Water Science and Technology Library

Volume 102

The aim of the *Water Science and Technology Library* is to provide a forum for dissemination of the state-of-the-art of topics of current interest in the area of water science and technology. This is accomplished through publication of reference books and monographs, authored or edited. Occasionally also proceedings volumes are accepted for publication in the series. *Water Science and Technology Library* encompasses a wide range of topics dealing with science as well as socio-economic aspects of water, environment, and ecology. Both the water quantity and quality issues are relevant and are embraced by *Water Science and Technology Library.* The emphasis may be on either the scientific content, or techniques of solution, or both. There is increasing emphasis these days on processes and *Water Science and Technology Library* is committed to promoting this emphasis by publishing books emphasizing scientific discussions of physical, chemical, and/or biological aspects of water resources. Likewise, current or emerging solution techniques receive high priority. Interdisciplinary coverage is encouraged. Case studies contributing to our knowledge of water science and technology are also embraced by the series. Innovative ideas and novel techniques are of particular interest.

Comments or suggestions for future volumes are welcomed.

Vijay P. Singh, Department of Biological and Agricultural Engineering & Zachry Department of Civil and Environment Engineering, Texas A&M University, USA
Email: vsingh@tamu.edu

All contributions to an edited volume should undergo standard peer review to ensure high scientific quality, while monographs should also be reviewed by at least two experts in the field.

Manuscripts that have undergone successful review should then be prepared according to the Publisher's guidelines manuscripts: https://www.springer.com/gp/authors-editors/book-authors-editors/book-manuscript-guidelines.

Aliva Nanda · Pankaj Kumar Gupta · Vivek Gupta ·
Prakash Kumar Jha · Swatantra Kumar Dubey
Editors

Navigating the Nexus

Hydrology, Agriculture, Pollution and Climate Change, Volume 1

Springer

Editors
Aliva Nanda
School of Civil and Environmental Engineering
Indian Institute of Technology Mandi
Mandi, Himachal Pradesh, India

Pankaj Kumar Gupta
Center for Rural Development and Technology
Indian Institute of Technology Delhi
New Delhi, Delhi, India

Vivek Gupta
School of Civil and Environmental Engineering
Indian Institute of Technology Mandi
Mandi, Himachal Pradesh, India

Prakash Kumar Jha
Mississippi State University
Starkville, MS, USA

Swatantra Kumar Dubey
Environmental Engineering
Seoul National University of Science and Technology
Seoul, Korea (Republic of)

ISSN 0921-092X ISSN 1872-4663 (electronic)
Water Science and Technology Library
ISBN 978-3-031-76531-5 ISBN 978-3-031-76532-2 (eBook)
https://doi.org/10.1007/978-3-031-76532-2

This Springer imprint is published by the registered company Springer Nature Switzerland AG
The registered company address is: Gewerbestrasse 11, 6330 Cham, Switzerland

Foreword

I am pleased to write this foreword for "*Navigating the Nexus—Hydrology, Agriculture, Pollution, and Climate Change.*" This timely and crucial book delves into the intricate relationships between the five elements of matter and life (earth, water, air, fire, and space). These relationships underpin global food, nutrition, climate, soil security, and environmental sustainability. Freshwater—the foundation of life—is a finite resource facing unprecedented pressures. Our planet's water cycle, intricately linked to agriculture, pollution, and climate change, has become a complex web of challenges demanding innovative solutions. This captivating book tackles these daunting tasks.

For decades, my research has focused on (a) understanding the impacts of climate change factors on the productivity of farming systems; (b) developing agricultural (e.g., crop, soil, water, nutrient) practices that improve efficiency and profitability; and (c) identifying climate-smart practices that improve adaptation, mitigation, and resilience to climate change. This book resonates deeply with my research interests and comprehensively explores the critical "nexus" where many of the elements of life and factors converge.

The editors have assembled a team of distinguished experts who dissect the complexities of water management, agricultural sustainability, pollution reduction, and adaptation and mitigation opportunities to minimize the impacts of climate change. Their diverse perspectives provide a holistic understanding of the challenges we face and identify potential pathways to address the challenges of food, nutrition, soil, and climate security.

This book's uniqueness is that it emphasizes integrated solutions that require transdisciplinary research, education, and outreach activities. By breaking down the silos and fostering collaborations between hydrologists, agricultural scientists, environmental engineers, and climate change experts, we can develop solutions that are practical and scalable to address our challenges. This book is not merely a scientific treatise but a call to action from all the key stakeholders, specifically researchers, policymakers, practitioners, and citizens. The nexus approach equips all stakeholders with the knowledge necessary to identify and advocate evidence-based solutions for sustainable agricultural practices and build the resilience of our farming systems and communities to climate change.

I wholeheartedly recommend this book to anyone seeking to navigate and learn about the complexities of water management in our interconnected world. Let us use the insights provided by the experts to design more sustainable, productive, profitable, and nature-positive agrifood systems for current and future generations.

P. V. Vara Prasad
Director, Feed the Future Sustainable Intensification Innovation Lab
Director, Center for Crops, Climate and Communities
University Distinguished Professor and R.O. Kruse Endowed Professor
Former President, Crop Science Society of America (2021)

Preface

In the era of accelerating climate change, the intricate connections between hydrology, agriculture, and environmental health are becoming increasingly critical to understand and manage. Our proposed book, "*Navigating the Nexus: Hydrology, Agriculture, Pollution, and Climate Change,*" delves into these complexities by integrating modern tools and applied research to provide a comprehensive overview of these interrelated fields. This book addresses the pressing impact of global warming on the hydrological cycle, particularly at headwater catchments and their downstream effects, crucial for water availability and agricultural productivity, soil-groundwater contaminations. Through advanced knowledge of agricultural water management and the application of crop modeling, it explores the challenges of soil-water pollution and proposes effective remediation strategies. With contributions from leading experts, the book spans five major domains:

- The hydrology of headwater catchments, acting as vital water sources for downstream communities.
- The impact of shifting precipitation and temperature regimes on irrigation planning and agricultural production.
- The remediation of soil-water pollution to mitigate climate crises.
- The application of new tools and technologies, including numerical modeling, machine learning, AI, and remote sensing, to address water and food security issues.
- The social aspects of hydrology, including transboundary water conflicts.

By integrating field experimentation, modeling approaches, and remote-sensing methods, the book aims to equip researchers, policymakers, and practitioners with the scientific insights needed to navigate and mitigate the adverse effects of climate change on water resources and agriculture. It highlights the necessity of understanding and managing the dynamic interactions between water availability, agricultural practices, and soil health in our changing world. As climate patterns become increasingly unpredictable, with uncertain rainfall and extreme weather events, the need for sustainable management of natural resources is more urgent than ever. This book offers solution-based insights and ground-based research to help readers grasp the intricate linkages between water availability, agricultural productivity, and soil-water-food security. We hope this book will serve as a valuable resource for hydrologists, environmental engineers, policymakers, and anyone involved in natural resource management. By enhancing their understanding of these interconnected domains, readers will be better equipped to address the challenges posed by climate change and ensure sustainable management of our vital natural resources.

Mandi, India — Aliva Nanda
New Delhi, India — Pankaj Kumar Gupta
Mandi, India — Vivek Gupta
Starkville, USA — Prakash Kumar Jha
Seoul, Korea (Republic of) — Swatantra Kumar Dubey

Contents

Part I Hydrology of the Headwater Catchments in Changing Climatic Conditions

1 Shift in Streamflow Regime in Headwater Catchments: Causes and Impacts 3
Sushant Mehan, Manoj Lamichhane, and Aditya Jha

2 Hydrology of the Upper Indus Basin under Changing Climate 33
Aftab Nazeer, Zakir Hussain Dahri, Faisal Baig, Muhammad Asif, and Muhammad Asif

3 The Open Global Glacier Model (OGGM) Based Assessment of Glacier Mass Variations in Chhota Shigri Glacier, Western Indian Himalaya 49
Tasso Yaka, Renji Remesan, and Akshay Pachore

4 Insight into the Impact of Land Use/Land Cover Change on Runoff Incorporating Hydrological Modelling to Analyse Their Impact on the Water Regime 71
Ankita Sharma, Radha Krishan, Kuljit Kour, Bhaskar R. Nikam, and Dhirendra Singh Bagri

Part II Efficient Irrigation and Agricultural Planning for Food Security

5 Divulging Insights into Irrigation Scheduling of *Allium cepa* (Onion) Using *Cocos nucifera* (Coconut) Husk Mulch 99
Yash Shukla, Vijay Shankar, and Vivek Gupta

6 Impact of Climatic Extreme Indices on Crop Productivity 117
Ayush Kumar, Ashok Mishra, and Idhaya Chandhiran Ilampooranan

7 Pathways to Accelerate Agricultural Production System Technologies in India: A Vision for New Generation Watersheds 137
Deep Banerjee, Krishna Reddy Kakumanu, and Swatantra Kumar Dubey

8 Analysing the Impact of Climate Change on Crop Yield and Its Consequences for the Water-Energy-Food Nexus in an Indian River Basin 173
Krishna Mondal, Vishal Gupta, Chandranath Chatterjee, and Rajendra Singh

Part III Hydro-chemical Environment under Climate Change

9 Exploring the Effects of Contemporary Agricultural Techniques on Soil and Water Pollution in Rice Fields 197
Ankita Paul and Arti Bhatia

10 Engineered Bioremediation for Decontamination of Aquifers 239
Bhavna Nigam, Dheeraj Rathore, Anuj Kumar Yadav, Ratna Chauhan, and Indra Jeet Chaudhary

11 Groundwater Remediation Using Meshless Local Petrov Galerkin (MLPG) Simulation Coupled with Differential Evolution (DE) 257
Sanjukta Das and T. I. Eldho

12 Transport of Nitrate, Phosphate and Potassium in Vadose Zone with Varying Rainfall Scenarios for a Semi-arid Cotton Growing Region of South India 275
Deepak Kumar, P. Vijay Kumar, Pankaj Kumar Gupta, and Suri L. Naidu

13 Health Risk Assessment of Heavy Metal Toxicity in Groundwater—A Comparative Study on the Industrial Corridors of Integrated Vellore District, Tamil Nadu 297
K. Loganathan and L. Vignesh Rajkumar

Part IV Advanced Tools and Techniques to Provide Climate Solutions

14 Assessing the Impacts of Climate Change on Hydroclimatic Regimes in Beas River Basin 325
Siddik Barbhuiya, Salil Sharma, Ashish Pathania, and Vivek Gupta

15 Precision Agriculture Technologies for Climate-Resiliency and Water Resource Management 351
Gaurav Jha, Udit Debangshi, Sindhu Palla, Farshina Nazrul, Sourajit Dey, Wri Dutta, and Sangeeta Bansal

16 Hydrometeorological Water Budget and It's Relationship with Extreme Over Alaknanda River Basin: An Application of ERA5-Land 383
Pramod Kumar, Sudheer Yadav, Khushboo Sharma, Prem Kumar, Jaya Rai, and Ravindra Pratap Singh

17 Bayesian Model Averaging for Multi-model Ensemble Streamflows of the Godavari Basin 409
Rajas Surlikar, Akshay Pachore, and Renji Remesan

Part V Socio-hydrological Impacts and Policy Implications in a Changing Climate

18 Application of AI/ML in Water Resource Management to Resolve Transboundary Water Conflict 431
Sayantan Sarkar and Prakash Kumar Jha

19 Suitability and its Physicochemical Characterization for Deciphering Surface Water Quality Using Entropy (E) and Fuzzy (F)-AHP Optimization Model in Mahanadi River Basin (MRB), Odisha (India) 457
Abhijeet Das

20 A Hydrologic Modeling Assessment of Future Water Scarcity in the Baitarani River Basin 499
Sushree Swagatika Swain, Ashok Mishra, Chandranath Chatterjee, and Morgan C Levy

21 Summary Chapter 527
Aliva Nanda, Pankaj Kumar Gupta, Vivek Gupta, Prakash Kumar Jha, and Swatantra Kumar Dubey

Part I
Hydrology of the Headwater Catchments in Changing Climatic Conditions

Chapter 1
Shift in Streamflow Regime in Headwater Catchments: Causes and Impacts

Sushant Mehan, Manoj Lamichhane, and Aditya Jha

Abstract The hydrological cycle within a basin is a complex interaction of climate patterns, basin characteristics, and human activities. Significant impacts on water distribution and yield have been noticed lately due to climate variability leading to global warming and shifting precipitation patterns, coupled with anthropogenic interventions. Changes in the river flow, primarily due to human water use and, in some cases, climate change, have been significant in arid and semi-arid regions. Assessing the impact of climate and land use change on streamflow involves various approaches, such as trend analysis, hydrological indicator-based analysis, and hydrological modeling. However, each approach comes with its strengths and limitations. This book chapter explores the pivotal role of headwater catchments globally, highlighting the impact of climate change, human activities, and land use changes in modifying streamflow regimes in the headwater catchments. Along with the foundational physics of different processes in headwater catchment, a meta-synthesis from diverse case studies and scientific research, the chapter also explores strategies for adaptation and mitigation, emphasizing the need for interdisciplinary collaboration and effective policy measures. By providing a comprehensive overview of the causes and impacts of streamflow regime shifts in headwater catchments, this chapter is an essential resource for researchers, practitioners, and policymakers seeking to better understand and address this critical issue in a changing environment.

Keywords Headwater catchments · Streamflow · Climate change

S. Mehan (✉) · M. Lamichhane
Agricultural and Biosystems Engineering Department, South Dakota State University, Brookings, SD 57006, USA
e-mail: Sushant.Mehan@sdstate.edu

A. Jha
Civil Engineering Department, Kansas State University, Manhattan, KS 66506, USA

A. Nanda et al. (eds.), *Navigating the Nexus*, Water Science and Technology Library 102, https://doi.org/10.1007/978-3-031-76532-2_1

1.1 Introduction

Headwater catchments play a crucial role in the broader hydrological cycle by influencing the movement and quality of water within river systems. The export regimes of headwater catchments have been studied to understand their contributions to downstream reaches. These studies have highlighted the importance of examining multiple solutes and time scales to compare headwater export regimes to the integral signal measured in downstream reaches (Dupas et al. 2017). Additionally, the hydrological response of headwater-dominated catchments has been analyzed to understand the relationship between aggregated and distributed hydrological responses, emphasizing the significance of inflow from headwater catchments in larger watershed systems (Özgen-Xian et al. 2023). Studies have shown that as catchments scale up from headwater to larger mesoscale systems, there is an increase in contributions from deeper groundwater with longer hydrological pathways flowing through mineral soils (Tiwari et al. 2014). This suggests that water from headwater catchments can be transferred to downstream rivers via groundwater flow paths, increasing the groundwater input in larger catchment areas (Bouaziz et al. 2018).

Moreover, the geological effects on scaling relationships of groundwater contributions in forested watersheds have challenged assumptions about impermeable bedrock in headwater catchments, revealing significant water and solute infiltration into bedrock (Iwasaki et al. 2021). This infiltration can lead to water leaving headwater catchments undetected at the weir, impacting downstream water quality and quantity. Studies have also explored the hydrological responses of headwater catchments to a warming climate, demonstrating the importance of these catchments in understanding long-term hydrological changes under future climate scenarios (Zheng et al. 2021). The variability of dominant hydrological processes across spatial scales is evident in the diverse spatiotemporal vegetation dynamics in headwater catchments, influenced by complex terrain, microclimate variation, disturbance history, and hydrological processes (Hwang et al. 2012).

Global streamflow regime changes are critical to hydrological systems, significantly affecting water resource management and ecosystem health. Studies have shown that alterations in streamflow components, such as high and low-flow components, directly impact river biodiversity and ecological functions (Li et al. 2022). These changes can manifest as decreasing trends in annual streamflow, with abrupt shifts detected in specific years affecting the timing and magnitude of streamflow (Sun et al. 2019). Additionally, changes in streamflow regimes following vegetation alterations in catchments can occur due to modifications in hydrological processes like rainfall interception and evapotranspiration (Zhao et al. 2011). Furthermore, the impact of climate change and human activities on streamflow has been a subject of research globally. Studies have highlighted the need to assess alterations in streamflow regimes to understand the variations in hydrological characteristics, such as seasonal patterns, magnitude, frequency, and duration of streamflow (Zhang et al. 2012). These alterations in hydrologic regimes may necessitate adjustments in water resource planning and management to adapt to streamflow seasonality, magnitude, and timing (Shin et al. 2016).

The spatial variability of streamflow responses to climate change and land use/ cover change has been emphasized, indicating the importance of studying hydrological responses at different scales within river basins (Liu et al. 2020). Changes in streamflow regimes can lead to increased probabilities of floods and droughts, highlighting the necessity to comprehend long-term variations in annual streamflow regimes for effective water management (Seyam and Othman 2014). Analyzing trends in streamflow characteristics and identifying changing streamflow regimes globally can provide valuable insights for sustainable water resource management and decision-making strategies (Zhang et al. 2015).

In this chapter, we aim to address the following research questions: How can changes in hydrological components in headwater catchments be accurately assessed? What are the capabilities and uncertainties of current hydrological models in predicting streamflow? What are the most influential factors that affect streamflow timing, magnitude, and duration? What human interventions affect streamflow in headwater catchments, and how can their impacts be mitigated? What are the ecological consequences of changes in streamflow, and how can these consequences be reduced? What are the infrastructure vulnerabilities associated with changes in streamflow in headwater catchments, and what are the potential mitigation strategies? How do changes in streamflow affect recreation and tourism in headwater catchments?

To answer these questions, the subsequent sections of this chapter are described in the following order: Section 1.2 provides fundamentals of different hydrological processes in headwater catchments, examining the foundational elements that influence water flow in these critical areas. Section 1.3 presents various modeling approaches for predicting changes in streamflow and associated uncertainties with these models. Section 1.4 explores the implications of climate change on streamflow dynamics, while Sect. 1.5 assesses the impacts of human activities and land use changes on these water systems. Section 1.6 discusses the ecological consequences of altered streamflow, reflecting on biodiversity and ecosystem health. Section 1.7 looked at infrastructure vulnerabilities and the strategies available to mitigate these challenges in the face of changing water regimes. Section 1.8 covers the effects on recreation and tourism, industries heavily dependent on stable and predictable water flows. Section 1.9 focuses on adaptation and mitigation strategies that can be employed to cope with the evolving challenges. Finally, Sect. 1.10 summarizes the discussion and outlines future research directions and the long-term outlook on managing and adapting to the environmental changes brought by dynamics in the streamflow regime in the headwater catchment.

1.2 Hydrological Processes in Headwater Catchments

The hydrological process comprises several components that govern water's appearance, distribution, and flow within a watershed (Fig. 1.1). This is particularly true for headwater catchments in the upper reaches of rivers, as they significantly influence the basin's hydrology. These headwater catchments involve a complex interaction of precipitation, evaporation, transpiration, infiltration, interception, runoff, and groundwater flow. These interactions directly affect the quality and quantity of water available further downstream. Extreme events in the headwater catchments can lead to numerous problems downstream, including flooding, drought, debris flow, and negative impacts on aquatic life habitats. Therefore, managing these catchments is critical to ensure water resources' sustainability and the environment's overall health. This section provides a concise overview of each hydrological component and its influence on headwater catchments.

Precipitation: Precipitation encompasses any moisture that reaches the Earth's surface. It includes rainfall and other forms such as snowfall, hail, dew, drizzle, and sleet. Snowfall plays a crucial role in headwater catchments, significantly impacting downstream flow as it melts and produces a continuous stream of water. The impact of climate change and variability is hardly felt at headwater catchments, resulting from the accelerated melting of snowpacks and consequently altering the flow patterns of rivers.

Interception and depression losses: When precipitation occurs, plants and other physical structures intercept a portion before reaching the Earth's surface, a phenomenon known as interception loss. This loss is eventually returned to the atmosphere through evaporation. Before any overland flow begins, all depressions in the landscape must fill up, and water stored in these depressions is typically lost through the infiltration process, referred to as depression loss. In hydrology, the combined effect of interception and depression loss is known as initial losses.

Evaporation and Transpiration: Evaporation refers to transforming water from liquid to vapor, primarily driven by the sun's heat energy. On the other hand, transpiration is the release of water from plants during photosynthesis, occurring through tiny openings known as stomata. The amalgamation of these two processes—evaporation and transpiration—is termed evapotranspiration. In headwater catchments, evaporation occurs from various surfaces, including soil, water bodies such as streams and lakes, and snow or ice. This process plays a pivotal role in returning water to the atmosphere. Transpiration regulates water flow and maintains the local microclimate, which is equally crucial in these catchments.

Runoff: Runoff, the water movement across the land surface typically from precipitation, plays a crucial role in headwater catchments by impacting water quantity and quality. It contributes to the volume of water in rivers and streams. In excessive volumes, runoff potentially causes flooding and transports pollutants that degrade water quality. Runoff also leads to erosion, affecting aquatic ecosystems through

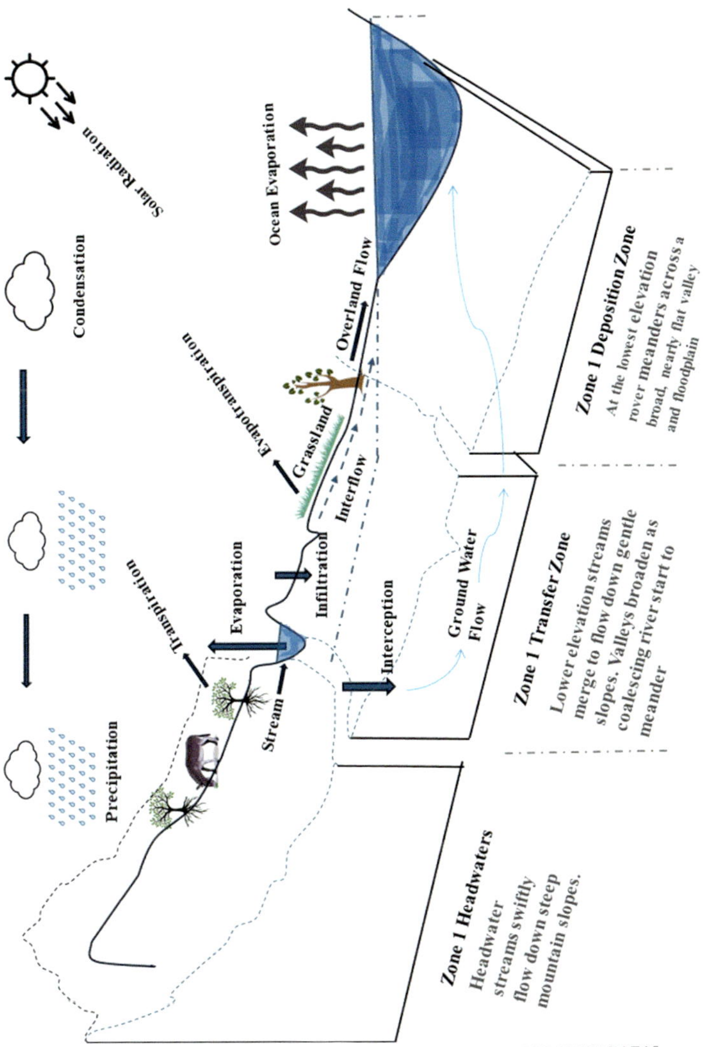

Fig. 1.1 Conceptual diagram representing headwater catchment, its components, and interactions with the streamflow, depicting major hydrologic processes. The diagram illustrates the hydrological cycle, showcasing the intricate connections between the atmosphere, land, and bodies of water. It begins with ocean evaporation, contributing to atmospheric moisture and evapotranspiration from land vegetation. This moisture then condenses, falling as Precipitation onto the landscape. Upon reaching the ground, some water is lost through Interception. At the same time, the remainder contributes to Overland Flow, Interflow, and Groundwater Flow, each with its time delay before returning to streams and rivers. These components, from evaporation to groundwater movement, encapsulate the continuous and complex water flow through the ecosystem, highlighting the cyclical nature of the hydrological processes within a head catchment

sedimentation and changes in water characteristics. In urban areas, impervious surfaces increase runoff, reducing groundwater recharge and exacerbating hydrological extremes due to climate change. Managing land use and implementing sustainable practices are essential in mitigating these impacts and ensuring the health and balance of headwater catchments and their ecosystems.

The literature emphasizes that understanding different hydrological processes in headwater catchments is crucial for controlling water movement, storage, and flow dynamics. Hydrological processes in headwater catchments control the recharge of subsurface water stores, flow paths, and residence times of water throughout landscapes. Any change in the surface–subsurface exchange fluxes triggers different flow regimes, connectivity dynamics, and runoff generation mechanisms inside the catchment, affecting the distributed hydrological response (Özgen-Xian et al. 2023). Responsive hydrological pathways associated with peat and shallow alpine soils in headwater sub-catchments produce seasonally variable runoff with short mean residence times (Rodgers et al. 2005a, b). Hydrological processes influence the spatial and temporal variations in streamwater chemistry, with land cover and downstream observations affecting organic carbon levels in headwater streams (Temnerud et al. 2016). Scale effects on headwater catchment runoff timing, flow sources, and groundwater-streamflow relations can vary with catchment size and antecedent moisture conditions (McGlynn et al. 2004). Saturated and unsaturated infiltration from soil to bedrock is a dominant hydrological process at the soil–bedrock interface, influencing runoff generation in headwater catchments (Kosugi et al. 2006).

Implementing high-frequency in situ monitoring with sensors can provide detailed and real-time data on hydrological parameters such as water levels, flow rates, and water quality, allowing for the characterization of short-term variations in hydrological processes. Long-term environmental monitoring programs incorporating high-frequency in situ sensors can help improve the mechanistic understanding of hydrological responses to extreme events and changes in catchment hydrology over time. Hydrological models, in conjunction with in situ measurements, can help simulate the dominant hydrologic processes (e.g., changes in water storage dynamics and runoff), and their responses to hydro-climatic variability, and estimate mean residence times of runoff in headwater catchments (Nauditt et al. 2016). Studying water isotope concentration signals through in situ monitoring can provide insights into episodic runoff generation processes and the dynamics of water sources in headwater catchments (Votrubova et al. 2017). Analyzing distributed surface–subsurface interactions through in situ monitoring can help constrain hydrograph analysis and provide valuable local information on hydrological responses (Özgen-Xian et al. 2023). Remote sensing can be used to monitor changes in vegetation cover, land use, and land cover types in headwater catchments, providing insights into the impact of these changes on hydrological processes (Buytaert et al. 2014). Besides, a summary of streamflow data sources adapted from Chen et al. (2023) is presented in Table 1.1.

The measured streamflow from in situ measurements has some characteristics that are widely used to define indices in ecological studies. Some selected streamflow characteristics are related to baseflow, peak flow, the flow duration curve (FDC) shape, and the mean water balance or flow timing. Table 1.2, adapted from Beck

Table 1.1 Different streamflow data sources in the world

Database	Spatial coverage	Time span	Included catchments	Data access	Quality flag
Global Streamflow Indices and Metadata Archive (GSIM)	Globe	1806–2016	30,959	https://doi.pangaea.de/10.1594/PANGAEA.887470 and Do et al. (2018): The Global Streamflow Indices and Metadata Archive—Part 1: Station catalog and Catchment boundary (pangaea.de)	–
Global River Discharge Centre (GRDC)	Globe	1806–2022	8458	https://www.bafg.de/GRDC/EN/Home/homepage_node.html	–
U.S. Geological Survey (USGS) National Water Information System	US and Canada	1857–2022	19,269	https://waterdata.usgs.gov/nwis	A: value has been validated to be published Ae: value was estimated and validated to be published P and Pe: provisional data
National Water Data Archive (HYDAT)	Canada	1860–2022	5786	https://www.canada.ca/en/environment-climate-change/services/water-overview/quantity/monitoring/survey/data-products-services/national-archive-hydat.html	A: partial day (1) B: ice conditions (2) D: dry (3) E: estimated (4) S: sample(s) collected this day (5)
National Water Agency (ANA)	Brazil	1901–2022	3691	https://www.ufrgs.br/lsh/products/ana-data-acquisition/	–

(continued)

Table 1.1 (continued)

Database	Spatial coverage	Time span	Included catchments	Data access	Quality flag
The Chilean Center for Climate and Resilience Research (CCCRR)	Chile	1913–2018	767	https://www.cr2.cl/datos-de-caudales/	–
Arctic Great Rivers Observatory (ArcticGRO)	The largest Arctic rivers	1927–2022	17	https://arcticgreatrivers.org/	A: certified data P: Provisional data
China Hydrological Yearbooks (CHY)	China	1947–2020	30	No public access	–
India Water Resources Information System (WRIS)	India	1960–2021	161	https://indiawris.gov.in/wris/	–
Australia Water Data Online, Australian Bureau of Meteorology (BOM)	Australia	1881–2022	4977	http://www.bom.gov.au/waterdata/	A (10): best available B (90): compromised to represent the parameter C (110): Estimated value E (140): Quality is not known F (210): poor quality or missing value Indic "–" also presents to flag missing value
Department of Hydrology and Meteorology (DHM), Nepal	Nepal	1963–2023	103	https://www.dhm.gov.np/request-data	–
Japan Water Information System (MLIT)	Japan	1979–Present	1023	http://www1.river.go.jp/	–

et al. (2013), mentions some streamflow characteristics that are often used to define indices in ecological studies.

1.3 Modeling Approaches for Predicting Streamflow Changes

Predicting streamflow in headwater catchments presents a formidable challenge due to the complex physical processes and significant climatic influences. Developing countries, in particular, struggle to forecast streamflow changes within these catchments, largely because of the scarcity of observed data. Many modeling approaches have been used to anticipate these changes, including empirical, conceptual, and physically based models. More recently, machine learning-based models and hybrid models have gained popularity for their ability to capture complex relationships within catchments. Nonetheless, each method carries its own set of strengths and weaknesses in the context of predicting streamflow changes in headwater catchments.

Empirical models, for example, the Autoregressive Integrated Moving Average (ARIMA) (Box et al. 2015), rely on the observed relationships between streamflow and variables such as temperature, precipitation, land use, and soil type. These straightforward approaches demand less data, making them accessible for quick forecasting. For example, ARIMA models have been applied in the Ganges River basin to forecast monsoon flows, which is crucial for agricultural planning (Mujumdar and Nagesh Kumar 1990). However, several uncertainties are associated with using empirical models to predict streamflow. Primarily, these models often assume that past relationships between inputs and outputs will remain consistent in the future, an assumption that may not hold true under evolving climatic and landscape conditions. Furthermore, empirical models rely heavily on historical data, making them particularly sensitive to the quality and availability of such data. Additionally, due to their simplicity, these models are prone to overfitting the calibration data, resulting in poor performance when applied to new, unseen data.

Conceptual models offer a middle ground, employing simplified representations of the catchment to predict streamflow. Notable examples include the Hydrologiska Byråns Vattenbalansavdelning (HBV) model, which has been applied across various Scandinavian catchments for flood forecasting (Bergström 1995), and the Stanford Watershed Model (SWM), which has been widely used in the United States for water resource assessment (Crawford and Linsley 1966). Another widely recognized conceptual model is the Soil and Water Assessment Tool (SWAT), used to assess the impacts of land management on water yields in large river basins such as the Mississippi (Arnold et al. 1998). While more detailed than empirical models, these models come with several uncertainties when predicting streamflow. These models include multiple parameters that are not directly measurable and must be estimated, introducing potential errors. Additionally, conceptual models simplify many complex

Table 1.2 Different streamflow characteristics used to define indices in ecological studies

Characteristic	Unit	Description and computation
BFI1	–	Baseflow index, defined as the ratio of long-term base flow to total Q (Smakhtin 2001), computed from daily Q data using the recursive digital filter of (Van Dijk 2010) with the "window size" set to 5 days (Peña-Arancibia et al. 2010)
BFI2	–	Baseflow index computed from daily Q data following the local-min method described in Pettyjohn and Henning (1979) and Sloto and Crouse (1996) with the "duration of surface runoff" N set to 5 days
BFI3	–	Baseflow index computed from daily Q data using a 7-day moving min to derive base flow. This method is referred to as the 7-day sliding interval in Pettyjohn and Henning (1979) and Sloto and Crouse (1996)
BFI4	–	Baseflow index computed from daily Q data following the procedure described in Gustard et al. (1992), which takes the minima at 5-day nonoverlapping intervals and subsequently connects the valleys in this series of minima to generate base flow
k	day^{-1}	Computed from daily Q data as described in Beck et al. (2013), with the "window size" set to 5 days and days with zero flow ignored
Q1, Q5, Q10, Q20, Q50, Q80, Q90, Q95, and Q99	mm day^{-1}	Daily flow percentiles (exceedance probability) computed from daily Q data. The number refers to the percentage of time that the flow is exceeded
T50	–	The day of the water year marking the timing of the center of mass of Q (Stewart et al. 2004)
RC	–	Runoff coefficient is the ratio of mean annual Q to mean annual P, where P is precipitation. Used. It can be computed from daily as well as monthly Q data
QMEAN	mm yr^{-1}	Mean annual Q per unit area, which can be computed from daily or monthly Q data
Streamflow Drought Index (SDI)	–	A robust index that quantifies streamflow deficits during drought periods, integrating historical flow data (Vicente-Serrano et al. 2011)
Low flood pulse count	–	Number of annual occurrences during which the magnitude of flow remains below a lower threshold. Hydrologic pulses are defined as those periods within a year in which the flow drops below the 25th percentile (low pulse) of all daily values for the time period (George et al. 2021)
High flood pulse count	–	High pulse is defined as the 75th percentile (George et al. 2021)
Constancy	–	A measure of temporal invariance (George et al. 2021)

(continued)

Table 1.2 (continued)

Characteristic	Unit	Description and computation
Predictability of a flow	–	Two independent components: additive components: constancy (a measure of temporal invariance) and contingency (a measure of periodicity) (George et al. 2021)
Rise rate	–	Mean rate of positive changes in flow from one day to the next (George et al. 2021)
Fall rate	–	Mean rate of negative changes in flow from one day to the next (George et al. 2021)
Reversals	–	Number of negative and positive changes in water conditions from one day to the next (George et al. 2021)

hydrological processes, which can lead to inaccuracies in representing these interactions. The accuracy of these models also heavily depends on their parameterization and calibration, which affects their reliability across different scenarios. This calibration sensitivity underscores the challenges in ensuring consistent model performance under varying conditions.

Physically based models are the most detailed, incorporating mathematical representations of physical processes to capture spatial variability. Examples of these models include the Distributed Hydrology Soil Vegetation Model (DHSVM), which has been instrumental in understanding the hydrological impacts of logging in the Pacific Northwest (Wigmosta et al. 1994), and the System Hydrologique Europeen (Mike SHE), which has been used for comprehensive catchment management in Europe (Abbott et al. 1986). The Hydrological Simulation Program—Fortran (HSPF) has also been utilized extensively in the United States to construct watershed models to study sediment and pollutant transport (Bicknell et al. 1997). In addition to the discussed modeling approaches, integrating hybrid models offers a promising avenue to enhance streamflow predictions. Hybrid models combine the strengths of different modeling techniques, such as empirical, conceptual, and physically based models, to overcome individual limitations and improve overall performance. Like empirical and conceptual models, physical models also face several uncertainties in predicting streamflows. These models are computationally intensive and require detailed inputs, which can often be challenging. A key uncertainty arises from the mismatch between the scale at which data is available and the scale at which the model operates. Additionally, physical models may not fully or accurately represent all processes despite their complexity, leading to potential errors in streamflow predictions. This limitation highlights the difficulty in capturing every aspect of hydrological dynamics, which can impact the model's accuracy and reliability.

In the past few decades, machine learning-based models have been gaining popularity, and they can excel at capturing complex relationships among a vast array of input variables, enhancing the model's efficiency and reliability. These methods necessitate extensive datasets for training purposes. Researchers utilized various methodologies to investigate changes in hydrological processes in headwater catchments using Machine Learning (ML). Mosavi et al. (2018) employed ML models to predict floods in headwater catchments by analyzing historical data to forecast flood events and understand hydrological processes. Almikaeel et al. (2022) utilized ML techniques to forecast hydrological droughts in headwater catchments, offering insights into water availability and streamflow dynamics. Zhu et al. (2019) applied ML models in headwater catchments and found that ML models can aid in assessing how hydrological systems in headwater catchments respond to future climate change scenarios, enabling the prediction of streamflow changes and water resource availability. Additionally, Yang and Chang (2020) employed the ML algorithms and observed that ML models can be applied for streamflow forecasting in headwater catchments, capturing the relationships between input variables and streamflow dynamics. Li et al. (2020) used ML techniques and found that these algorithms can assist in parameterizing hydrological models in headwater catchments, enhancing model performance, and improving the representation of hydrological processes. Although ML models can effectively capture the non-linear and complex relationships between input and output variables without the need to understand the underlying physics, several uncertainties still challenge the accuracy of predicted streamflow. A major uncertainty is that these models may not adapt well to conditions not represented in the training data, such as new climatic regions or changes in land use. Furthermore, while ML models can excel with training data, they might not generalize well to data outside the training set, often due to overfitting. Additionally, models from the branch of ML known as deep learning are frequently considered "black boxes" because it is difficult to interpret how inputs are transformed into outputs. This opacity can affect the understanding and trust in the model's predictions.

1.4 Climate Change and Streamflow

Climate change significantly influences precipitation patterns, impacting streamflow in headwater catchments. Research has shown that anthropogenic climate change can change precipitation distribution into streamflow, evapotranspiration, and storage in forested headwater catchments (Gaertner et al. 2020). These alterations in precipitation patterns can affect the hydrological regime of headwater catchments by influencing streamflow timing, magnitude, and duration. Climate change can also impact headwater streams by thawing permafrost and altering hydrological regimes, vegetation distribution, physiognomy, and productivity. Studies have indicated that changes in precipitation due to climate change can have a nonlinear dynamic impact on streamflow in headwater regions. The response of streamflow to climate change is influenced by precipitation at decadal scales and modified by temperature and evapotranspiration within shorter timeframes (Wu et al. 2018). Additionally, the sensitivity of streamflow to climate variability and human activities in headwater catchments has been studied, with findings suggesting that land use changes may play a more significant role than climate in altering streamflow (Zhao et al. 2009). The impact of climate change on streamflow in headwater catchments is complex and multi-faceted. Factors such as precipitation intensity, catchment characteristics, and land surface changes need to be considered when assessing the effects of climate variability on streamflow (Lv and Ma 2018). Climate change can exacerbate the impact of other disturbances, such as wildfires, on catchment streamflow, leading to noticeable increases in mean annual streamflow in burnt areas (Guo et al. 2021). Understanding the complex interactions between climate change, precipitation, and streamflow is crucial for effective water resource management and ecosystem conservation in these sensitive environments.

Rising temperatures due to climate change significantly impact snowmelt and runoff processes in various regions. Understanding the complex interactions between temperature changes, snowmelt dynamics, and runoff responses is crucial for assessing the impacts of climate change on water resources and ecosystem dynamics in snowmelt-dominated regions. Studies have shown that increasing temperatures can lead to earlier snowmelt initiation, resulting in changes in the timing and volume of snowmelt runoff (Stigter et al. 2017, 2016; Lamichhane et al. 2024). The earlier onset of snowmelt due to rising temperatures can cause an increase in snowmelt runoff during the winter and early melt season while decreasing runoff during the late melt season (Stigter et al. 2017, 2016). Additionally, the influence of temperature rise on the average discharge of the entire snow season may not be immediately apparent, indicating a complex relationship between temperature and snowmelt runoff (Wheeler et al. 2016). The effects of warming temperatures on snowpack energy balance can shift the timing of snowmelt initiation earlier in the year. However, earlier spring snowmelt during periods of lower solar radiation can result in slower snowmelt rates under a warming climate (Harpold and Brooks 2018). This highlights the intricate relationship between temperature, solar radiation, and snowmelt dynamics in response to climate change. Some studies have indicated that changes in snowmelt runoff in response to climate change scenarios can lead to shifts in the timing of streamflow and alterations in runoff volume (Wheeler et al. 2016; Wang et al. 2010). Predicted changes in precipitation and air temperature patterns may result in significant modifications in the timing and volume of mountain snowmelt runoff, potentially leading to increased incidences of flooding and summer droughts (Molini et al. 2011).

Several studies offer valuable insights to help comprehend the impacts of climate change on streamflow in headwater catchments worldwide. For example, research by Chiew (2006) in Australia estimated the rainfall elasticity of streamflow in 219 catchments, showcasing the sensitivity of streamflow to changes in precipitation. Li et al. (2007) assessed the impact of climate variability and human activities on streamflow in the Wuding River basin in China, revealing a significant reduction in streamflow regime percentile flows. This study demonstrated the substantial influence of both climate variability and human activities on altering streamflow patterns in headwater catchments. Evans et al. (2015) investigated groundwater flow in mountainous headwater catchments with permafrost, suggesting that permafrost thawing could lead to a three-fold increase in groundwater contribution to streamflow under a warming scenario. Vlach et al. (2020) also indicated that increasing temperature, evaporation rates, and decreasing snow cover could intensify summer streamflow deficits in Central European headwaters. Climate change, driven partly by human activities, has also influenced streamflow regimes in headwater catchments. Wang et al. (2019) highlighted that increasing temperatures can lead to declining streamflow due to enhanced evaporation. Similarly, Hu et al. (2011) noted that decreasing precipitation, particularly in the wet season, coupled with rising temperatures, can significantly decrease water resources, posing challenges for downstream water uses.

Fischer et al. (2015) examined the effects of vegetation change and climate variability on streamflow, underscoring the substantial impact of vegetation cover on streamflow dynamics. Meriö et al. (2019) highlighted the pronounced effects of a warming climate on hydrology and catchment sensitivity concerning snow quantity and snow cover duration. Research by Zheng et al. (2009) in the Yellow River Basin highlighted that land use change was predominant in reducing streamflow, with climate change contributing to a lesser extent. This study emphasized the inverse relationship between precipitation and runoff coefficient, indicating that drier catchments are more sensitive to precipitation changes. Leach and Moore (2019) highlighted the importance of understanding the thermal regimes of headwater catchments in western North America due to their prevalence in the landscape and their significance as a salmonid habitat. Different mitigation strategies and their impacts are discussed in subsequent sections of the chapter.

1.5 Human Activities and Land Use Changes

Anthropogenic factors have significantly influenced streamflow regimes in headwater catchments. Studies have shown that land use changes, such as deforestation and urbanization, can substantially impact streamflow dynamics. Understanding these interactions is crucial for effective water resource management and ecosystem conservation in these environments. Zheng et al. (2009) found that land use change was responsible for over 70% of streamflow reduction in the Yellow River Basin. Patterson et al. (2013) highlighted the contributions of both anthropogenic and climate factors to shifts in streamflow. Molina et al. (2015) demonstrated that observed changes in streamflow were likely due to anthropogenic disturbances associated with land cover changes. Patterson et al. (2013) found that the extent of agricultural land within basins positively correlated with streamflow change, highlighting the influence of land use practices on streamflow alterations. Agricultural practices also significantly impact streamflow regimes in headwater catchments. Various studies have highlighted how agricultural activities influence streamflow dynamics. Research by Kim et al. (2022) suggested that changing grazing management practices can impact streamflow regimes by altering the magnitude of high and low flows, affecting flood and drought occurrences. Changes in land cover, such as deforestation, afforestation, and urbanization, can alter streamflow patterns in headwater catchments (Rogger et al. 2017; Buechel et al. 2022). Implementing soil conservation practices can influence streamflow by altering infiltration rates, erosion, and sediment transport (Gao et al. 2015). The establishment of forests through afforestation can impact streamflow by modifying evapotranspiration rates and interception of precipitation (Buechel et al. 2022). The intensity of land use practices, such as agricultural intensification or urban development, can lead to changes in streamflow regimes due to altered runoff patterns and water infiltration rates (Bai et al. 2023). Effective land use planning strategies considering streamflow dynamics can help maintain water resources, protect ecosystems, and support sustainable development in headwater catchments (Leh et al. 2011). Penna et al. (2011) found that antecedent wetness conditions influenced the catchment's response time, with streamflow reacting and peaking based on soil moisture levels. Williams et al. (2015) indicated that the rapid response of streamflow to precipitation inputs in headwater catchments could be attributed to various ditches and tile drains in the upper catchment, affecting streamflow dynamics.

Besides water quantity, agricultural activities were observed to affect the water quality. Agricultural fertilizers and livestock wastes can lead to nutrient loading in headwater streams, affecting water quality and downstream ecosystems (Alexander et al. 2007). Agricultural activities can impact groundwater discharge to headwater streams, influencing nutrient transport and stream health (Fischer et al. 2016). Agricultural management practices can enhance organic matter turnover, changing the export of dissolved organic matter to headwater streams (Heinz et al. 2015). Excessive nutrients from agriculture have been associated with persistent eutrophication in aquatic ecosystems, affecting stream water quality (Dupas et al. 2019). Herbicide pollution from agricultural practices can result in episodic peak concentrations in headwater streams, raising environmental concerns (Ammann et al. 2020). Managing pesticide contamination in surface water catchments is crucial, with restrictions on land treatment necessary to meet downstream water quality standards (Villamizar et al. 2020).

Implementing riparian buffers in agricultural landscapes can help decrease the delivery of agricultural pollutants to streams, thereby improving water quality (Tomer et al. 2021). Nunes et al. (2020) demonstrated that wildfire and post-fire land management practices in humid Mediterranean headwater catchments could temporarily alter sediment processes, potentially impacting downstream water quality.

Regulatory and policy considerations are crucial in managing human-induced alterations to streamflow in headwater catchments. By incorporating regulatory frameworks and policy measures, it is possible to mitigate the impacts of anthropogenic activities on streamflow dynamics. Here are some ways in which regulatory and policy considerations can help manage human-induced alterations to streamflow in headwater catchments. Implementing land use planning regulations can help control land development activities in headwater catchments, ensuring sustainable land use practices that minimize adverse impacts on streamflow regimes (Penna et al. 2011). Effective water resource management policies can help regulate water use, allocation, and conservation practices to maintain streamflow levels in headwater catchments (Gaertner et al. 2020). Enforcing environmental protection regulations can safeguard headwater catchments from degradation due to human activities, preserving natural hydrological processes and streamflow dynamics (Hatley et al. 2023). Policies aimed at climate change adaptation can also help address the impacts of changing climatic conditions on streamflow regimes in headwater catchments, ensuring resilience to future hydrological challenges (Biederman et al. 2015).

Regulations governing groundwater extraction and recharge practices can help maintain baseflow contributions to streams in headwater catchments, ensuring sustainable streamflow levels (Hellwig and Stahl 2018). Policies focusing on erosion control measures can help reduce sediment inputs to streams, preserving water quality and streamflow patterns in headwater catchments (Goulsbra et al. 2009). Regulations promoting wetland conservation and restoration can enhance water retention capacity in headwater catchments, influencing streamflow dynamics and reducing runoff variability (Freyberg et al. 2015). Involving local communities in decision-making through participatory approaches can help raise awareness about the importance of sustainable land management practices for streamflow in headwater catchments (Spence et al. 2010).

1.6 Ecological Consequences

It is commonly understood that the flow regime significantly influences the structure and function of rivers and their adjoining floodplain wetlands. Alterations to this flow pose a persistent and serious threat to freshwater ecosystems (Bunn and Arthington 2002). Streamflow alterations disrupt the delicate balance of aquatic ecosystems, impacting various ecological processes and functions. Changes in flow magnitude, frequency, and timing can have far-reaching consequences. Altered streamflow can modify the physical characteristics of aquatic habitats, such as channel morphology, substrate composition, and water depth. These changes can affect the availability of suitable habitats for aquatic organisms, leading to species distribution and abundance shifts. The duration and seasonal timing of associated low flow conditions, along with the reduced flow variability, strongly influence riverine organisms directly and via changes to habitat (Munasinghe et al. 2021). This inhibits access to spawning grounds, feeding areas, and essential habitats, leading to population declines and genetic isolation (Chen et al. 2023; Folke et al. 2004). Altered streamflow can fragment habitats by isolating populations and reducing connectivity between riverine habitats. This fragmentation limits gene flow and dispersal, increasing the vulnerability of fish populations to environmental stressors and reducing overall biodiversity (Marshall et al. 2021). Modifications of hydrologic regimes can indirectly change the composition, structure, or function of aquatic systems, affecting water quality, energy sources, physical habitat characteristics, and biotic interactions and damaging the ecological integrity of rivers (Richter et al. 1996). Any modifications in flow regimes can influence water quality parameters such as turbidity, nutrient concentrations, and dissolved oxygen levels, with implications for the health of aquatic biota (Nakulopa et al. 2024; Usman et al. 2021). Streamflow variations can influence primary production dynamics in aquatic ecosystems. Changes in flow regimes may alter nutrient availability, light penetration, and sediment deposition, consequently affecting the growth rates of aquatic plants and algae, which form the food web base

(Talbot et al. 2018). Streamflow influences various ecosystem processes, including nutrient uptake, organic matter processing, and ecosystem metabolism. Changes in streamflow regimes can disrupt these processes, impacting the overall functioning of aquatic ecosystems (Palmer and Ruhí 2019).

Riparian ecosystems play a crucial role in the health and functioning of aquatic ecosystems, particularly in headwater catchments. Streamflow regime changes can significantly impact riparian ecosystems, influencing biodiversity, habitat quality, and ecosystem processes. Various studies have highlighted the influence of rapid soil saturation in riparian zones on catchment runoff response, indicating the importance of soil moisture in riparian areas for runoff generation processes (Penna et al. 2011). Riparian zones in headwater streams contribute significantly to the biodiversity of river networks, with the diversity of life in headwater streams playing a crucial role in the overall biodiversity of aquatic ecosystems (Meyer et al. 2007). Changes in the dominant water sources between headwaters and larger streams can lead to a disconnection in which the persistent browning signal in headwaters, driven by riparian soils, is only occasionally transmitted downstream to larger rivers, indicating the complex dynamics of water sources in riparian ecosystems (Fork et al. 2020).

Changes in riparian and channel structures over time and space, influenced by mass movements and flow characteristics in headwater tributaries, lead to various patterns of biological assemblies within the channel networks, influencing the ecological responses in riparian ecosystems (Gomi et al. 2002). Riparian forests have been shown to mitigate the warming and environmental degradation of agricultural headwater streams, with forested reaches exhibiting lower water temperatures and temperature variations than open reaches (Turunen et al. 2021). Riparian zones, particularly in headwater lakes, play a critical role in regulating and influencing downstream discharge dynamics, which can significantly influence stream ecosystems and water quality downstream of lakes (Leach et al. 2017). Seasonal soil moisture patterns in riparian zones can control transit time distributions in forested headwater catchments, highlighting the importance of riparian areas in regulating water flow dynamics (Stockinger et al. 2014).

1.7 Infrastructure Vulnerabilities and Mitigation Strategies

The headwater stream originates from higher-slope areas and moves towards the mid-slope and finally towards the plains (Fig. 1.1). Meanwhile, in the transport process, water is used for different purposes, such as human civilization's well-being and the ecosystem's benefit. Any change in streamflow in headwater catchments can pose challenges to infrastructure and adaptation strategies in various ways. Changes in streamflow patterns can affect water availability for various uses, including agriculture, industry, and domestic consumption. Reduced streamflow can challenge water supply infrastructure and require adaptations to ensure water availability (Zheng et al. 2009). Altered streamflow regimes can challenge water resource management, requiring adjustments in water allocation strategies, reservoir operations, and water supply planning to cope with changing flow patterns (Penna et al. 2011). Changes in streamflow can impact the vulnerability of infrastructure such as dams, bridges, and flood protection systems, necessitating upgrades or modifications to infrastructure to withstand changing hydrological conditions (Evans et al. 2015; Fischer et al. 2015). Shifting streamflow patterns, including changes in flood magnitudes and frequencies, can challenge infrastructure resilience to extreme events (Vlach et al. 2020). Changes in streamflow regimes can affect water quality parameters such as sediment transport, nutrient concentrations, and pollutant levels (Leach et al. 2020).

Changes in streamflow can influence groundwater-surface water interactions, affecting water availability and quality (Camporese et al. 2014; Özgen-Xian et al. 2023). Changes in streamflow dynamics, mainly related to snowmelt processes, can impact water availability and timing (Sprenger et al. 2022). Increased frequency and intensity of extreme weather events, such as heavy rainfall, can overwhelm existing flood defense mechanisms, leading to an increased risk of flooding in urban and rural areas (Bhowmik et al. 2020). Extreme weather events, including heavy rainfall and floods, can exceed the capacity of wastewater treatment plants, leading to untreated or partially treated wastewater being released into the environment, posing risks to public health and ecosystems (Hughes et al. 2021). Variability in streamflow and changing precipitation patterns can lead to water scarcity, affecting the availability of water for irrigation and consequently impacting food security (Liu et al. 2023). Alterations in streamflow patterns can affect the generation capacity of hydropower facilities, impacting energy supply and reliability (Liu et al. 2023).

Changing streamflow affects water distribution among various users, leading to potential conflicts between agricultural, industrial, municipal, and environmental stakeholders. Managing these conflicts, especially under conditions of scarcity, is a growing challenge.

The different adaptation strategies need to consider the following criteria:

(a) Adaptation strategies may need to enhance infrastructure resilience to withstand more frequent or intense flood events.
(b) Infrastructure and adaptation strategies may need to address water quality concerns from altered streamflow patterns
(c) Infrastructure and adaptation strategies may need to account for shifts in snowmelt contributions to streamflow and associated water resource management challenges
(d) As flow patterns change, the structures may become less effective or inadequate, requiring significant investments in retrofitting or building new infrastructure.
(e) Water rights and allocation systems that are based on historical streamflow patterns may no longer be appropriate as those patterns change. This necessitates legal and regulatory adjustments, which can be contentious and time-consuming.
(f) The economic implications of changing streamflow patterns are significant. They affect water treatment and distribution costs, agricultural productivity, energy generation, and overall economic growth in regions dependent on water resources.

Adapting infrastructure to the challenges posed by changing streamflow, climate change, and human activities requires a multifaceted approach that includes technological, regulatory, and behavioral changes. Upgrading and retrofitting existing water infrastructure to withstand extreme weather events and changing hydrological conditions. This includes reinforcing flood defenses, expanding the capacity of stormwater management systems, and enhancing the robustness of water supply and treatment facilities (Climate-Resilient Infrastructure 2018). Utilizing climate models and projections to inform the design and operation of water-related infrastructure ensures that future climate variability and trends are considered in the planning stage (Brockway and Dunn 2020). Implementing water-saving technologies and practices in agricultural, industrial, and municipal sectors to reduce water demand and increase efficiency. This includes adopting drip irrigation, recycling wastewater, and promoting water conservation measures among consumers (Levidow et al. 2014). Adopting IWRM (Integrated Water Resource Management) approaches that consider the entire watershed and encourage cross-sectoral coordination and cooperation among different water users. This helps balance human needs with ecosystem requirements and manage water resources in a holistic manner. Implementing NBS (Nature-Based Solutions), such as wetland restoration, green infrastructure, and reforestation, to enhance water retention, improve water quality, and reduce flood risks in a sustainable manner. Developing flexible water management policies and regulations that can be adapted in response to new information and changing conditions. This includes establishing water allocation priorities, pricing mechanisms, and incentives for water conservation and efficiency. Engaging communities and stakeholders in the planning and decision-making processes to raise awareness about water-related challenges and encourage collective action towards sustainable water management. By implementing these adaptation strategies, communities and governments can enhance the resilience of water-related infrastructure, mitigate the impacts of climate change and human activities, and ensure the sustainable management of water resources for future generations.

1.8 Recreation and Tourism

Changes in streamflow in headwater catchments can significantly impact recreation and tourism, affecting the availability of water-based activities, scenic beauty, and overall visitor experience. Altered streamflow regimes can impact the availability and quality of water for recreational activities such as fishing, boating, and swimming, and may affect visitor numbers, revenue generation, and the overall economic viability of tourism operations in headwater catchments (Gronsdahl et al. 2019). Changes in streamflow patterns may affect the suitability and safety of these activities, potentially reducing visitor satisfaction (Zheng et al. 2009). Changes in streamflow can influence the visual appeal and aesthetic value of headwater catchments, impacting the scenic beauty of rivers, waterfalls, and natural landscapes. Altered streamflow regimes may affect the attractiveness of these areas for tourists and outdoor enthusiasts (Biederman et al. 2015). Changes in streamflow can affect riparian habitats and wildlife populations, potentially impacting opportunities for wildlife viewing and nature-based tourism. Shifts in streamflow patterns may alter the presence and behavior of wildlife species, influencing visitor experiences (Brantley et al. 2014). Variations in streamflow patterns may affect the timing and availability of recreational opportunities, influencing visitor preferences and visitation patterns (Fischer et al. 2015). Streamflow changes can influence local communities' resilience, dependent on recreation and tourism.

Altered streamflow regimes may require adjustments to recreational infrastructure such as boat ramps, fishing docks, and hiking trails to accommodate changing water levels and flow conditions. Infrastructure adaptation may be necessary to ensure visitor safety and access (Hu et al. 2011). Adapting to altered streamflow regimes may require community engagement, planning, and investment in sustainable tourism practices (Zhao et al. 2010).

1.9 Adaptation and Mitigation

Mitigating streamflow shifts is crucial for reducing their adverse effects and ensuring the sustainability of basin hydrology. This complex process goes beyond simple watershed management and encompasses many factors and strategies. Among these, sustainable agricultural practices (Janker et al. 2018), the creation of riparian buffers (Sargac et al. 2021), the development of artificial wetlands (Nottingham and Messer 2021), the implementation of rainwater harvesting systems (Ayling et al. 2021), the conceptualization of sponge cities (Zha et al. 2021), and the application of various other ecohydrological approaches play a significant role. These methods are increasingly integrated into watershed management practices to address the multifaceted challenges posed by streamflow shifts. Effective mitigation requires a holistic approach that combines watershed management, climate change mitigation, urban planning and green infrastructure, adaptive management, and robust policy and governance frameworks. Together, these strategies comprehensively respond to the complex issues surrounding streamflow changes. Sustainable agriculture, for instance, helps reduce runoff and enhance groundwater recharge, while riparian buffers act as natural filters, improving water quality. Artificial wetlands store floodwater and release it during droughts, contributing to water balance in the ecosystem. Rainwater harvesting and sponge cities are innovative urban strategies that manage water locally to reduce pressure on natural water bodies and enhance urban resilience. Integrating these ecohydrological approaches within the broader context of watershed management, climate action, and policy development is critical for effectively mitigating the effects of streamflow shifts. This underscores the importance of adopting multifaceted and interconnected strategies to ensure the long-term health and sustainability of water basins.

Water resource systems designed for sustainability are built and managed to be more adaptable, strong, and able to withstand uncertain and changing conditions. These systems need to work well even when supplies, management goals, and demands change. Like any system, they might fail, but they should be able to get back to normal without too much cost. The failure to meet diverse water needs often shows problems in planning, controlling, and decision-making at various levels. Effective management of water resources, ensuring enough, affordable, and lasting water supplies of good quality for people and the environment, depends on recognizing and addressing social and economic issues like poverty, population growth, corruption, and lack of education.

There are two general approaches to sustainable planning and management. One is from the top down, often called command and control. The other is called the grassroots approach. Working together, both approaches can lead to an integrated plan and management policy (Loucks and Van Beek 2017).

Top-Down Planning and Management

Headwater sustainability depends on how it is monitored and managed for the people's and the environment's welfare. This approach assumes that one or more institutions have the ability and authority to develop and implement a plan for monitoring and managing the coordinated development and operation of basin activities impacting the basin's surface and ground waters.

Bottom-Up Planning and Management

Sustainable water resources directly rely on the participation of the local stakeholders, often related to citizens who are directly affected by any steps implied during water resource management. Thus, stakeholder participation improves the chance that the managed system will be sustainable. This is a bottom-up strategy.

Implementing Integrated Water Resource Management (IWRM) approaches involves coordinating water management across different sectors and stakeholders and considering social, economic, and environmental dimensions to achieve sustainable and equitable water use (Meran et al. 2021).

1.10 Conclusion—Future Prospects and Research Needs

In conclusion, the dynamics of headwater catchments are intricately linked to various environmental, climatic, and anthropogenic factors. The alterations in streamflow observed in these regions have profound impacts not only on the hydrological cycle but also on the ecosystems, infrastructures, and communities dependent on it. From this comprehensive analysis, the following impacts, along with their adaptation and mitigation strategies, are outlined.

- Changes in surface and subsurface fluxes lead to variations in flow regime connectivity, runoff generation mechanisms, and dynamics within the catchment, affecting the hydrological response across the area. These changes in hydrological processes in headwater catchments can be assessed using in-situ monitoring. Furthermore, combined with in-situ measurements, hydrological models can help evaluate the predominant runoff processes, understand the hydrological reactions to hydro-climatic variability, and estimate the mean residence times of runoff in these catchments. Additionally, advancements in remote sensing allow us to monitor changes in vegetation cover, land use, and land cover types in headwater catchments, providing insights into the impact of these changes on hydrological processes.
- To assess the shift in streamflow within headwater catchments, various models ranging from empirical to physically based hydrological models and more complex machine learning algorithms are used. Each model has its own advantages and uncertainties. The choice of model depends on the availability and quality of data. Combining traditional physical and empirical models with machine learning algorithms is beneficial for gaining a deeper understanding of streamflow shifts in headwater catchments.
- The impact of climate change on streamflow volume and frequency in headwater catchments is complex and multifaceted. Climate change significantly influences precipitation patterns, permafrost thaw, and snowmelt due to rising temperatures, and altering vegetation distribution. Several studies have examined the sensitivity of streamflow to climate variability and human activities in these areas, with findings suggesting that land use changes may play a more significant role than climate in altering streamflow. These changes in climatic variables and vegetation patterns can affect the hydrological regime of headwater catchments by influencing streamflow timing, magnitude, and duration.
- Human activities have significantly impacted streamflow regimes in headwater catchments. These anthropogenic factors encompass changes in land cover, such as deforestation, afforestation, urbanization, agricultural practices, and soil conservation measures. Collectively, these factors can modify infiltration rates and evapotranspiration, leading to alterations in streamflow patterns. Effective land use planning considering streamflow dynamics can preserve water resources, protect ecosystems, and promote sustainable development in these critical areas. Additionally, regulatory measures and policy considerations are essential for managing human-induced changes to streamflow in headwater catchments.
- Alterations in streamflow disrupt the balance of aquatic ecosystems, impacting habitat characteristics such as channel morphology, substrate composition, and water depth. These changes can isolate populations and affect species distribution and abundance. To mitigate these effects, restoration of natural flow regimes, the implementation of riparian buffers, and adaptive management practices are essential. These strategies help maintain biodiversity and support the health of aquatic environments.

- Water infrastructure, such as dams, reservoirs, and treatment plants, was traditionally designed based on historical streamflow data. However, shifting streamflow patterns, including changes in the magnitude and frequency of floods, pose challenges to the resilience of this infrastructure against extreme events. Key strategies to mitigate vulnerabilities under changing hydrological conditions and extreme weather include upgrading and retrofitting existing infrastructure, using climate models to predict future climatic variables for informed design, implementing water-saving technologies, and adopting Integrated Water Resource Management (IWRM) approaches. These measures collectively enhance infrastructure resilience and sustainability.
- Changes in streamflow in headwater catchments can significantly impact recreation and tourism, affecting the availability of water-based activities, scenic beauty, and overall visitor experience. Altered streamflow regimes may require adjustments to recreational infrastructure such as boat ramps, fishing docks, and hiking trails to accommodate changing water levels and flow conditions. To mitigate these challenges, it is essential to develop adaptive management strategies that regularly monitor water conditions, update infrastructure plans, and engage with stakeholders to align recreational offerings with the changing environment.

Future research on headwater catchments should address the impacts of climate change, urbanization, and pollution by developing strategies for climate resilience, exploring integrated water management approaches, and investigating pollution control methods to improve water quality. Studies should also focus on ecosystem conservation, assess socio-economic impacts to inform policy, and embrace technological innovations for monitoring and management. Additionally, efforts should include climate change mitigation to reduce these impacts and promote collaborative management practices for shared water resources. By integrating these diverse research directions, we can ensure the sustainable management of headwater catchments, supporting ecological health and community well-being in the face of environmental challenges. To be precise, a few potential research questions can be: What are the primary drivers of changes in streamflow regimes globally, and how do natural factors such as climate variability and human activities contribute to these shifts? How do changes in streamflow regimes impact water availability, water quality, and water management practices in different regions, and what are the implications for sustainable water resource planning? What are the potential ecological consequences of streamflow regime shifts, and how do alterations in streamflow patterns affect aquatic ecosystems, biodiversity, and riparian habitats? How can advanced modeling techniques, such as machine learning and hidden Markov models, be utilized to analyze and predict changes in streamflow regimes, and what are the implications for water resource management?

References

Abbott MB et al (1986) An introduction to the European Hydrological System—Systeme Hydrologique Europeen, "SHE", 2: Structure of a physically-based, distributed modelling system. J Hydrol 87(1–2):61–77

Alexander R, Boyer E, Smith R, Schwarz G, Moore R (2007) The role of headwater streams in downstream water quality. Jawra J Am Water Resources Assoc 43(1):41–59. https://doi.org/10.1111/j.1752-1688.2007.00005.x

Almikaeel W, Čubanová L, Šoltész A (2022) Hydrological drought forecasting using machine learning—Gidra river case study. Water 14(3):387. https://doi.org/10.3390/w14030387

Ammann L, Doppler T, Stamm C, Reichert P, Fenicia F (2020) Characterizing fast herbicide transport in a small agricultural catchment with conceptual models. J Hydrol 586:124812. https://doi.org/10.1016/j.jhydrol.2020.124812

Arnold JG, Srinivasan R, Muttiah RS, Williams JR (1998) Large area hydrologic modeling and assessment part I: model development. JAWRA J Am Water Resources Assoc 34(1):73–89. https://doi.org/10.1111/j.1752-1688.1998.tb05961.x

Ayling SM, Phillips N, Bunney S (2021) Allotments in the future: Building resilience to climate change through improved site design and efficient water practices. Water (Switzerland) 13(11):1457. https://doi.org/10.3390/W13111457/S1

Bai X, Zhao W, Liu H, Zhang Y, Yang Q, Liu J, Chang X (2023) Effects of precipitation changes and land-use alteration on streamflow: a comparative analysis from two adjacent catchments in the Qilian mountains, arid northwestern China. Front Environ Sci 11. https://doi.org/10.3389/fenvs.2023.1097049

Beck HE, Bruijnzeel LA, Van Dijk AIJM, McVicar TR, Scatena FN, Schellekens J (2013) The impact of forest regeneration on streamflow in 12 mesoscale humid tropical catchments. Hydrol Earth Syst Sci 17(7):2613–2635. https://doi.org/10.5194/HESS-17-2613-2013

Bergström S (1995) The HBV model. Comp Models Watershed Hydrol:443–476

Bhowmik R, Ng T, Wang J (2020) Understanding the impact of observation data uncertainty on probabilistic streamflow forecasts using a dynamic hierarchical model. Water Resources Res 56(4). https://doi.org/10.1029/2019wr025463

Bicknell BR, Imhoff JC, Kittle Jr JL, Donigian Jr AS, Johanson RC (1997) Hydrological simulation program—FORTRAN user's manual for version 11. Environmental Protection Agency Report No. EPA/600/R-97/080. US Environmental Protection Agency, Athens, GA. https://p2infohouse.org/ref/07/06417.pdf

Biederman J, Somor A, Harpold A, Gutmann E, Breshears D, Troch P, Brooks P (2015) Recent tree die-off has little effect on streamflow in contrast to expected increases from historical studies. Water Resour Res 51(12):9775–9789. https://doi.org/10.1002/2015wr017401

Bouaziz L, Weerts A, Schellekens J, Sprokkereef E, Stam J, Savenije H, Hrachowitz M (2018) Redressing the balance: quantifying net intercatchment groundwater flows. Hydrol Earth Syst Sci 22(12):6415–6434. https://doi.org/10.5194/hess-22-6415-2018

Box GEP et al (2015) Time series analysis: forecasting and control. John Wiley and Sons

Brantley S, Miniat C, Elliott K, Laseter S, Vose J (2014) Changes to southern Appalachian water yield and stormflow after loss of a foundation species. Ecohydrology 8(3):518–528. https://doi.org/10.1002/eco.1521

Brockway AM, Dunn LN (2020). Weathering adaptation: Grid infrastructure planning in a changing climate. Clim Risk Manage 30. https://doi.org/10.1016/j.crm.2020.100256

Buechel M, Slater L, Dadson S (2022) Hydrological impact of widespread afforestation in great britain using a large ensemble of modelled scenarios. Commun Earth Environ 3(1). https://doi.org/10.1038/s43247-021-00334-0

Bunn S, Arthington A (2002) Basic principles and ecological consequences of altered flow regimes for aquatic biodiversity. Environ Manage 30(4):492–507. https://doi.org/10.1007/s00267-002-2737-0

Buytaert W, Zulkafli Z, Grainger S, Acosta L, Alemie T, Bastiaensen J, Zhumanova M (2014) Citizen science in hydrology and water resources: opportunities for knowledge generation, ecosystem service management, and sustainable development. Front Earth Sci 2. https://doi.org/10.3389/feart.2014.00026

Camporese M, Penna D, Borga M, Paniconi C (2014) A field and modeling study of nonlinear storage-discharge dynamics for an Alpine headwater catchment. Water Resour Res 50(2):806–822. https://doi.org/10.1002/2013wr013604

Chen X, Jiang L, Luo Y, Liu J (2023) A global streamflow indices time series dataset for large-sample hydrological analyses on streamflow regime (until 2021). https://doi.org/10.5194/essd-2023-49

Chiew F (2006) Estimation of rainfall elasticity of streamflow in Australia. Hydrol Sci J 51(4):613–625. https://doi.org/10.1623/hysj.51.4.613

Climate-resilient Infrastructure (2018)

Crawford NH, Linsley RK (1966) Digital simulation in hydrology. Stanford Watershed Model 4

Dupas R, Musolff A, Jawitz J, Rao P, Jäger C, Fleckenstein J, Borchardt D (2017) Carbon and nutrient export regimes from headwater catchments to downstream reaches. Biogeosciences 14(18):4391–4407. https://doi.org/10.5194/bg-14-4391-2017

Dupas R, Abbott B, Minaudo C, Fovet O (2019) Distribution of landscape units within catchments influences nutrient export dynamics. Front Environ Sci 7. https://doi.org/10.3389/fenvs.2019.00043

Evans S, Ge S, Liang S (2015) Analysis of groundwater flow in mountainous, headwater catchments with permafrost. Water Resour Res 51(12):9564–9576. https://doi.org/10.1002/2015wr017732

Fischer B, Rinderer M, Schneider P, Ewen T, Seibert J (2015) Contributing sources to baseflow in pre-alpine headwaters using spatial snapshot sampling. Hydrol Process 29(26):5321–5336. https://doi.org/10.1002/hyp.10529

Fischer B, Stähli M, Seibert J (2016) Pre-event water contributions to runoff events of different magnitude in pre-alpine headwaters. Hydrol Res 48(1):28–47. https://doi.org/10.2166/nh.2016.176

Folke C, Carpenter S, Walker B, Scheffer M, Elmqvist T, Gunderson L, Holling C (2004) Regime shifts, resilience, and biodiversity in ecosystem management. Annu Rev Ecol Evol Syst 35(1):557–581. https://doi.org/10.1146/annurev.ecolsys.35.021103.105711

Fork M, Sponseller R, Laudon H (2020) Changing source-transport dynamics drive differential browning trends in a boreal stream network. Water Resources Res 56(2). https://doi.org/10.1029/2019wr026336

Freyberg J, Rao P, Radny D, Schirmer M (2015) The impact of hillslope groundwater dynamics and landscape functioning in event-flow generation: a field study in the Rietholzbach catchment, Switzerland. Hydrogeol J 23(5):935–948. https://doi.org/10.1007/s10040-015-1238-1

Gaertner B, Fernández R, Zégre N (2020) Twenty-first century streamflow and climate change in forest catchments of the central Appalachian mountains region, US. Water 12(2):453. https://doi.org/10.3390/w12020453

Gao Z, Zhang L, Zhang X, Cheng L, Potter N, Cowan T, Cai W (2015) Long-term streamflow trends in the middle reaches of the Yellow river basin: detecting drivers of change. Hydrol Process 30(9):1315–1329. https://doi.org/10.1002/hyp.10704

George R, McManamay R, Perry D, Sabo J, Ruddell BL (2021) Indicators of hydro-ecological alteration for the rivers of the United States. Ecol Ind 120:106908

Gomi T, Sidle R, Richardson J (2002) Understanding processes and downstream linkages of headwater systems. Bioscience 52(10):905. https://doi.org/10.1641/0006-3568(2002)052[0905:upadlo]2.0.co;2

Goulsbra C, Lindsay J, Evans M (2009) A new approach to the application of electrical resistance sensors to measuring the onset of ephemeral streamflow in wetland environments. Water Resources Res 45(9). https://doi.org/10.1029/2009wr007789

Gronsdahl S, Moore R, Rosenfeld J, McCleary R, Winkler R (2019) Effects of forestry on summertime low flows and physical fish habitat in snowmelt-dominant headwater catchments of the pacific northwest. Hydrol Process 33(25):3152–3168. https://doi.org/10.1002/hyp.13580

Guo Y, Zhang L, Zhang Y, Wang Z, Zheng H (2021) Estimating impacts of wildfire and climate variability on streamflow in Victoria, Australia. Hydrol Process 35(12). https://doi.org/10.1002/hyp.14439

Gustard A, Bullock A, Dixon JM (1992) Low flow estimation in the United Kingdom. Institute of Hydrology

Harpold A, Brooks P (2018) Humidity determines snowpack ablation under a warming climate. Proc Natl Acad Sci 115(6):1215–1220. https://doi.org/10.1073/pnas.1716789115

Hatley C, Armijo B, Andrews K, Anhold C, Nippert J, Kirk M (2023) Intermittent streamflow generation in a merokarst headwater catchment. Environ Sci Adv 2(1):115–131. https://doi.org/10.1039/d2va00191h

Heinz M, Graeber D, Zak D, Zwirnmann E, Gelbrecht J, Pusch M (2015) Comparison of organic matter composition in agricultural versus forest affected headwaters with special emphasis on organic nitrogen. Environ Sci Technol 49(4):2081–2090. https://doi.org/10.1021/es505146h

Hellwig J, Stahl K (2018) An assessment of trends and potential future changes in groundwater-baseflow drought based on catchment response times. Hydrol Earth Syst Sci 22(12):6209–6224. https://doi.org/10.5194/hess-22-6209-2018

Hughes J, Cowper-Heays K, Olesson E, Bell R, Stroombergen A (2021) Impacts and implications of climate change on wastewater systems: a New Zealand perspective. Clim Risk Manage 31. https://doi.org/10.1016/j.crm.2020.100262

Hu Y, Maskey S, Uhlenbrook S, Zhao H (2011) Streamflow trends and climate linkages in the source region of the Yellow river, China. Hydrol Process 25(22):3399–3411. https://doi.org/10.1002/hyp.8069

Hwang T, Band L, Vose J, Tague C (2012) Ecosystem processes at the watershed scale: hydrologic vegetation gradient as an indicator for lateral hydrologic connectivity of headwater catchments. Water Resources Res 48(6). https://doi.org/10.1029/2011wr011301

Iwasaki K, Nagasaka Y, Nagasaka A (2021) Geological effects on the scaling relationships of groundwater contributions in forested watersheds. Water Resources Res 57(7). https://doi.org/10.1029/2021wr029641

Janker J, Mann S, Rist S (2018) What is sustainable agriculture? Critical analysis of the international political discourse. Sustainability 10(12):4707. https://doi.org/10.3390/SU10124707

Kim J, Ale S, Teague W, Wang T (2022) Evaluating hydrological components and streamflow characteristics under conventional and adaptive multi-paddock grazing management. River Res Appl 38(4):776–787. https://doi.org/10.1002/rra.3948

Kosugi K, Katsura S, Katsuyama M, Mizuyama T (2006) Water flow processes in weathered granitic bedrock and their effects on runoff generation in a small headwater catchment. Water Resources Res 42(2). https://doi.org/10.1029/2005wr004275

Lamichhane M, Phuyal S, Mahato R, Shrestha A, Pudasaini U, Lama SD, Chapagain AR, Mehan S, Neupane D (2024) Assessing climate change impacts on streamflow and baseflow in the Karnali River Basin, Nepal: a CMIP6 multi-model ensemble approach using SWAT and web-based hydrograph analysis tool. Sustainability 16(8):3262. https://doi.org/10.3390/su16083262

Leach J, Moore R (2019) Empirical stream thermal sensitivities may underestimate stream temperature response to climate warming. Water Resour Res 55(7):5453–5467. https://doi.org/10.1029/2018wr024236

Leach J, Lidberg W, Kuglerová L, Peralta-Tapia A, Ågren A, Laudon H (2017) Evaluating topography-based predictions of shallow lateral groundwater discharge zones for a boreal lake-stream system. Water Resour Res 53(7):5420–5437. https://doi.org/10.1002/2016wr019804

Leach J, Buttle J, Webster K, Hazlett P, Jeffries D (2020) Travel times for snowmelt-dominated headwater catchments: influences of wetlands and forest harvesting, and linkages to stream water quality. Hydrol Process 34(10):2154–2175. https://doi.org/10.1002/hyp.13746

Leh M, Bajwa S, Chaubey I (2011) Impact of land use change on erosion risk: an integrated remote sensing, geographic information system and modeling methodology. Land Degrad Dev 24(5):409–421. https://doi.org/10.1002/ldr.1137

Levidow L, Zaccaria D, Maia R, Vivas E, Todorovic M, Scardigno A (2014) Improving water-efficient irrigation: prospects and difficulties of innovative practices. Agric Water Manag 146:84–94. https://doi.org/10.1016/j.agwat.2014.07.012

Li L, Zhang L, Wang H, Wang J, Yang J, Jiang D, Du Q (2007) Assessing the impact of climate variability and human activities on streamflow from the Wuding river basin in China. Hydrol Process 21(25):3485–3491. https://doi.org/10.1002/hyp.6485

Li L, Sullivan P, Benettin P, Cirpka O, Bishop K, Brantley S, Kirchner J (2020) Toward catchment hydro-biogeochemical theories. Wiley Interdiscip Rev Water 8(1). https://doi.org/10.1002/wat2.1495

Li M, Gu H, Wang H, Wang Y, Chi B (2022) Separating the impacts of climate variability and human activities on the discharge and component of streamflow in Taoer river basin, China. https://doi.org/10.21203/rs.3.rs-2212197/v1

Liu Z, Cuo L, Li Q, Liu X, Ma X, Liang L, Ding J (2020) Impacts of climate change and land use/cover change on streamflow in beichuan river basin in Qinghai province, China. Water 12(4):1198. https://doi.org/10.3390/w12041198

Liu J, Fu Z, Liu W (2023) Impacts of precipitation variations on agricultural water scarcity under historical and future climate change. J Hydrol 617. https://doi.org/10.1016/j.jhydrol.2022.128999

Loucks DP, Van Beek E (2017) Water resource systems planning and management: An introduction to methods, models, and applications. Springer.

Lv M, Ma Z (2018) Effects of climate/land surface changes on streamflow with consideration of precipitation intensity and catchment characteristics in the Yellow river basin. J Geophys Res Atmos 123(4):1942–1958. https://doi.org/10.1002/2017jd027625

Marshall JC, Lobegeiger JS, Starkey A (2021) Risks to fish populations in dryland rivers from the combined threats of drought and instream barriers. Front Environ Sci 9. https://doi.org/10.3389/fenvs.2021.671556

Meran G, Siehlow M, Von Hirschhausen C (n.d.) Springer water the economics of water rules and institutions. http://www.springer.com/series/13419

McGlynn B, McDonnell J, Seibert J, Kendall C (2004) Scale effects on headwater catchment runoff timing, flow sources, and groundwater-streamflow relations. Water Resources Res 40(7). https://doi.org/10.1029/2003wr002494

Meran G, Siehlow M, von Hirschhausen C (2021) The economics of water: rules and institutions. Springer.

Meriö L, Ala-aho P, Linjama J, Hjort J, Kløve B (2019) Snow to precipitation ratio controls catchment storage and summer flows in boreal headwater catchments. Water Resour Res 55(5):4096–4109. https://doi.org/10.1029/2018wr023031

Meyer J, Strayer D, Wallace J, Eggert S, Helfman G, Leonard N (2007) The contribution of headwater streams to biodiversity in river networks. Jawra J Am Water Resources Assoc 43(1):86–103. https://doi.org/10.1111/j.1752-1688.2007.00008.x

Molina A, Vanacker V, Brisson E, Mora D, Balthazar V (2015) Multidecadal change in streamflow associated with anthropogenic disturbances in the tropical andes. Hydrol Earth Syst Sci 19(10):4201–4213. https://doi.org/10.5194/hess-19-4201-2015

Molini A, Katul G, Porporato A (2011) Maximum discharge from snowmelt in a changing climate. Geophys Res Lett 38(5). https://doi.org/10.1029/2010gl046477

Mosavi A, Öztürk P, Chau K (2018) Flood prediction using machine learning models: literature review. Water 10(11):1536. https://doi.org/10.3390/w10111536

Mujumdar PP, Nagesh Kumar D (1990) Stochastic models of streamflow: some case studies. Hydrol Sci J 35(4):395–410

Munasinghe DSN, Najim MMM, Quadroni S, Musthafa MM (2021) Impacts of streamflow alteration on benthic macroinvertebrates by mini-hydro diversion in Sri Lanka. Scient Rep 11(1). https://doi.org/10.1038/s41598-020-79576-5
Nakulopa F, Bärlund I, Borchardt D (2024) How a reservoir modulates downstream water quality under declining upstream loading and progressing climate change. Sci Tot Environ 912. https://doi.org/10.1016/j.scitotenv.2023.169460
Nauditt A, Birkel C, Soulsby C, Ribbe L (2016) Conceptual modelling to assess the influence of hydro-climatic variability on runoff processes in data scarce semi-arid Andean catchments. Hydrol Sci J 62(4):515–532. https://doi.org/10.1080/02626667.2016.1240870
Nottingham ER, Messer TL (2021) A literature review of wetland treatment systems used to treat runoff mixtures containing antibiotics and pesticides from urban and agricultural landscapes. Water (Switzerland) 13(24):3631. https://doi.org/10.3390/W13243631/S1
Nunes J, Bernard-Jannin L, Rodríguez-Blanco M, Boulet A, Santos J, Keizer J (2020) Impacts of wildfire and post-fire land management on hydrological and sediment processes in a humid mediterranean headwater catchment. Hydrol Process 34(26):5210–5228. https://doi.org/10.1002/hyp.13926
Özgen-Xian I, Molins S, Johnson R, Xu Z, Dwivedi D, Loritz R, Steefel C (2023) Understanding the hydrological response of a headwater-dominated catchment by analysis of distributed surface–subsurface interactions. Scient Rep 13(1). https://doi.org/10.1038/s41598-023-31925-w
Palmer M, Ruhí A (2019) Linkages between flow regime, biota, and ecosystem processes: implications for river restoration. Science 365(6459). https://doi.org/10.1126/science.aaw2087
Patterson L, Lutz B, Doyle M (2013) Climate and direct human contributions to changes in mean annual streamflow in the south Atlantic, USA. Water Resour Res 49(11):7278–7291. https://doi.org/10.1002/2013wr014618
Peña-Arancibia JL, Van Dijk AIJM, Mulligan M, Bruijnzeel LA (2010) The role of climatic and terrain attributes in estimating baseflow recession in tropical catchments. Hydrol Earth Syst Sci 14(11):2193–2205. https://doi.org/10.5194/HESS-14-2193-2010
Penna D, Meerveld I, Gobbi A, Borga M, Fontana G (2011) The influence of soil moisture on threshold runoff generation processes in an alpine headwater catchment. Hydrol Earth Syst Sci 15(3):689–702. https://doi.org/10.5194/hess-15-689-2011
Pettyjohn WA, Henning R (1979) Preliminary estimate of ground-water recharge rates, related streamflow and water quality in Ohio. Water Resources Centre Project Completion Rep. 552, Ohio State University, 323 pp
Richter B, Baumgartner J, Powell J, Braun D (1996) A method for assessing hydrologic alteration within ecosystems. Conserv Biol 10(4):1163–1174. https://doi.org/10.1046/j.1523-1739.1996.10041163.x
Rodgers P, Soulsby C, Waldron S (2005a) Stable isotope tracers as diagnostic tools in upscaling flow path understanding and residence time estimates in a mountainous mesoscale catchment. Hydrol Process 19(11):2291–2307. https://doi.org/10.1002/hyp.5677
Rodgers P, Soulsby C, Waldron S, Tetzlaff D (2005b) Using stable isotope tracers to assess hydrological flow paths, residence times and landscape influences in a nested mesoscale catchment. Hydrol Earth Syst Sci 9(3):139–155. https://doi.org/10.5194/hess-9-139-2005
Rogger M, Agnoletti M, Alaoui A, Bathurst J, Bodner G, Borga M, Blöschl G (2017) Land use change impacts on floods at the catchment scale: challenges and opportunities for future research. Water Resour Res 53(7):5209–5219. https://doi.org/10.1002/2017wr020723
Sargac J, Johnson RK, Burdon FJ, Truchy A, Rîşnoveanu G, Goethals P, McKie BG (2021) Forested riparian buffers change the taxonomic and functional composition of stream invertebrate communities in agricultural catchments. Water (Switzerland) 13(8):1028. https://doi.org/10.3390/W13081028/S1
Seyam M, Othman F (2014) Long-term variation analysis of a tropical river's annual streamflow regime over a 50-year period. Theoret Appl Climatol 121(1–2):71–85. https://doi.org/10.1007/s00704-014-1225-9

Sloto RA, Crouse MY (1996) Hysep: a computer program for streamflow hydrograph separation and analysis
Smakhtin VU (2001) Low flow hydrology: a review. J Hydrol 240(3–4):147–186. https://doi.org/10.1016/S0022-1694(00)00340-1
Shin M, Eum H, Kim C, Jung I (2016) Alteration of hydrologic indicators for korean catchments under cmip5 climate projections. Hydrol Process 30(24):4517–4542. https://doi.org/10.1002/hyp.10948
Spence C, Guan X, Phillips R, Hedstrom N, Granger R, Reid B (2010) Storage dynamics and streamflow in a catchment with a variable contributing area. Hydrol Process 24(16):2209–2221. https://doi.org/10.1002/hyp.7492
Sprenger M, Carroll R, Siirila-Woodburn E, Newcomer M, Brown W, Williams K (2022) Variability of snow and rainfall partitioning into evapotranspiration and summer runoff across nine mountainous catchments. Geophys Res Lett 49(13). https://doi.org/10.1029/2022gl099324
Stewart IT, Cayan DR, Dettinger MD (2004) Changes in snowmelt runoff timing in western North America under a "business as usual" climate change scenario. Clim Change 62(1–3):217–232. https://doi.org/10.1023/B:CLIM.0000013702.22656.E8/METRICS
Stigter E, Wanders N, Saloranta T, Shea J, Bierkens M, Immerzeel W (2016) Climate sensitivity of snow water equivalent and snowmelt runoff in a himalayan catchment. https://doi.org/10.5194/tc-2016-216
Stigter E, Wanders N, Saloranta T, Shea J, Bierkens M, Immerzeel W (2017) Assimilation of snow cover and snow depth into a snow model to estimate snow water equivalent and snowmelt runoff in a Himalayan catchment. Cryosphere 11(4):1647–1664. https://doi.org/10.5194/tc-11-1647-2017
Stockinger M, Bogena H, Lücke A, Diekkrüger B, Weiler M, Vereecken H (2014) Seasonal soil moisture patterns: controlling transit time distributions in a forested headwater catchment. Water Resour Res 50(6):5270–5289. https://doi.org/10.1002/2013wr014815
Sun Y, Liu Q, Yuan X, Yan S (2019) Assessment of streamflow regime alterations in tang river, China. E3s Web Conf 136:04003. https://doi.org/10.1051/e3sconf/201913604003
Talbot CJ, Bennett EM, Cassell K, Hanes DM, Minor EC, Paerl H, Raymond PA, Vargas R, Vidon PG, Wollheim W, Xenopoulos MA (2018) The impact of flooding on aquatic ecosystem services. Biogeochemistry 141(3):439–461. https://doi.org/10.1007/s10533-018-0449-7
Temnerud J, Brömssen C, Fölster J, Buffam I, Andersson J, Nyberg L, Bishop K (2016) Map-based prediction of organic carbon in headwater streams improved by downstream observations from the river outlet. Biogeosciences 13(2):399–413. https://doi.org/10.5194/bg-13-399-2016
Tiwari T, Laudon H, Beven K, Ågren A (2014) Downstream changes in doc: inferring contributions in the face of model uncertainties. Water Resour Res 50(1):514–525. https://doi.org/10.1002/2013wr014275
Tomer M, Porter S, James D, Horn J, Niemi J (2021) Comparing riparian buffer design classification data among watersheds representing iowa landscapes. Agrosyst Geosci Environ 4(2). https://doi.org/10.1002/agg2.20159
Turunen J, Elbrecht V, Steinke D, Aroviita J (2021) Riparian forests can mitigate warming and ecological degradation of agricultural headwater streams. Freshw Biol 66(4):785–798. https://doi.org/10.1111/fwb.13678
Usman M, Ndehedehe C, Ahmad B (2021) Modeling implications of climate induced streamflow changes on the fish species of the soan river, Pakistan. Model Earth Syst Environ 8(3):2965–2975. https://doi.org/10.1007/s40808-021-01272-8
Van Dijk AIJM (2010) Climate and terrain factors explaining streamflow response and recession in Australian catchments. Hydrol Earth Syst Sci 14(1):159–169. https://doi.org/10.5194/HESS-14-159-2010
Vicente-Serrano SM, López-Moreno JI, Beguería S, Lorenzo-Lacruz J, Azorin-Molina C, Morán-Tejeda E (2011) Accurate computation of a streamflow drought index. J Hydrol Eng 17(2):318–332. https://doi.org/10.1061/(ASCE)HE.1943-5584.0000433

Villamizar M, Stoate C, Biggs J, Morris C, Szczur J, Brown C (2020) Comparison of technical and systems-based approaches to managing pesticide contamination in surface water catchments. J Environ Manage 260:110027. https://doi.org/10.1016/j.jenvman.2019.110027

Vlach V, Ledvinka O, Matoušková M (2020) Changing low flow and streamflow drought seasonality in central european headwaters. Water 12(12):3575. https://doi.org/10.3390/w12123575

Votrubova J, Dohnal M, Vogel T, Šanda M, Tesař M (2017) Episodic runoff generation at central european headwater catchments studied using water isotope concentration signals. J Hydrol Hydromech 65(2):114–122. https://doi.org/10.1515/johh-2017-0002

Wigmosta MS, Vail LW, Lettenmaier DP (1994) A distributed hydrology-vegetation model for complex terrain. Water Resour Res 30(6):1665–1679

Wang J, Li H, Hao X (2010) Responses of snowmelt runoff to climatic change in an inland river basin, northwestern China, over the past 50 years. https://doi.org/10.5194/hessd-7-493-2010

Wang W, Dong Z, Lall U, Dong N, Yang M (2019) Monthly streamflow simulation for the headwater catchment of the Yellow river basin with a hybrid statistical-dynamical model. Water Resour Res 55(9):7606–7621. https://doi.org/10.1029/2019wr025103

Wheeler J, Cortés A, Sedlacek J, Karrenberg S, Kleunen M, Wipf S, Rixen C (2016) The snow and the willows: earlier spring snowmelt reduces performance in the low-lying alpine shrub salix herbacea. J Ecol 104(4):1041–1050. https://doi.org/10.1111/1365-2745.12579

Williams M, Buda A, Elliott H, Singha K, Hamlett J (2015) Influence of riparian seepage zones on nitrate variability in two agricultural headwater streams. Jawra J Am Water Resources Assoc 51(4):883–897. https://doi.org/10.1111/1752-1688.12335

Wu Y, Zhang G, Shen H, Xu Y (2018) Nonlinear response of streamflow to climate change in high-latitude regions: a case study in headwaters of nenjiang river basin in China's far northeast. Water 10(3):294. https://doi.org/10.3390/w10030294

Yang S, Chang L (2020) Regional inundation forecasting using machine learning techniques with the internet of things. Water 12(6):1578. https://doi.org/10.3390/w12061578

Zha X, Luo P, Zhu W, Wang S, Lyu J, Zhou M, Huo A, Wang Z (2021) A bibliometric analysis of the research on Sponge city: current situation and future development direction. Ecohydrology 14(7):e2328. https://doi.org/10.1002/ECO.2328

Zhang L, Zhao F, Brown A (2012) Predicting effects of plantation expansion on streamflow regime for catchments in Australia. Hydrol Earth Syst Sci 16(7):2109–2121

Zhang J, Zhang T, Lei Y, Zhang X, Li R (2015) Streamflow regime variations following ecological management on the loess plateau, China. Forests 7(12):6. https://doi.org/10.3390/f7010006

Zhao F, Xu Z, Zhang L, Zuo D (2009) Streamflow response to climate variability and human activities in the upper catchment of the Yellow river basin. Sci China Technol Sci 52(11):3249–3256. https://doi.org/10.1007/s11431-009-0354-3

Zhao F, Zhang L, Xu Z, Scott D (2010) Evaluation of methods for estimating the effects of vegetation change and climate variability on streamflow. Water Resources Res 46(3). https://doi.org/10.1029/2009wr007702

Zhao F, Xu Z, Zhang L (2011) Changes in streamflow regime following vegetation changes from paired catchments. Hydrol Process 26(10):1561–1573. https://doi.org/10.1002/hyp.8266

Zheng H, Zhang L, Zhu R, Liu C, Sato Y, Fukushima Y (2009) Responses of streamflow to climate and land surface change in the headwaters of the Yellow river basin. Water Resources Res 45(7). https://doi.org/10.1029/2007wr006665

Zheng W, Lamačová A, Yu X, Krám P, Hruška J, Zahradníček P, Farda A (2021) Assess hydrological responses to a warming climate at the lysina critical zone observatory in central europe. Hydrol Process 35(9). https://doi.org/10.1002/hyp.14281

Zhu R, Yang L, Liu T, Wen X, Zhang L, Chang Y (2019) Hydrological responses to the future climate change in a data scarce region, northwest China: application of machine learning models. Water 11(8):1588. https://doi.org/10.3390/w11081588

Chapter 2
Hydrology of the Upper Indus Basin under Changing Climate

Aftab Nazeer, Zakir Hussain Dahri, Faisal Baig, Muhammad Asif, and Muhammad Asif

Abstract The recent climate change accelerations intensify future water availability concerns from high-altitude mountainous regions. The Upper Indus Basin (UIB) is a highly snow-fed and glacierized basin from the High Mountains of Asia; thus, an accelerated melt due to climate change may severely affect these frozen water resources. Therefore, increasing knowledge of the future water availability in the Indus Basin under climate scenarios is vital. This research was designed to analyse the future water availability from the Hunza River Basin (sub-basins of the UIB) under the Coupled Model Intercomparison Project Phase 6 (CMIP6) climatic projections. The projected precipitation and temperature from two General Circulation Models (GCMs) were analysed. The Distance Distribution Dynamics (DDD), a data parsimonious rainfall-runoff model was set up using these projections. The simulations were performed for the historical (1991–2010), middle of the century (2041–2060) and end of the century (2081–2100) period under Shared Socioeconomics Pathways (SSP) SSP1, SSP2 and SSP5 emission scenarios. Relative to the baseline temperature, the future projections showed an increase between 1.1 and 8.6 °C based on both GCMs. The 12–32% annual precipitation increase is expected by the end of the twenty-first century, relative to the baseline precipitation. Furthermore, reduced precipitation as snow and increased rainfall are expected together with the changes

A. Nazeer (✉)
Department of Agricultural Engineering, Bahauddin Zakariya University (BZU), Multan 60800, Pakistan
e-mail: aftabnazeerihe@gmail.com

Department of Water Management, Delft University of Technology, P.O. Box 5048, 2600 GA Delft, The Netherlands

Z. H. Dahri
Pakistan Agricultural Research Council (PARC), Islamabad 45500, Pakistan

F. Baig
National Water and Energy Center, United Arab Emirates University, Al Ain, Abu Dhabi, UAE

M. Asif
School of Engineering, University of Basilicata, 85100 Potenza, Italy

Department of Mechanical and Aerospace Engineering, Mechatronics Laboratory, Politecnico di Torino, Turin, Italy

A. Nanda et al. (eds.), *Navigating the Nexus*, Water Science and Technology Library 102, https://doi.org/10.1007/978-3-031-76532-2_2

in precipitation cycles. Relative to the baseline glacier melt, a 38–265% glacier melt increase is expected in the river flow. An overall river flow increase of 23–126% is expected by the end of the twenty-first century. This research presented climate change associated future water availability for the study area. Thus, policy-makers and stakeholders could use these findings for reservoir development, modifying hydropower and reservoir development policy, sustainable agriculture, and efficient water resources management.

Keywords CMIP6 · Distance distribution dynamics (DDD) · General circulation models (GCMs) · Hunza basin · Upper Indus basin (UIB)

2.1 Introduction

Due to the unceasing greenhouse gas emissions and global warming, the climate change threat is continuously increasing. Evidence shows climate change impacts are most substantial for natural systems (Masson-Delmotte et al. 2021). The climate change acceleration further intensifies concerns about water availability from mountainous catchments at the pace of climate change (Hasson 2016). This may alter the changing frequency, and quantity of the climatic variables, for instance, precipitation (Hasson 2016). Yet, there are significant uncertainties in the projected and observed climate, particularly precipitation. These uncertainties prevent reliable assessment of the changes and variability at the global and regional scales for current and future water cycles.

The Upper Indus Basin (UIB) exhibited significantly increasing precipitation trends at several stations between 1961 and 1990 (Archer and Fowler 2004). For the same stations, a rise in temperature of 0.76 °C during the last four decades is reported. Akhtar et al. (2008), in their regional climatic modelling using the SRES A2 scenario from PRECIS (2071–2100), projected that by the end of this century, UIB will see a 16% increase in annual mean precipitation and a 4.8 °C rise in temperature. Khattak et al. (2011) reported temperature decline for monsoon and minor warming for winter and spring. Another recent finding concludes that the Shigar River basin (central Karakoram) reported increased flows (Mukhopadhyay and Khan 2014). The authors argued that with temperature and precipitation increase, the glacier's contribution is increasing with the neutral mass balance (Lutz et al. 2016). Bashir et al. (2017) also reported glacier's anomalous behavior of surge with a rise in humidity. However, because of the lack of in-situ data, the complicated terrain, and the local climatic circulations, there are still a lot of unknowns when it comes to accurately reflecting the hydrological and climatic dynamics of the basin. (Lutz et al. 2014; Dahri et al. 2021). The accuracy of future climatic projections is also exposed to variabilities and spread in the outcomes of the general circulation models (GCM) (Lutz et al. 2016). Moreover, these projections have structural limitations in their modelling and reliable observed data is missing for bias corrections and validations.

Hence, significant changes are expected concerning the hydro-climatic regime of the Indus Basin (Dahri et al. 2021).

The novelty of the current study lies in attempting the simulation of the future projections and hydro-climatic regimes by analysing the newly released projections from the Couple Model Intercomparison Project phase 6 (CMIP6) (Eyring et al. 2015) for the Indus basin. Moreover, these projections are forced to a relatively new modelling framework that uses the energy balance (EB) approach for simulating snow and glacier melt. Moreover, as per our knowledge, only a few studies have been conducted on the CMIP6 projections for the UIB. The main objective of this research is to investigate the future runoff regime and water availability under different glacier presence scenarios. The findings of this current study may improve the understanding of the current and future hydro-climatic regime and may provide helpful information concerning future meltwater contributions.

2.2 Data and Methods

2.2.1 Study Area

The Hunza Basin lies in the Karakoram mountains in the HKH region. The total basin area is 13,713 km^2. The basin is starched between 2300 and 7889 masl, with 31% of its entire surface permanently covered by glaciers. Karakoram's frozen freshwater resources are predominantly nourished by the western disturbances (Hasson 2016) (Fig. 2.1). The Hunza basin's precipitation is received with a westerly dominant source (Bookhagen and Burbank 2010). The summer monsoon contributes 27% of the annual precipitation.

2.2.2 Historical Data

The observed runoff was used to calibrate the simulated runoff and GCM historical projections based simulations (1991–2010). The few in-situ gauges in the Hunza basin with a limited record period discourage using this data for such a crucial study (Nazeer et al. 2022). The Hunza basin in the Upper Indus is particularly the most scarcely monitored part. Therefore, in this intricate and orographically impacted high-mountain environment, the current gauges ineffectively measure the precipitation (Dahri et al. 2016). Precipitation estimates by European reanalysis 5 Land gridded dataset developed by the Copernicus Climate Change Service (C3S) were reasonable for the HKH region (Nazeer et al. 2022).

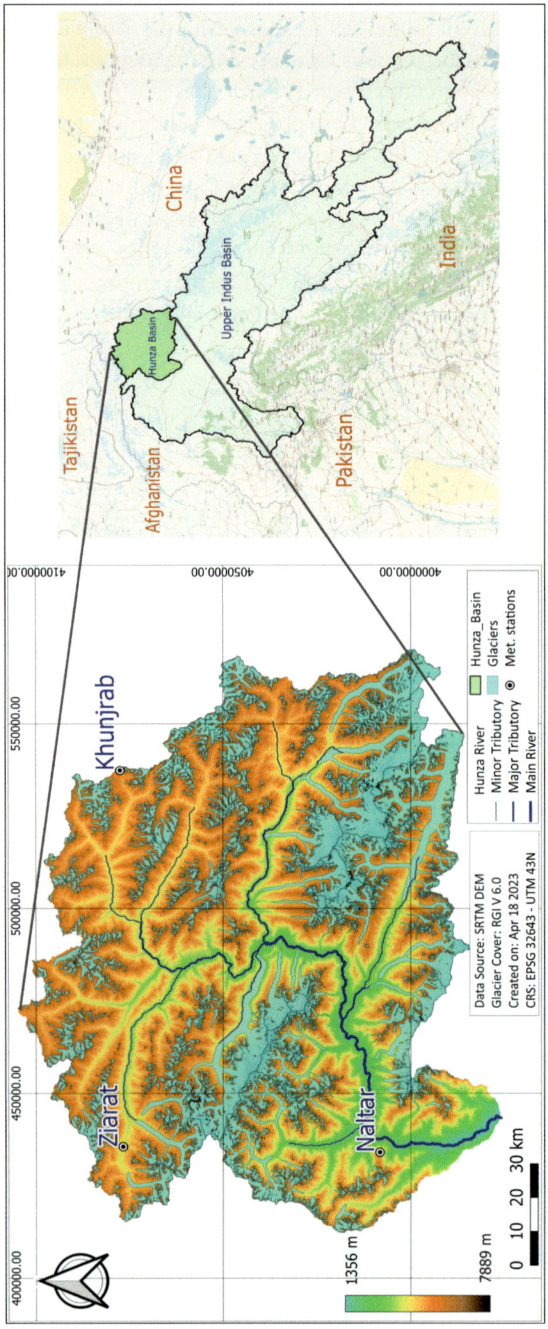

Fig. 2.1 Hunza Basin with glacier cover, river network and meteorological stations

2.2.3 CMIP6 GCM Data

Coupled Models Intercomparison Project Phase 6 (CMIP6) was established by the EC-Earth consortium (Hazeleger et al. 2010; Massonnet et al. 2020). EC-Earth3 with grid size of $0.7^\circ \times 0.7^\circ$ and has global spatial coverage for CMIP6. Because of its comparatively good resolution and performance based on the primary analysis, EC-Earth3 data is chosen for this investigation. The data set is used to derive Hunza's precipitation and temperature from the historical periods of 1991–2010 and respective future periods (2041–2060, and 2081–2100) for Shared Socioeconomic Pathways (SSP) scenarios of SSP1, SSP2 and SSP5. CMIP6 based MPI-ESMI data is used for this study due to its relatively fine resolution and performance based on the primary analysis. The data set is used for temperature and precipitation for the Hunza basin for the historical periods of 1991–2010 and future periods (2041–2060 and 2081–2100) for SSP1, SSP2 and SSP5 scenarios.

2.2.4 Modelling Framework

For this study, the semi-distributed rainfall-runoff Distance Distribution Dynamics (DDD) model was employed. Skaugen and Onof (2014) of the Norwegian Water Resources and Energy Directorate (NVE) derived the model. River flow, SWE, SCA, subsurface water storage, GM, and actual evapotranspiration (ET) are all simulated by the model (Skaugen and Weltzien 2016). The model is data parsimonious and best suited for data-scarce regions having inputs like precipitation and temperature (Nazeer et al. 2021). For data-scarce regions like the Indus, the model is highly reliable for its data parsimonious nature and due to the majority of the model's parameters are not calibrated against runoff but are obtained from digital/GIS maps.

Many uncertainties and model biases affect the accuracy of GCM projections when compared to observed data. Thereby, GCM projections often need bias correction before being applied to future climate and hydrological studies. However, global precipitation products can be good alternatives for such regions' observed data. For the precipitation estimates of data-poor regions, ERA5-Land precipitation is proven as a good alternative (Dahri et al. 2021; Nazeer et al. 2021).

This data, however, overestimates the precipitation and needs correction before applying it as observed data. EAR5-Land precipitation corrected data is an alternative to the Hunza basin's observed data. Hence, three steps of correction include: (i) setting up the model using ERA5-Land and getting the corrected precipitation based on the observed river runoff, (ii) using bias-corrected ERA5 precipitation to bias-correct the elevation-distributed GCM precipitation, and (iii) setting up the model using elevation-distributed bias-corrected GCM's projections. In the model setup using the GCMs projections, the GCMs elevation-distributed, bias-corrected temperature, and precipitation projections are used for the historical period (1991–2010). The model is used in the validation mode. The calibrated parameters are kept similar to those for the observed period simulations.

The accuracy and sound evaluation are assessed using the KGE and NSE efficiency standards. The KGE (Eq. 2.1) is a goodness-of-fit given by KGE (Gupta et al. 2009) is employed for model evaluation, and has applicability negative infinity to 1. On the other hand (Nash and Sutcliffe 1970) (Eq. 2.2) assess the relative magnitude of variance compared with the variance in observed data and subtracted from unity.

$$KGE = 1 - \sqrt{(r-1)^2 + \left(\frac{\sigma_{sim}}{up\sigma_{obs}} - 1\right)^2 + \left(\frac{\mu_{sim}}{\mu_{obs}} - 1\right)^2} \tag{2.1}$$

$$NSE = 1 - \frac{\sum_{i=1}^{n}(Q_{obsi} - Q_{simi})^2}{\sum_{i=1}^{n}\left(Q_{obsi} - \overline{Q_{obs}}\right)^2} \tag{2.2}$$

where r, σ, μ and Q show the linear correlation, standard deviation, mean, and flow, respectively. Subscripts *obs* and *sim* show the observed and simulated data, respectively.

2.3 Results

2.3.1 Historical Simulations

In the first modelling setup, the runoff was simulated using ERA5-Land based precipitation on calibration (1997–2005) and validation (2006–2010) for the Basin. The station temperature data were used as model input. For the higher elevations, the highest station (4730 masl) data was used as a reference together with the temperature lapse rate derived from the APHRODITE data. The slightly revised subroutine to simulate glacier melt changed the results, but these changes are less significant and discussed later.

The ECE3 based simulation showed KGE as 0.80 and NSE as 0.74, with an overestimated runoff of 5.1%. In comparison, the ESMI based simulation showed KGE as 0.83 and NSE as 0.77, with a slightly overestimated runoff of 8.1%. With the increase in temperature in April, the runoff starts increasing, initially due to the snowmelt from lower elevations. The sharp rise and decline in temperature generate immediate snowmelt. The observed runoff of the Hunza River indicates short-term multiple flow peaks. The GCM data simulated similar multiple peaks but could not reproduce all of them and the flow peaks due to flooding in the river (for instance, 1994 and 2010). The ESMI based GCM simulation performed relatively better in this regard. There is a slight increasing trend in runoff from 1991 to 2010, both in modelled and gauged flow for the basin.

The ECE3 based 1991–2010 simulation showed the maximum basin-scale yearly SCA as 70.3% for 1992 and at least 60.2% in 1996, with 63.7% for the Hunza basin. The ESMI simulation showed its maximum as 70.5% for 1994, least as 59.4%, with an average of 63.6%. These SCA estimates agree with MODIS data (Muhammad et al. 2019), showing 60% (average) area under snow with an average of 233 mm per year, the ECE3-based SWE displayed a maximum of 278 mm in 2007 and a minimum of 174 mm in 2010. By contrast, the simulation based on ECE3 revealed a minimum of 174 mm in 2010 and a high of 337 mm in 1993.

The observed data-based simulations and elevation-distributed estimations are in agreement where the lowest elevations are the least snow-covered, while the higher altitudes are primarily under snow for the year. Both observed and GCM-based simulations indicate that 14–17% area of the lowest elevation (a1) is covered with snow on an annual mean scale. While the highest altitude (a10) is 97–100% covered with snow. The SWE is least simulated (6 mm) from a1 and maximum (43–45 mm) simulated from GCM-based simulations.

The glacier melt simulations are shown in Fig. 2.2. Using two GCM inputs and an energy balance technique, the DDD model simulates the melting of glaciers. The maximum annual glacier melt is simulated as 620 mm for 2008 and 679 mm for 2007, the minimum as 288 mm for 1995 and 351 mm for 1994, respectively, based on ECE3 and ESMI inputs. The annual mean glacier melt) is projected as 512 mm from ECE3 based simulations and 553 mm from ESMI based simulations. The glacier melt shows a slightly increasing trend from 1991 to 2010. Based on GCM and observed data simulations, the lowest elevation with less than 2% of total glacier cover contributes 28–29 mm of water. At the same time, the highest elevation, with about 27% of the total glacier melt, contributes 1–13 mm. Because of the higher temperature in lower elevations, more melt is generated and vice versa for higher elevations. The elevation zone a9 generates the maximum melt, followed by a8 and a2.

2.3.2 *Future Simulations*

The GCM based bias-corrected temperature estimates for the Hunza basin are discussed in this section. Results further show a comparison with respective GCM's bias-corrected historical temperatures. Based on all scenarios, the mean monthly temperature increase is slightly higher in ECE3 GCM than ESMI. This future temperature increases from historical end-century periods and from year to year within twenty years of these periods.

The mean annual basin's temperature indicates all 20 years of the historical period below 0 °C in both GCMs. The SSP1 indicates a slight temperature rise for the mid-century period but then a temperature drops. ECE3 indicates a mean annual temperature below 0 °C for a few years for mid-century, but the temperature is closer to 0 °C. In comparison, ESMI indicates below 0 °C for all years of the mid-century period. ECE3's SSP2 indicates only a few years with a temperature below 0 °C than

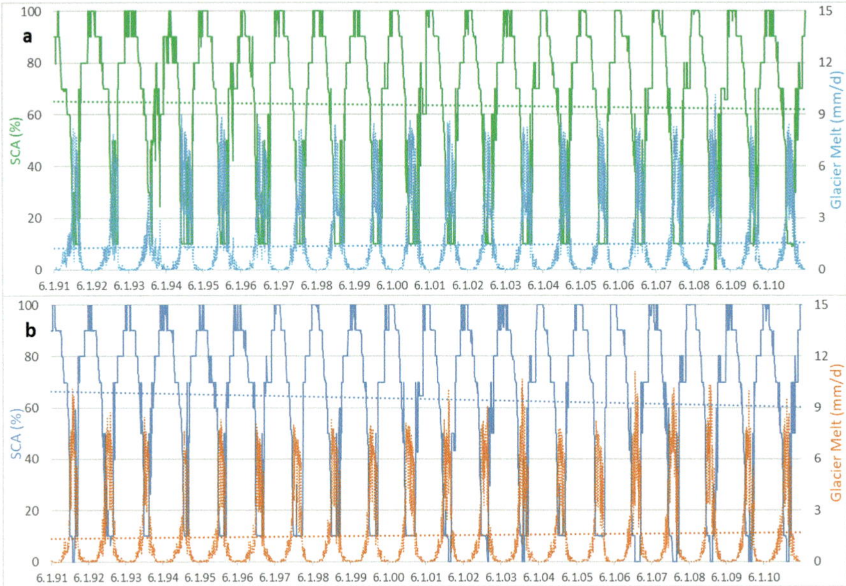

Fig. 2.2 Time series of daily snow cover area (SCA) and glacier melt using **a** ECE3 **b** ESMI GCM

ESMI's SSP2, showing more years with a temperature above 0 °C. For SSP5, both GCMs indicate temperatures above 0 °C for all years of the end-century period. The basin-scale annual temperature averaged for all 20 years of the historical period indicates around − 3 °C for both GCMs.

Relative to the historical period, the annual precipitation seems to increase for all selected scenarios for both GCMs. The post-monsoon receives the least precipitation in the basin. These monthly estimates for historical periods are similar. The GCM's monthly changes in precipitation relative to their respective historical periods are shown in Table 2.1. The precipitations are increasing annually, but data shows uncertain monthly precipitation changes relative to the historical period. For ECE3 data, the maximum increase in precipitation is expected in May for all scenarios, and reduced precipitation is expected in April. The maximum precipitation increase could be 133% for May for SSP2-end-century followed by 130% for June for SSP5-end-century, and the maximum decrease could be 26% for April followed by 19% for December for SSP5-end century.

The ECE3 based historical estimates indicate a minimum annual precipitation of 312 mm for 2010 and a maximum of 498 mm for 2007, with an annual mean of 428 mm. For mid-century, ECE3 indicated the minimum annual precipitation of 407 mm for 2057 and a maximum of 760 mm for 2059 in SSP5. The mean annual precipitation is estimated at 509, 508 and 541 mm for the mid-century period for SSP1, SSP2, and SSP5 scenarios, respectively. For the end-century, ECE3 indicated the minimum annual precipitation of 323 mm for 2088 and a maximum of 773 mm for 2094 for SSP5. For the end-century period, the mean annual precipitation is

Table 2.1 Changes in monthly (mean value) precipitation (%) versus historical data

		Jan	Feb	Mar	Apr	May	Jun	Jul	Aug	Sep	Oct	Nov	Dec
ECE3-Baseline (mm)		*44*	*48*	*34*	*30*	*18*	*21*	*30*	*35*	*27*	*9*	*11*	*35*
ECE3 - SSP1	mid-century	18	18	16	-15	74	16	56	30	-11	12	41	6
	end-century	35	15	16	23	90	-10	20	12	27	7	74	11
	Avg	26	17	16	4	82	3	38	21	8	9	58	8
ECE3 - SSP2	mid-century	8	15	22	5	88	3	28	5	3	26	44	4
	end-century	27	17	18	-11	133	17	13	3	-1	21	89	10
	Avg	17	16	20	-3	111	10	21	4	1	24	66	7
ECE3 - SSP5	mid-century	58	9	22	-9	111	11	58	-4	15	38	17	26
	end-century	31	43	8	-26	96	130	59	-14	25	120	-8	-19
	Avg	45	26	15	-17	103	70	59	-9	20	79	4	3
ESMI-Baseline (mm)		*45*	*52*	*36*	*31*	*17*	*20*	*29*	*33*	*24*	*10*	*12*	*36*
ESMI - SSP1	mid-century	32	18	19	4	13	-6	59	31	-8	38	43	32
	end-century	18	25	1	6	-7	-21	13	-30	-10	41	78	23
	Avg	25	22	10	5	3	-13	36	0	-9	40	61	28
ESMI - SSP2	mid-century	42	5	13	4	8	-27	43	-29	-45	14	27	25
	end-century	40	6	7	-6	23	-16	22	-12	-11	52	37	15
	Avg	41	5	10	-1	15	-22	32	-21	-28	33	32	20
ESMI - SSP5	mid-century	15	2	4	-12	24	0	36	9	5	19	66	14
	end-century	28	29	-1	-20	80	-32	16	-62	-47	21	61	40
	Avg	21	15	1	-16	52	-16	26	-27	-21	20	64	27

estimated at 539, 530, and 562 mm for SSP1, SSP2 and SSP5 scenarios. The ESMI based historical estimates indicate a minimum annual precipitation of 318 mm for 2002 and a maximum of 567 mm for 1993, with an annual mean of 427 mm. For mid-century, ESMI indicated the minimum annual precipitation of 303 mm for 2057 for SSP2 and a maximum of 775 mm for 2053 for SSP1. The mean annual precipitation is 547, 479, and 493 mm for the mid-century SSP1, SSP2 and SSP5 scenarios. For the end-century, ESMI indicated the minimum annual precipitation of 303 mm for 2082 and a maximum of 694 mm for 2094 for SSP1.

2.3.3 Future Runoff and Water Availability

This section presents future runoff simulations based on GCMs and all selected scenarios. Runoff determined by past forecasts is also contrasted with future runoff based on GCM. The future runoff simulations are further presented in the following section under three different glacier presence scenarios of the Hunza basin. The whole glacier cover is considered intact (100% present) for all selected scenarios in this scenario. The simulated runoffs under different scenarios are compared with historical data-driven runoff and intact glaciers. The runoff with intact glacier scenario is increased compared with historical simulated runoff for all scenarios of both future periods. The runoff increases from year to year within twenty years of these periods and from historical to future periods.

The ECE3-based simulation for the historical period showed a minimum monthly flow of 24 m^3/sec for February and a maximum of 836 m^3/s for July, with a mean of 306 m^3/s. The ECE3-based mid-century period has the minimum mean monthly flow of 51 m^3/s for February for SSP2 and a maximum of 1080 m^3/s for July for SSP1. In comparison, ECE3 based end-century period has the minimum mean monthly flow of 60 m^3/s for SSP1 for Jan and a maximum of 1092 m^3/sec for July for SSP5. The ESMI-based simulation for the historical period showed a minimum mean monthly flow of 20 m^3/sec for February and a maximum of 917 m^3/sec for August, with a mean of 315 m^3/sec. The mid-century period showed the minimum mean monthly flow of 40 m^3/sec in February for SSP5 and a maximum of 1194 m^3/sec in July for SSP1.

In the glaciers reduced to elevation above 4000 masl scenarios, the mean monthly simulated runoff ECE3 based inputs are shown in Fig. 2.3. The simulated runoffs are compared with historical data driven runoff with reduced glacier cover. However, the runoff using the same inputs and scenarios with reduced glacier cover is reduced than with intact glacier scenarios. But it is still higher than the runoff simulated using historical data from the respective GCMS for future periods and all scenarios. The runoff increases from year to year within twenty years of these periods and from historical to mid-century to end-century periods.

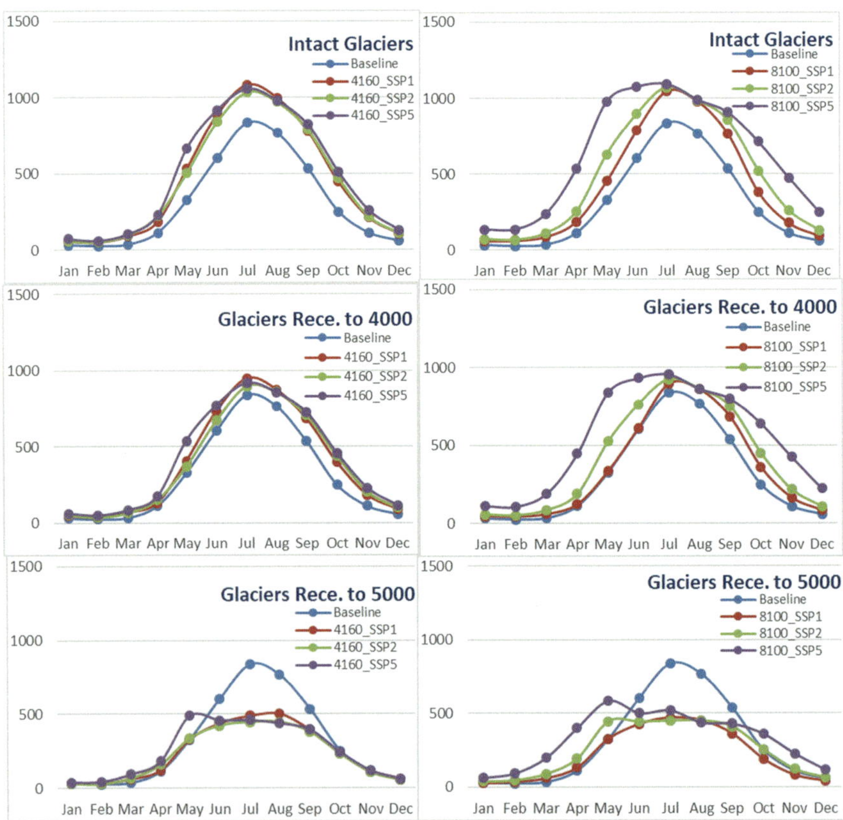

Fig. 2.3 Mean monthly simulated runoff for future periods vs historical period using ECE3 GCM

In this scenario, the glaciers are reduced to elevation above 5000 for all selected scenarios of future periods. The mean monthly simulated runoff with further reduced glacier cover and ESM based inputs are shown in Fig. 2.4. The simulated runoffs are compared with historical data-driven runoff, reducing glacier cover. There is a significant reduction in runoff relative to the historical runoff. Also, runoff is significantly reduced compared with the runoff produced with glacier presence scenarios 1 and 2. These runoff reduction patterns with further reduced glacier cover are consistent for all selected scenarios of both GCMs. The elevation-distributed least melting contributions difference is observed in the lowest elevation, with more differences relative to the historical period melt. The least melting difference could be 2–3 mm for the lowest elevation (a1) for all scenarios for both future periods.

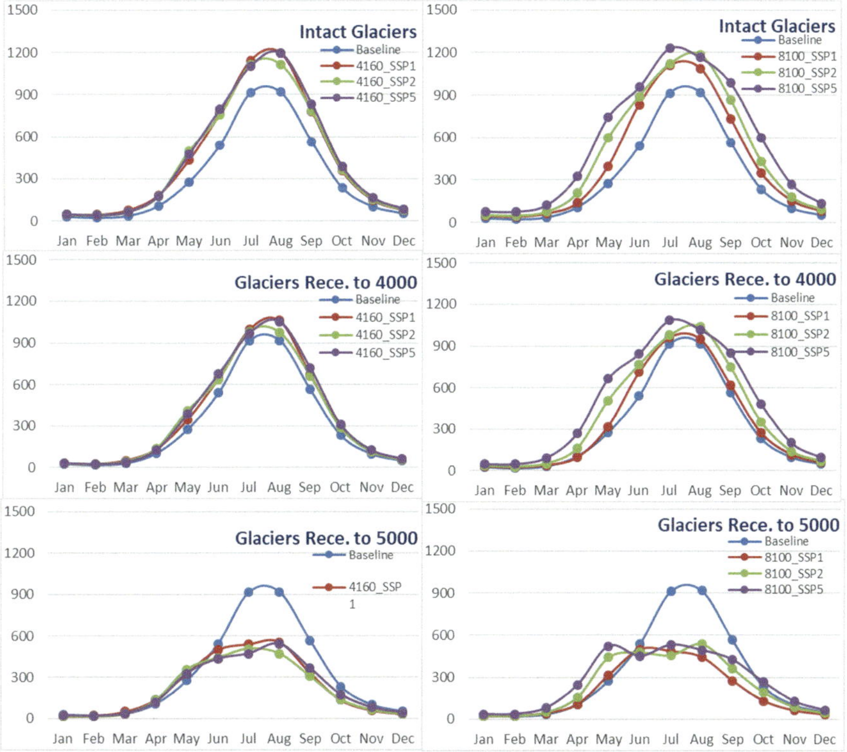

Fig. 2.4 Mean monthly simulated runoff for future periods vs historical period using ESMI GCM

2.4 Discussions

Temperature is the primary indicator of sensitive climate change indicators with significant impacts on the basin's hydrology. The CMIP6 GCM of ECE3 and ESMI-based scenarios for SSP1, SSP2, and SSP5 indicate a highly uncertain future climate for the Hunza basin. But all scenarios based on both GCMs show that the future will warm, with the end-of-century SSP5 scenario showing a more significant warming. A temperature above 0 °C may significantly affect the hydroclimatic regime and runoff dynamics. A temperature of about 0 °C is a threshold for snow or rain. So, this higher temperature immediately may change the precipitation dynamics with more rain than snow. This will ultimately influence the runoff regime as more rainfalls will generate more direct runoff, and less snow will be stored. The year-to-year increase is also associated with year-to-year temperature increases. The temperature rise in all future scenarios is evident. Elevation-distributed glaciers in the Hunza basin and mean annual future glacier melt are presented in Table 2.2. The melt dependent Hunza basin has proved to be very sensitive to temperature. Consequently, higher year-to-year and overall will produce higher runoff because of more melt contribution.

The increased future runoff is primarily associated with a surge in precipitation and temperature. The increased precipitation will generate direct river runoff if it falls as rainfall. Assume that the precipitation falls in the form of snow. Due to temperature rise in future periods, precipitation as rainfall will be increased than snow. This will generate more base flow, and the stored snow will start melting earlier than it melts. So, significant changes are likely in the base flow regime with

Table 2.2 Future glaciers melt from different elevations in the Hunza basin

		GM1	GM2	GM3	GM4	GM5	GM6	GM7	GM8	GM9	GM10
Glacier cover (%)		*1.6*	*4.7*	*5.4*	*6.0*	*6.8*	*8.1*	*10.2*	*13.4*	*17.2*	*26.7*
ECE3 Baseline (mm)		*29*	*67*	*65*	*54*	*51*	*50*	*56*	*66*	*73*	*1*
ECE3 -	mid-century	31	99	75	65	67	72	85	107	126	107
SSP1	end-century	31	97	72	61	60	62	72	89	101	67
(mm)	Avg	31	98	73	63	63	67	79	98	114	87
ECE3 -	mid-century	32	100	77	66	67	71	86	106	124	93
SSP2	end-century	33	103	80	72	72	78	92	118	137	128
(mm)	Avg	33	101	78	69	70	74	89	112	130	110
ECE3 -	mid-century	32	100	77	69	71	77	92	118	139	115
SSP5	end-century	36	119	95	89	93	104	125	165	200	254
(mm)	Avg	34	110	86	79	82	90	109	141	170	184
ESMI Baseline (mm)		*28*	*66*	*65*	*55*	*53*	*53*	*61*	*74*	*84*	*13*
ECE3 -	mid-century	31	94	74	63	62	67	80	97	114	56
SSP1	end-century	32	94	75	65	66	70	83	101	118	60
(mm)	Avg	32	94	75	64	64	68	82	99	116	58
ECE3 -	mid-century	32	95	77	68	70	72	87	108	127	81
SSP2	end-century	33	98	81	72	75	81	98	124	146	120
(mm)	Avg	33	96	79	70	72	77	93	116	136	100
ECE3 -	mid-century	32	95	78	69	70	75	88	110	130	92
SSP5	end-century	36	108	93	87	90	98	120	152	186	205
(mm)	Avg	34	102	86	78	80	86	104	131	158	148

more runoff and reduced base flow duration. The increased temperature will increase the energy available and, ultimately, the accelerated melt process. The increased melt contributions will increase the river runoff.

The hydrological regimes and processes in the HKH region are complex and challenging to investigate, so the underestimated runoff peaks are reported in the previous investigations. Tahir et al. (2011) for the Hunza Basin simulated underestimated peak flows and suggested the precipitation input is responsible. The changes in different hydrological components for scenarios and both GCMS are presented in Table 2.3. These changes are presented as percentages relative to the historical period in all components. The Hunza's mean runoff for the intact glaciers scenario is expected to increase between 31 and 126%. The Hunza's mean simulated runoff for glaciers presence scenario two is expected to increase for mid-century between 15–62% and 29–66%, and end-century between 14–109% and 18–119% for ECE3 and ESMI based simulations, respectively. The runoff for the glacier presence third scenario is expected to be reduced significantly relative to the historical period with intact glaciers and the second glacier presence scenario, except for ECE3 based SSP5-end-century. The runoff is expected to decrease for mid-century between 11–26% and 20–39%, and for end-century, between 10–26% and 14–40% for ECE3 and ESMI based simulations. For ECE3 based SSP5-end-century simulations, there could be a slight increase (14–27%).

The glacier melt is likely to increase for all scenarios for both GCMs. The Hunza's mean glacier melt for intact glaciers scenario is expected to increase between 53 and 265% for ECE3 based simulations. For ESMI based simulations, this glacier melt is expected to surge between 38 and 172%. The glacier melt also increases for

Table 2.3 Percentage variations in the future water availability under all future climate change scenarios relative to the historical period

Runoff comp.	SSP's	mid-century intact glaciers			end-century intact glaciers			mid-century glac. recession to 4000 m			end-century glac. recession to 5000 m			mid-century glac. recession to 4000 m			end-century glac. recession to 5000 m		
		Min	Max	Median	Min	Max	Median	Min	Max	Median	Min	Max	Median	Min	Max	Median	Min	Max	Median
Qsim - ECE3	SSP1	72	39	47	63	31	38	45	21	25	33	14	16	-11	-26	-24	-22	-26	-23
	SSP2	73	34	46	84	58	59	43	15	24	49	37	35	-17	-27	-22	-21	-10	-19
	SSP5	93	48	58	126	95	104	62	29	35	109	73	77	-11	-18	-15	27	14	7
Glacier melt - ECE3	SSP1	140	63	63	92	53	39	87	41	33	11	30	10	-16	-25	-21	-71	-33	-48
	SSP2	86	70	61	149	75	78	37	41	31	96	50	48	-49	-31	-35	-21	-11	-16
	SSP5	141	66	74	265	172	150	88	39	44	204	141	115	-35	-19	-25	60	51	23
Rainfall - ECE3	SSP1	64	58	49	103	43	40	64	58	49	103	43	40	64	58	49	103	43	40
	SSP2	39	52	46	87	62	54	39	52	46	87	62	54	39	52	46	87	62	54
	SSP5	81	47	59	154	101	117	81	47	59	154	101	117	81	47	59	154	101	117
Snowmelt - ECE3	SSP1	1	16	2	-1	35	14	1	16	2	-1	35	14	1	16	2	-1	35	14
	SSP2	-25	17	0	-18	25	1	-25	17	0	-18	25	1	-25	17	0	-18	25	1
	SSP5	-28	25	8	23	22	-24	-28	25	8	23	22	-24	-28	25	8	23	22	21
ET - ECE3	SSP1	61	30	35	27	28	22	59	29	33	25	25	18	51	22	27	24	24	18
	SSP2	45	37	35	70	39	44	43	33	31	69	39	43	43	31	29	58	35	39
	SSP5	65	39	42	120	96	86	66	36	40	119	94	84	62	33	37	115	91	81
Qsim - ESMI	SSP1	50	34	39	49	23	33	-31	-20	-28	-40	-27	-35	-31	-20	-28	-40	-27	-35
	SSP2	49	40	38	69	43	53	-39	-29	-33	-29	-24	-24	-39	-29	-33	-29	-24	-24
	SSP5	62	30	43	99	60	78	-36	-28	-30	-14	-17	-13	-36	-28	-30	-14	-17	-13
Glacier melt - ESMI	SSP1	38	38	34	64	42	38	-2	17	8	28	20	12	-71	-38	-50	-55	-39	-47
	SSP2	48	72	48	110	62	68	7	51	21	62	37	40	-68	-18	-41	-26	-28	-27
	SSP5	110	47	52	172	104	112	67	23	25	128	75	82	-24	-34	-38	28	-1	0
Rainfall - ESMI	SSP1	26	30	21	-1	17	17	26	30	21	-1	17	17	26	30	21	-1	17	17
	SSP2	15	15	14	-31	18	9	15	15	14	-31	18	9	15	15	14	-31	18	9
	SSP5	1	9	5	-10	35	3	1	9	5	-10	35	3	1	9	5	-10	35	3
Snowmelt - ESMI	SSP1	35	32	27	34	-22	-6	35	32	27	34	-22	-6	35	32	27	34	-22	-6
	SSP2	-50	3	-1	35	1	15	-50	3	-1	35	1	15	-50	3	-1	35	1	15
	SSP5	22	43	26	71	-8	18	22	43	26	71	-8	18	22	43	26	71	-8	18
ET - ESMI	SSP1	27	15	15	37	13	17	26	14	14	40	14	18	24	12	11	34	9	14
	SSP2	31	35	24	61	29	34	30	38	25	59	29	34	28	28	19	56	21	29
	SSP5	43	21	25	88	55	59	45	22	25	91	56	62	39	16	20	85	47	55

glaciers presence scenario 2, except for ESMI based SSP1-mid-century. The mean glacier melt is expected to increase for mid-century between 39–88% and 7–23% and for end-century between 11–204% and 20–128% for ECE3 and ESMI based simulations, respectively. For the third glacier presence scenario, glacier melt is significantly reduced relative to all presence scenarios except for SSP5-end-century. The mean glacier melt is expected to decrease for mid-century between 19–49% and 18–71%, and for end-century, between 11–71% and 1–55% for ECE3 and ESMI based simulations, respectively.

The future precipitation inputs (rainfall and snow) are the same for all three glacier presence scenarios. The precipitation increases relative to the historical period for all future scenarios, but rainfall and snowmelt contribution changes are different in both GCMs. The ECE3 based precipitation increases rainfall from 39 to 154% for future periods. The snowmelt contribution increases in a few scenarios between 1 and 35% and decreases in others between 1 and 28%. For the end-century period, rainfall increases for a few scenarios and decreases for others, but the mean value still indicates rainfall increases from 5 to 21%. The snowmelt contributions are increasing between 1 and 71% and decreasing in a few scenarios between 1 and 50%. The increase in snowmelt contribution in a few scenarios could be associated with a change in the precipitation seasonality with more precipitation in winter. Temperature rise can cause the same precipitation as now in some basin elevations. The decline in snowmelt in a few scenarios contribution to the future scenario could be for two reasons. One is less precipitation to fall as snow due to higher temperature (remember the threshold temperature that decides precipitation as snow or rain). This will reduce overall snow cover and stored snow to contribute to the following melt season. Second is the basin's increased temperature relative to the historical period temperature, so snow may melt immediately after the snow event.

Although the current modelling employs an energy balance-based approach to simulate the glacier melt, proxy equations are temperature-based. Hence, more temperature refers to more energy available and more melt (Nazeer et al. 2023). For historical period simulations, the lowest elevation (a1) with the maximum temperature generates the second least melt, and the second-highest elevation (a9) generates the maximum melt. The temperature is extremely low at the highest altitude (a10) due to the extremely high elevation (5550–7889 m) with a mean of 6719 m), even though the glacier cover is the most substantial (27% of total glacier cover). Thus the melt from the glaciers at these altitudes is negligible.

For future simulations, the significant melt is generated from higher elevations due to high enough temperature to melt from all elevations. With about 30% of permanent glacier cover, the Hunza basin receives approximately 50% of its flow from glaciers. However, as this cover shrinks, runoff and the amount of water from glacial melt will also decrease. Hasson (2016) suggested that the glacier melt regime be enhanced under warmer climates for the far future scenario and, subsequently, the overall water availability rise. He further suggested this trend may continue until glaciers deplete or are reduced to a tipping point with significantly less melt contribution, and reduce the overall water availability.

2.5 Conclusion

The following few straightforward outcomes from the current study;

- The temperature rise is evident for all investigated scenarios, with a maximum increase of 10 °C.
- A surge in precipitation is also evident in future scenarios. Also, changes in their cycles, with reduced snow and increased rainfall, are expected.
- The glacier presence could significantly affect the hydrological regime with almost 50% flow contribution from glacier melt for current runoff. These contributions could increase up to 265% for the intact glacier scenario for ECE3 based SSP5-end-century scenario relative to the historical period glacier melt. For the glacier recession to 5000 m scenario, the melt contribution could decrease to 71% for ECE3 based SSP1-end-century.
- The base flow period could only be reduced to a few months with increased flow. The peak flow period could be extended.
- The future river runoff projections are uncertain, but increasing trends are evident overall. For the glacier recession to 5000 masl scenarios, runoff could decrease to 40% for ESMI based SSP1-end-century.
- Significant changes in land cover are expected under the projected warming and reduced glacier cover—which can severely affect the vulnerable populations living in downstream plains.

References

Akhtar M, Ahmad N, Booij MJ (2008) The impact of climate change on the water resources of Hindukush–Karakorum–Himalaya region under different glacier coverage scenarios. J Hydrol 355(1–4):148–163

Archer DR, Fowler HJ (2004) Spatial and temporal variations in precipitation in the Upper Indus Basin, global teleconnections and hydrological implications. Hydrol Earth Syst Sci

Bashir F, Zeng X, Gupta H, Hazenberg P (2017) A hydrometeorological perspective on the Karakoram anomaly using unique valley-based synoptic weather observations. Geophy Res Lett 44(20):10–470

Bookhagen B, Burbank DW (2010) Toward a complete Himalayan hydrological budget: spatiotemporal distribution of snowmelt and rainfall and their impact on river discharge. J Geophys Res: Earth Surface 115(F3)

Dahri ZH, Ludwig F, Moors E, Ahmad B, Khan A, Kabat P (2016) An appraisal of precipitation distribution in the high-altitude catchments of the Indus basin. Sci Total Environ 548:289–306

Dahri ZH, Ludwig F, Moors E, Ahmad S, Ahmad B, Ahmad S, Riaz M, Kabat P (2021) Climate change and hydrological regime of the high-altitude Indus basin under extreme climate scenarios. Sci Total Environ 768:144467

Eyring V, Bony S, Meehl G, Senior C, Stevens B, Stouffer R, Taylor K (2015) Overview of the coupled model intercomparison project phase 6 (CMIP6) experimental design and organisation. Geosci Model Dev Discussions 8(12)

Gupta HV et al (2009) Decomposition of the mean squared error and NSE performance criteria: implications for improving hydrological modelling. J Hydrol 377:80–91

Hasson SU (2016) Future water availability from Hindukush-Karakoram-Himalaya Upper Indus Basin under conflicting climate change scenarios: Climate 4(3):40

Hazeleger W, Severijns C, Semmler T, Ştefănescu S, Yang S, Wang X, Wyser K, Dutra E, Baldasano JM, Bintanja R (2010) EC-Earth: a seamless earth-system prediction approach in action. Bull Am Meteor Soc 91(10):1357–1364

Khattak MS, Babel MS, Sharif M (2011) Hydro-meteorological trends in the upper Indus River basin in Pakistan. Climate Res 46(2):103–119

Lutz A, Immerzeel W, Shrestha A, Bierkens M (2014) Consistent increase in High Asia's runoff due to increasing glacier melt and precipitation. Nat Clim Chang 4(7):587–592

Lutz AF, Immerzeel WW, Kraaijenbrink PD, Shrestha AB, Bierkens MF (2016) Climate change impacts on the Upper Indus hydrology: sources, shifts and extremes. PLoS ONE 11(11):e0165630

Masson-Delmotte V, Zhai P, Pirani A, Connors SL, Péan C, Berger S, Caud N, Chen Y, Goldfarb L, Gomis MI, Huang M, Leitzell K, Lonnoy E, Matthews JBR, Maycock TK, Waterfield T, Yelekçi O, Yu R, Zhou B (eds) (2021) Climate change 2021: The physical science basis. In: Contribution of working group I to the Sixth assessment report of the intergovernmental panel on climate change. IPCC. https://doi.org/10.1017/9781009157896

Massonnet F, Ménégoz M, Acosta M, Yepes-Arbós X, Exarchou E, Doblas-Reyes FJ (2020) Replicability of the EC-Earth3 Earth system model under a change in computing environment. Geosci Model Dev 13(3):1165–1178

Muhammad S, Tian L, Khan A (2019) Early twenty-first century glacier mass losses in the Indus Basin constrained by density assumptions. J Hydrol 574:467–475

Mukhopadhyay B, Khan A (2014) Rising river flows and glacial mass balance in central Karakoram. J Hydrol 513:192–203

Nash JE, Sutcliffe JV (1970) River flow forecasting through conceptual models part I—a discussion of principles. J Hydrol 10:282–290

Nazeer A, Maskey S, Skaugen T, McClain ME (2021) Simulating the hydrological regime of the snow fed and glaciarised Gilgit Basin in the Upper Indus using global precipitation products and a data parsimonious precipitation-runoff model. Sci Tot Environ:149872

Nazeer A, Maskey S, Skaugen T, McClain ME (2022) Changes in the hydro-climatic regime of the Hunza Basin in the Upper Indus under CMIP6 climate change projections. Sci Rep 12(1):21442

Nazeer A, Maskey S, Skaugen T, McClain ME (2023) Analysing the elevation-distributed hydrological regime of the highly glaciated and snow-fed Hunza Basin in the Hindukush Karakoram Himalaya (HKH) region. Front Earth Sci 2023:1215878

Skaugen T, Onof C (2014) A rainfall-runoff model parameterized from GIS and runoff data. Hydrol Process 28(15):4529–4542

Skaugen T, Weltzien IH (2016) A model for the spatial distribution of snow water equivalent parameterized from the spatial variability of precipitation. Cryosphere 10(5):1947–1963

Tahir AA, Chevallier P, Arnaud Y, Neppel L, Ahmad B (2011) Modeling snowmelt-runoff under climate scenarios in the Hunza River basin, Karakoram Range, Northern Pakistan. J Hydrol 409(1–2):104–117. https://doi.org/10.1016/j.jhydrol.2011.08.035

Chapter 3
The Open Global Glacier Model (OGGM) Based Assessment of Glacier Mass Variations in Chhota Shigri Glacier, Western Indian Himalaya

Tasso Yaka, Renji Remesan, and Akshay Pachore

Abstract Indian Himalayan glaciers are significant indicators of regional climate change and future water availability. Due to its inaccessible location and difficult terrain, observational data is scarce. Hence, it is imperative to prioritize modeling glaciers in the Indian context to enhance our comprehension of how glaciers are reacting to the current climate changes. This paper utilizes the Open Global Glacier Model (OGGM) to simulate the changes in glacier mass and volume over time, focusing on the initial state of the Chhota Shigri Glacier. OGGM is an open source, global, modular, community driven numerical framework which utilizes the publicly available data [e.g., geodetic data and World Glacier Monitoring Service (WGMS)] for calibration and validation purposes. The study has employed baseline climate data (W5E5) and Randolph Glacier Inventory (RGI v6.0) outlines as input for the modeling workflow in the Chhota Shigri Glacier, a bench mark glacier located in the Western Indian Himalaya. The study has involved detailed sensitivity analysis of three model parameters (viz. melt factor, precipitation factor and temperature bias) on the modeled glacier evolutions. The study outcomes during the year 2003 to 2019 would give generalized understanding on impacts of glacier evolution, other glaciological and hydrological processes in Western Indian Himalayan River basins.

Keywords Indian Himalayan glaciers · OGGM · Monthly temperature index model · WGMS · RGI

3.1 Introduction

The towering Himalayan peaks situated in northern India displays a crucial part in shaping and sustaining the climate across the entire northern region of the Indian subcontinent supplying vital meltwater to the most of the rivers in the region.

T. Yaka · R. Remesan (✉) · A. Pachore
School of Water Resources, Indian Institute of Technology Kharagpur, Kharagpur, India
e-mail: renji.remesan@swr.iitkgp.ac.in

A. Nanda et al. (eds.), *Navigating the Nexus*, Water Science and Technology Library 102, https://doi.org/10.1007/978-3-031-76532-2_3

Studying the historical temperature and precipitation patterns and its effects on glacier modeling holds immense importance in the Himalayan regions (Wang et al. 2024). It serves as the foundational step in understanding climate change and its visible impacts, notably evident in the glaciers of the area. The meteorological and glacier information available for glacier dominated regions especially the Indian Himalayan Region (IHR) is constrained due to challenges in establishing weather stations at high altitudes and ensuring the durability of these stations in harsh weather conditions (Engelhardt et al. 2017; Cogley 2011). Glaciers stand out as notable characteristics in mountainous terrains; however, their considerable impact on the volume and distribution of catchment runoff is frequently disregarded. The worldwide attention on glacier research and especially on glacier mass balance modeling is increasing because glaciers serve as sensitive indicators of climate change (Bolch et al. 2019). Affected directly by changes in climate, glaciers are experiencing significant melting as a result of projected global warming, leading to a notable increase in global sea levels as reported in many researches (Bajracharya et al. 2008; Romshoo et al. 2022). Glaciers, primarily located in the union territory of Ladakh and the Indian states of Himachal Pradesh, Uttarakhand, Sikkim, and Arunachal Pradesh, are pivotal for comprehending the impacts of climate change. Therefore, understanding changes in glacier mass, volume, and other parameters through glacier modeling is essential.

Measuring glacier mass balance involves utilizing diverse methodologies and models, including advanced glacio-hydrology models (Song et al. 2020; Gupta et al. 2019), which incorporate elements such as glaciological, geodetic, hydrological methods, and modeling techniques. One instance of a modern model is the Open Global Glacier Model (OGGM), offering a modular and openly available numerical framework for simulating the past and anticipated changes in glaciers worldwide (Maussion et al. 2019). OGGM has diverse applications, ranging from analyzing ice–climate interactions over millennial timescales to assessing glaciers' contributions to historical and projected sea-level changes (Parkes and Goosse 2020; Pesci et al. 2023). As a model with the capacity to develop into a self-sustaining, community-driven system, OGGM serves as a valuable instrument for understanding both global and regional glacier dynamics. Its capability to discern the glaciers mass balance is pivotal for evaluating the pace of the glacier's retreat and deducing the ramifications of climate change (Eis et al. 2021). The critical assessment of parameter sensitivity in glacier models, as shown by Schuster et al. (2023), elucidates the impact of model design variations stemming from parameter choice on present and future glacier volumes.

As of now, there are limited studies available for the Indian Himalayas (Koul and Ganjoo 2010; Azam et al. 2014). Hence, there is an emergency to enhance understanding of glacier variations in reaction to climatic alterations in the Indian part of the Himalayas (Koldunov et al. 2016). Comprehending the rates of snow and ice surface melting holds significant importance for both assessing and managing water resources within the country, while also aiding in the evaluation of cryospheric changes associated with climate variation. In our study, we specifically examined the Chhota Shigri Glacier as a case study region, employing OGGM for analysis. The objective of this research is to assess glacier mass balance and glacier volume in the

Indian state of Himachal Pradesh using OGGM given emphasis on sensitivity analysis of three model parameters (viz. melt factor, precipitation factor and temperature bias).

3.2 Study Area

The Chhota Shigri Glacier, located at coordinates 32.28° N and 77.58° E, is of notable significance within the Chandra River basin (Fig. 3.1). The glacier is situated within the Pir Panjal Mountain Range of the Western Himalayas, in Himachal Pradesh, India. Specifically, it is oriented towards the north and lies within a valley, nestled amidst the rugged peaks and terrain of this prominent mountain range. Encompassing approximately 16 km^2 and with a catchment area of around 35 km^2 (Wagnon et al. 2007; Ramanathan 2011), nearly half of the research area is under ice and glacier coverage. Stretching around 9 km in length, the glacier ranges in elevation from 4050 to 5830 m above sea level. Its surface exhibits slopes ranging from 10° to 45°, with its terminus located within a couloir. Water from the basin streams into a sole fore-glacial watercourse, adding to the Chandra River, a tributary of the Chenab River, which is part of the extensive Indus River system. Located in the transitional area between monsoon and arid climates, the Chhota Shigri Glacier encounters two different precipitation regimes: the Indian Summer Monsoon (ISM) and the Mid-Latitude Westerlies (MLW) (Azam et al. 2014). Most of the precipitation falls throughout the winter months (October–May) due to the influence of the MLW.

Starting from 2002, measurements of surface mass balance have been conducted consistently at Chhota Shigri Glacier (Ramanathan 2011) using the ground-based methods (glaciological method), establishing it as the longest uninterrupted sequence of mass-balance data accessible throughout the Himalayan region. The individual mass balance ranges from 1.5 m to − 2.5 m water equivalent per year, with a rise in individual mass balance below an altitude of about 4400 m attributed to the existence of debris cover.

3.3 Data and Methods

3.3.1 Datasets

OGGM requires a range of data files to initiate its operations. It depends on openly accessible data sources that users can readily access and automatically download. However, before use, this data must undergo extraction and pre-processing tailored to each specific glacier. This model creates the interface where it downloads the data, and uses it for further analysis.

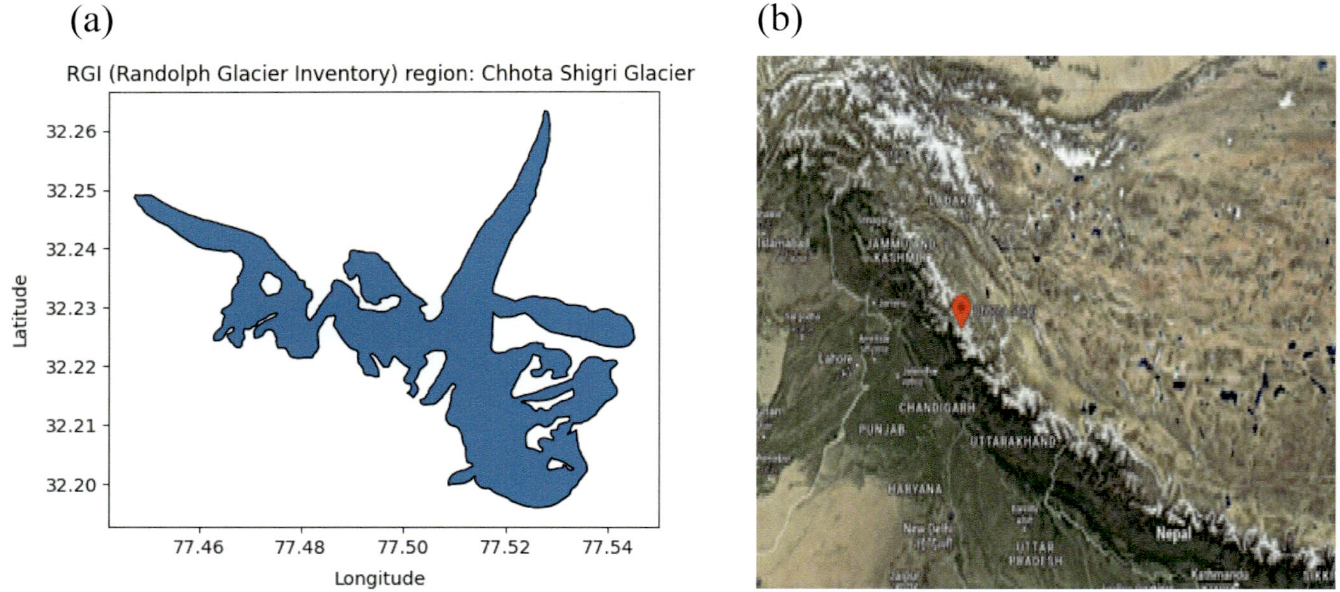

Fig. 3.1 **a** Chhota Shigri Glacier catchment map generated using OGGM simulation; **b** location (google map) of Chhota Shigri Glacier in Indian Himalayan Region

3.3.1.1 Data on Climate: Temperature and Precipitation

In the present investigation, GSWP3-W5E5 dataset for 1901 – 2019 is used as baseline climate data which utilizes total monthly precipitation (kg/m^2) and 2 m temperature (°C) at reference height, i.e., 5384 m a.s.l. Starting from version 1.6, OGGM utilizes GSWP3-W5E5 as its standard baseline climate dataset across all preprocessed directories. GSWP3-W5E5 combines W5E5 v2.0 data spanning from 1979 to 2019 with GSWP3 v1.09 data, which has been adjusted to match the W5E5 standards from 1901 to 1978. The GSWP3-W5E5 offers a global resolution of 0.5 degrees spatially. W5E5 utilizes the version 2 of the WATCH Forcing Data (WFD) methodology applied to ERA5 data for terrestrial regions (WFDE5; [Cuchi_2020]). Precipitation in W5E5 is derived from WFDE5 rainfall and snowfall data, followed by bias-adjustment employing the version 2.3 of the monthly precipitation dataset from the Global Precipitation Climatology Project (GPCP). The W5E5 is selected for all preprocessed directories motivated by the fact that the climate input data for the impact assessments performed in Coupled Model Intercomparison Project Phase 6 (CMIP6) Global Climate Models (GCMs) of phase 3b of the Inter-Sectoral Impact Model Intercomparison project (ISIMIP3b) have undergone bias correction using this dataset. Typically, bias correction of the GCMs is required to align them approximately with the climate dataset employed to calibrate the model. By using W5E5 for model calibration and ISIMIP3b GCMs for projections, OGGM eliminates the need for supplementary bias adjustments. The GCMs from ISIMIP3b, downscaled statistically at a 0.5° resolution, undergo internal bias adjustment to W5E5 for the period from 1979 to 2014. This approach offers a significant benefit as the quantile-mapping bias correction method employed by ISIMIP3b provides greater reliability for outliers in contrast to the "delta-method" commonly utilized in OGGM.

3.3.1.2 Data on the Mass Balance Within Glaciers

(i) ***Traditional in-situ mass balance data***

OGGM utilizes in-situ mass balance data for the calibration and validation of its initial iteration mass balance model. In the present study, mass balance observations sourced from the World Glacier Monitoring Service (WGMS). The WGMS functions under the International Association of Cryospheric Sciences (IACS) within the International Union of Geodesy and Geophysics (IUGG). It is affiliated with the World Data System of the International Science Council (ISC, previously known as ICSU) and receives backing from the United Nations Environment Programme (UN Environment), UNESCO and the World Meteorological Organization (WMO). In a tight partnership with the U.S. National Snow and Ice Data Center (NSIDC) and the Global Land Ice Measurements from Space (GLIMS) initiative, the WGMS oversees the Global Terrestrial Network for Glaciers (GTN-G) to support the objectives of the United Nations Framework Convention on Climate Change (UNFCCC). The Fluctuations of Glaciers (FoG) database holds yearly mass balance measurements for numerous glaciers globally, except those ending in water bodies and time series with fewer than five years of data. For our study, we have used WGMS specific mass balance data from 2003 to 2019 as reference data and utilized for calibration of mass balance model. The problem with in-situ WGMS mass balance time series is that there are often gaps and so calibration requires to be done for the same modeled time period. The WGMS specific mass balance averaged over the years 2003–2019 is –463.12 kg/m^2.

(ii) ***Geodetic mass balance data***

OGGM includes a geodetic mass balance table containing mass balance data for all glaciers worldwide, sourced from Hugonnet et al. (2021). Geodetic data for each glacier were categorized into three periods: 2000–2020, 2000–2010, and 2010–2020. These geodetic mass balance data were averaged for each time period. The averaged data (2000–2020) was employed for calibration through dynamic spin-up to simulate volume changes between 2003 and 2019.

3.3.1.3 Additional Data Sets

This study utilized the OGGM model, incorporating glacier outlines sourced from the Randolph Glacier Inventory v6 (RGI v6), a comprehensive inventory of global glacier outlines released in 2017. The dataset merge outlines from GLIMS, DCW, and WGI datasets, along with numerous other (frequently unpublished) sources, selecting the highest quality version for each region. The outlines, supplied by the Global Land Ice Measurements from Space (GLIMS), were used for the initial topographic processing in the model. Topographical data is chosen automatically, depending on the glacier's location. The topographical data is chosen as per the list: NASADEM, COPDEM90, GIMP, REMA, TANDEM and MAPZEN. However, NASADEM and COPDEM90 are found to be sufficient for all glaciers. The gridded topography undergoes re-projection and resampling to match the glacier map, which follows a Transverse Mercator (TM) projection centered on the glacier. The spatial resolution of this customized grid varies according to the glacier's size. By default, OGGM employs the following rule:

$$\Delta \text{x} = d_1\sqrt{S} + d_2 \tag{3.1}$$

Here, Δx signifies the grid's spatial resolution (measured in meters), S denotes the glacier's area (measured in square kilometers), while d_1 and d_2 are specific parameters (assigned values of 14 and 10, respectively). If the selected spatial resolution exceeds 200 m (S $\geq$ 185 km^2), it is adjusted to this value.

Moreover, elevation-band flowlines were generated using the NASADEM dataset obtained in February 2000. The NASADEM dataset, available at an approximate resolution of one arc-second (about 30 m), can be acquired from the Land Processes Distributed Active Archive Center (LP DAAC) via NASA Earth data. After downloading, the data was interpolated onto the nearby grid, aligning with the exact positions and shapes of the glaciers. NASADEM, an upgraded iteration of the DEM extracted from the Shuttle Radar Topography Mission (SRTM) data, offers higher resolution compared to other elevation datasets, rendering it highly valuable for projects requiring detailed elevation data.

3.3.2 Methods

3.3.2.1 Open Global Glacier Model

The model effectively simulates glacier dynamics by using the shallow ice approximation, which allows for the calculation of depth-integrated ice flux along various flow channels. Moreover, it has the capability to replicate crucial characteristics of glaciers, including mass balance (the equilibrium between ice accumulation and loss), determine ice thickness, and accurately depict the changing dynamics of glaciers over a period of time. The latest version utilized for the current study, OGGM v-1.6, incorporates an enhanced calibration method compared to earlier versions. First, OGGM comes with some pre-processed directories where the data needed for our study region is already stored such as reference data that can be easily accessed and requires no further processing. In our study, we have conducted calibration of the monthly temperature index model with WGMS specific mass balance data (2003–2019) and average geodetic mass balance data (2000–2020). In both the calibration procedures, we have followed three step calibration approach wherein, we can take upto maximum of three calibration parameters, i.e., parameter 1, parameter 2 and parameter 3 (Fig. 3.2). We tried placing melt_f, prcp_fac and temp_bias individually first in parameter 1 then in 2 and 3 (Fig. 3.2a). After completing the single trials, we tried playing with combination of two parameters such as taking parameter 1 and 2 or 2 and 3 or 1 and 2 (Fig. 3.2b). In the first combination, i.e., parameter 1 and 2, we tried placing melt_f, prcp_fac or temp_bias in both the parameters. In the similar manner, it was done with other combinations. And lastly, after completing the combination of two parameters, we proceeded to play with combination of three parameters in the similar manner. We found that some combinations gave the same set of parameter values ignoring which only five set of parameter values were unique and different for WGMS and geodetic mass balance data. We considered parameter set of WGMS data as it is more reliable which when analyzed for specific mass balance using correlation, standard deviation and Nash–Sutcliffe coefficient (NSC) indicated that one of these five sets showed agreement with these two tests, i.e., standard deviation and Nash–Sutcliffe coefficient. Hence, we got the best parameter set combination and this set of parameter values was utilized to model glacier mass balance, i.e., modeled specific mass balance. On the other hand, the spin-up was followed to get the volume evolution of the glacier from 2003 to 2019 which is explained in Sect. 3.2.3. Spin-up helps in getting the approximate past state of glacier so that future states can be correctly estimated. Spin-up was conducted with fixed geometry, dynamic spin-up with matching criteria as area and volume so that whichever showed higher correlation with the calculated volume from WGMS is considered as modeled volume.

3.3.2.2 Monthly Temperature Index Model

The glacier mass balance component within OGGM represents an enhanced iteration of the temperature index model originally devised by Marzeion et al. (2012). Equation (3.2) was employed to estimate the monthly glacier mass balance B_i at a given elevation z.

$$B_i(z) = P_i^{Solid}(z) - d_f \max(T_i(z) - T_{Melt}, 0) \tag{3.2}$$

P_i^{Solid} (z) denotes the monthly solid precipitation (kg/m^2) at altitude z; $T_i(z)$ represents the monthly air temperature (°C) at altitude z; and T_{Melt} is the monthly temperature threshold at which ice begins to melt, typically set at − 1 °C. The solid precipitation proportion was quantified as a percentage of the entire precipitation. When the temperature (T_i) was equal to or below the solid temperature threshold (T_{Solid}) of 0 °C, it was considered 100% solid; conversely, when the temperature was equal to or above the liquid temperature threshold (T_{liquid}) of 2 °C, it was considered 0% solid. The proportion of solid precipitation was estimated based on the monthly mean temperature. The percentage of solid precipitation was linearly interpolated for temperatures between T_{solid} and T_{liquid}. Precipitation was presumed to be completely solid when the temperature was below 0 °C and completely liquid when the temperature was above 2 °C, with a gradual change in state in between. The overall precipitation is derived from the climate dataset and then adjusted by a precipitation correction factor. The parameter d_f, i.e., degree-day factor of glaciers signifies the glacier's temperature sensitivity (in kg/m^2/K/month) and requires calibration. The model calculates the temperature and precipitation at the elevation of the glacier grid points. By default, a constant lapse rate of − 6.5 K/km is applied for temperature, while no gradient is assumed for precipitation.

3.3.2.3 Model Spin-Up

OGGM provides a feature for recreating a glacier's recent history through a process known as "dynamic spinup", wherein the glacier's past state is estimated iteratively. Multiple potential past states, dating back to 1979, are evaluated, and the first one that accurately reflects the current glacier geometry after simulating forward from 1979 to the RGI date is selected. We can select either area or volume as the matching criterion at the RGI date. Moreover, this dynamic spinup can be integrated with a dynamic calibration process for adjusting the temperature sensitivity parameter. This calibration endeavor aims to align the model with geodetic mass balance (MB) data, while also considering the evolving glacier geometry. Dynamic initialization was implemented

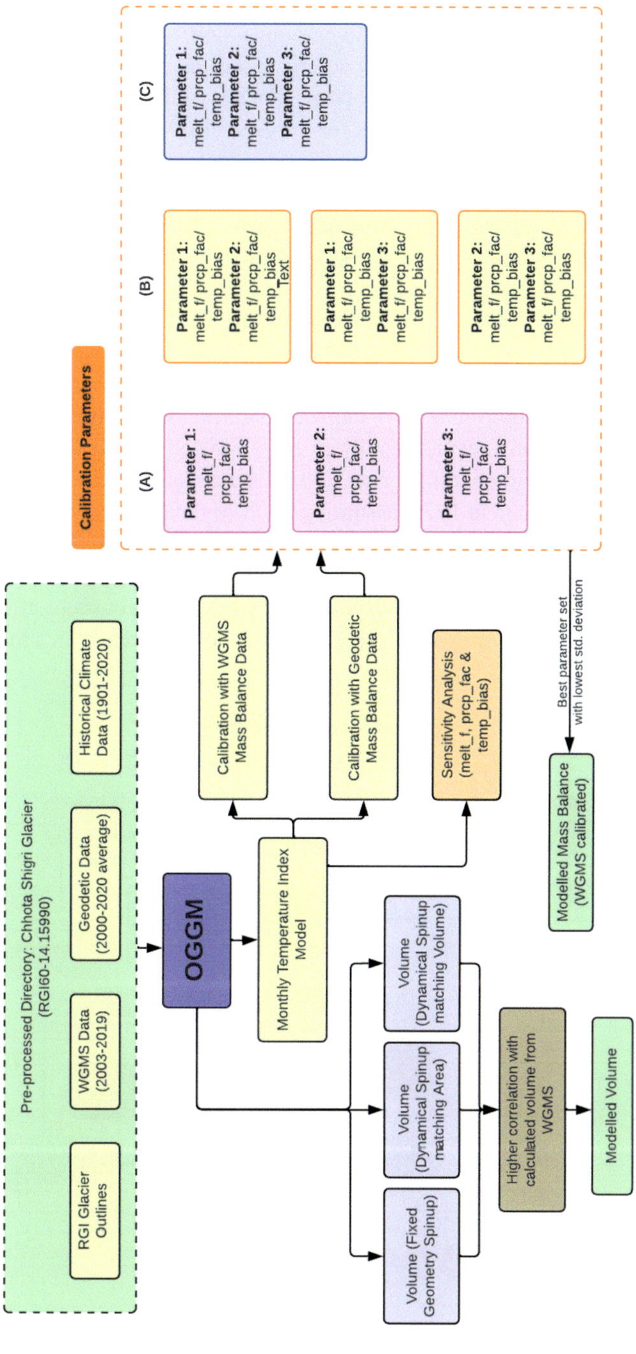

Fig. 3.2 The methodology diagram showing dynamical spin-up and calibration procedures in OGGM

to address significant assumptions embedded in OGGM's default settings. These assumptions include considering glaciers to be in dynamic equilibrium at the specified glacier outline date, crucial for ice thickness inversion, and calibrating the mass balance melt factor parameter ("melt_f") solely based on geodetic mass balance, without considering the dynamically evolving glacier geometry. The subsequent section elaborates on the main approaches for dynamical spin-up (Fig. 3.3).

Initially, a glacier sensitivity parameter (d_f) was determined by matching the modeled mass balance with geodetic mass balance data provided by Hugonnet et al. (2021). This was followed by establishing the apparent mass balance by utilizing a static glacier geometry and baseline climate conditions during the geodetic mass balance duration. A mass balance residual was then identified and added to the mass balance profile between 2000 and 2020 to ensure an average mass balance of 0, maintaining the equilibrium assumption. Subsequently, the apparent mass balance was utilized to invert the underlying glacier bed, defining the creep parameter A, with resulting glacier volume matching the regional volume estimate by Farinotti et al. (2009). This inversion method followed the approach outlined by Maussion et al. (2019), with the sliding parameter set to 0. A dynamic spin-up was conducted to determine the glacier state in 1979, assuming the RGI data were representative of the year 2000. The establishment of different glacier states in the past was achieved by introducing a temperature bias to the baseline climate during a 20-year period of constant mass balance simulation. Progress was made if this stage effectively identified a previous glacial state that closely matched the RGI area, with a margin of

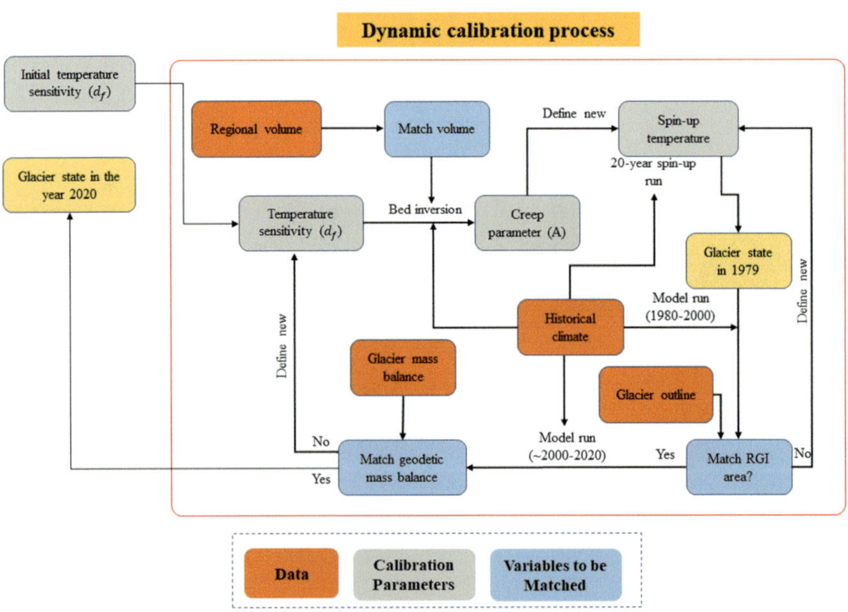

Fig. 3.3 Flowchart showing dynamical spin-up step of OGGM model

1 km^2. After determining the previous condition of the glacier, a dynamic simulation was carried out from 1979 to 2020. This simulation used climate data as a reference and started from the previously determined state of the glacier. The model ultimately computed the mass balance for the period of 2000–2020 and compared it to the geodetic measurements provided by Hugonnet et al. (2021). If the results aligned, the calibration process concluded, and the resulting glacier state in 2020, creep parameter A, and d_f were used for projection runs. If not, the newly defined d_f was utilized to repeat the process. Upon completion of the model spin-up, the past state of glaciers and their state in 2020 were obtained.

3.4 Results and Discussions

3.4.1 Calibration of prcp_fac, temp_bias and melt_f Parameters Using In-Situ Geodetic Data

In this study, we have used the three-step calibration method for calibrating against WGMS in-situ specific mass balance data. This method employs three major parameters for glacier temperature index model such as precipitation factor (prcp_fac), temperature bias (temp_bias), and melt factor or degree-day factor (melt_f) to see how these parameters influence the model on calibration. The snowmelt amount is determined by the positive air temperature using the temperature index model. The proportionality factor in this model is referred to as the melt factor (melt_f). Temperature bias refers to the adjustment made to the input temperature data in order to compensate for any consistent inaccuracies or biases present in the temperature records. OGGM incorporates temperature bias modifications to the input climatic data in order to enhance the alignment between the modeled and actual glacier mass balance. OGGM use the precipitation factor to correct the measurement inaccuracies in the snow input. Pesci et al. (2023) has adopted the similar approach by employing the above mentioned three parameters (prcp_fac, temp_bias, and melt_f) for the OGGM calibration. Hanus et al. (2023) also followed the similar calibration method with the focus of coupling the large-scale glacier model OGGM (v1.5.3) and the large-scale hydrological model CWatM (V1.08). When OGGM model is calibrated for geodetic data, it is tried with all three parameters and then its combinations to match with its average geodetic specific mass balance of − 242 kg/m^2 (2000–2020). Out of all the trials conducted, only five calibrations gave unique and different results. First calibration was successfully completed when temp_bias was used as calibration parameter 1. Second and third calibration was with prcp_fac and temp_bias as calibration parameter 2. Fourth and fifth calibration was conducted with combination of parameters keeping prcp_fac and melt_f in one set and prcp_fac and temp_bias in another as parameters 1 and 2. The simulations maintained the temperature bias parameter ("temp_bias") within a range of − 15 to 15 °C, and the temperature sensitivity parameter (“melt_f”) ranged from 1.5 to 17 °C. The parameter “prcp_fac” demonstrated reasonable consistency between the model simulation data and the measured data within the range of 0.1–10 kg/m^2. The Fig. 3.4 shows the glacier specific mass balance between geodetic mass balance and the modeled values between 2000 and 2019 using different major parameters combinations during calibration. The period 2000–2019 was chosen as geodetic specific mass balance is available for three periods: 2000–2010, 2010–2020 and 2000–2020. So, we have considered the specific mass balance of the period 2000–2020 for calibration.

The Fig. 3.4 shows that the average mass balance over the entire period of 2000–2019 is matched well with the observed mass balance as we calibrate it to different combinations of parameters, but considerable interannual mass-balance variability is observed. When analyzed the calibration results (as shown in Table 3.1), it implies that one of these five modeled mass balance set of values (when applied temp_bias as calibration parameter 2 (“melt_f” = 1.5, “prcp_fac” = 2.66 and “temp_bias” = – 3.19) have highest correlation of 0.97 and least root mean square error (RMSE) of 95.32 when calibrated against geodetic specific mass balance of – 242 kg/m^2 between 2000 and 2020. The study done by Schuster et al. (2023), has adopted the similar average geodetic mass balance data along with the temperature index calibration model over the 88 considered glaciers. Results from this research shows that using higher resolution data in terms of both time and space improves the depiction of mass balance gradients because of better calibration.

3.4.2 Sensitivity Analysis of Parameters

Though through several manual calibration efforts, we could identify a reasonably suitable calibration parameter sets, it is important to know how the temperature-index model’s parameter choices display variable sensitivities to climate anomalies in the Chhota Shigri Glacier. For which, we have indicted isolated calibration experiments (sensitivity analysis) for each parameter for defined set of ranges as shown in the Fig. 3.5 to isolate these differences for “prcp_fac”, “temp_bias” and “melt_f” parameters. The Fig. 3.5a shows the change in specific mass balance when melt factor is varied on keeping the precipitation factor as 0.1 and temperature bias as 0. It indicates that specific mass balance is indirectly proportional to melt factor. Higher values of melt factor will lead to smaller mass balance, i.e., negative mass balance. This study focused on annual values, more detailed evaluations with more focus on solid winter and summer precipitation anomalies as reported in Schuster et al. (2023) would give more detailed understanding of model sensitivities with temperature-index parameter choices. Similarly, Fig. 3.5b shows the change in specific mass balance with change in temperature bias on keeping precipitation factor as 0.1 and melt factor as 0. It indicates that higher negative temperature bias values lead to higher constant specific mass balance while on the other hand, the higher positive temperature bias values lead to zero specific mass balance after a certain temperature bias. Temperature bias can be seen to be proportional with melt factor up to certain degree. Schuster et al. (2023)

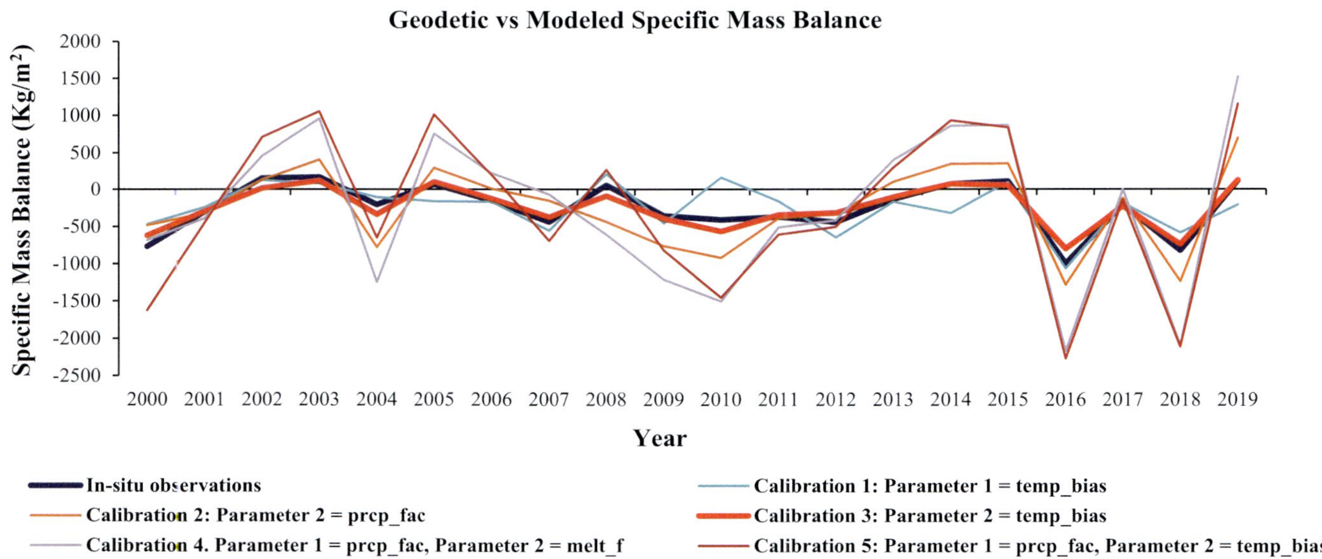

Fig. 3.4 The variations in the specific mass balance in comparison to in-situ observations with different combinations of parameters for calibration of the Chhota Shigri Glacier

Table 3.1 Results of calibration with in-situ geodetic specific mass balance data and its correlation and RMSE w.r.t modeled data

Calibration	Parameter(s)	melt_f (ranges: 1.5–17)	prcp_fac (ranges: 0.1–10)	temp_bias (ranges: – 15–15)	SMB ((Ref = – 242 (kg/m^2))	Correlation	RMSE
1	1: temp_bias	5	2.66	– 7.85	– 242	0.77	216.96
2	2: prcp_fac	1.5	5.35	0	– 242	0.83	323.82
3	2: temp_bias	1.5	2.66	– 3.19	– 242	0.97	95.32
4	1: prcp_fac 2: melt_f	2.7	10	0	– 242	0.83	755.46
5	1: prcp_fac 2: temp_bias	5	10	– 3.09	– 242	0.96	718.6

has reported the similar observation for the temperature anomalies, mentioning that, solid precipitation and specific mass balance decreases with the increasing temperature for average results over 2000 to 2019 for 217 glaciers considered in the study. In the same manner, Fig. 3.5c. shows the change in specific mass balance with change in precipitation factor by keeping both melt factor and temperature bias as 0. In this figure, the specific mass balance varies directly proportional to precipitation factor with least value as 0.1 necessarily.

3.4.3 Calibration of prcp_fac, temp_bias and melt_f Parameters Using WGMS Data

In the similar manner as in Sect. 3.4.1, we have done the calibration against WGMS specific mass balance data which is available continuously from 2003 to 2019. This uses the fixed values from earlier calibrations for all parameters except for the parameter we calibrate. The calibration is done using the above same values for other two parameters and tries to match with the average mass balance. We have conducted multiple trials calibrating parameters and its different combinations. Only five calibrations gave different and valid results. The calibration using prcp_fac and melt_f as parameters 1 and 2 was the closest one to the WGMS specific mass balance data. From Table 3.2 we can infer that in calibration 1, the met_f and prcp_fac have same values as that of calibration with geodetic data, only the temp_bias is changed, which is the calibration parameter in the first trial. Similarly, other four calibrations also followed in the same manner. Out of these five calibrations, only the third calibration shows high relation (Fig. 3.6).

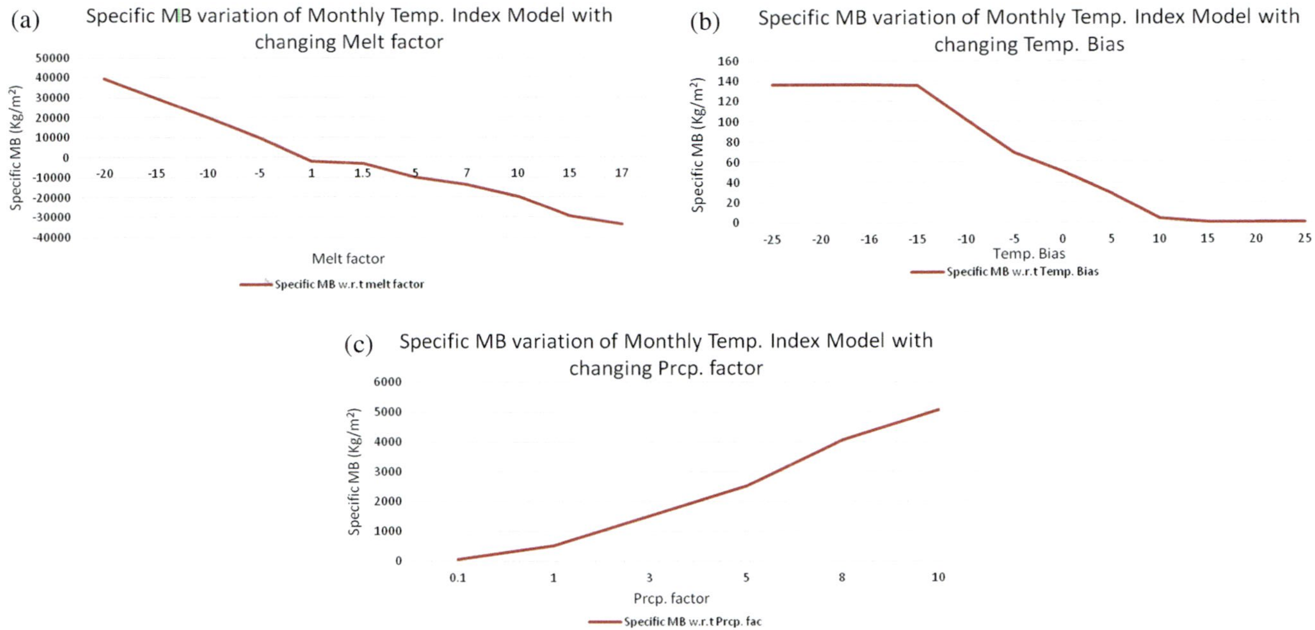

Fig. 3.5 The sensitivity of specific mass balance with parametrization in the OGGM model in the Chhota Shigri Glacier **a** with respect to changes in melt factor **b** with respect to changes in temperature bias and **c** with respect to changes in precipitation factor

Table 3.2 Results of calibration with WGMS specific mass balance data and its correlation, std. deviation, RMSE, and NSC w.r.t modeled data

Calibration	Parameter(s)	melt_f (ranges: 1.5–17)	prcp_fac (ranges: 0.1–10)	temp_bias (ranges: – 15–15)	SMB (Ref = – 463.12 (kg/m^2))	Correlation	Std. deviation	RMSE	NSC
1	1: temp_bias	5	2.66	– 7.46	– 572.84	0.032	324.174	676.125	– 0.276
2	2: prcp_fac	1.5	4.68	0	– 586.42	0.096	528.145	759.521	– 0.61
3	2: temp_bias	1.5	2.66	– 2.42	– 547.74	0.044	300.949	659.845	– 0.215
4	1: prcp_fac 2: melt_f	2.96	10	0	– 759.31	0.1	1118.199	1221.952	– 3.168
5	1: prcp_fac 2: temp_bias	5	10	– 2.62	– 864.24	0.051	1096.166	1259.039	– 3.425

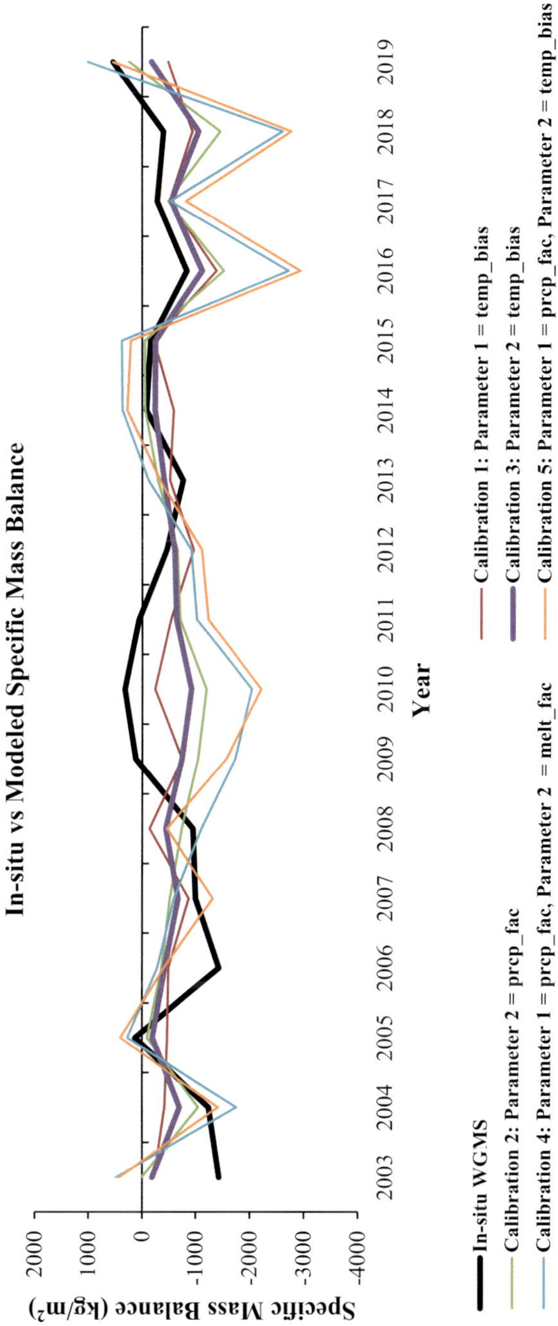

Fig. 3.6 The variations in the specific mass balance in comparison to WGMS data with different combinations of parameters for calibration of the Chhota Shigri Glacier

3.4.4 *Modeled Glacier Mass Balance and Glacier Volume*

To investigate the mass balance changes in the region, the OGGM has driven during 2003 to 2019 (Fig. 3.7) using the WGMS data for which the above reported best calibration parameters viz. melt factor, temperature bias, and precipitation factor having values of 1.5, 2.66, and – 2.42. Figure 3.7 shows that the mass balance of glaciers in Chhota Shigri region, Western Indian Himalaya, during the study period from 2003 to 2019, was calculated to be – 0.547 ± 0.572 m of water equivalent per year (m w.e. a^{-1}). Further analysis focused on the glacier volume changes over the period 2003–2006 and 2016–2019 as shown in the Fig. 3.8 revealing a volume change of – 0.0069 km^3 a^{-1} from 2003 to 2006 and – 0.0134 km^3 a^{-1} from 2016 to 2019. The analysis shown that the modeling results from OGGM is inconsistent with that observed data in Glacier Mass Balance and Glacier volume (– 0.017 km^3 a^{-1} from 2003 to 2006 and – 0.0042 km^3 a^{-1} for 2016–2019) with a significant negative mass balance of 0.659 m w.e. a^{-1} of the Chhota Shigri Glacier, indicating substantial ice loss over the research period 2003 to 2019. Similar findings on glacier loss were reported in the study done by Singh et al. (2018), highlighting the negative mass balance trend over Indian Himalayan Region showing greater glacier loss under the influence of changing climate.

The Fig. 3.8 shows the graph of OGGM modeled volume and calculated volume using WGMS specific mass balance data. There are three spinups viz., fixed geometry spinup, dynamic spinup matching area, and dynamic spinup matching volume. The modeled volume is generated using fixed geometry spinup. The first spinup, i.e., fixed geometry spinup gave the highest correlation of 0.987 with the calculated volume. The calculated volume is derived using WGMS's 'area' data which is given for only six years (2003, 2004, 2005, 2006, 2016, 2017, 2018 and 2019) and multiplied to its specific mass balance divided by density (900 kg/m^3).

The investigation of glacial mass balance is highly significant due to its close connection with climate change. However, the number of glaciers that have been examined for mass balance is rather less in comparison to the total number of glaciers worldwide. In the present study, attempt has been made to model the Chhota Shigri Glacier of the Indian Himalayan Region using OGGM model, though the results of present analysis can be improved significantly with enhanced calibration strategies and in-situ datasets (Schuster et al. 2023).

3.5 Conclusion

This study examined the sensitivity of the Chhota Shigri Glacier through a series of experiments conducted from 2003 to 2019, focusing on the parameters prcp_fac, temp_bias and melt_f using in-situ WGMS and geodetic data. Changes in glacier mass balance were assessed by adjusting a single parameter at a time: temperature bias (temp_bias) ranging from – 15 to 15 °C, melt factor (melt_f) ranging from 1.5

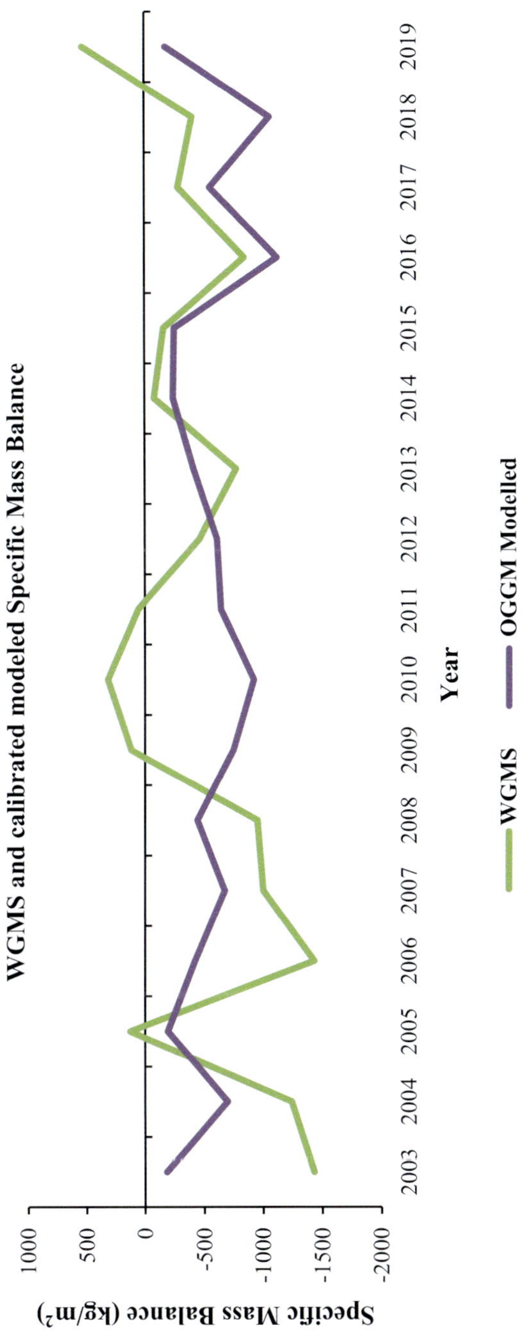

Fig. 3.7 WGMS specific mass balance and modeled specific mass balance (2003–2019)

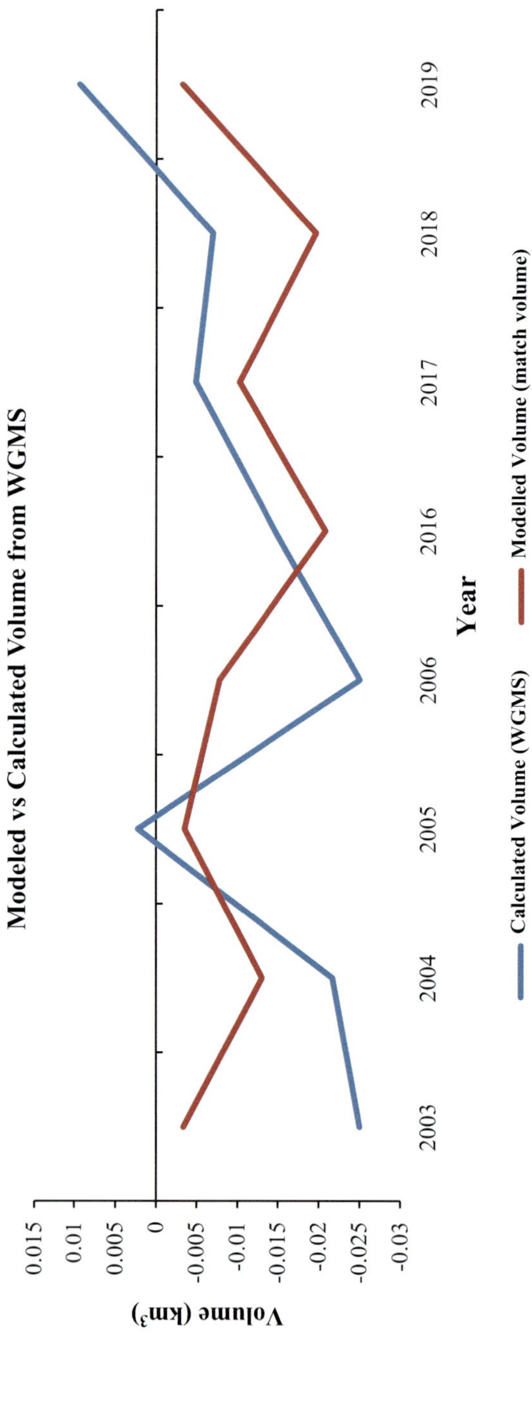

Fig. 3.8 Modeled volume and volume originated from WGMS specific mass balance

to 17 °C, and precipitation (prcp_fac) ranging from 0.1 to 1 kg/m^2. By controlling these parameters, changes in glacier mass balance and volume were evaluated. The study identified parameter values of 'prcp_fac' = 1.5, 'temp_bias' = 2.66 and 'melt_f' = − 2.42 that aligned with in-situ glacier mass balance for the period from 2003 to 2019. The modeled results showed that the average mass balance of the Chhota Shigri Glacier from 2003 to 2019 was − 0.547 ± 0.572 m w.e. a^{-1}, with the highest loss occurring in 2016 (− 1.12 m w.e. a^{-1}), corresponding to a glacier volume change of − 0.020 km^3 a^{-1} in 2016. These initial modeling findings suggest that after proper uncertainty assessment and evaluation over extended periods, the model setup can offer valuable insights for developing effective river basin strategies in glacier-dominated catchments in the western Himalayan regions of India. Furthermore, this model could be applied with future GCM projections to assess future trends in glacier dynamics for the Chhota Shigri Glacier, using bias-corrected CMIP6 model data to drive the OGGM model.

References

Azam MF, Wagnon P, Vincent C, Ramanathan A, Linda A, Singh VB (2014) Reconstruction of the annual mass balance of Chhota Shigri glacier, Western Himalaya, India, since 1969. Ann Glaciol 55(66):69–80

Bajracharya SR, Mool PK, Shrestha BR (2008) Global climate change and melting of Himalayan glaciers. Melting glaciers and rising sea levels: impacts and implications, pp 28–46

Bolch T, Shea JM, Liu S, Azam FM, Gao Y, Gruber S, Immerzeel WW, Kulkarni A, Li H, Tahir AA, Zhang G (2019) Status and change of the cryosphere in the extended Hindu Kush Himalaya region. The Hindu Kush Himalaya assessment: mountains, climate change, sustainability and people, pp 209–255

Cogley JG (2011) Present and future states of Himalaya and Karakoram glaciers. Ann Glaciol 52(59):69–73

Eis J, Van der Laan L, Maussion F, Marzeion B (2021) Reconstruction of past glacier changes with an ice-flow glacier model: proof of concept and validation. Front Earth Sci 9:595755

Engelhardt M, Ramanathan AL, Eidhammer T, Kumar P, Landgren O, Mandal A, Rasmussen ROY (2017) Modelling 60 years of glacier mass balance and runoff for Chhota Shigri Glacier, Western Himalaya, Northern India. J Glaciol 63(240):618–628

Farinotti D, Huss M, Bauder A, Funk M, Truffer M (2009) A method to estimate the ice volume and ice-thickness distribution of alpine glaciers. J Glaciol 55(191):422–430

Gupta A, Kayastha RB, Ramanathan AL, Dimri AP (2019) Comparison of hydrological regime of glacierized Marshyangdi and Tamor river basins of Nepal. Environ Earth Sci 78:1–15

Hanus S, Schuster L, Burek P, Maussion F, Wada Y, Viviroli D (2023) Coupling a large-scale glacier and hydrological model (OGGM v1. 5.3 and CWatM V1. 08)-Data Set

Hugonnet R, McNabb R, Berthier E, Menounos B, Nuth C, Girod L, Farinotti D, Huss M, Dussaillant I, Brun F, Kääb A (2021) Accelerated global glacier mass loss in the early twenty-first century. Nature 592(7856):726–731

Koldunov NV, Kumar P, Rasmussen R, Ramanathan AL, Nesje A, Engelhardt M, Tewari M, Haensler A, Jacob D (2016) Identifying climate change information needs for the Himalayan Region: results from the GLACINDIA stakeholder workshop and training program. Bull Am Meteorol Soc 97(2):ES37–ES40

Koul MN, Ganjoo RK (2010) Impact of inter-and intra-annual variation in weather parameters on mass balance and equilibrium line altitude of Naradu Glacier (Himachal Pradesh), NW Himalaya, India. Clim Change 99(1):119–139

Maussion F, Butenko A, Champollion N, Dusch M, Eis J, Fourteau K, Gregor P, Jarosch AH, Landmann J, Oesterle F, Recinos B (2019) The open global glacier model (OGGM) v1. 1. Geoscien Model Dev 12(3):909–931

Marzeion B, Jarosch AH, Hofer M (2012) Past and future sea-level change from the surface mass balance of glaciers. Cryosphere 6(6):1295–1322

Parkes D, Goosse H (2020) Modelling regional glacier length changes over the last millennium using the Open Global Glacier Model. Cryosphere 14(9):3135–3153

Pesci MH, Schulte Overberg P, Bosshard T, Förster K (2023) From global glacier modeling to catchment hydrology: bridging the gap with the WaSiM-OGGM coupling scheme. Front Water 5:1296344

Ramanathan AL (2011) Status report on Chhota Shigri glacier (Himachal Pradesh). Sci. Eng. Res. Counc. Department of Science and Technology, Government of India, pp 1–89

RGI Consortium (2017) Randolph glacier inventory—a dataset of global glacier outlines, version 6 [Indicate subset used]. NSIDC: National Snow and Ice Data Center, Boulder, Colorado USA. https://doi.org/10.7265/4m1f-gd79

Romshoo SA, Murtaza KO, Shah W, Ramzan T, Ameen U, Bhat MH (2022) Anthropogenic climate change drives melting of glaciers in the Himalaya. Environ Sci Pollut Res 29(35):52732–52751

Schuster L, Rounce DR, Maussion F (2023) Glacier projections sensitivity to temperature-index model choices and calibration strategies. Ann Glaciol:1–16

Singh S, Kumar R, Dimri AP (2018) Mass balance status of Indian Himalayan glaciers: a brief review. Front Environ Sci 6:30

Song L, Wang L, Li X, Zhou J, Luo D, Jin H, Qi J, Zeng T, Yin Y (2020) Improving permafrost physics in a distributed cryosphere-hydrology model and its evaluations at the upper Yellow River Basin. J Geophys Res: Atmos 125(18):e2020JD032916

Wagnon P, Linda A, Arnaud Y, Kumar R, Sharma P, Vincent C, Pottakkal JG, Berthier E, Ramanathan A, Hasnain SI, Chevallier P (2007) Four years of mass balance on Chhota Shigri Glacier, Himachal Pradesh, India, a new benchmark glacier in the western Himalaya. J Glaciol 53(183):603–611

Wang L, Liu H, Bhlon R, Chen D, Long J, Sherpa TC (2024) Modeling glacio-hydrological processes in the Himalayas: a review and future perspectives. Geogr Sustain

Chapter 4
Insight into the Impact of Land Use/Land Cover Change on Runoff Incorporating Hydrological Modelling to Analyse Their Impact on the Water Regime

Ankita Sharma, Radha Krishan, Kuljit Kour, Bhaskar R. Nikam, and Dhirendra Singh Bagri

Abstract Land and water are the two essential natural resources for human survival. Land Use/Land Cover (LULC) is an important parameter governing the hydrology of a region. Increased human interventions have caused different modifications in the land cover, directly or indirectly affecting the hydrological cycle and processes in the long run. Raising urban/rural growth leads to an increase in the runoff by diminishing the perviousness of the region. Distributed hydrological modelling provides an efficient way to evaluate such changes by allowing quantification of streamflow patterns and distribution of hydrological components throughout watersheds. Remote sensing has the potential to be a more accurate tool for collecting spatial and temporal databases for studying hydrological processes. This study helps researchers to convey their ideas to quantify the influence of alterations in LULC patterns over time on the runoff of the Asan watershed, India from 1980 to 2016 and opted for the Variable Infiltration Capacity (VIC) model to understand broader-scale hydrological effects on the watershed. Datasets related to temperature, precipitation and discharge are required in order to generate inflow-outflow statistics. The outcome shows that, with a relative increase in 2011, mean annual runoff increased from 509.8 cusec in 1980 to 549.5 cusec in 2016. Annual discharge was highest in July–September and lowest

A. Sharma (✉)
Department of Geology, HNBGU, Chauras Campus, Srinagar, Uttarakhand 246174, India
e-mail: ankita.geors@gmail.com

R. Krishan
Department of Civil Engineering, Government of Engineering College, Banka 813102, India

K. Kour
Department of RS and GIS, University of Jammu, BR Ambedkar Road, Jammu, Jammu & Kashmir 180006, India

B. R. Nikam
EDPO, ISRO, Bangalore, Karnataka 560096, India

D. S. Bagri
Department of Geology, HNBGU, S.R.T. Campus, New Tehri, Uttarakhand 249001, India

A. Nanda et al. (eds.), *Navigating the Nexus*, Water Science and Technology Library 102, https://doi.org/10.1007/978-3-031-76532-2_4

in December–February. Urbanized areas increased from 9.55 km^2 in 1980 to 36.50 km^2 in 2016, leading to an increase in impervious surfaces and runoff. The decrease in natural land cover, including water bodies, and intensive agriculture contributed to higher runoff and lower groundwater recharge. Overall, the study emphasizes the importance of hydrological modelling in understanding water balance components and simulating hydrological responses under LULC alterations. Also, the adoption of appropriate workflows for assembling multiple datasets and extraction techniques and analysis of the time-series impact of urbanisation on the watershed's runoff potential.

Keywords LULC · Hydrological modelling · Watershed · VIC · Rainfall/runoff · Urbanisation

4.1 Introduction

In geologic strata, waters present on the surface of the earth in the form of vapour, snow/ice and underground water which keep on circulating from the surface to the atmosphere and vice versa. Physical, chemical, and biological processes and anthropogenic activities affect the hydrological cycle and thus greatly impact changing climatic patterns (Tricart 1999). Thus, studying hydrology is important because it helps to understand its components, viz., evapotranspiration, precipitation, infiltration, runoff, baseflow, soil depth and moisture content, etc., to quantify, manage and preserve water resources in the region effectively (Asrar and Dozier 1994; Berhanu et al. 2015). To analyse the implications of disturbance and alteration of these components, hydrological modelling acts as a good aid at a regional scale. Hydrological modelling is a mathematical procedure for representing the natural process influencing the water balance components of a watershed or basin (Liang et al. 1996). It is crucial for understanding hydrological processes, their spatial variability and water resources-related problems that occur within the hydrological cycle as well as simulating the behaviour of these processes under various conditions e.g., Climate change or LULC change (Singh and Marcy 2017).

The selection of a perfect hydrological model for any research is core dependent upon the objectives taken. Cunderlik (2003) proposed four generic fundamental criteria, which include the model's capacity, the availability of input data, the overall cost of the model and the significance of the output generated by the model. This study examines the theoretical framework of the VIC-hydrological model, describing its distinct hydrological properties and the mathematical relationships that are used to simulate different hydrological processes, including flow, evaporation, rainfall/runoff, sediment load and the water cycle (Hamman et al. 2018). The Variable Infiltration Capacity (VIC) model was chosen for this study because it effectively simulates major hydrological processes in the study area, which justifies the research objectives. In addition, it provides flexibility on required datasets, and long-term simulation capabilities and is freely available. Developed collaboratively by the Universities of

Washington and Princeton, the model operates in both water and energy balance modes within river flow regimes and accommodates the effects of multiple vegetation types (Zhao et al. 1980). Initially, the model was discussed in detail by Liang et al. (1994, 1996).

Parameter identification and approximation pose challenges in models like VIC, which integrate both conceptual and physically based elements. The model uses empirical estimation to simulate runoff, rainfall and infiltration processes. The grid-based hydrologic model is being discussed here, and it operated in the water balance mode of the river flow regime using physical watershed components (e.g. land cover, soil and topography) and meteorological forcing files (e.g. precipitation, temperature and wind speed). It incorporates hydrological fluxes for each grid cell from the watershed quantifying influential hydro-meteorological processes at the surface of the area (Witham 2015). The parameterization of surface runoff and baseflow is based on the storage and infiltration capacity of soil moisture representing sub-grid variability (Srivastava et al. 2017). Each grid cell takes vegetation coverage, soil properties and infiltration capacity into consideration. The model was initially created for UNIX and LINUX platforms. On Windows, UNIX emulators such as Cygwin are often used for compiling and simulating the dataset.

The VIC model generates a complex interaction between precipitation and LULC aids in predicting the climate change's impact on the hydrological response of the area (Yeo et al. 2007). Models such as VIC, alongside others like MIKE-SHE, SWAT, HEC-HMS, and HEC-RAS, are commonly used to assess the impact of LULC changes on basin hydrology (Liang et al. 1994, 1996; Nijssen et al. 1997). The VIC model helps in studying complex terrains/topographies and soil properties and their effects on several hydrological components, such as construction of hydrological datasets, water balance, base flow, forecasting, evapotranspiration (ET), coupled climate modelling and runoff, which are significantly influenced by LULC and climatic changes in the area (Gao et al. 2009; Zhai and Tao 2017). However, it's important to recognize that each model operates based on its own assumptions and input data.

Various factors, such as climatic variability, topographic changes, altitude, slope, the soil erosion coefficient, conservation policies and socioeconomic factors, such as distance from the river, major roads, settlement, and population density, are responsible for exerting large impacts on changing LULC patterns (Bronstert et al. 2002; Elfert and Bormann 2010). Changes in LULC affect watershed conditions, such as the water-holding capacity of soil and surface runoff, which in turn changes the river's path. Runoff is the product of precipitation covering the Earth's surface. It resided in the watershed and was extracted from the outlet. The distribution of precipitation is uneven throughout the watershed, whereas that of runoff is uniform. Utilizing Remote Sensing (RS) and Geographical Information System (GIS) techniques enabled us to offer timely and precise insights into evolving Land Use and Land Cover (LULC) patterns over time. This, in turn, greatly facilitated numerous hydrological investigations, particularly in forecasting surface runoff and assessing the effectiveness and efficiency of hydrologic models based on watershed characteristics. The presence of inherent errors and uncertainty in the modelling procedure is examined when calibrating the model (Gupta et al. 1998; Seibert and McDonnell 2002). A better calibration procedure is responsible for providing accurate hydrological variables. These potential errors may affect the results of the simulation. The calibrated models are significant under various water management scenarios.

Altogether the present study addresses the setup of a macroscale semi-distributed hydrological VIC model using observed discharge data and quantifies the influence of changes in LULC from 1980 to 2016 on the runoff of the Asan watershed, Uttarakhand, India. The long list of input parameters is assembled from various sources, and the entire setup is executed in the ArcMap interface (Nikam 2015). The VIC model is useful for studying complex topographies and their effects on water cycle components (Gao et al. 2009). Runoff is the portion of precipitation that occurs at an outlet of the watershed. The total runoff of the basin is the collective term used for surface runoff, subsurface runoff and groundwater. For the present study, the runoff was calculated every month and measured in units of Cubic foot/feet per second (Cusec). Sharma and Bagri (2020) reported that increased urbanization/industrialization and deforestation occurring over time in the study area are responsible for gradually changing LULC and subsequently hydrological processes in the long run. By considering all the literature reviews and research-based facts, we consider these objectives for our study: (i) To quantify the impact of altercation in LUC patterns on the runoff between 1980 to 2016. (ii) To access how well the VIC hydrological model simulates the impact of LULC modification on the runoff (iii) To create a foundation for upcoming modelling initiatives aimed at comprehending the broader-scale hydrological effect of LULC change on the watershed.

4.2 Study Area

The Asan River watershed is located between geographic latitudes 30°14′ 14″ N to 30° 29′ 54″ N and longitudes 77° 39′ 42″ E to 78° 05′ 30″ E in Dehradun district, Uttarakhand State, India and ranges in elevation from 345 to 2188 m above mean sea level (MSL). It covers an area of about 712.34 km^2 (Fig. 4.1). It falls under 53F/11, 53F/15, 53J/3 and 53F/16 toposheet of Survey of India, scale (RF)-1: 50,000. It is among the most significant watersheds in Doon valley, located at the foothills of the Siwalik ranges with the Asan river flowing through the central part of the valley from the north-east to south-west direction. The watershed extends for a length of approximately 40 km with an approximate widest width of 18 km. It spans approximately 43 km in length. As the river moves toward the southwest, it carries water from several streams flowing southwards from the Lesser Himalayan range and northwards from the Siwaliks. In the lower reaches, a relatively flat section flows that feeds the Asan barrage. It is located in the vicinity of the Asan and Yamuna Rivers. The water from the barrage later drains into the Yamuna at Paonta Sahib.

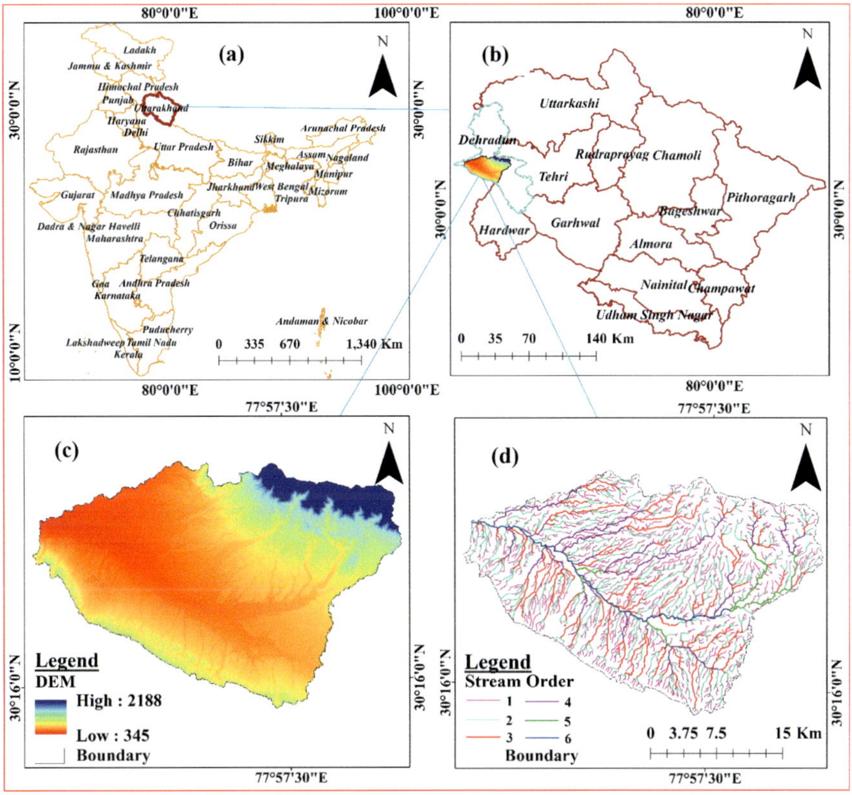

Fig. 4.1 Location of the study area; **a** depiction of Uttarakhand state in the map of India; **b** depiction of study area in the Dehradun district of Uttarakhand; **c** DEM used in the study; **d** stream Order distribution in the area

4.3 Data and Software Used

The details of the software and datasets used in the study are provided in Table 4.1.

Table 4.1 Input datasets and software used in the study

Data	Software
• Cartosat-DEM (10 m)	• ArcGIS 10.3
• LULC maps (Scale-1:250,000)	• QGIS
• Hydro-meteorological data	• ERDAS Imagine 9.3
• Discharge data	• Cygwin
• Soil Map (Scale-1:250,000)	• Python script

4.4 Detailed Methodology Adopted

The methodology described the procedure for generating various datasets and input parameter files of the study area to simulate the VIC hydrological model. The model is executed in three steps: Simulation, Calibration and Validation with a necessary setting on all the required input files. Figure 4.2, depicts the flow chart of the methodology used for the study.

To carry out the study, a grid network is generated over the study area using QGIS and the parameters are derived using DEM, meteorological and other remote sensing data as input.

4.4.1 Grid Generation Over the Watershed

The preparation of the grid map requires the extent of 1 km * 1 km to be generated for the watershed's area by using QGIS (Fig. 4.3). Each grid has been assigned a number starting from the top left corner of the watershed to the whole extent. The total number of grids is 1242, with 27 and 46 being the total number of rows and

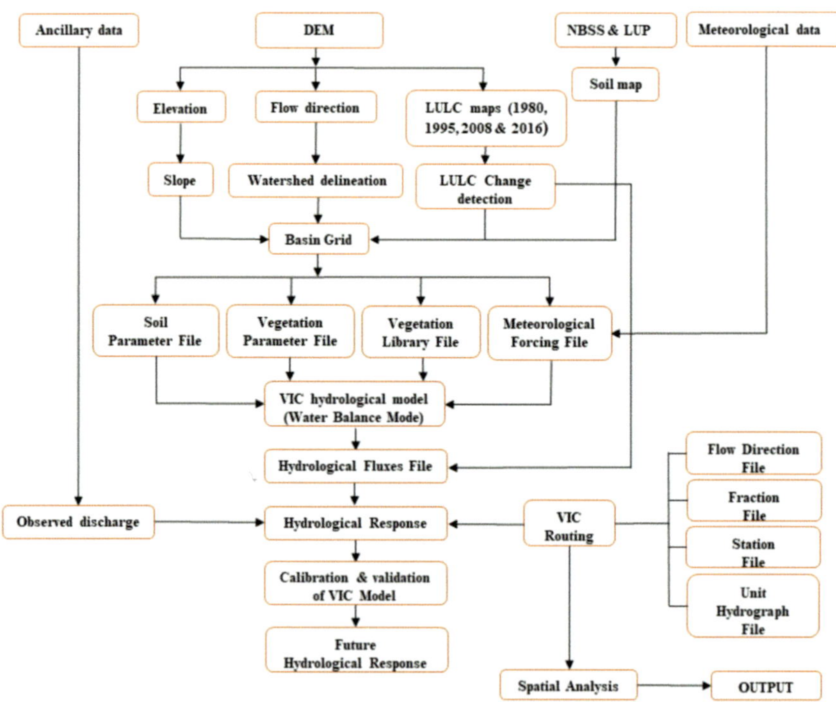

Fig. 4.2 Flowchart represents the methodology of the VIC hydrological model

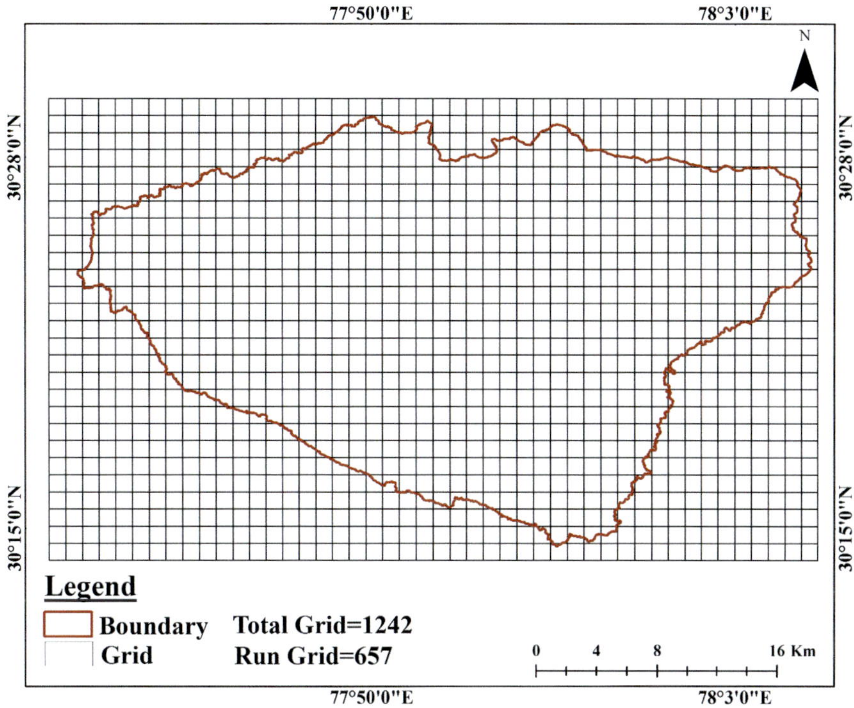

Fig. 4.3 Grid file used in the hydro-processing of the model

columns respectively. Depending upon the intersecting watershed's boundary with the grid, 657 'Run grids' are generated which are used for further analysis. The 'Run grid' is represented by the value 1 if the grid lies inside the watershed boundary and 0 if outside. The grid map was generated with ID, XMIN, XMAX, YMIN and YMAX fields. Further, different fields of rainfall, soil, slope, elevation, and run grids are added to the grid file.

4.4.2 *Global Parameter File for Model Simulation*

To operate in the Cygwin environment, a global parameter file is used, which contains all the information needed for the input and output path locations as well as other details like the number of soil layers, time step, duration, meteorological data, etc. It is formatted in ASCII. In order to run the VIC model for four distinct LULCs while maintaining the same level of meteorological forcing on the water balance model, a global parameter file that generates flux files for hydrological components has been created.

4.4.3 *Input Parameter Files*

The input database requires four major input parameter files namely: vegetation parameter, soil, vegetation library and meteorological forcing parameter files. The information in these files is stored in ASCII format. The process of preparing these input files is explained in the sequential sections.

(a) **Vegetation Parameter File**

The primary input required for creating a vegetation parameter file is the LULC map. Together with the grid map, it specifies the number of grid cells, vegetation composition and their fractional area. LULC also has an impact on hydrological fluxes because the amount of precipitation that falls on the ground is generally determined by the type and density of vegetation. The vegetation parameter file includes Leaf Area Index (LAI), rooting depth and rooting fraction for LULC type (vegetation cover) and it computed infiltration, moisture fluxes and soil layer. These database files are created by using the International Geosphere-Biosphere Programme (IGBP) LULC classification scheme. The mentioned link contains a detailed description of the vegetation parameter file: (https://www.hydro.washington.edu/Lettenmaier/Models/VIC/Documentation/VegParam.html).

(b) **Vegetation Library File**

It describes the land cover and vegetation type generated by a LULC map of the area. It contains detailed land cover characteristics on a monthly average basis. To analyse the hydrological changes throughout the watershed in a spatial context, some grid cells are considered insignificant and require changes at the pixel level, eventually responsible for the increase or decrease in the runoff. The VIC-hydrological modelling tool is used to create a comparison of vegetation cover parameter files from different classes for different LULC scenarios. The detailed description of the vegetation library file can be seen at the link mentioned: (https://www.hydro.washington.edu/Lettenmaier/Models/VIC/Documentation/VegLib.html).

(c) **Soil Parameter File**

It defines the soil layer's characteristics such as its identification number, geographic information, mean elevation, grid cell number and soil moisture condition for each grid cell. The digitised soil texture map is prepared from NBSS & LUP at a scale of 1:10,00,000 (Fig. 4.4). Soil texture IDs have been assigned to each grid cell to identify them. The input file contains the fraction of each grid that covers a specific soil. To extract the dominant soil type in each active/run grid, the soil texture map is rasterized and overlaid with the grid map. The link provides a more detailed description: (https://www.hydro.washington.edu/Lettenmaier/Models/VIC/Documentation/SoilParam.html).

The soil map shows that the watershed contains the highest proportion of loamy soil and is distributed almost all over the basin. The western portion of the watershed is occupied by loamy skeletal soil in significant quantities along with some proportion

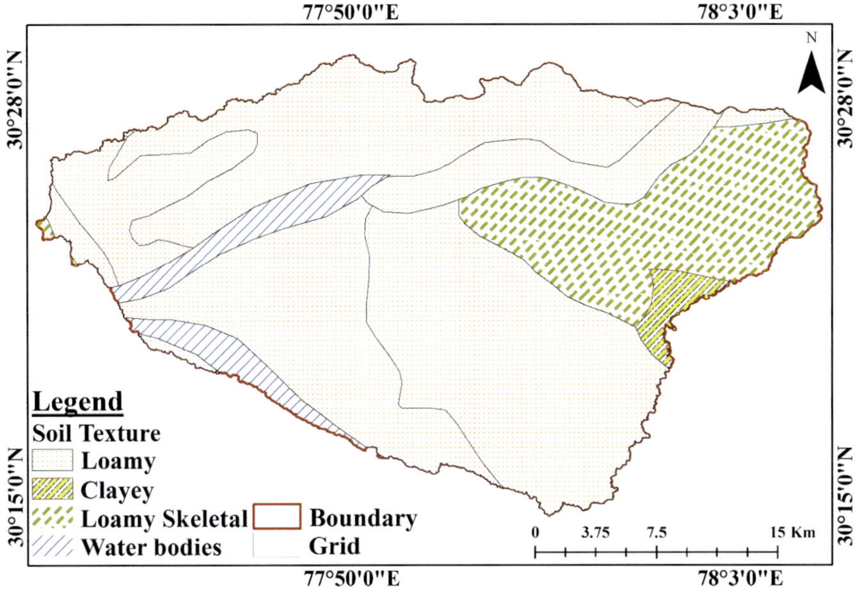

Fig. 4.4 Soil map of the study area. *Source* NBSS-LUP

of clayey soil. The information in the soil layer is specified in the user-defined soil parameter file (Zhao et al. 1980).

(d) **Meteorological Forcing File**

This file is prepared using a long-term gridded dataset of the Indian Meteorological Department (IMD) and generated for each grid cell. It contains four significant meteorological variables: minimum temperature (Tmin), Maximum temperature (Tmax), precipitation and wind speed. These are daily data and saved in ASCII format. Forcing files is one of the important model inputs for creating runoff and simulating the hydrological cycle. The model has been run for 36 years from 1980 to 2016 with four different scenarios each represented by decadal LULCs of 1980, 1995, 2008 and 2016 (Sharma and Bagri 2020, 2023; Sharma et al. 2021), and all the input parameters are set to investigate the effect of change in LULC pattern on runoff with meteorological forcing remains constant for 1951 to 2013.

Further, the next step includes model calibration has to be done from 1996 to 2003 and validated from 2004 to 2011 using the same meteorological forcing files obtained from IMD from 1951 to 2013 and observed discharge data (1996–2011). This method of calibration and validation provided a positive match between simulated and observed discharge. Now, the model is run again using the same routing process as above taking three other LULC scenarios of 1980, 1995 and 2016, using an IDL code average values of runoff.

4.4.4 Global Parameter File for Model Calibration and Validation

The global parameter file used for the routing model contains the location path of the input and output folder, details of the routed month and year, and necessary input files such as flow direction, fraction, station and UH for routing. The fluxes generated after VIC simulation are also used as input.

4.4.5 Routing Model of Surface Flow

Lohmann et al. developed the routing model in (1998a, b) (I; II). In this study, we implemented the model within the Cygwin environment to route the fluxes for the active grid of the watershed. Subsequently, the model was routed for the LULC data of 2008 at the outlet of the watershed using various input files. The code for the routing model is accessible in the source code section of the download page via the provided link: (https://www.hydro.washington.edu/Lettenmaier/Models/VIC/SourceCode/Download.html).

Further, compilation will be done using edit '*rout.f*' from the extracted zipped file. This is the executable file for the model which is extracted into a folder '*Rout_*'. Put all input layers (*Flow direction file, Station location file, Fraction file and UH*) together with fluxes and global parameter file into a similar place. To initiate the model in the Cygwin platform following steps are taken:

- Updating the Global parameter file by adding all relevant information and paths.
- Open a window on the Linux/Cygwin environment and give the path where the folder to be routed is kept.
- At the command prompt, type:./rout <input file name> .

4.4.6 Input Parameter Files

For routing the model, essential elements like flow direction, station data, fraction information, and Unit Hydrograph (UH) are necessary. These files, prepared in ASCII format by the user, along with fluxes, global parameter files, and discharge data, are vital for executing the routing of the total surface runoff model at the designated discharge station (Asan Barrage). Below, we discuss all the input files required for this routing process:

(a) **Flow Direction File (*Fdr*)**

The Flow direction map is extracted from Cartosat-1 (spatial resolution-10 m) using the Hydro tool in the GIS environment by using the resampling tool to convert the resolution of DEM similar to the extent of the grid file (1 km * 1 km). The value is

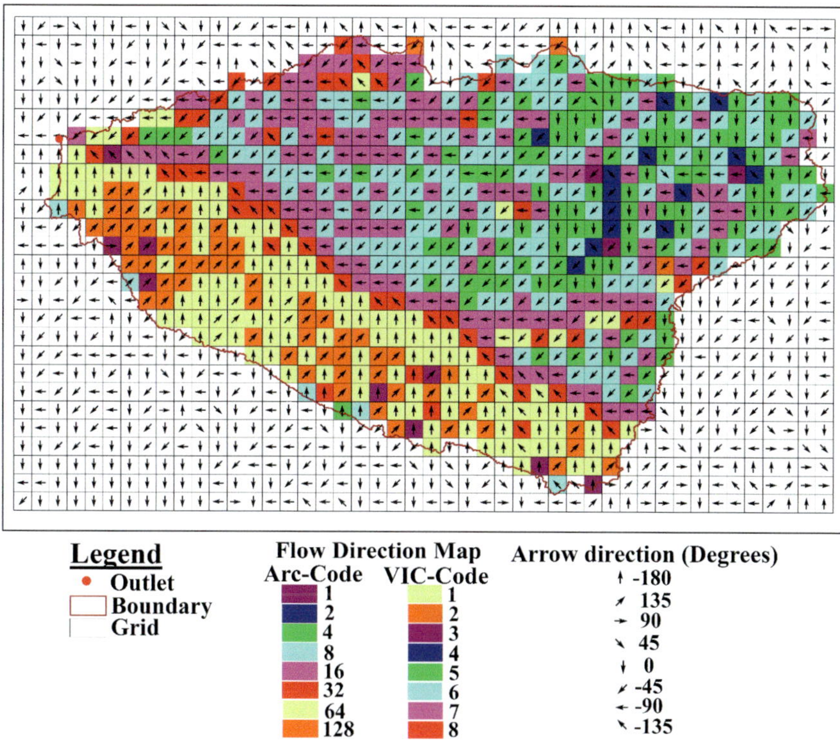

Fig. 4.5 Flow direction map used in the study

assigned for each grid cell and creates a streamflow network throughout the watershed (Fig. 4.5).

The output reflects arrows for each grid representing the direction of streamflow. As *'Fdr'* is very significant for generating runoff there is some anomaly present in it. The stream network used for the study is overlapped within the grid to improve the orientation of the arrows and is further manually edited which are not pointing towards the right direction of the stream (Fig. 4.6).

Further, the flow direction codes for ArcGIS are replaced and defined with related VIC codes (Table 4.2).

The LULC maps are generated through Landsat 3 MSS, Landsat 5 TM and LISS-IV images through visual interpretation for four different years i.e. 1980, 1995, 2008 and 2016. Different LULC maps are classified into eight different classes viz., the built-up (urban and rural), cropland, fallow land, plantation, evergreen needle leaf forest, deciduous broadleaf forest, mixed forest, and water bodies (Fig. 4.7).

The assignment of VIC code represents eight LULC classes selected for the study out of twenty different classes of LULC map are shown in Table 4.3.

ArcGIS begins counting from the east and works clockwise by multiplying the number by two. In VIC, counting begins in the north and proceeds clockwise by

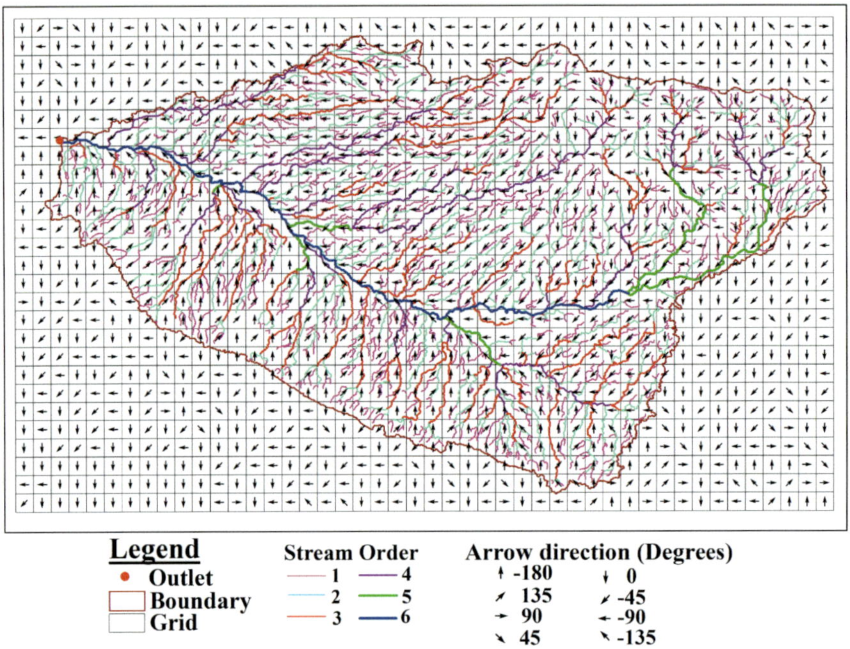

Fig. 4.6 Orientation of arrows within the watershed boundary represents each grid is directed toward the flow path

Table 4.2 Flow direction code

Direction	ArcGIS	VIC
East	1	3
South-East	2	4
South	4	5
South-West	8	6
West	16	7
North-West	32	8
North	64	1
North-East	128	2

adding one. The conversion of ArcGIS into VIC codes of flow direction is done by the raster calculator tool of ArcGIS. The output raster file is then converted into ASCII format for input in route.

(b) **Fraction File**

The fraction file specifies the percentage of each grid cell that generates streamflow into the watershed being routed. This file is generated using an active grid map by the user and it is in ASCII format (Fig. 4.8). Each grid represents 0 or 1 numeric

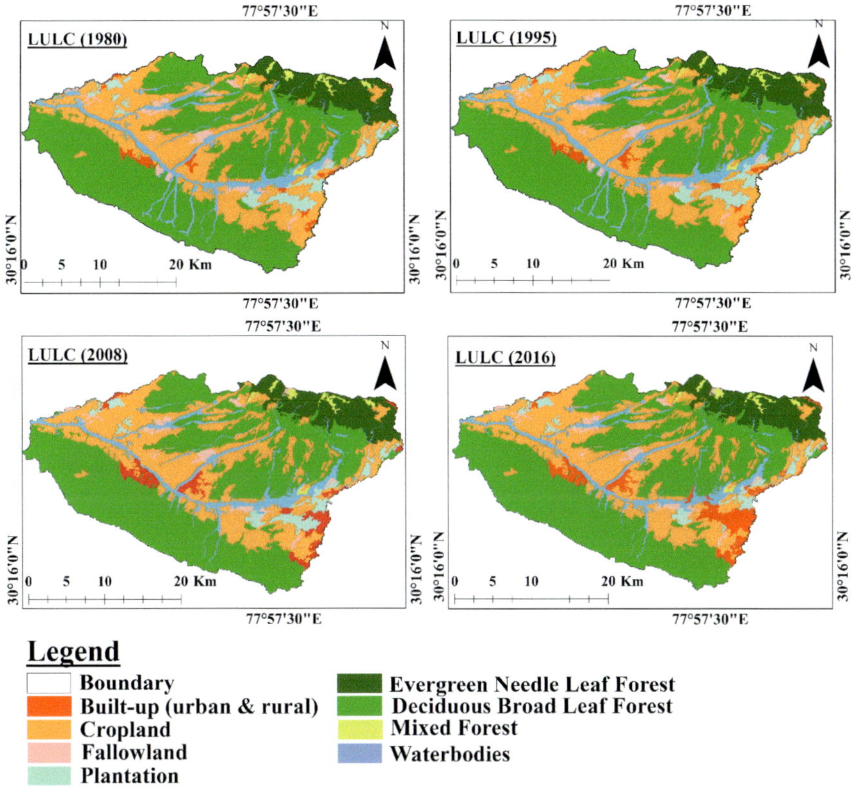

Fig. 4.7 Land use/land cover map of the study area for 1980–2016. *Source* Sharma and Bagri (2020, 2023), Sharma et al. (2021)

value. The grids outside the watershed's boundary are assigned a value of 0, which is considered NO DATA, and grids within the boundary represent a value of 1. The following steps are taken into account for the creation of the fraction file:

- Conversion of grid file and boundary into a feature class
- Selection of run grids
- The area of each grid cell is determined by intersecting the watershed boundary and the resultant 8
- The area field is further divided into 0.0625 to determine the fraction of each grid (area of the square grid).

(c) Station Location File

It specifies which station in the watershed should be routed (flow data generated from grid cells). The file stored the information about the discharge station e.g., its name, depending upon their state of being active or non-active, mention the number 1 or 0 respectively. The routing model must generate a UH file for the mentioned station

Table 4.3 Representation of VIC codes, class names and designation of LULC classes

VIC code	Class name	Designation	VIC code	Class name	Designation
1	Built-up (urban and rural)	BU	11	Grassland	GL
2	Crop Land	CL	12	Mangrove Forest	MF
3	Fallow Land	FL	13	Water Bodies	WB
4	Plantation	PL	14	Permanent Wetland	PW
5	Evergreen Needle Leaf Forest	ENLF	15	Aquaculture	AQ
6	Evergreen Broad Leaf Forest	EBLF	16	Salt pans	SP
7	Deciduous Needle Leaf Forest	DNLF	17	Snow and Ice	SI
8	Deciduous Broad Leaf Forest	DBLF	18	Barren land	BL
9	Mixed Forest	MF	19	Wasteland	WL
10	Shrubland	SL	20	Mining	MN

Source Nikam (2015)

Fraction.txt - Notepad

File Edit Format View Help

```
ncols         48
nrows         29
xllcorner     77.630
yllcorner     30.210
cellsize      0.01
NODATA_value  0
0 0 0 0 0 0 0 0 0 0 0 0 0 0 0 0 0 0 0 0 0 0 0 0 0 0 0 0 0 0 0 0 0 0 0 0 0 0 0 0 0 0 0 0 0 0 0 0
0 0 0 0 0 0 0 0 0 0 0 0 0 0 0 0 0 0 0 0 0 0 0 0 0 0 0 0 0 0 0 0 0 0 0 0 0 0 0 0 0 0 0 0 0 0 0 0
0 0 0 0 0 0 0 0 0 0 0 0 0 0 0 0 0 0 0 0 0 0 0 0 0 0 0 0 0 0 0 0 0 0 0 0 0 0 0 0 0 0 0 0 0 0 0 0
0 0 0 0 0 0 0 0 0 0 0 0 0 0 0 0 0 0 0 1 1 1 1 0 0 0 0 0 0 0 0 1 0 0 0 0 0 0 0 0 0 0 0 0 0 0 0 0
0 0 0 0 0 0 0 0 0 0 0 0 0 0 0 0 0 1 1 1 1 1 1 1 0 0 0 0 1 1 1 1 1 1 0 0 0 0 0 0 0 0 0 0 0 0 0 0
0 0 0 0 0 0 0 0 0 0 0 0 0 1 1 1 1 1 1 1 1 1 1 1 1 1 1 1 1 1 1 1 1 1 1 1 1 1 1 1 0 0 0 0 0 0 0 0
0 0 0 0 0 0 0 0 0 0 1 1 1 1 1 1 1 1 1 1 1 1 1 1 1 1 1 1 1 1 1 1 1 1 1 1 1 1 1 1 1 1 1 1 1 0 0 0
0 0 0 0 0 0 0 1 1 1 1 1 1 1 1 1 1 1 1 1 1 1 1 1 1 1 1 1 1 1 1 1 1 1 1 1 1 1 1 1 1 1 1 1 1 0 0 0
0 0 0 0 1 1 1 1 1 1 1 1 1 1 1 1 1 1 1 1 1 1 1 1 1 1 1 1 1 1 1 1 1 1 1 1 1 1 1 1 1 1 1 1 1 0 0 0
0 0 0 0 1 1 1 1 1 1 1 1 1 1 1 1 1 1 1 1 1 1 1 1 1 1 1 1 1 1 1 1 1 1 1 1 1 1 1 1 1 1 1 1 1 1 0 0
0 0 0 0 1 1 1 1 1 1 1 1 1 1 1 1 1 1 1 1 1 1 1 1 1 1 1 1 1 1 1 1 1 1 1 1 1 1 1 1 1 1 1 1 1 1 0 0
0 0 0 0 0 1 1 1 1 1 1 1 1 1 1 1 1 1 1 1 1 1 1 1 1 1 1 1 1 1 1 1 1 1 1 1 1 1 1 1 1 1 1 1 0 0 0 0
0 0 0 0 0 1 1 1 1 1 1 1 1 1 1 1 1 1 1 1 1 1 1 1 1 1 1 1 1 1 1 1 1 1 1 1 1 1 1 1 1 1 1 0 0 0 0 0
0 0 0 0 0 0 0 1 1 1 1 1 1 1 1 1 1 1 1 1 1 1 1 1 1 1 1 1 1 1 1 1 1 1 1 1 1 1 1 1 1 1 0 0 0 0 0 0
0 0 0 0 0 0 0 1 1 1 1 1 1 1 1 1 1 1 1 1 1 1 1 1 1 1 1 1 1 1 1 1 1 1 1 1 1 1 1 1 0 0 0 0 0 0 0 0
0 0 0 0 0 0 0 0 1 1 1 1 1 1 1 1 1 1 1 1 1 1 1 1 1 1 1 1 1 1 1 1 1 1 1 1 1 1 0 0 0 0 0 0 0 0 0 0
0 0 0 0 0 0 0 0 0 1 1 1 1 1 1 1 1 1 1 1 1 1 1 1 1 1 1 1 1 1 1 1 1 1 1 1 1 0 0 0 0 0 0 0 0 0 0 0
0 0 0 0 0 0 0 0 0 0 1 1 1 1 1 1 1 1 1 1 1 1 1 1 1 1 1 1 1 1 1 1 1 1 1 1 1 0 0 0 0 0 0 0 0 0 0 0
0 0 0 0 0 0 0 0 0 0 0 0 0 1 1 1 1 1 1 1 1 1 1 1 1 1 1 1 1 1 1 1 1 1 1 1 1 0 0 0 0 0 0 0 0 0 0 0
0 0 0 0 0 0 0 0 0 0 0 0 0 0 0 1 1 1 1 1 1 1 1 1 1 1 1 1 1 1 1 1 1 1 1 1 1 0 0 0 0 0 0 0 0 0 0 0
0 0 0 0 0 0 0 0 0 0 0 0 0 0 0 0 0 1 1 1 1 1 1 1 1 1 1 1 1 1 1 1 1 1 1 1 1 0 0 0 0 0 0 0 0 0 0 0
0 0 0 0 0 0 0 0 0 0 0 0 0 0 0 0 0 0 1 1 1 1 1 1 1 1 1 1 1 1 1 1 1 1 1 1 0 0 0 0 0 0 0 0 0 0 0 0
0 0 0 0 0 0 0 0 0 0 0 0 0 0 0 0 0 0 0 0 0 1 1 1 1 1 1 1 1 1 1 1 1 1 1 1 0 0 0 0 0 0 0 0 0 0 0 0
0 0 0 0 0 0 0 0 0 0 0 0 0 0 0 0 0 0 0 0 0 0 0 0 1 1 1 1 1 1 1 1 1 1 1 0 0 0 0 0 0 0 0 0 0 0 0 0
0 0 0 0 0 0 0 0 0 0 0 0 0 0 0 0 0 0 0 0 0 0 0 0 0 0 0 0 1 1 1 1 1 1 0 0 0 0 0 0 0 0 0 0 0 0 0 0
0 0 0 0 0 0 0 0 0 0 0 0 0 0 0 0 0 0 0 0 0 0 0 0 0 0 0 0 0 0 0 1 1 1 0 0 0 0 0 0 0 0 0 0 0 0 0 0
0 0 0 0 0 0 0 0 0 0 0 0 0 0 0 0 0 0 0 0 0 0 0 0 0 0 0 0 0 0 0 0 0 0 0 0 0 0 0 0 0 0 0 0 0 0 0 0
0 0 0 0 0 0 0 0 0 0 0 0 0 0 0 0 0 0 0 0 0 0 0 0 0 0 0 0 0 0 0 0 0 0 0 0 0 0 0 0 0 0 0 0 0 0 0 0
0 0 0 0 0 0 0 0 0 0 0 0 0 0 0 0 0 0 0 0 0 0 0 0 0 0 0 0 0 0 0 0 0 0 0 0 0 0 0 0 0 0 0 0 0 0 0 0
```

Fig. 4.8 Fraction file used in the routing model

based on its position in the '*column*' from the left, and '*row*' from the bottom (by default set to NONE). The number of stations may vary for the watershed, here in the present study we use one discharge station (Asan Barrage). This file is generated using an active grid map by the user and it is also in ASCII format.

(d) **Unit Hydrograph (UH) File**

The grid impulse response function with a monthly summation of 1 is described by the Unit Hydrograph (UH-file). The basin characteristics are used to calculate total runoff at discharge stations and the unit hydrograph is derived from them.

(e) **Discharge Data**

The discharge data is acquired from the office of the Uttarakhand Jal Vidyut Nigam Limited (UNJVL) situated at the Asan Barrage (outlet's location) from the year January 1996 to December 2011 to validate the VIC model. The output (runoff) is stored daily and monthly streamflow in Cusec (a unit of water flow, Cubic foot per second).

4.5 Results and Analyses

The simulation period is divided into two parts: one is calibration and the other is validation. The VIC model was developed from 1996 through 2011 over the Asan watershed. For this period, the simulation from 1996 to 2003, is selected for calibration and 2004–2011 is selected for validation. This is the widely used method for assessing the model. The detailed steps for VIC model calibration and validation are as follows:

4.5.1 Simulation of the Model

The VIC 4.0.5 source code can be downloaded from the link: (https://www.hydro.washington.edu/Lettenmaier/Models/VIC/Development/ArchievedVersions.html). The VIC model was originally designed for LINUX and UNIX platforms, but to use it in Windows, a UNIX emulator called 'Cygwin' is required. The compilation of the VIC model is done using the *'gcc'* compiler on LINUX and the code compilation is made using the *'make'* command prompt type. The executable file *'vicNL'* is generated through compiler code. To initiate the model in the Cygwin platform following steps are taken:

(a) Updating the Global parameter file by adding all relevant information
(b) Open a window on Linux/Cygwin environment, and give the path where *'src'* folder under route VIC tool.
(c) Then, again at Cygwin, give the path of the source code, where *'vicNL'* is kept.

(d) Type: vicNL—g Global parameter, at the command prompt,

As previously discussed, the model is run in water balance mode and thus the generated output fluxes for each grid carry information about the major components such as ET, base flow, runoff and soil moisture of each soil layer. The generated fluxes have no header and for removing the *'nan'* value, Python script is used to convert these values into zero. Further, Simulation of streamflow at an outlet is done through the Routing model.

4.5.2 Method Development for the Calibration and Validation of the Model

Every hydrological model is dependent upon the inputs and it may contain some uncertainties at grid and watershed scale. For the detection and removal of uncertainties, calibration and validation are necessary and significant steps (Alvarenga et al. 2020). In Calibration, simulated and observed data values are compared for evaluation, whereas validation is a continuation of the calibration process in which the model properly evaluates all parameters and conditions. Out of 10 calibrating parameters of soil, the model considers 4, which will affect the output considerably. So, the calibration has been done by adjusting the parameter's value using the trial and error method till getting an acceptable result. The four adopted ranges of calibrating parameters of soil include:

- **Fraction of maximum soil moisture (*Ws*)**: Its value ranges from > 0 to 1; a higher '*Ws*' will raise the amount of water needed for exponentially growing, non-linear base flow (*Ds*) thereby delaying runoff peaks. It's the percentage of moisture in the soil.
- **Fraction of maximum velocity of base flow (D_{Smax})**: Its value ranges from > 0 to 1; a higher value of non-linear baseflow '*Ds*' (fraction of D_{Smax}) at lower moisture content in the underlying soil.
- **Variable infiltration curve parameter (*Infilt*)**: The shape of the variable infiltration curve is determined by this parameter. The accessible infiltration capacity is expressed as a function of the relative saturate area of the grid cell. Its value ranges from > 0 to ~ 0.4; a higher '*Infilt*' value indicates that there is less infiltration and more surface runoff.
- **Depth of soil number 2 (*D2*)**: Soil depth affects the variable of many models. Overall, thicker soil depth reduces seasonal flow peaks (which are dominated by base flow) and boosts evapotranspiration losses. Furthermore, as a function of the relatively saturated grid cell area, the storage capacity of soil moisture is instantaneously calculated by changes in soil thickness, which results in low runoff (Gao et al. 2009). Its value is typically varying from 0.1 to 1.5 m.

Using altered values of calibrated soil parameters: '*Ws*' = 1.0, '*Ds*' = 1.0, '*D2*' = 1.0, '*Infilt*' = 0.2; the model gives the best match of observed and simulated

Fig. 4.9 **a** Location of discharge station (30° 26′ 06.21″ N; 77° 40′ 49.66″ E) (Yamuna Hydel Canal); **b** course of the canal flowing toward the western direction from an outlet of the area

discharge values after running it multiple times using trial and error methodology. The simulated discharge values are calculated using the VIC model and observed data collected from the discharge station (Fig. 4.9a, b).

4.5.3 Calibration of the Model

The input parameters of the model may contain certain uncertainties and errors at the grid or watershed scale. The calibration of the model is required to measure the degree of uncertainties. It can be done by minimizing the difference or by adjusting the values of input parameters till the model's performance likely matches the observed data values of the watershed on an annual basis. The calibration of the model is dependent upon the observations and reference availability. The observation includes the observed and simulated data of runoff/discharge and the initial values of input parameters as a reference and adjusted them till it makes it suitable for calibration. The first calibration of observed and simulated flow was done from January 1996 to December 2003 for the LULC scenario of 2008. This value suggests the model is calibrated very well and used for further analysis. Here, the Coefficient of determination (R^2) and Nash Sutcliffe model Efficiency (NSE) are used as the model performance evaluation measure. R^2 and NSE coefficients for the calibration period are 0.82 and 0.9 respectively. The value of NSE closer to 1, refers to that model simulated very well with the observed data (Nash and Sutcliffe 1970). Although it is expected that a mean will not accurately capture the extreme high and low values, it is clear that the simulated data has a high bias with the peak of observed inflows (Fig. 4.10a). A scatterplot of these flows shows the variance between the monthly simulated and observed flows with statistics (Fig. 4.10b).

Once the calibration process is done, the model is simulated for the rest of the period using the calibrated soil parameter. To validate the model and compute quality checking with observed streamflow data from 2004 to 2011. The validated streamflow for the Asan watershed is as follows:

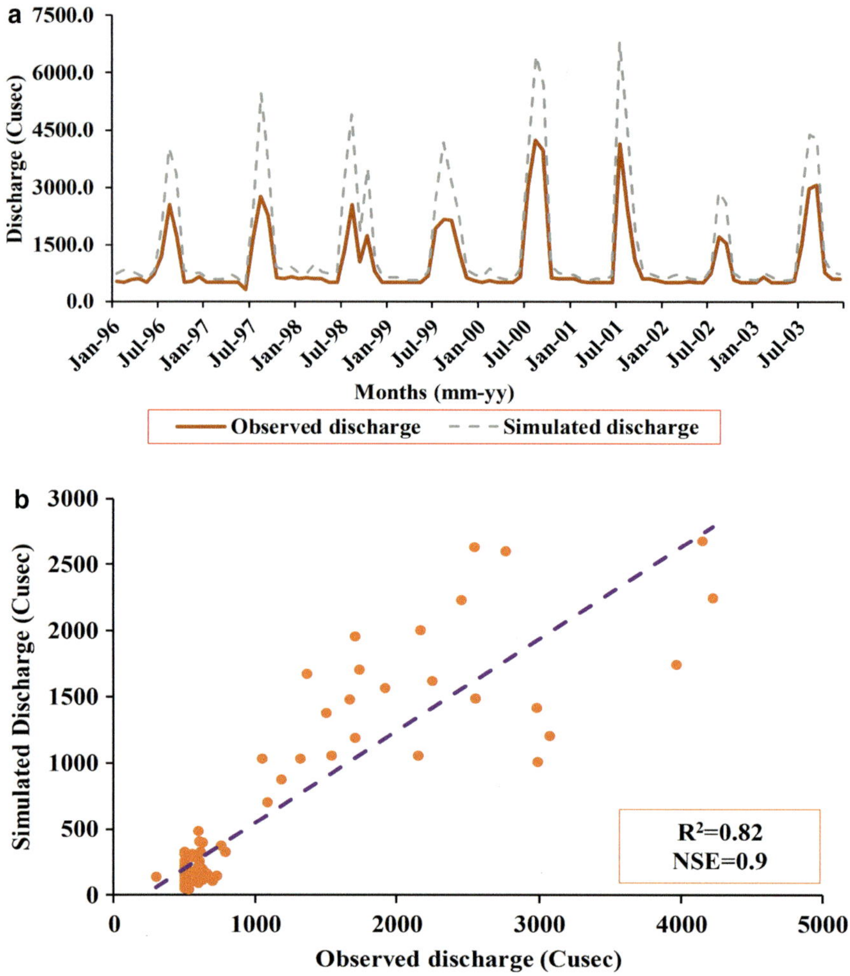

Fig. 4.10 **a** Graphical comparison of simulated and observed discharge data values for the years 1996 to 2003 of Scenario 3, **b** scatterplot shows the variance between simulated and observed discharge with statistics for the years 1996 to 2003 of Scenario 3

4.5.4 Validation

The extension of the model calibration process has been done in the validation period. This gives a basic assurance that in the calibration period, the model assesses all the input variables and is set up effectively which in turn, strongly affects the model results. After completion of the calibration period, the simulation has to be done (with similar inputs) for the remaining time duration i.e. January 2004 to December 2011 for Scenario 3 (Fig. 4.11a). This process of split-calibration is recommended for the study of watershed modelling. A scatterplot of these flows shows the variance between the monthly simulated and observed flows with statistics (Fig. 4.11b). For discharge simulations, the R^2 and NSE criterion suggests that the model meets the satisfactory result with observed data are 0.80 and 0.8 respectively.

For further analysis, the validation result is evaluated for the entire time window of 1980 through 2016 (Fig. 4.12). The entire process is carried out through calibration of the model and then the verification would be done for a longer period. Furthermore, based on the entire simulation period (1996 to 2011) and using different years of LULC's scenarios, station-wise streamflow (runoff) has been generated. Since the available observed data is in Cubic foot per second (Cusec), the output of the route model is also converted into the same unit.

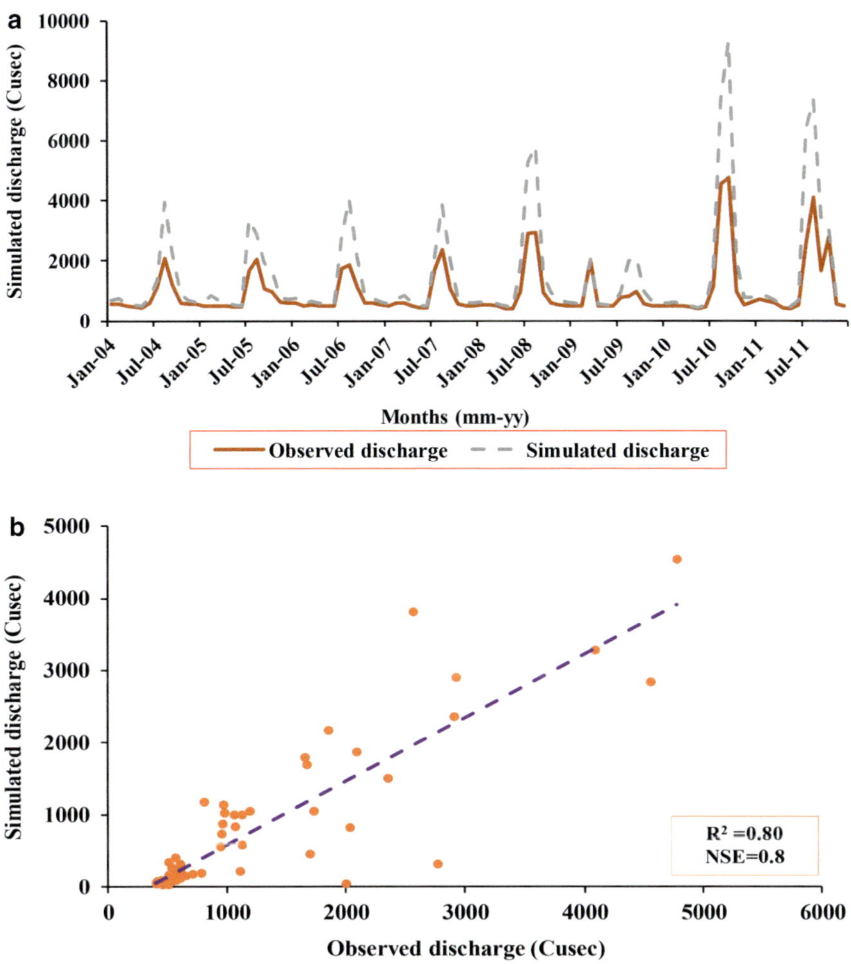

Fig. 4.11 **a** Graphical comparison of simulated and observed discharge data values for the years 2004 to 2011 of Scenario, **b** scatterplot shows the variance between simulated and observed discharge with statistics for the years 2004 to 2011 of Scenario 3

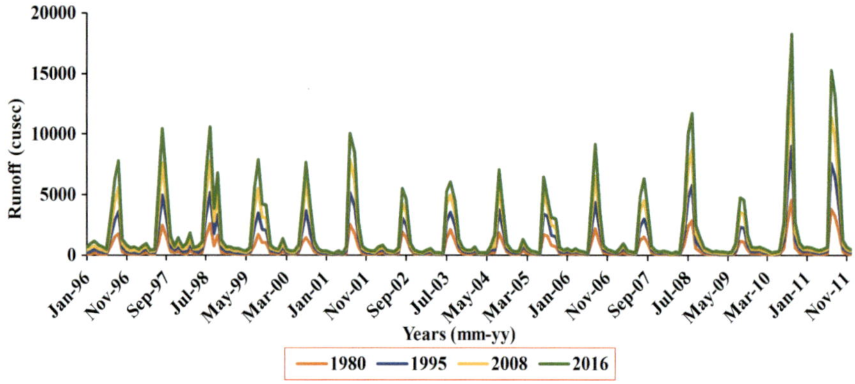

Fig. 4.12 Graphical representation of the simulated discharge (runoff) data values as a water balance component for the period of 1996 to 2011 of different scenarios of LULC

The performance of the model is evaluated through the comparison of monthly observed and simulated discharge at the watershed's outlet (Fig. 4.12). It is evident, that the simulated (peak) inflow for different scenarios is more or less fitted each other because of the variance of discharge values. However, the overall results generally indicated that the model performs sufficient flow.

4.6 Impact of LULC Change on the Hydrological Component (Runoff)

LULC plays a pivotal role in shaping the hydrology of a region. The increased human footprint has triggered numerous transformations in land cover, directly or indirectly impacting hydrological processes and the hydrological cycle, especially prevalent in developing countries. Urbanization, in particular, escalates runoff rates. For the analysis, the outlet was selected based on the availability of observed discharge data. Comparing the simulated and observed inflow of the Asan watershed involved evaluating the monthly mean discharge for decadal LULCs taken for the study (1980, 1995, 2008, and 2016). The comparison reveals that the alteration in LULC appears to be the primary driver of the runoff variation observed between the periods 1980 to 2016 (Fig. 4.13).

To scrutinize these variations, distributed hydrological modelling emerges as an effective approach for assessing changes, enabling the quantification of streamflow patterns and distribution of hydrological components across the basin. Remote sensing stands out as a valuable tool for collecting spatial and temporal data, enhancing the accuracy of studying hydrological processes.

The mean annual observed discharge at the outlet of the study area is 959.8 cusec whereas the value of the simulated one is 509.8 cusec for 1980. Further, for the years

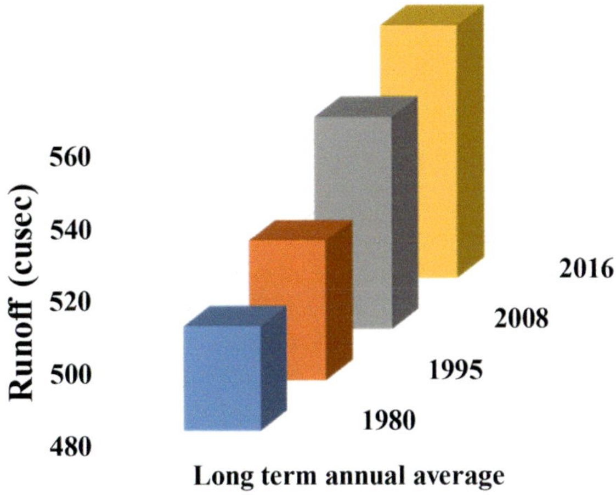

Fig. 4.13 Graphical representation of mean annual runoff under different LULCs

1995 and 2008, the mean annual discharge of simulated value is 518.4 and 538.3 cusec respectively (Fig. 4.14). From 1980 to 2016, the urbanized area increased from 9.55 to 36.50 km^2 (Table 4.4), resulting in an increase in the mean annual discharge from 509.8 cusecs (1980) to 549.5 cusecs (2016).

At the basin scale, the changing LULC does not seem to affect the hydrology that much as overall it appears as a minor change for the basin. This is because an

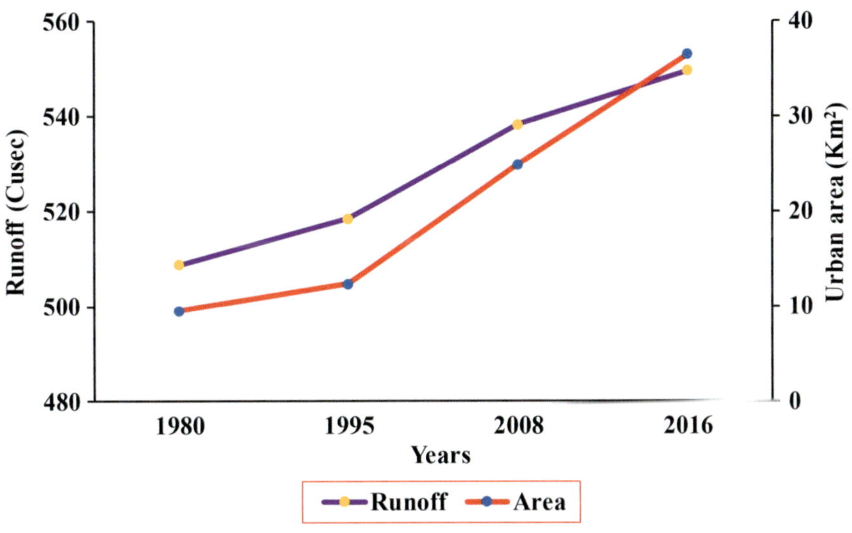

Fig. 4.14 Relationship between 'urban area' and 'mean annual runoff' during 1980–2016

Table 4.4 Urban growth in the study area (1980–2016)

Year	Urban area (km^2)	Urban area change (%)	Increase (km^2)
Till 1980	9.55	11.45	–
1980–1995	12.39	14.86	2.84
1995–2008	24.91	29.88	12.52
2008–2016	36.50	45.79	11.59

increase in runoff (Positive change) at some places in the basin gets cancelled out by the decrease in runoff (Negative change) at other places.

4.7 Limitations and Future Aspects

- Choosing the right hydrological model in recent scenarios is crucial and often requires an extensive literature review to align with research objectives. The current study relies on the Variable Infiltration Capacity (VIC) hydrological model, acknowledging the inherent assumptions and limitations associated with any modelling approach.
- The study involves utilizing multiple datasets from various sources to initiate the model. Managing these diverse datasets can be challenging. Also, the outcome of the study is heavily depending on the quality and precision of input datasets particularly the meteorological data, soil parameters, and land cover information.
- The simulation and calibration process of the model entails tweaking parameters in the global parameter files until reaching the desired outcome. This often involves multiple attempts, as the model operates through a trial-and-error method, making it a time-consuming and tedious task. Adjusting the appropriate parameter values leads to the creation of multiple files that require organization and compilation, potentially introducing subjectivity and biases into the results.
- The research outcomes pertain specifically to the Asan River watershed in India, which constrains their applicability to other areas with distinct characteristics and conditions.
- Future studies could focus on improving hydrological models, like VIC, to better represent the complexities of changing LULC patterns and how they affect hydrological processes.
- Long-term observation of LULC changes and the hydrological effects they have can provide important new information about the long-term patterns and dynamics of watershed systems.
- A more comprehensive understanding of the interactions between land use, climate dynamics, and hydrology can be achieved by taking into account the impact of climate change in addition to LULC modifications.

- The outcome of the study can inform strategies for managing water resources and land use, highlighting the importance of implementing sustainable land management techniques to mitigate unfavourable hydrological effects.

4.8 Conclusion

The correlation between rainfall and runoff is influenced by climate and physiography, governing various runoff mechanisms such as interflow, groundwater flow, and surface runoff. Mathematical modelling studies aid in comprehending watershed management, atmospheric interactions, and climate change by providing quantitative estimations of the spatial and temporal distribution of various hydrologic processes.

This study utilized the Variable Infiltration Capacity (VIC) hydrological model to quantify the impact of land use and land cover (LULC) changes on runoff in the Asan River watershed, Uttarakhand, India from 1980 to 2016. Further investigation should compare simulated and observed discharge from the Asan barrage utilizing the VIC model in water balance mode. The VIC model was set up, calibrated, and validated using observed discharge data from 1996 to 2011. Discharge results based on LULC were analysed across four different scenarios representing the years 1980, 1995, 2008, and 2016. Calibrated models like VIC are valuable for assessing the baseline implications of diverse water management strategies. Calibration and validation of the VIC model involve comparing observed and simulated streamflow, with the calibration period set from January 1996 to December 2003. The process involved repeatedly adjusting soil parameter values within the global parameter file until the model produced the most accurate match between observed and simulated discharge values. The model performance was evaluated using the coefficient of determination (R^2) and Nash–Sutcliffe efficiency (NSE) criteria, which showed satisfactory results with observed data. However, the study is limited by its reliance on the VIC model, which is subject to inherent assumptions and limitations. The accuracy of the findings is also influenced by the quality of input data, particularly meteorological data, soil parameters, and LULC information. The calibration process involves adjusting parameters through trial and error, which can introduce subjectivity and potential biases in the results. LULC have a direct impact on regional hydrology. The study demonstrated that the watershed's runoff characteristics were greatly impacted by changes in LULC throughout time. Also, urbanization, industrialization, and deforestation over time typically result in heightened surface runoff, decreased evapotranspiration, diminished baseflow and groundwater recharge.

The VIC model underwent thorough calibration and validation. Runoff measurements were taken monthly in '*Cusec*', yielding calibration period coefficients of $R^2 = 0.82$ and NSE = 0.9, and validation period coefficients of $R^2 = 0.80$ and NSE = 0.8. Furthermore, a similar calibration was conducted for the broader period spanning 1980 to 2016. Analysis unveiled a rise in annual discharge in 2011 across all LULC scenarios, with peak discharge typically occurring from July to September and lows from December to February. A comparison between the 1980 and 2016 LULC maps revealed a notable increase in settlement density, significantly impacting runoff. The research suggested that the decline in natural water cover, expansion of urban areas, and intensification of agriculture have heightened surface runoff while diminishing groundwater recharge.

Future research should prioritize refining hydrological models to better capture the complexities of LULC changes and their impact on hydrology. Long-term monitoring of these changes and their hydrological effects can provide valuable insights into long-term trends and dynamics. Integrating the influence of climate change alongside LULC changes can offer a more comprehensive understanding of the interactions between land use, climate, and hydrology. These findings underscore the significance of adopting sustainable land management practices to mitigate adverse hydrological consequences. Changes in LULCs were identified as the primary driver of runoff variability over the study period, emphasizing the need for a comprehensive framework to understand the hydrological impact of LULC changes on a broader scale in future modelling efforts. The study's results can inform land use planning and water resource management strategies to ensure sustainable development in the Asan River watershed and similar regions.

Author's Contribution Conceptualization and design of the article; acquisition, methodology, interpretation and validation of the data, formal analysis, writing and original draft preparation-A.S.; Validation and formal analysis-R.K.; Methodology-K.K.; Visualization, writing-review and editing-D.S.B; Writing-review-B.R.N.

Conflict of Interest I testify on behalf of my co-authors that our research article is already published on Research Square with a https://doi.org/10.21203/rs.3.rs-3664860/v2. Other than that there is no conflict of interest associated with this research work.

Data Availability The author used datasets from the United States Geological Survey, USGS (https://earthexplorer.usgs.gov/); National Remote Sensing Centre, Bhuvan-NRSC (https://bhuvan.nrsc.gov.in/home/index.php); ICAR-National Bureau of Soil Survey and Land Use Planning (https://nbsslup.icar.gov.in/), Uttarakhand Jal Vidyut Nigam Limited (UNJVL), Indian Meteorological Department (IMD) are duly acknowledged.

Funding Statement The authors received no specific funding or research grant for this work from any agency in the public, commercial, or not-for-profit sectors, that could have influenced its outcome.

References

Alvarenga LA, Carvalho VSO, Oliveira VAD, Mello CRD, Colombo A, Tomasella J, Melo PA (2020) Hydrological simulation with SWAT and VIC models in the Verde river watershed, Minas Gerais. Revista Ambiente & Água 15(4):e2492. https://doi.org/10.4136/ambi-agua.2492

Asrar G, Dozier J (1994) EOS-science strategy for the earth observing system. AIP Press, Woodbury, NY, p 199

Berhanu B, Seleshi Y, Demisse SS, Melesse AM (2015) Flow regime classification and hydrological characterization: a case study of Ethiopian rivers. Water 7(6):3149–3165. https://doi.org/10.3390/w7063149

Bronstert A, Niehoff D, Burger G (2002) Effects of climate and land-use change on storm runoff generation: present knowledge and modelling capabilities. Hydrol Process 16:509–529

Cunderlik J (2003) Hydrologic model selection for the CFCAS project: assessment of water resources risk and vulnerability to changing climatic conditions. Department of Civil and Environmental Engineering, The University of Western Ontario: Canadian Foundation for Climatic and Atmospheric Sciences (CFCAS) project. Water Resources Research Report, 9

Elfert S, Bormann H (2010) Simulated impact of past and possible future land-use changes on the hydrological response of the Northern German lowland 'Hunte' catchment. J Hydrol 383:245–255

Gao H, Tang Q, Shi X, Zhu C, Bohn T, Su F, Wood E (2009) Water budget record from variable infiltration capacity (VIC) model. In: Algorithm theoretical basis document for terrestrial water cycle data records version 1.2, pp 120–173

Gupta HV, Sorooshian S, Yapo PO (1998) Toward improved calibration of hydrologic models: multiple and non-commensurable measures of information. Water Resour Res 34:751–763

Hamman JJ, Nijssen B, Bohn TJ, Gergel DR, Mao Y (2018) The variable infiltration capacity model version 5 (VIC-5): infrastructure improvements for new applications and reproducibility. Geoscien Model Dev 11(8):3481–3496

Liang X, Lettenmaier DP, Wood EF, Burges SJ (1994) A simple hydrologically based model of land surface water and energy fluxes for general circulation models. J Geophys Res: Atmos 99:14415–14428

Liang X, Wood EF, Lettenmaier DP (1996) Surface soil moisture parameterization of the VIC-2L model: evaluation and modification. Glob Planet Change 13:195–206

Lohmann D, Raschke E, Nijssen B, Lettenmaier DP (1998a) I. Regional scale hydrology: I. Formulation of the VIC-2L model coupled to the routing model. Hydrol Sci J 43:131–141

Lohmann D, Raschke E, Nijssen B, Lettenmaier DP (1998b) II. Regional scale hydrology: II. Application of the VIC-2L model to the Weser River, Germany. Hydrol Sci J 43:143–158

Nash JE, Sutcliffe JV (1970) River flow forecasting through conceptual models part I—a discussion of principles. J Hydrol 10:282–290

Nijssen B, Lettenmaier DP, Liang X, Wetzel SW, Wood EF (1997) Streamflow simulation for continental-scale river basins. Water Resour Res 33:711–724

Nikam VV (2015) Assimilation of remote sensing derived parameters in hydrological model for Ganga Basin. M.Tech, a thesis submitted to Andhra University, Visakhapatnam, 2015

Seibert J, McDonnell JJ (2002) On the dialog between experimentalist and modeller in catchment hydrology: use of soft data for multicriteria model calibration. Water Resour Res 38:23–31

Sharma A, Bagri DS (2020) Monitoring of urban sprawl and spatio-temporal variability of landuse/landcover change in Asan watershed, Doon valley, Uttarakhand, India. In: AGU fall meeting 2020. AGU. https://doi.org/10.1002/essoar.10505204.1

Sharma A, Bagri DS (2023) Integrative assessment of geomorphic indices and land cover dynamics for sustainable development: a case study of the Asan watershed, Doon Valley, Uttarakhand, India. Environ Earth Sci Res J 10(4):132–150. https://doi.org/10.18280/eesrj.100402

Sharma A, Nikam BR, Bagri DS, Kumar D (2021) Geo-environmental investigations in the Asan river watershed of Doon valley, Uttarakhand. PhD thesis submitted to HNB Garhwal University. IIRS-ISRO Dehradun. https://shodhganga.inflibnet.ac.in:8443/jspui/handle/10603/401114

Singh SK, Marcy N (2017) Comparison of simple and complex hydrological models for predicting catchment discharge under climate change. AIMS Geosci 3(3):467–497

Srivastava A, Sahoo B, Raghuwanshi NS, Singh R (2017) Evaluation of variable-infiltration capacity model and MODIS-terra satellite-derived grid-scale evapotranspiration estimates in a River Basin with Tropical Monsoon-Type climatology. J Irrig Drain Eng 143(8). https://doi.org/10.1061/(ASCE)IR.1943-4774.0001199

Tricart JLF (1999) Hydrological cycle. In: Environmental geology, encyclopedia of earth science. Springer, pp 337–338

Witham JD (2015) Calibration, verification, and diagnosis of a season-ahead drought prediction model: limits to predictability in central Texas. Doctoral dissertation, Michigan Technological University

Yeo IY, Guldmann JM, Gordon SI (2007) A hierarchical optimization approach to watershed land use planning. Water Resources Res 43

Zhai R, Tao F (2017) Contributions of climate change and human activities to runoff change in seven typical catchments across China. Sci Total Environ 605:219–229

Zhao RJ, Zhuang YL, Fang LR, Liu XR, Zhang QS (1980) The Xinanjiang model. In: Hydrological forecasting (proceedings of the Oxford symposium). IAHS Publ. 129. IAHS Press, Wallingford, U.K., pp 351–356

Part II
Efficient Irrigation and Agricultural Planning for Food Security

Chapter 5
Divulging Insights into Irrigation Scheduling of *Allium cepa* (Onion) Using *Cocos nucifera* (Coconut) Husk Mulch

Yash Shukla, Vijay Shankar, and Vivek Gupta

Abstract To manage water resources for agricultural output in areas with limited water resources, innovation and sustainable methods must be used. The aim of this book chapter is to delve into the synergistic effects of hydrothermal regime, growth characteristics and consumption of water, by onion crop over different mulch layer of Coconut husk. Within an altered crop growing environment, the application of mulch exhibits beneficial impacts on the retained soil moisture, crop growth and productivity of crop. In order to assess the potency of readily available *Cocos nucifera* husk in retaining soil moisture field trials have been carried out. The onion crops utilized a variety of application rates and revised watering schedule accordingly. The study's conclusions demonstrate that the use of coconut husk mulch reduced the maximum soil temperature, promoted plant development, and decreased the overall amount of irrigation needed. The study also revealed for higher permitted depletion having Maximum Allowable Depletion (MAD) 70% the overall water requirement is low while a higher value of water need for a lower permitted depletion of MAD value 30%.

Keywords Hydrothermal regime · Coconut husk · Soil moisture · Irrigation scheduling · Maximum allowable depletion

Y. Shukla (✉) · V. Gupta
Indian Institute of Technology, Mandi, India
e-mail: yashshuklakt55@gmail.com

V. Gupta
e-mail: vivekgupta@iitmandi.ac.in

V. Shankar
National Institute of Technology, Hamirpur, India

A. Nanda et al. (eds.), *Navigating the Nexus*, Water Science and Technology Library 102, https://doi.org/10.1007/978-3-031-76532-2_5

5.1 Introduction

The term "mulch" probably originates from the German word "molsch," which means "soft to decay." It seems that this term referred to the application of straw and leaves as mulch by gardeners on the ground. Mulching minimizes weed infestation, controls water evaporation, and prevents runoff and soil loss, all of which contribute to the reduction of soil deterioration. As a result, it increases soil moisture retention, aids in temperature regulation, boosts the physical, chemical, and biological qualities of the soil, adds nutrients to the soil, and eventually promotes crop development and production (Ray and Biswasi 2016). By applying various materials to the soil's surface, farmers and horticulturists can improve the quality of agricultural soils through a process called mulching. Better plant production is correlated with improved soil physical conditions (Vandenberg and Tiessen 1972; Bot and Benites 2005; Chakraborty et al. 2008). Mulch application can improve the soil's organic matter content, inhibit the growth of weeds, and minimise or completely stop erosion (Bot and Benites 2005). By applying various materials to the soil's surface, farmers and horticulturists can improve the quality of agricultural soils through a process called mulching. Better plant production is correlated with improved soil physical conditions (Vandenberg and Tiessen 1972; Bot and Benites 2005; Chakraborty et al. 2008). Mulch application can improve the soil's organic matter content, inhibit the growth of weeds, and minimise or completely stop erosion (Bot and Benites 2005). Many types of organic mulches are frequently used in landscaping to improve plant health and manage weeds (Tiquia et al. 2002).

Mulch materials are commonly categorized into two groups: organic or biodegradable, which comprise organic components, and inorganic, which are primarily composed of polymers. Organic mulch offers the benefit of using easily accessible materials in the field that break down over time and add to the soil's organic matter. The effect of organic mulches on agricultural productivity varies. Mulching enhances plant growth, yield, and quality. Mulching resulted in higher yields for early season crops. According to Lal (1974), mulching can increase grain output by lowering soil temperature and improving moisture levels. Certain mulches, such as straw, peat, and sawdust, can deplete soil nitrogen due to their high C:N ratio (Johnson et al. 2004; Sønsteby et al. 2004). Gruber and Claupein (2008) found that mulching with wood chips had no influence on crop output. Kar and Kumar (2007) found that plots with straw mulch produced higher potato yields and improved crop growth. Potato yields were comparable between mulched and unmulched plots, whereas watermelon yields were higher in straw mulch plots (Johnson et al. 2004).

Plastic mulches reduce water evaporation from the soil surface by 50–80%. Inorganic mulches are long-lasting, but decay over time and may cause disposal issues, while organic mulches are biodegradable and decompose faster, and need to be renewed more regularly than inorganic mulches. Because of reduced evaporation, soil mulching promotes water conservation, particularly in the topsoil, reduces the evapo-concentration of salts in irrigation water and soil solution, and reduces soil salinization and sodication (Rahman et al. 2005).

5.2 Various Mulching Materials in Agriculture

Mulching with organic matter dates back to ancient agriculture, although lithic (stone) mulches have also been employed for centuries. Thousands of years ago, farmers in both the Old and New Worlds employed lithic materials to boost crop productivity and prevent dryness in parched regions. Materials used include stones, gravel, pebbles, volcanic ash, cinder, and other lithic materials. The approach reduces evaporation, wind erosion, and surface runoff in fields (Lightfoot 1996). In the 1920s, pineapple (*Ananas comosus* (L.) Merr.) production in the Hawaiian Islands relied on asphalt paper mulch to suppress weeds and maintain soil moisture. Plastics are widely used for mulching, with an estimated 700,000 t used annually in China alone.

Mulching materials utilised in modern crop production can vary. Farmers and horticulturists can use crop leftovers, polymers, biodegradable films, or paper mulches (coated or uncoated). The impact of mulch type on crop development varies depending on production procedures, growing circumstances, and crop species (Ashworth and Harrison 1983).

Additionally, by competing with weeds for light, nutrients, and water, living mulches can be employed to inhibit and minimize weed growth. They inhibit the growth of weeds by generating substances known as allelopathic, which lessen the need to apply herbicides to crops. Table 5.1 lists the different crops that were mulched in various areas and the outcomes that were obtained.

5.3 The Beneficial Effects of Mulching

1. *Soil Moisture Conservation*

Mulching minimizes evaporation from the soil surface, increasing water availability for transpiration in water-limited settings and maintaining plant water status (Chakraborty et al. 2008). As a result, mulch-covered areas require less watering (Ranjan et al. 2017). Over time, some mulches, such as those made of organic materials, can help the soil's structure (Blanco-Canqui and Lal 2007). A better soil structure increases the capacity of water infiltration and retention, lowering runoff and improving plant availability. So, organic mulches improve soil structure and raise the amount of organic matter in the soil, which improves water retention (Wang et al. 2021).

2. *Controlling Erosion*

Mulch decreases the kinetic energy of rain and slows its passage, preventing runoff while providing the soil more time to absorb it. Additional moisture promotes plant root growth, which contributes to soil stabilisation. Mulch also protects soil against wind erosion (Ranjan et al. 2017).

Table 5.1 Mulching treatment in different crops

Crop	Mulch used	Outcomes	References
Wheat	Plastic and straw mulch	Increase in yield; mulch was found to be more effective than straw	Qin et al. (2015)
Maize	Plastic and straw mulch	60% increases in yield; plastic-mulch is more effective than organic mulch	Qin et al. (2015)
Brinjal	Straw mulching, 30 micron Bi-colored silver, Black plastic mulch	Highest yield-480.24 quintal per hectare was obtained from 30 micron bi-colored silver mulch	Kumar et al. (2019)
Cauliflower	Polythene mulch with different concentrations of OPE (open pan evaporation), RD of nitrogen and potassium	282.53 quintal per hectare was obtained with bilayer polythene mulch when 24.96 L of OPE was supplemented through drip system per plant; 125% of fertigation of nitrogen and potash was done	Bhoutekar et al. (2017)
Carrot	Sugarcane straw mulch, black polythene mulch, leave mulch, blue polythene mulch, paddy straw mulch, grass mulch, white polythene mulch	54.69 tonnes per hectare of yield was obtained using black polythene mulch; it is the maximum yield obtained among different mulched plots	Jaysawal et al. (2018)
Lemon	Bajra straw, maize straw, grasses, brankad, farmyard manure, black polythene	Maximum soil content and yield (1848 kg per hectare) was observed with the use of black polythene mulch, followed by farmyard manure mulch (yield: 1780 kg per hectare) and brankad mulch and (yield: 1744 kg per hectare)	Kumar et al. (2015)
Sugarcane	No straw, straw, and straw burned over the soil	75 and 100% straw-mulch showed highest suppression of weeds	Hoshino et al. (2017)

3. *Decrease Soil Temperature*

Mulching regulates soil temperature throughout the year, preventing extremes. This may mitigate water loss from evaporation and heat stress. When compared to bare soil, mulching decreased soil temperature swings by 5–10 °C, which decreased water loss. Also, Wheat straw mulch increased soil temperature by 2–3 °C during peak winter season.

4. *Lowering Weed Growth*

Mulch slows seed germination by blocking sunlight from reaching the top layer of soil. Plastic films and landscape fabrics prevent weeds from germinating by establishing a protective layer on the soil (Sharma and Bhardwaj 2017).

5. *Quality and Yield Enhancement*

Mulch protects fruits from soil rot, cracking, and blossom end rot. Fruits typically have softer skin and less scars. Properly laid plastic mulch prevents dirt from splashing onto plants after rainfall, reducing grading time (Ray and Biswasi 2016).

6. *Reduce Soil Salinity*

Increased soil salinity is caused by a shallow groundwater table, high evaporation rate, and limited rainfall. Mulching the soil surface with a variety of materials (such as crop residue, plastic film, sand, or gravel) might lessen evaporative water loss and aid in preventing salt buildup in the shallow soil layer.

5.4 *Cocos nucifera* (Coconut) Husk as a Mulch Material

It is recommended to apply organic mulch, especially when using organic farming methods. Compost, sawdust, straw, coconut husk, and other organic wastes are among the often used organic materials as mulch. Because of its high rate of water absorption, enhanced air circulation, and low rate of evaporation, coconut husk makes a valuable planting medium (Udayana et al. 2017). A range of dietary fibers, including cellulose, hemicellulose, lignin, and pectin, are abundant in coconut husks (Mittal and Chaudhary 2018). It produces potassium-rich soaking water when submerged in water, which can be used in place of organic fertilizer (Evans et al. 1996).

Onions are produced commercially in more than a hundred different countries worldwide. India comes in second place with an output share of 8% of the world total. In India, onions were produced on 1.22 million hectares of land in 2020–2021, yielding 22.8 million metric tonnes annually. Therefore, field crop studies on onions were conducted from December 2022 to March 2023.

Mulching materials can be made from a variety of materials, but their effects on various crops and ecosystems can differ (Monks et al. 1997). This variation has been noted in terms of the varying influence on the hydrothermal regimes of the soil. Tuber yield is affected by the interaction of mulching and irrigation (Xing et al. 2012). Thus, studies on the capacity of mulch materials to hold onto moisture, their role in providing nutrients that are beneficial to the soil, their impact on soil moisture retention, and the resulting variations in irrigation requirements between mulched and unmulched environments are crucial (Goel et al. 2019). Thus, in the research region situated in the northwest mid-hills of India, this case study investigated the effects of coconut husk on irrigation scheduling and crop growth characteristics of onion crop.

5.5 Case Study

5.5.1 *Study Area and Data*

The study was carried out at the Hamirpur, India, field experimental site of the National Institute of Technology, which is situated in the Himachal Pradesh state. The coordinates of the study region are 31° 42′ 40″ North, 76° 31′ 33″ East, or 738 m above sea level. This region, which is a portion of the western Himalayas, experiences sub-humid sub-tropical weather (Kottek et al. 2006).

Hamirpur's average yearly temperature is 24.46 °C, and the city enjoys pleasant, mild, and warm weather. A major factor influencing the growth, yield, and quality of crops throughout the growing season is the weather. It also affects the soil's biotic phase. Thus, it is imperative that climate variables be incorporated into this study. The daily temperatures and any precipitation that falls are recorded by the all-weather station (AWS) at NIT Hamirpur. Figure 5.1 illustrates the 159.14 mm total precipitation corresponding to the onion crop grown from December 2022 to February 2023.

Before beginning the design of the experiment, a field sample of the soil was gathered and assessed to establish a baseline for comparison. Table 5.2 provides

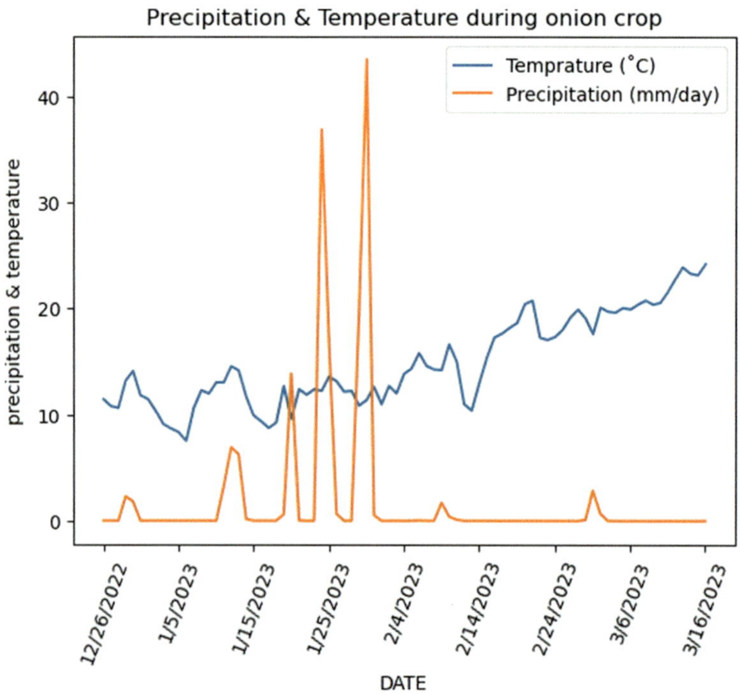

Fig. 5.1 Rainfall and temperature data during potato crop

Table 5.2 Soil physical characteristics from the field experiments

S. No.	Properties	Value	Method
1	Sand (%)	77.64	Sieve analysis
2	Silt (%)	14.32	
3	Clay (%)	8.04	
4	Textural class	Sandy loam	USDA triangle
5	Bulk density (g/cm^3)	1.57	Core cutter method
6	Particle density (g/cm^3)	2.65	Core cutter method
7	Porosity (%)	40	USDA handbook 60 method no 40
8	Specific gravity	2.51	Density bottle
9	Field capacity (m^3/m^3)	0.23	Pressure plate apparatus
10	Permanent wilting point (m^3/m^3)	0.08	

specifics about the initial physical properties of the soil samples that were collected at the experiment site.

5.5.2 *Field Experiments*

Between December 5, 2022, and February 27, 2023, field trials were carried out for the onion crop. It is possible to evaluate the mulches' efficacy as soil moisture-retaining materials by looking at how they affected the soil's hydrothermal and plant properties. Because of this, the field trails featured three coconut husk treatments: no mulch (NM), a layer that was reproduced three times using a randomized block design, and layers that were 1 cm thick (CH1), 2 cm thick (CH2), and 3 cm thick (CH3). The area was tilled and raised beds approximately 1 m wide by 3 m long were constructed at a distance of 1.2 m apart in order to prepare it for onion planting (Table 5.3).

Table 5.3 Information on the individual interventions along with their symbol

Treatment	Symbol
Main plot treatment (irrigation schedule)	
(i) Irrigation at 30% MAD	I30
(ii) Irrigation at 50% MAD	I50
(iii) Irrigation at 70% MAD	I70
Subplot treatment	
(i) Coconut husk mulch @ 1 cm height	C_1
(ii) Coconut husk mulch @ 2 cm height	C_2
(iii) Coconut husk mulch @ 3 cm height	C_3
(iv) No mulch	N

5.5.3 Establishing the Parameters for the Treatment Assessment

Conventional techniques were used to gather information on crop metrics and the hydrothermal properties of the soil in the field and the lab in order to assess and compare each treatment. The amount of water in the pores of the soil is determined by the Soil Moisture Content (SMC). Soil water content is represented as a proportion of the overall volume of soil (Hillel 2013). In order to measure SMC in the crop root zone, the current study used a soil moisture probe (Diviner 200; Sentek Pty Ltd; Stepney, South Australia). Because of the capacitance probe, measurements of the SMC's capacitance can be made; nevertheless, local calibration of the device produced the best results. The calibration technique employed a gravimetric SMC estimate, and the outcomes were contrasted with the probe's findings. The SMC measured in the field and the SMC measured by the probe are matched by the calibration. Using the lab method described by Shukla (2013), the samples' SMC was ascertained gravimetrically.

With the aid of a core cutter, samples of undisturbed soil with a volume of around 100 cm^3 were obtained in order to calculate the particle density and the bulk density. After being dried for enough period and weighed once more, it was found that the sample's weight had stabilized. The porosity (η) of the soil samples was determined by using their particle density and their dry bulk density.

$$\text{Porosity}(\eta) = 1 - \frac{\textit{Bulk Density}}{\textit{Particle Density}} \tag{5.1}$$

5.5.4 Crop Parameter Observation

One of the most important metrics for assessing crop development and canopy structure is the Leaf Area Index (LAI). Stated differently, it is an estimation of the quantity of foliage that a plant has. It is the ratio of the entire one-sided photosynthetic tissue area to the total area of the ground surface, according to Watson (1947). It also affects yield and biomass, which are both influenced by the quantity of water the plant consumes (Firouzabadi et al. 2015). It had previously described the methodology that was used in the current investigation to determine the LAI. To measure the amount of LAI, randomly selected plants from each duplication under every condition were tagged. One plant was removed from each location during each visit as part of the disruptive sampling procedure. A digital planimeter was used to measure the leaf area of each plant after one plant was removed from each section, and the plant's outer leaf border was sketched on paper.

Understanding the depth within which roots have developed is essential to understanding soil moisture dynamics, according to Bohm (1979). Only when the root depth is understood can the depth of water treatment at each certain soil water depletion be precisely approximated. In this investigation, root depth was assessed once a week. To measure the depth of the roots, plants were removed and submerged in water. Using forceps to straighten the roots, they were positioned on a flat glass plate so that they did not overlap. The root-depth readings were then recorded using a scale. Since plant height reflects the height of the plants, it serves as a stand-in for crop development. Over the course of the trial, average crop heights were monitored once a week. For this reason, a small sample of randomly selected plants was measured for height. In terms of statistics, the average height was noted.

5.5.5 *Methodology for Assessing Irrigation Schedules*

Experiments were designed to examine the effects of switching the onion crop's irrigation schedule to one of three different patterns, each of which applied varying amounts of coconut husk mulch. In the present study, a soil moisture content assessment was conducted with the aim of distributing irrigation water in compliance with MAD (maximum acceptable depletion) values. According to Ortola and Knox (2015), the watering schedule for onions was set between 30 and 70% MAD values. Mulch made from coconut husk was applied at 1, 2, and 3 cm. The control plots were the ones with no mulch (NM). A total of twelve treatment combinations were used in the study, with varying amounts of mulch applied at different watering levels. The plot size for each treatment combination is 3 × 1 m. A split plot design was used to set up the experiment, with three different rates of mulch application—no mulch or coconut husk mulch—in the sub plots and an irrigation schedule in the main plot.

The percentage of soil water in the effective root zone that had been utilised was computed using the formula below (Martin et al. 1990).

$$\text{Depletion } \% = 100 \times \frac{1}{n} \sum_{i=1}^{n} \frac{Fci - \text{Ø}i}{Fci - PWPi} \tag{5.2}$$

The soil moisture absorption at field capacity of the ith layer is represented by this equation, where n is the total number of layers of the effective rooting depth used for the sampling, PWP is the soil moisture at permanent wilting point, and $\text{Ø}i$ is the volumetric soil moisture in the ith layer before irrigation. The following formula (Michael 1978) was used to calculate the depth of water delivered at a given MAD:

$$\text{d} = \frac{1}{IRE} \sum_{i=1}^{n} \frac{MAD(Fci - PWPi) \times Rzi}{100} \tag{5.3}$$

where IRE is the irrigation efficiency, d is the targeted irrigation water application depth, and Rzi is the depth of the soil layer within the effective rooting depth.

5.5.6 Results and Discussion

5.5.6.1 Effect of Mulch on Soil Moisture

Crops with shallow roots include onions. While most roots were detected in the first 0.18 m of soil, very few were found below 0.31 m, according to Drinkwater and Janes' (1955) greatest root penetration measurement of 0.76 m. It is unlikely that irrigation water that falls below 0.76 m will get to the crop of onions. For the crop, this meant that daily measurements of soil moisture were taken at depths of 10 cm, 20 cm, and 30 cm. Irrigation was conducted once the moisture content was measured and reached the designated permitted depletion. Using the measured moisture content as a baseline, the percentage of moisture retention for the other treatments were computed.

The soil moisture content rose in all treatments at a depth of 10 cm as the January rains intensified, and the onion crop's soil moisture retention hierarchy was CH1 > CH2 > CH3 > NM. In comparison to NM, the soil moisture content in CH1, CH2, and CH3 was 35.6%, 30.4%, and 29.3% greater, respectively. Similarly, the pattern of soil moisture holding at 20 cm depth was comparable to that found at 10 cm depth; however, in CH1, CH2, and CH3, the increase in soil moisture found at this level was 23.5%, 10.3%, and 9.3%, respectively, with respect to NM. At 30 cm, the soil moisture content rose 32.5% for CH3, 9.8% for CH2, and 7.0% for CH1, compared to the no mulch condition. Organic mulches such as rice, wheat, and peanut straw can save around 20% of the necessary water application without reducing yield (El-Metwally et al. 2022).

5.5.6.2 Effect of Mulch on Soil Temperature

The practical use of mulch resulted in a significant shift in the soil's temperature, based on the data collected on soil temperature. The low conducting surface of mulch materials explains why mulches have a cooling influence on temperatures. As a result, the soil in the mulched area continues to have a low temperature because less sun radiation can penetrate it. According to the current study, CHM lowers soil temperature in the onion case by 1–2 °C. According to the researcher, mulching the soil's surface creates a layer of insulation between solar and heat radiation from the earth. The movement of heat and water between the two radiation sources is impeded by this layer. The enhanced water retention in the soil that took place beneath the layer of mulch made of coconut husks contributed to the lower soil temperature. According to a study on onion crops done in West Hebei Oasis (Deng et al. 2021), the soil temperature changes most significantly in the 5 cm soil layer, where external influences had a greater impact, and in the 10 and 20 cm soil layers, where external influences had less of an impact.

5.5.6.3 Effect of Mulches on Plant Parameters

Mulching helps soil hold and release water for longer periods of time, which enables plants to absorb more of the water and nutrition they require to thrive (Kumar and Srivastava 1998). This has a positive influence on plant growth. Plants get more from the nutrients when mulching helps the soil hold and distribute water for longer periods of time.

The highest root depth in the onion crop ranged from 31.4 to 33.07 cm at a MAD of 30%. C1I30 demonstrated the largest increase in root depth, measuring 5.31% higher than NI30. The maximum root depth at the 50% MAD level increased by 13% from 45.22 in NI50 to 51.1 cm in C1I50. At 70% MAD, C1I70's root depth of 58.22 was 3.33 percentage points lower than that of the unmulched control. The impact of different mulch rate treatments and watering times on the root depth of onion crops is shown graphically in Fig. 5.2.

Every two weeks, starting on the day of transplantation and concluding on the day of harvest, the height of the onions plants was measured. The effects of different mulch rate treatments on onion crop plant height at different growth stages are shown in Fig. 5.3. With a maximum plant height of 24.65 cm to 27.9 cm at a MAD of 30%, the onion crop showed the largest increase in plant height, with C1I30 showing a 13.18% increase over NI30. Maximum plant height at 50% MAD level ranged from 23 cm in NI50 to 24.4 cm in C1I50, representing a 6.08% gain over the former. C3I70 was 9.63% points shorter than the unmulched control, with a plant growth of

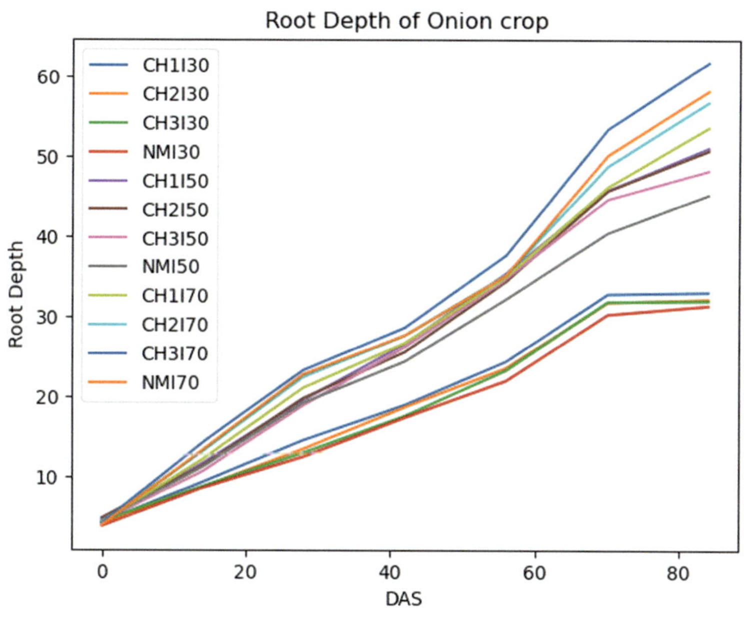

Fig. 5.2 Root depth at different MAD levels for onion crop

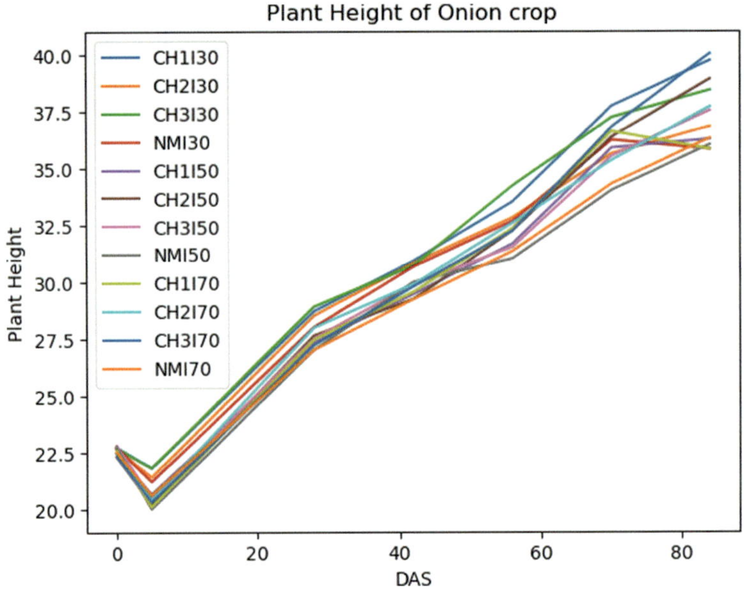

Fig. 5.3 Plant height at different MAD levels for onion crop

23.27 cm at 70% MAD. A study was conducted in Eastern Ethiopia using plastic and straw mulch on onion crops, and the results show that there is a relative tardiness of maturity, increased plant height, number of leaves per plant, marketable bulb output, and higher water productivity (Sali et al. 2022).

5.5.6.4 Effect of Varying Mulch Rates in Irrigation Schedules

For the onion crop, the watering schedule was determined using the MAD values of 30%, 50%, and 70%. The volume of irrigation water that should be applied under each MAD schedule was determined by taking into account the root penetration in different soil layers and the crop age (days after sowing). The average daily soil moisture for the onion crop in 2022–2023 under different treatment combinations is shown in Figs. 5.4, 5.5 and 5.6. This information is taken into account when scheduling irrigation.

Figure 5.4 shows the highest amount of soil moisture that can be achieved for an onion crop in 2022 is 30%. The graph shows that throughout these treatments, the soil's daily average moisture level was maintained between 18.5 and 23%. Irrigation was started to restore the soil to its pre-drying levels as soon as the moisture content dropped to 18.5% or less. At the years 2022 and 2023, Fig. 5.5 illustrate that the daily average moisture of the soil at 50% MAD varied from 15.5 to 22%. Figure 5.6 illustrates how SMC changed from 15.5 to 23% in 2022–2023; however, MAD remained below 70% in all regimens.

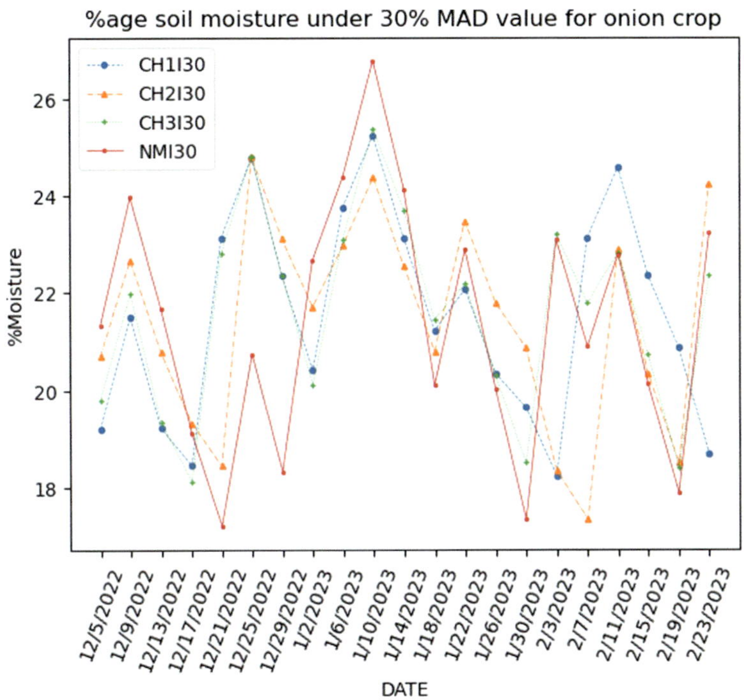

Fig. 5.4 Percentage soil moisture under 30% MAD value for onion crop

Table 5.4 shows that for the onion crop, the unmulched treatment with NI30 needed seven irrigations to maintain a MAD level of 30% during the growth season, while the mulched treatments only needed five irrigations. As a result, depending on the type and quantity of mulch applied, mulching may assist reduce the need for up to two irrigations at a 30% MAD level. The unmulched treatment needed five or six irrigations to reach 50% MAD, whereas the mulched treatments only needed four or five irrigations to achieve the same MAD level. This suggests that, at the 50% MAD level, the mulched treatments needed one or two fewer irrigation than the unmulched control NI50. The amount of irrigation that occurs when the MAD is 70% is not significantly different from the unmulched stage.

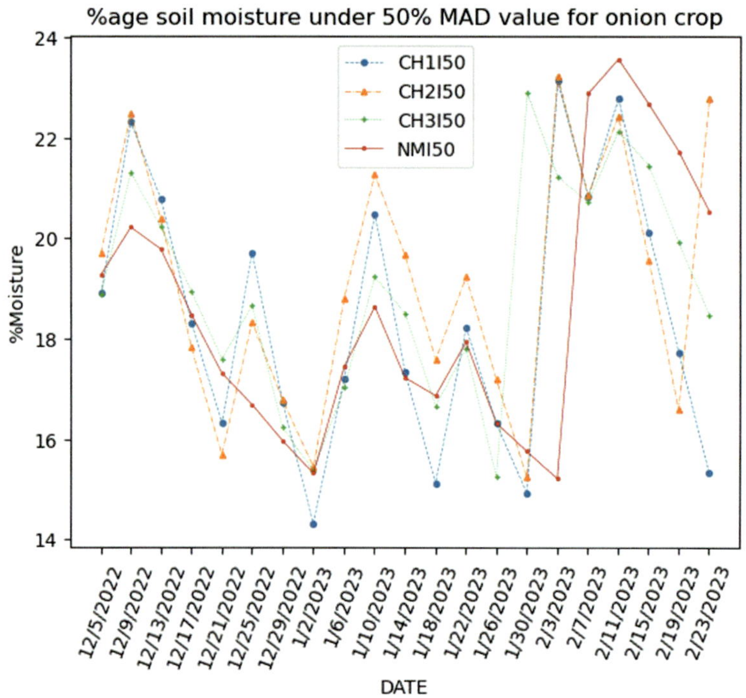

Fig. 5.5 Percentage soil moisture under 50% MAD value for onion crop

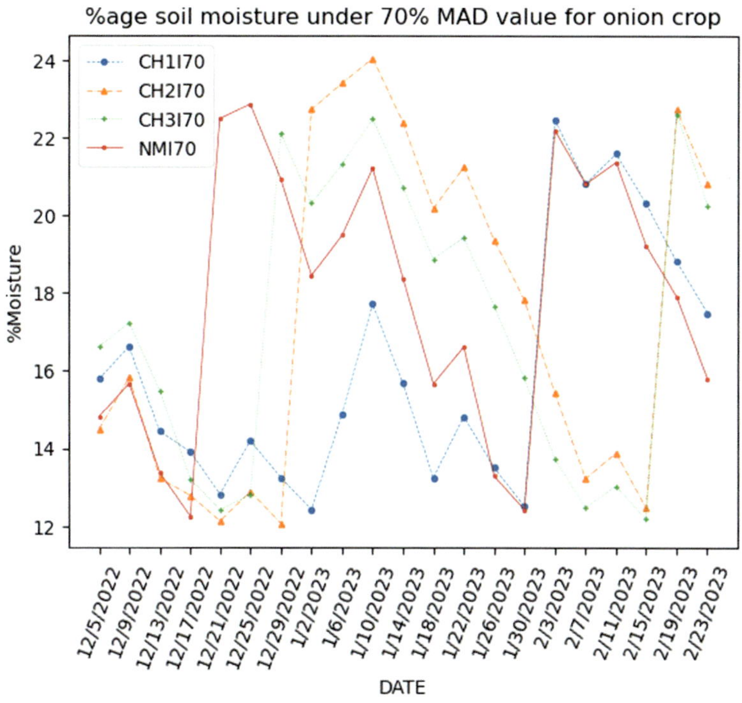

Fig. 5.6 Percentage soil moisture under 70% MAD value for onion crop

Table 5.4 Irrigation and water usage for onion crops in various combinations

S. No.	Treatment combination	Year	Numbers of irrigations applied	Water added through irrigation	Water added through rainfall during crop period (mm)	Total amount of water applied during crop period (mm)
1	NI_{20}	2022–23	13	232	159.14	323.31
2	NI_{30}	2022–23	10	232	159.14	323.31
3	NI_{40}	2022–23	7	232	159.14	323.31
4	C_1I_{20}	2022–23	10	195	159.14	286.31
5	C_1I_{30}	2022–23	8	220	159.14	311.31
6	C_1I_{40}	2022–23	7	212	159.14	303.31
7	C_2I_{20}	2022–23	10	201	159.14	292.31
8	C_2I_{30}	2022–23	7	195	159.14	286.31
9	C_2I_{40}	2022–23	7	212	159.14	303.31
10	C_3I_{20}	2022–23	10	182	159.14	273.31
11	C_3I_{30}	2022–23	8	195	159.14	286.31
12	C_3I_{40}	2022–23	6	195	159.14	286.31

5.6 Conclusion

While irrigation regimens guarantee a sufficient supply of water for productivity. The most popular method for preserving soil moisture in agricultural production systems is mulching.

These are the conclusions drawn from the investigation.

1. Mulching has a significant effect on soil hydrothermal properties and plant growth parameters for onion crops. The CHM creates an optimal soil thermodynamic atmosphere for plant growth, controls soil temperature, and enhances soil moisture retention. The results of the study show that a 3 cm layer of CHM significantly lowers soil temperature when compared to the other two CHM conditions and the absence of mulch.
2. Mulching has a significant effect on soil moisture. For onions, depending on the amount of mulch used, CHM decreases 1–2 irrigations between 30 and 70% MAD levels, while at 50% MAD, there is no change.
3. Mulch has a tremendous impact on plant growth. It enhances root depth, plant height, and leaf area index.

References

Ashworth S, Harrison H (1983) Evaluation of mulches for use in the home garden. HortScience 18(2):180–182

Bhoutekar S, Luchon S, Bonti G, Sonbeer C (2017) Fertigation level and mulching in Cauliflower (Brassica oleraceae L. var. botrytis) cv. Snowball White. Int J Agric Sci 9:4226–4228

Blanco-Canqui H, Lal R (2007) Soil structure and organic carbon relationships following 10 years of wheat straw management in no-till. Soil Tillage Res 95(1–2):240–254

Bohm W (1979) Root parameters and their measurement. In: Methods of studying root systems. Springer, Berlin, Heidelberg, pp 125–138. https://doi.org/10.1007/978-3-642-67282-8_12

Bot A, Benites J (2005) The importance of soil organic matter: Key to drought-resistant soil and sustained food production (No. 80). Food & Agriculture Organisation

Chakraborty D, Nagarajan S, Aggarwal P, Gupta VK, Tomar RK, Garg RN, Kalra N (2008) Effect of mulching on soil and plant water status, and the growth and yield of wheat (Triticum aestivum L.) in a semi-arid environment. Agric Water Manage 95(12):1323–1334

Deng H, Wang P, Zhang Y, Xiao R, Pan X, Li X, Chen T, et al (2021) Effect of different irrigation patterns and covering methods on soil temperature of onions under membrane drip irrigation. IOP Conf Ser: Earth Environ Sci 643(1):012107

Drinkwater WO, Janes BE (1955) Effects of irrigation and soil water on maturity, yield and storage of two onion hybrids. Proc Am Soc Hortic Sci 66:267–279

El–Metwally I, Geries L, Saudy H (2022) Interactive effect of soil mulching and irrigation regime on yield, irrigation water use efficiency and weeds of trickle–irrigated onion. Arch Agron Soil Sci 68(8):1103–1116

Evans MR, Konduru S, Stamps RH (1996) Source variation in physical and chemical properties of coconut coir dust. HortScience 31(6):965–967. https://doi.org/10.21273/HORTSCI.31.6.965

Firouzabadi AG, Raeini-Sarjaz M, Shahnazari A, Zareabyaneh H (2015) Non-destructive estimation of sunflower leaf area and leaf area index under different water regime managements. Arch Agron Soil Sci 61(10):1357–1367. https://doi.org/10.1080/03650340.2014.1002776

Goel L, Shankar V, Sharma RK (2019) Investigations on effectiveness of wheat and rice straw mulches on moisture retention in potato crop (Solanum tuberosum L.). Int J Recyc Org Waste Agric 8:345–356
Gruber S, Claupein W (2008) Wood chips from hedgerows–biomass potential for on-farm mulching and bioenergy?
Hillel D (2013) Fundamentals of soil physics. Academic Press
Hoshino AT, Hata FT, Aquino GSd, Junior AdOM, Ventura MU, Medina CdC (2017) Mulching with sugarcane straw reduces weed density in sugarcane field. Int J Agric Biol 19:121–124
Jaysawal N, Singh G, Kanojia A, Debbarma B (2018) Effect of different mulches on growth and yield of carrot (Dacus carota L.). Int J Chem Stud 6(4): 381–384
Johnson JM, Hough-Goldstein JA, Vangessel MJ (2004) Effects of straw mulch on pest insects, predators, and weeds in watermelons and potatoes. Environ Entomol 33(6):1632–1643
Kar G, Kumar A (2007) Effects of irrigation and straw mulch on water use and tuber yield of potato in eastern India. Agric Water Manag 94(1–3):109–116
Kottek M, Grieser J, Beck C, Rudolf B, Rubel F (2006) World Map of the Köppen-Geiger climate classification updated. Meteorol Z 15:259–263
Kumar R, Srivastava BK (1998) Influence of different mulch materials on flowering and fruit setting of winter tomato. Crop Res-Hisar 15:196–198
Kumar V, Bhatt AK, Sharma V, Gupta N, Sohan P, Singh VP (2015) Effect of different mulches on soil moisture, growth and yield of Eureka Lemon under rainfed conditions. Indian Journal of Dryland Agriculture Research and Development 30(1):83–88
Kumar P, Kumar S, Kumari M, Kumar V (2019) Effect of mulching on brinjle cultivation. Int J Sci Environ Technol 8(3):624–629
Lal R (1974) Soil temperature, soil moisture and maize yield from mulched and unmulched tropical soils. Plant Soil 40(1):129–143
Lightfoot DR (1996) The nature, history, and distribution of lithic mulch agriculture: an ancient technique of dryland agriculture. Agric Hist Rev:206–222
Martin DL, Stegman EC, Fereres E (1990) Irrigation scheduling principles. In: Management of farm irrigation systems. American Society of Agricultural Engineers, St. Joseph, MI, p 155–203, 19 fig., 9 tab, 81 ref
Michael AM (1978) Irrigation theory and practice. Vikash Publishing House Private Limited, New Delhi, Reprint 1999, pp 448–584
Mittal M, Chaudhary R (2018) Experimental study on the water absorption and surface characteristics of alkali treated pineapple leaf fibre and coconut husk fibre. Int J Appl Eng Res 13(15):12237–12243
Monks CD, Monks DW, Basden T, Selders A, Poland S, Rayburn E (1997) Soil temperature, soil moisture, weed control, and tomato (Lycopersicon esculentum) response to mulching. Weed Technol 11(3):561–566
Ortola MP, Knox JW (2015) Water relations and irrigation requirements of onion (*Allium cepa* L.): a review of yield and quality impacts. Exp Agric 51(2):210–231
Qin W, Hu C, Oenema O (2015) Soil mulching significantly enhances yields and water and nitrogen use efficiencies of maize and wheat: a meta analysis. Sci Rep 5:16210
Rahman MA, Chikushi J, Saifizzaman M, Lauren JG (2005) Rice straw mulching and nitrogen response of no-till wheat following rice in Bangladesh. Field Crop Res 91(1):71–81
Ranjan P, Patle GT, Prem M, Solanke KR (2017) Organic mulching-a water saving technique to increase the production of fruits and vegetables. Curr Agric Res J 5(3)
Ray M, Biswasi S (2016) Impact of mulching on crop production: A review. Trends Biosci 9:757–767
Sali AM, Alemayehu Y, Hordofa T (2022) Effects deficit irrigation and mulching on yield and water productivity of furrow irrigated onion (*Allium cepa* L.) under Haramaya condition, Eastern Ethiopia. Turk J Agric Food Sci Technol 10(2):360–367
Sharma R, Bhardwaj S (2017) Effect of mulching on soil and water conservation-a review. Agric Rev 38(4):311–315
Shukla MK (2013) Soil physics: an introduction. CRC Press, pp 505–517

Sønsteby A, Nes A, Måge F (2004) Effects of bark mulch and NPK fertilizer on yield, leaf nutrient status and soil mineral nitrogen during three years of strawberry production. Acta Agriculturae Scandinavica, Sect B-Soil Plant Sci 54(3):128–134

Tiquia SM, Lloyd J, Herms DA, Hoitink HA, Michel FC Jr (2002) Effects of mulching and fertilization on soil nutrients, microbial activity and rhizosphere bacterial community structure determined by analysis of TRFLPs of PCR-amplified 16S rRNA genes. Appl Soil Ecol 21(1):31–48

Udayana SK, Naorem A, Singh NA (2017) The multipurpose utilization of coconut by-products in agriculture: prospects and concerns. Int J Curr Microbiol Appl Sci 6(6):1408–1415. https://doi.org/10.20546/IJCMAS.2017.606.165

Vandenberg J, Tiessen H (1972) Influence of Wax-coated and Polyethylene-coated Paper Mulch on Growth and Flowering of Tomato1. HortScience 7(5):464–465

Wang B, Niu J, Berndtsson R, Zhang L, Chen X, Li X, Zhu Z (2021) Efficient organic mulch thickness for soil and water conservation in urban areas. Sci Rep 11(1):6259

Watson DJ (1947) Comparative physiological studies on the growth of field crops: I. Variation in net assimilation rate and leaf area between species 180 and varieties, and within and between years. Ann Botany 11(41):41–76

Xing Z, Toner P, Chow L, Rees HW, Li S, Meng F (2012) Effects of Hay Mulch on soil properties and potato tuber yield under irrigation and non-irrigation in New Brunswick, Canada. J Irrig Drain Eng 138(8):703–714

Chapter 6
Impact of Climatic Extreme Indices on Crop Productivity

Ayush Kumar, Ashok Mishra, and Idhaya Chandhiran Ilampooranan

Abstract Climate change is associated with rising temperatures and changing rainfall patterns, which can degrade soil and reduce crop productivity. In this study, we assess the effects of rainfall and temperature-based extreme climatic indices on rice and wheat crops across all 38 Bihar districts for the historical period (1998–2019). Results indicate that rice and wheat yield trends are increasing except for Arwal, where a decrease in wheat yield was observed. Increasing yield is because of improved agricultural practices like high-yielding seeds, fertilizer, tillage, and modern technologies. Trend analysis for indices shows a decreasing trend in the number of colder days during the kharif season (TX10P_KHARIF, TN10_KHARIF) and consecutive wet days (CWD) in most districts, whereas an increase in warmer days and higher daily temperature extremes, attributable to erratic rainfall and droughts. Analyzing the relationship between climate extremes and crop yields, positive correlations were observed with precipitation indices, which boost plant growth and soil microbial activities, enhancing productivity. Specifically, wheat benefited from increased precipitation. However, specific temperature indices, like summer days (SU), days when minimum temperature > 90th percentile (TN90P), and days when maximum temperature > 90th percentile (TX90P), negatively impacted yields due to elevated temperatures causing early plant maturity and affecting kernel weight. These findings offer valuable insights for decision-makers aiming to formulate effective adaptation and mitigation policies to address the impacts of climate change on crop yield in Bihar. More studies are needed to understand how these climate extremes affect plants' phenological stages and yield.

Keywords Trend analysis · Crop yield · Climate change · Extreme rainfall indices · Extreme temperature indices

A. Kumar · I. C. Ilampooranan
Water Resources Development and Management Department, Indian Institute of Technology Roorkee, Roorkee, Uttarakhand 247667, India

A. Mishra (✉)
Agricultural and Food Engineering Department, Indian Institute of Technology Kharagpur, Kharagpur, West Bengal 721 302, India
e-mail: amishra@agfe.iitkgp.ac.in

A. Nanda et al. (eds.), *Navigating the Nexus*, Water Science and Technology Library 102, https://doi.org/10.1007/978-3-031-76532-2_6

6.1 Introduction

Climate change has significantly impacted human and environmental processes (Allan et al. 2021). It is associated with temperature rise, variability in rainfall patterns, and extreme climatic events, such as floods, droughts, and landslides, potentially degrading soil and reducing crop productivity (Skendžić et al. 2021). These effects are particularly challenging for smallholder farmers in developing countries like India. Earth has become considerably hotter in these two three decades compared to an earlier decadal period (Dahal et al. 2018), which leads to climate change, consequently affecting the hydrological cycle and resulting in more frequent flooding, droughts, and other extreme occurrences (Ur et al. 2021). Recent studies have highlighted spatiotemporal variations in the temperature and precipitation extremes based on the Expert Team on Climate Change Detection and Indices (ETCCDI) (Yuan et al. 2021; Zhou et al. 2016). These studies not only track the occurrence of extreme events but also examine how the frequency and intensity of these extremes change seasonally and annually (Dos Santos et al. 2011; Zhang et al. 2016). The fluctuation of climatic variables such as air humidity, rainfall, solar radiation, temperature, and wind speed regularly impact the global distribution of livestock and crop productivity (Adisa et al. 2018; Vogel et al. 2019). Due to this, food production has become an increasingly important challenge with the growing global population. As a result, variability and climate change have immediate and long-term impacts on crop productivity. Most simulation studies suggested that climate change directly affects crops because of increasing chances of occurrence of extreme events such as rainfall intensity, decreasing rainy days, temperature rise, the occurrence of flood and drought, and humidity change (Birthal et al. 2014; Mall et al. 2006; Tirkey et al. 2018; Vogel et al. 2019). New research indicates that the impact of climate change on crop output has undeniably started. Rising temperatures and erratic precipitation during the critical stages can adversely affect crops (Hatfield et al. 2011), and challenges that these climatic variables introduce to the agricultural sector are quickly becoming more pressing and evident (Hatfield 2013). The most significant climate variables, precipitation and temperature, directly impact grain output quality, quantity, and market value (Dai et al. 2016). Higher temperatures shorten the time it takes for crops to develop and, as a result, reduce plant photosynthesis. In contrast, the crop matures faster at the right low temperature and accumulates more crop biomass, resulting in higher crop productivity (Tian et al. 2017). If extremes and average annual temperatures continue to rise, then the quality and quantity of grain may be lowered (Högy et al. 2013). Subash et al. (2013) studied the trend of temperature-based indices and their effect on rice and wheat productivity for two districts of Bihar (i.e., Patna and Samastipur). Their results suggested that temperature affects crop productivity. Tesfaye et al. (2017) analyzed that heat stress from rising minimum temperatures during the rabi (winter) season, elevated minimum and maximum temperatures in the spring, and extended periods of dry days in the Kharif season will be the main climate threats to crop output in Bihar. Although it is widely predicted that variability in climate has a significant impact on agriculture, these studies have examined the effect of climatic indices on crop yield and have yet to consider some essential index, such as summer days (SU), maximum length of a dry spell (CDD), consecutive wet days (CWD), and Tropical nights (TR). Therefore, assessing the risk of changing climate on agricultural production has come up as a challenging task for researchers to identify appropriate adaptation options as it is the most critical environmental hazard to agricultural productivity on a global scale. This study aims to examine the trends of extreme indices and crop yield and the impact of extreme climate indices on crop yields using long-term data from 1998 to 2019.

6.2 Materials and Methods

6.2.1 Study Area

The study area is Bihar, India's twelfth-largest state situated in the eastern part of India. Geography and socio-economic conditions make it highly susceptible to climate change. It falls within latitudes 24° 20′ 10″ N and 27° 31′ 15″ N and longitudes 83° 19′ 50″ E and 88° 17′ 40″ E. The average annual rainfall is 1205.6 mm. The geographical area of Bihar covers 94,163 km^2, with an average elevation of approximately 150 m above sea level. It is divided into 38 (Fig. 6.1) districts, classified into three agro-climatic zones: North-East, North-West, and South. The North-West zone comprises 13 districts with annual rainfall ranging from 1040 to 1450 mm and predominantly consisting of loam and sandy loam soils. On the other hand, the northeast zone consists of eight districts with rainfall varying from 1200 to 1700 mm and is primarily characterized by loam and clay loam soils. Lastly, the South Zone includes 17 districts with annual rainfall between 990 to 1300 mm and is dominated by sandy loam, loam clay, and clay loam soils (Tesfaye et al. 2017). It is in a subtropical temperate zone with a humid subtropical climate. It has rich alluvial soil and plenty of water resources, particularly groundwater. Agriculture is the state's most important industry, accounting for 77.3% of total land use and a rural–urban divide (about six times the urban area) significantly larger than the national average. The main cereal crops are rice, wheat, and maize. Rice is Bihar's principal monsoon crop, and it is grown in all the state's districts. The state is highly agriculture-centric, with a significant portion of its residents depending on farming. Most farmers own small plots and use traditional methods (Gathala et al. 2021). The location and landscape make it vulnerable to climate change effects like inconsistent rains, regular floods, especially from rivers like the Ganges and Kosi, and droughts in areas like the south (Misra et al. 2007). These climate-induced changes pose severe challenges to the state's agricultural productivity and the livelihoods of millions who depend on it (Aryal et al. 2020).

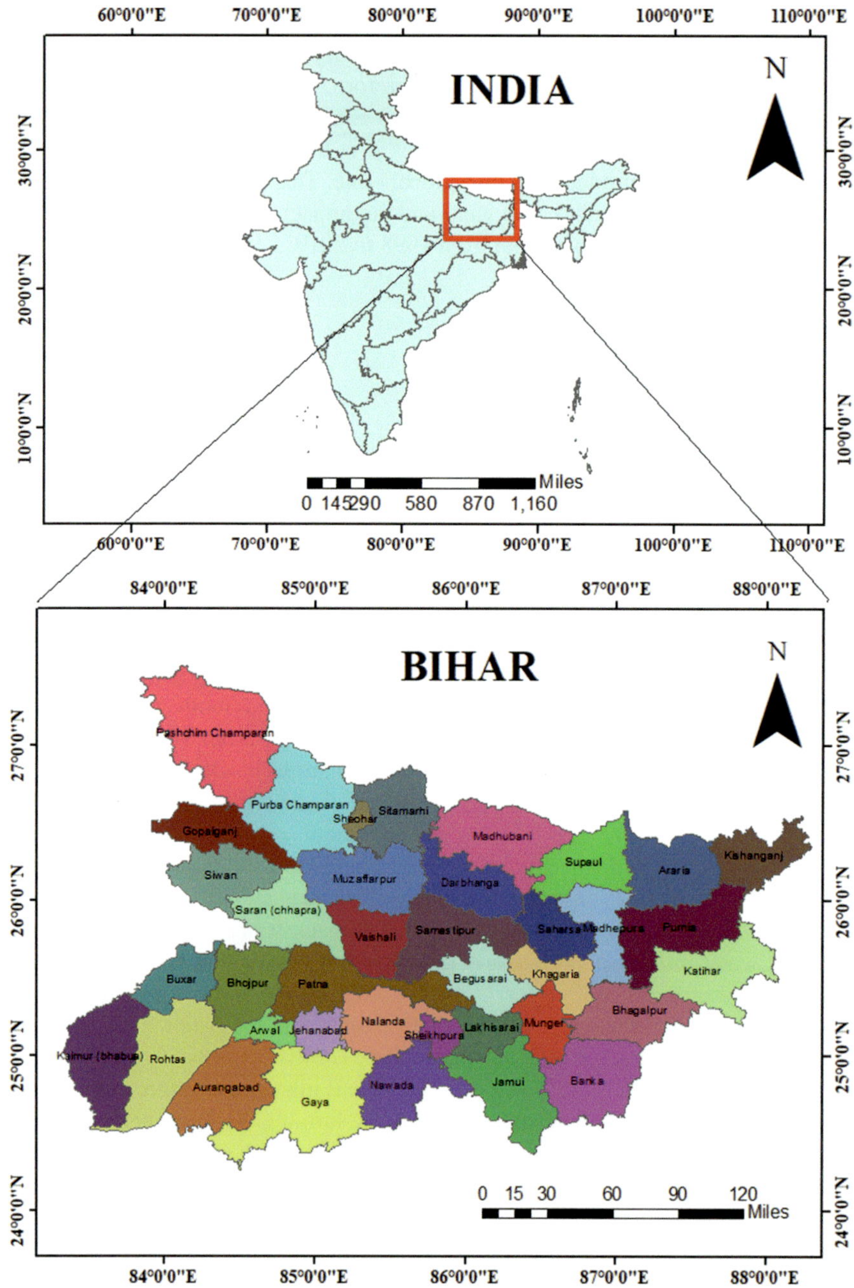

Fig. 6.1 The administrative boundary of Bihar shows 38 districts

Table 6.1 Details of the climatic and yield data

Sr. No.	Data type	Duration	Resolution	Source
1	Rainfall	1998–2019	0.25° × 0.25°	IMD
2	Maximum temperature	1998–2019	1° × 1°	IMD
3	Minimum temperature	1998–2019	1° × 1°	IMD
4	Yield (rice and wheat)	1998–2019		aps.dac.gov.in

IMD—India Meteorological Department

6.2.2 Data Used and Acquisition

Data required to achieve the objectives of this study were collected from various sources, as given in the Table 6.1.

6.2.3 Data Collection

The data used for this study are precipitation at 0.25° × 0.25°, maximum temperature, and minimum temperature at 1° × 1° resolution, which were collected from the India Meteorological Department (IMD), and crop yield data (Rice and Wheat) of 38 districts of Bihar on an annual scale from the government website, Crop Production Statistics Information System (DES, accessed 13 July 2021) from 1998 to 2019. Name of the districts and corresponding weather stations' location are shown in Table 6.2.

6.3 Methodology

6.3.1 Climate Change Extreme Indices

The Expert Team on Climate Change Detection and Indices (ETCCDI) website comprehensively compiles 27 recognized climate change indicators related to temperature and rainfall. It is accompanied by detailed explanations and equations for their calculation (ETCCDI). This study considered nine precipitation-based indices (Table 6.3) and ten temperature-based indices (Table 6.4). Many temperature indices rely on counting the days exceeding certain thresholds, which can be either fixed absolute values or variable percentiles. The threshold values for maximum and minimum temperature are 35 °C and 28 °C, respectively, and for precipitation, 2.5 mm is used. Any negative value in the data was eliminated, and the missing value was replaced with code 99.9 (Tank et al. 2009; Zhang and Yang 2004). The Mann–Kendall trend test (Chatfield and Collins 2018) demonstrates robustness and remains unaffected

Table 6.2 Weather stations for 38 districts of Bihar

Sr. No.	District	Latitude	Longitude
1	Arwal	26.1311	87.4528
2	Aurangabad	24.7500	84.3700
3	Banka	24.8874	86.9198
4	Begusarai	25.5184	86.1752
5	Bhagalpur	25.3478	86.9824
6	Bhojpur	25.4662	84.5222
7	Buxar	25.4988	84.1435
8	Darbhanga	26.0811	86.0122
9	Purba Champaran (East Champaran)	24.7213	84.8568
10	Gaya	26.4514	84.3963
11	Gopalganj	24.9200	86.2200
12	Jamui	25.0426	83.6056
13	Kaimur (Bhabua)	25.5335	87.5837
14	Katihar	25.4838	86.7014
15	Khagaria	26.3077	87.7763
16	Kishanganj	25.1861	86.1345
17	Lakhisarai	25.9247	86.7945
18	Madhepura	26.4628	86.1752
19	Madhubani	25.3708	86.4734
20	Munger	26.0667	85.5200
21	Muzaffarpur	25.2622	85.4788
22	Nalanda	24.7426	85.5200
23	Nawada	27.1543	84.3542
24	Patna	25.4840	85.0233
25	Purnia	26.6438	84.9040
26	Rohtas	25.7771	87.4753
27	Saharsa	25.0686	84.0167
28	Samastipur	25.8603	86.6208
29	Saran (Chhapra)	25.7471	85.8896
30	Sheikhpura	25.8560	84.8568
31	Sheohar	25.0848	85.8896
32	Sitamarhi	26.5146	85.2942
33	Siwan	26.6322	85.5200
34	Supaul	26.1689	84.3963
35	Vaishali	26.2899	86.8220
36	Pashchim Champaran (West Champaran)	25.6838	85.3550
37	Araria	25.2391	84.6696
38	Jehanabad	25.1516	84.9818

Table 6.3 Precipitation-based extreme indices

Sr. No.	Indices	Unit	Definition
1	PRCPTOT	mm	Yearly total rainfall on wet days
2	R99p	mm	Yearly total rainfall when $RR > 99$th percentile
3	R10mm	No. of days	Yearly count of days when $PRCP \geq 10$ mm
4	CDD	No. of days	Maximum length of dry spell: maximum number of consecutive days with daily rainfall, $RR < 2.5$ mm
5	SDII	mm/day	Simple Daily Intensity Index Let RR_{wj} be the daily rainfall amount on wet days, w ($RR \geq$ 2.5 mm) in period j. If w denotes the number of wet days in j, then; $SDIIi = \sum RR_{wj}/w$ $w = 1$
6	CWD	No. of days	Consecutive Wet Day Let $RRij$ be the daily precipitation amount on day i in period j. Count the largest number of consecutive days where $RRij \geq 2.5$ mm
7	R95p	No. of days	Yearly total PRCP when $RR > 95p$
8	R20mm	No. of days	Yearly count of days when $PRCP \geq 20$ mm
9	R25mm	mm	Yearly count of days when $PRCP \geq 25$ mm

by missing values in the time series. Additionally, Sen's slope estimator, another non-parametric method, is employed to estimate the trend's slope in a time series and to calculate the rate of change of extreme indicators (Theil 1950; Sen 1968). Mann–Kendall and Sen's slope tests assess climatic variability and identify trends (Data 2009).

6.3.2 *Methodological Flowchart*

Methodology followed, to achieve the objective of the study, is shown in flowchart (Fig. 6.2).

6.3.3 *Trend Analysis*

The data was analyzed using the Mann–Kendall trend test to detect monotonic trends over time, whether increasing or decreasing. This test is non-parametric, making it applicable to various distributions without requiring normality assumptions. However, ensuring that the data does not exhibit serial correlation is essential.

Table 6.4 Temperature-based extreme indices

Sr. No.	Indices	Unit	Definition
1	TXN	°C	Monthly minimum value of daily maximum temperature
2	TXX	°C	Monthly maximum value of daily maximum temperature
3	TNN	°C	Monthly minimum value of daily minimum temperature
4	TNX	°C	Monthly maximum value of daily minimum temperature
5	TN10p	% of days	Cold nights Number of days TN < 10th percentile calculated for each calendar day (for a 30-year climate reference period) of a 5-day running window
6	TX90p	% of days	Warm days Number of days TX > 90th percentile calculated for each calendar day (for a 30-year climate reference period) of a 5-day running window
7	SU	No. of days	Number of summer days Yearly count of days when TX (daily maximum temperature) > 35 °C
8	TR	No. of days	Number of tropical nights Yearly count of days when TN (daily minimum temperature) > 28 °C
9	TX10P	% of days	Number of days when TX < 10th percentile
10	TN90P	% of days	Number of days when TN > 90th percentile

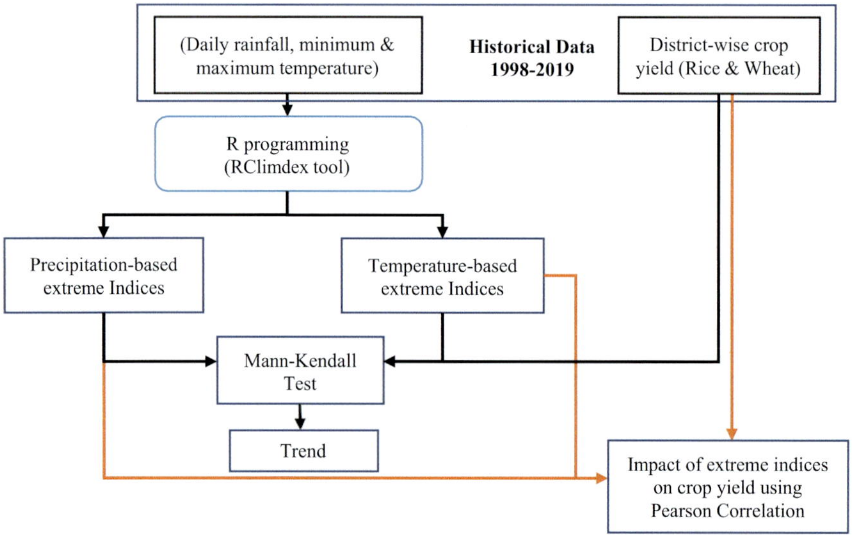

Fig. 6.2 Methodological flowchart

Sen's slope estimator was also employed to investigate the trends and magnitudes of rainfall and temperature.

Mann–Kendall statistic (S) for series R_1, R_2, and R_n is given by:

$$\mathrm{S} = \sum_{i=1}^{n-1} \sum_{j=i+1}^{n} \mathrm{sgn}\left(R_j - R_i\right) \tag{6.1}$$

$$\mathrm{sgn}\left(R_j - R_i\right) = \begin{cases} 1 & \textit{if } R_j > R_i \\ 0 & \textit{if } R_j = R_i \\ -1 & \textit{if } R_j < R_i \end{cases} \tag{6.2}$$

Variance of the Mann–Kendall statistic.

$$\mathrm{var(S)} = \frac{1}{18}\left[\mathrm{n(n-1)(2n+5)} - \sum_{\mathrm{p-1}}^{\mathrm{g}} \mathrm{t_p}\left(\mathrm{t_p} - 1\right)\left(2\mathrm{t_p} + 5\right)\right] \tag{6.3}$$

Test statistic z

$$Z = \begin{cases} \frac{s-1}{\sqrt{\mathrm{var}(s)}} & \textit{if } s > 0 \\ 0 & \textit{if } s = 0 \\ \frac{s+1}{\sqrt{\mathrm{var}(s)}} & \textit{if } s < 0 \end{cases} \tag{6.4}$$

Hypothesis: $\mathrm{H_0}$ = Null hypothesis trend rejected,

$\mathrm{H_1}$ = Alternate Hypothesis of no trend rejected if $|Z| > Z_{1-\alpha/2}$ $\alpha = 10\%$ significant level $Z_{1-\alpha/2}$ is 1.64.

The trend of time series data using Sen's slope is calculated as

$$\beta = Median\left(\frac{x_j - x_i}{j - i}\right), \quad j > i \tag{6.5}$$

where β is Sen's slope estimate and $\beta > 0$ indicates an increasing trend. Otherwise, the data series presents a downward trend during the period.

6.4 Results and Discussion

6.4.1 Trends of Rice and Wheat Yield

Figure 6.3 shows the trend analysis of crop yields (i.e., Rice and Wheat) for Bihar state. The different colors show the magnitude of Sen's slope; the upward triangle represents a significant positive trend, the downward triangle represents a significant negative trend, and the dark circle shows no trend. The analysis indicates that yield has a strong positive trend in some districts and no trend in many districts of Bihar. As for the rice yield trend (Fig. 6.3a), a significant positive trend is analyzed for all districts except the few districts like Sheohar, Sitamarhi, Gaya, Nalanda, Sheikhpura, Jamui, Munger, Samastipur, and Vaishali which do not show any significant trend. The magnitude of the trend for rice yield varies from 0.00 to 0.13 ton/ha/yr. Similarly, for wheat, all districts have a significant positive trend except for Arwal (Fig. 6.3b). The districts that do not show any significant trend for wheat are Katihar, Jamui, Munger, Gaya, Nalanda, Sitamarhi, Vaishali, Nalanda, and Patna. Arwal has shown a significant negative trend, and the remaining districts have shown a significant positive trend. The magnitude of the trend varies from − 0.02 to 0.10 ton/ha/yr. Regarding the percentage distribution, about 23.68% of the area shows no trend, and the remaining 76.32% resulted in a strong positive trend for rice yield. In contrast, for wheat, 73.69% of the area shows a significant positive trend, 23.68% shows no trend, and 2.63% has a significant negative trend. The reasons for showing an increasing trend for most districts may be high-yielding seeds, excessive use of fertilizers, over-irrigation, increased soil tillage, and modern agricultural technologies and implements.

6.4.2 Trends of Precipitation-Based Extreme Indices

Figure 6.4 shows the yearly trends of the precipitation-based extreme indices for India's Bihar state. Regardless of the different indexes, only a few indices show a significant positive trend for a few districts. Except for these indices, all others either show a considerable negative or no trend. This is because of the spatial–temporal variability of rainfall and the erratic behavior of rains due to climate change. SDII showed a positive trend for seven districts (Aurangabad, Darbhanga, Gaya, Purba Champaran, Samastipur, Saran (Chhapra), and Sheikhpura) and a negative trend for three districts (Bhagalpur, Katihar, and Madhepura). The only index showing a significant positive trend is because of a decrease in wet days and rainfall variability. A positive trend in precipitation indices, such as the SDII, in agricultural regions may hint towards potential benefits, like enhanced crop yield in certain districts. However, the negative trends and variability in rainfall can pose substantial threats to crop production, as consistent water supply is pivotal for crop growth. This variability can result in periods of water scarcity or excess, leading to drought conditions, which are detrimental to agriculture and the livelihoods of smallholder farmers.

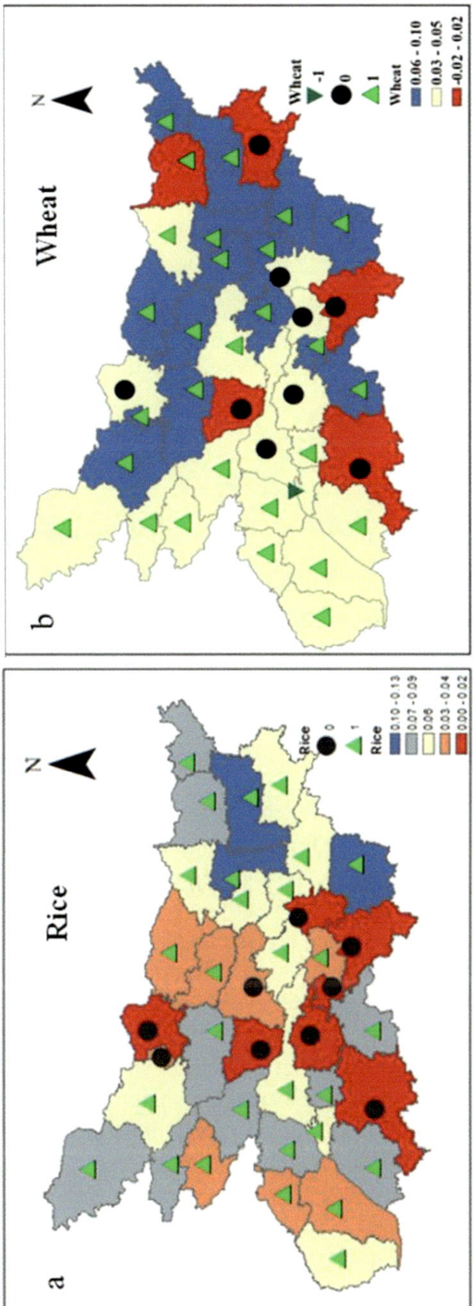

Fig. 6.3 Trends of rice and wheat yield

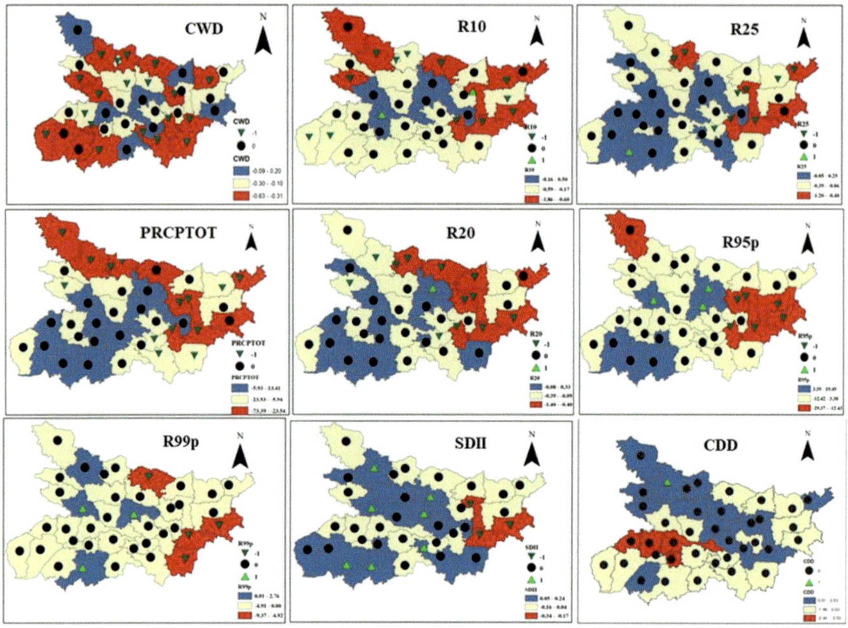

Fig. 6.4 Trends of precipitation-based extreme indices

6.4.3 Trends of Maximum Temperature-Based Extreme Indices

Figure 6.5 shows the trend of maximum temperature-based indices for both seasons (i.e., Kharif and Rabi). Results for the Rabi season-based extremes show no trend. In contrast, Kharif season-based indices show either a significant positive trend or no trend except the TX10P_Kharif. TX90P_Kharif and TXX_Kharif show a significant positive trend. This is due to increased emissions of harmful gases such as carbon dioxide, nitrous oxide, and methane because anthropogenic activities cause global warming, rainfall variability, and extreme events such as drought. Sen's slope also indicated that the magnitude indices have an increasing trend. This increase in the trend of warmer days directly impacts agriculture. In simple terms, if temperatures continue to rise during crucial growing seasons, it could threaten crop yields, affecting the food supply and the livelihoods of those dependent on farming.

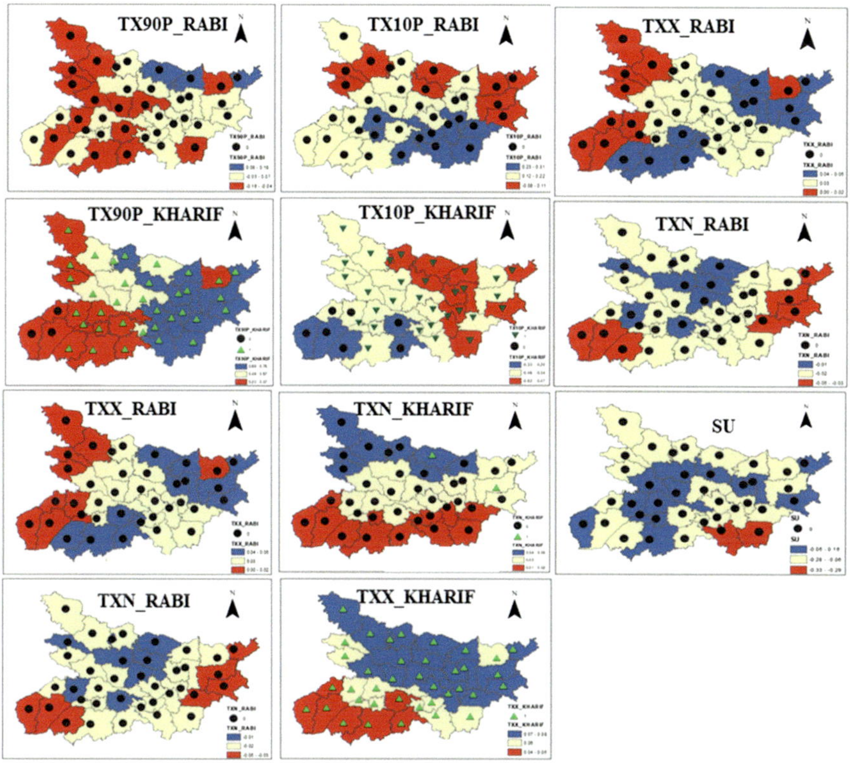

Fig. 6.5 Trends of maximum temperature-based extreme indices

6.4.4 Trends of Minimum Temperature-Based Extreme Indices

Figure 6.6 shows the trend of minimum temperature-based indices for both seasons (i.e., Kharif and Rabi). Results indicated that the maximum value of the minimum temperature-based index, such as TNX and TN90P, show a significant positive trend, and the minimum value of the minimum temperature-based indices shows no trend except TN10P, which shows a significant negative trend. This is because of heat waves, global warming, deforestation, land use change, and extreme weather events, which increase the diurnal temperature and decrease the cold nights. Higher nighttime temperatures can affect crop growth, germination, and yield.

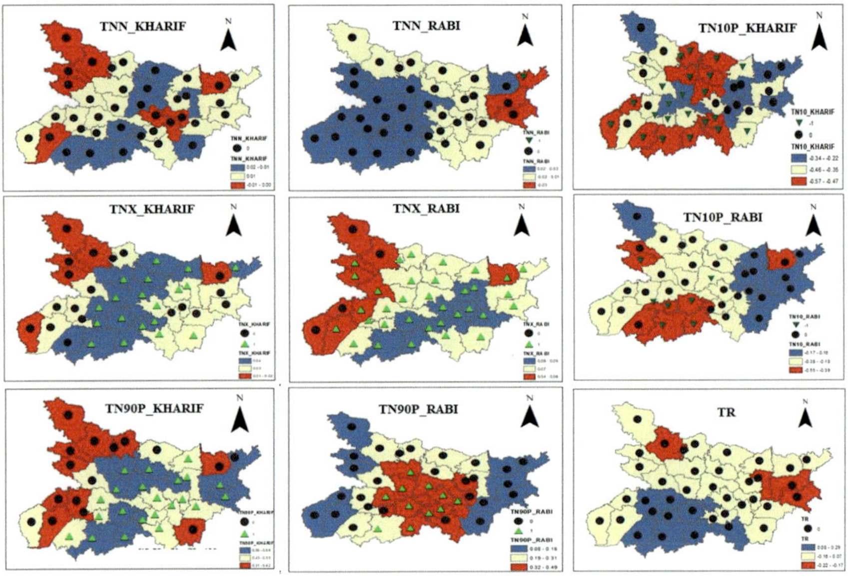

Fig. 6.6 Trends of minimum temperature-based extreme indices

6.4.5 *Correlation Between Rice Yield with Extreme Climatic Indices*

Figure 6.7 shows the correlation between rice yield and climatic extremes indices for the historical period (1998–2019) varying from − 0.5 to 0.7. Here, the blue color indicates a negative correlation, and the red color shows a positive correlation. PRCPTOT, R10, R20, R25, and R95 indices strongly correlate with various districts of Bihar. Districts such as Aurangabad, Banka, Bhojpur, Nalanda, Buxar, Kaimur, Kishanganj, and Purnia strongly correlate with temperature and precipitation-based indices. Some of these indices are positively correlated, and some of these are negatively correlated with rice yield. The precipitation-based indices show a strong positive correlation with the rice yield for most of the districts of Bihar because the rice crop requires standing water for its growth and development, and these rainfall indices help with it. Banka district shows a strong positive correlation because of the rivers Badua, Chir, and Chandan, which allow irrigation. In contrast, Kishanganj shows a negative correlation with the precipitation-based indices because the district is significantly affected by rivers such as Mahananda, Kankai, Mechi, Donk, Rauta, and Ramzan, which consistently pose a significant challenge as they frequently lead to flooding in the area. The temperature-based indices generally do not correlate well with rice yields for various districts, with some districts, such as Madhubani, showing a positive correlation. However, crop yields of Bhojpur, Buxar, and Kaimur show a negative correlation with most of the temperature-based indices. Positive correlations of precipitation-based and colder days indices such as TR, TN10P, and TX10P increase rice yield because they help boost plant growth and soil microbial activities. This benefits farmers in districts like Banka, where increased rainfall or river water aids in rice cultivation. Conversely, negative correlations imply that increased climatic indices can decrease rice yield. For example, in Kishanganj, more rainfall often results in floods, harming crops.

6.4.6 *Correlation Analysis of Wheat Yield with Climatic Extreme Indices*

Figure 6.8 presents a correlation analysis between wheat yield and climatic extremes for the historical period (1998–2019). CWD, PRCPTOT, R10, R20, R25, R95p, and R99p indices strongly correlate with the wheat yield for most districts. Wheat yield of districts such as Bhagalpur, Kaimur, and Lakhisarai show a high correlation (positive or negative) with the temperature and precipitation-based indices. Weak correlations with extreme indices are mainly obtained in Araria, Arwal, Begusarai, Jehanabad, and Madhepura. The precipitation-based indices show a strong positive correlation with the wheat yield for most of the districts of Bihar. Wheat crops rely heavily on water availability during the growing season. Adequate rainfall can provide the necessary moisture for the wheat plants to grow and develop properly, resulting in higher yields. Results also indicated that temperature-based indices negatively correlate with wheat yields for the various districts. High temperatures can lead to heat stress, affecting multiple stages of wheat development, including flowering, grain filling, and maturation. This can reduce grain size, lower grain number, and decrease wheat yield. For farmers in districts with strong positive correlations to rainfall, it indicates that their wheat yields are highly dependent on rainfall amounts. To ensure consistent yields, they might need to adopt water-conservation practices or irrigation systems during drier years. In contrast, in areas with a negative correlation with temperature, farmers might consider adopting heat-resistant wheat varieties or adjusting their planting schedules to avoid peak heat periods.

6.5 Conclusion

The study examined the spatiotemporal variations of precipitation and temperature-based extreme indices and their effects on crop yield in Bihar for the period 1998 to 2019. Initially, we assessed the long-term trends in extreme weather indicators by utilizing climate data (i.e., precipitation, maximum and minimum temperature) on a daily temporal scale for 38 districts. Then, the correlation between extremes and crop yield was developed. The trend analysis for the historical period (1998–2019) showed a significant positive trend for both crops in most districts. However, the districts that didn't show a significant trend also had a positive trend. For climatic

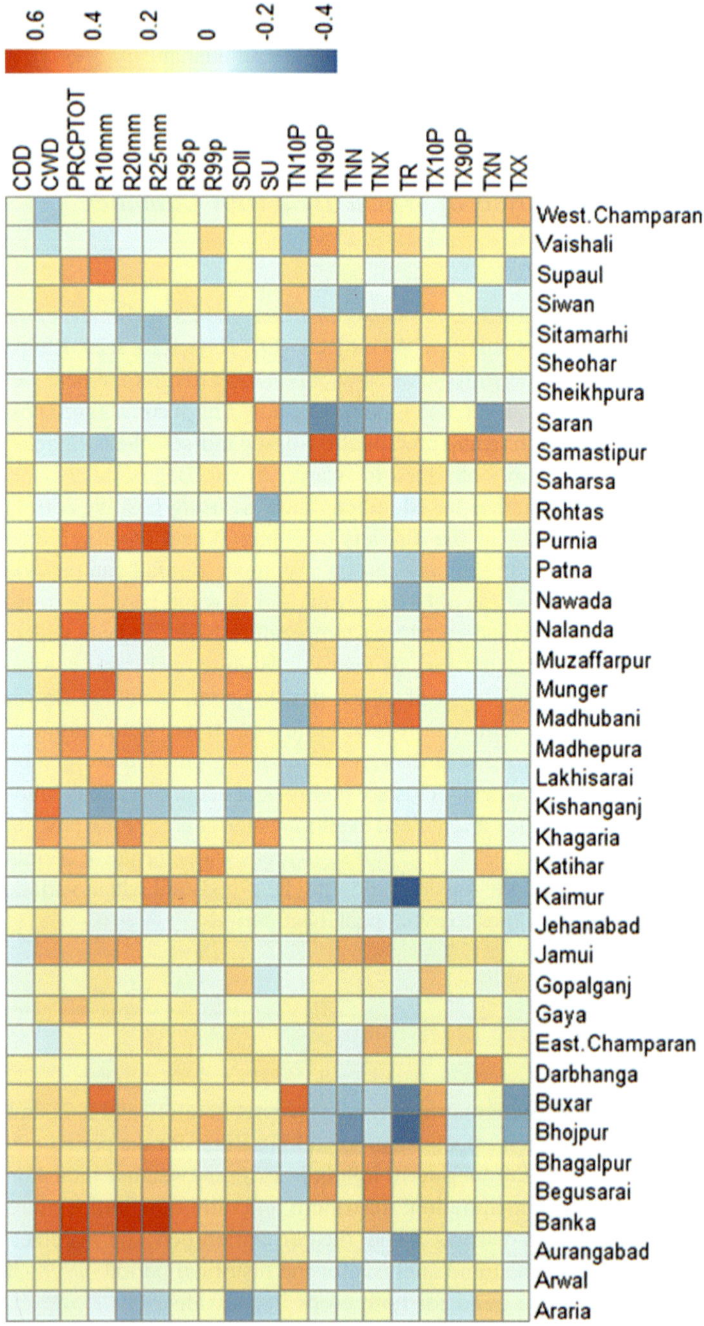

Fig. 6.7 Correlation between rice yield and Kharif season extremes indices

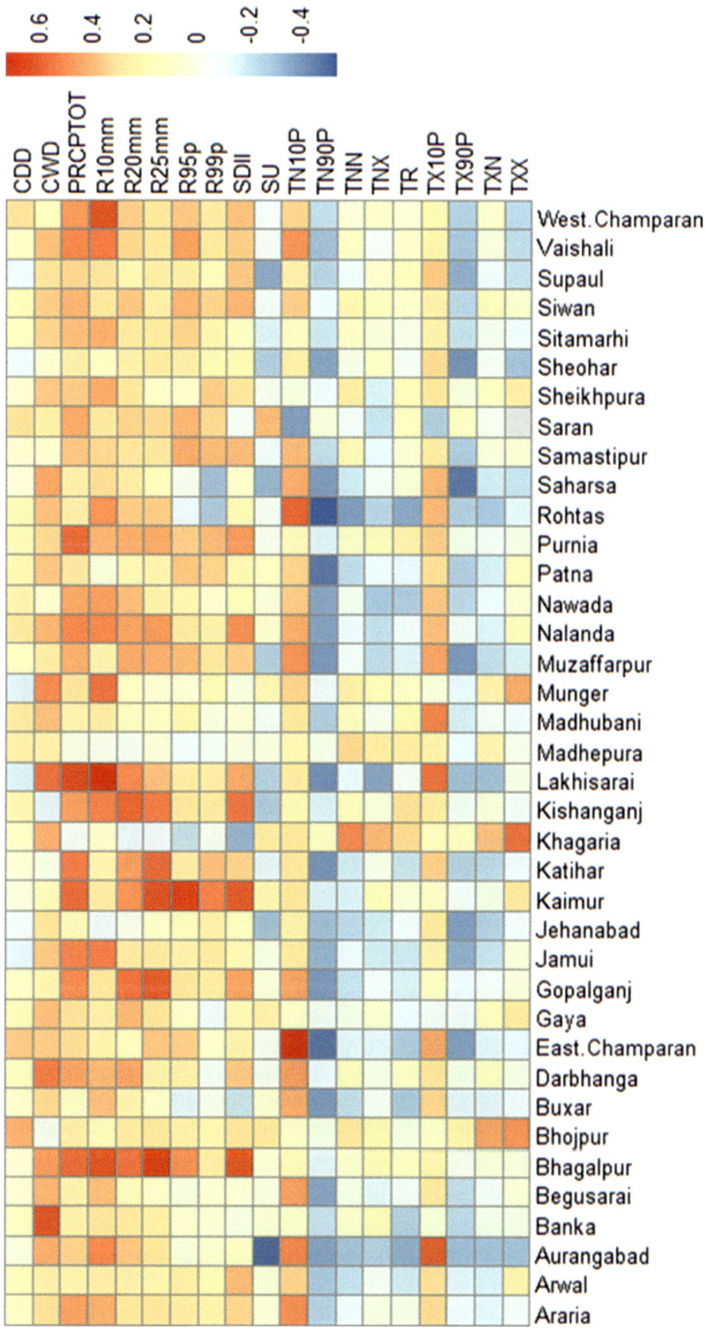

Fig. 6.8 Correlation between wheat yield and Rabi season extremes indices

extremes analysis, it was observed that most precipitation-based indices show a significant negative trend and temperature-based indices show a significant positive trend. The findings indicate the need to develop thermo-insensitive varieties to sustain and increase crop productivity and meet future food demand. The results of this study will help the decision-makers develop appropriate adaptation and mitigation policies to reduce the effects of climate change on crop yield in Bihar. Further research is required to study how this extreme precipitation and temperature-based indices will affect the stages of plant phenology and, therefore, crop productivity.

Acknowledgements I sincerely thank the Department of Agricultural and Food Engineering, IIT Kharagpur, and the Department of Water Resources Development and Management, IIT Roorkee, for making all the necessary facilities available to execute this research. I would also like to thank Mr. Uday Partap Singh Bhadoriya, Mr. Rajesh Ranjan, and Mr. Shashi Bhushan for helping with this research. Lastly, I thank my supervisors, Prof. Ashok Mishra and Prof. Idhaya Chandhiran Ilampooranan, for their guidance and mentorship throughout this research.

Declaration of Competing Interest The authors declare that they have no identified financial or personal conflicts that could have potentially influenced the research findings in this study.

References

Adisa OM, Botai CM, Botai JO, Hassen A, Darkey D, Tesfamariam E, Adisa AF, Adeola AM, Ncongwane KP (2018) Analysis of agro-climatic parameters and their influence on maize production in South Africa. Theor Appl Climatol 134(3–4):991–1004. https://doi.org/10.1007/S00704-017-2327-Y

Allan RP, Hawkins E, Bellouin N, Collins B (2021) IPCC, 2021: summary for policymakers. Cambridge University Press. https://doi.org/10.1017/9781009157896.001

Aryal JP, Sapkota TB, Khurana R, Khatri-Chhetri A, Rahut DB, Jat ML (2020) Climate change and agriculture in South Asia: adaptation options in smallholder production systems. Environ Dev Sustain 22(6):5045–5075

Birthal P, Khan T, Negi DS, Agarwal S (2014) Impact of climate change on yields of major food crops in India: implications for food security. indianjournals.com. https://doi.org/10.5958/0974-0279.2014.00019.6

Chatfield C and Collins (2018) Introduction to multivariate analysis. In: Introduction to multivariate analysis. CRC Press, pp 1–248. https://doi.org/10.1201/9780203749999/INTRODUCTION-MULTIVARIATE-ANALYSIS-CHRIS-CHATFIELD

Dahal V, Gautam S, Bhattarai R (2018) Analysis of the long-term precipitation trend in Illinois and its implications for agricultural production. mdpi.com. https://doi.org/10.3390/w10040433

Dai S, Shulski MD, Hubbard KG, Takle ES (2016) A spatiotemporal analysis of Midwest US temperature and precipitation trends during the growing season from 1980 to 2013. Int J Climatol 36(1):517–525

Data C (2009) Guidelines on analysis of extremes in a changing climate in support of informed decisions for adaptation. World Meteorological Organization

Dos Santos CA, Neale CM, Rao TV, Da Silva BB (2011) Trends in indices for extremes in daily temperature and precipitation over Utah, USA. Int J Climatol 31(12):1813–1822

Gathala MK, Laing AM, Tiwari TP, Timsina J, Rola-Rubzen F, Islam S, Maharjan S, Brown PR, Das KK, Pradhan K, Chowdhury AK, Kumar R, Datt R, Anwar M, Hossain S, Kumar U, Adhikari S, Magar DBT, Sapkota BK, Shrestha HK, Islam R, Rashid M, Hossain I, Hossain A, Brown B,

Gerard B (2021) Improving smallholder farmers' gross margins and labor-use efficiency across a range of cropping systems in the eastern gangetic plains. World Dev 138:105266

Hatfield JL (2013) North American perspectives on potential climate change and agricultural responses. In: Handbook of climate change and agroecosystems, pp 33–55

Hatfield JL, Boote KJ, Kimball BA, Ziska LH, Izaurralde RC, Hatfield JL, Boote KJ (2011) Climate impacts on agriculture: implications for crop production. Wiley Online Library 103(2):351–370. https://doi.org/10.2134/agronj2010.0303

Högy P, Poll C, Marhan S, Kandeler E, Fangmeier A (2013) Impacts of temperature increase and change in precipitation pattern on crop yield and yield quality of barley. Food Chem 136(3–4):1470–1477

Tank AMGK, Zwiers FW, Zhang X (2009) Guidelines on analysis of extremes in a changing climate in support of informed decisions for adaptation. Clim data Monit WCDMPNo. 72, WMO-TD No. 1500:56

Mall RK, Singh R, Gupta A, Srinivasan G, Rathore LS (2006) Impact of climate change on Indian agriculture: a review. Clim Change 78(2–4):445–478. https://doi.org/10.1007/S10584-005-9042-X

Misra AK, Saxena A, Yaduvanshi M, Mishra A, Bhadauriya Y, Thakur A (2007) Proposed river-linking project of India: a boon or bane to nature. Environ Geol 51:1361–1376

Sen PK (1968) Estimates of the regression coefficient based on Kendall's Tau. J Am Stat Assoc 63(324):1379–1389. https://doi.org/10.1080/01621459.1968.10480934

Skendžić SS, Zovko M, Pajač I, Živković Ž, Lešić VL, Lemić DL (2021) The impact of climate change on agricultural insect pests. mdpi.com. https://doi.org/10.3390/insects12050440

Subash N, Singh SS, Priya N (2013) Observed variability and trends in extreme temperature indices and rice-wheat productivity over two districts of Bihar, India-a case study. Theor Appl Climatol 111(1–2):235–250. https://doi.org/10.1007/S00704-012-0665-3

Tesfaye K, Aggarwal PK, Mequanint F, Shirsath PB, Stirling CM, Khatri-Chhetri A, Bahadur Rahut D (2017) Climate variability and change in Bihar, India: challenges and opportunities for sustainable crop production. mdpi.com. https://doi.org/10.3390/su9111998

Theil H (1950) A rank-invariant method of linear and polynomial regression analysis. Indag Math 12(85):173

Tian J, Liu J, Wang J, Li C, Nie H, Yu F (2017) Trend analysis of temperature and precipitation extremes in major grain producing area of China. Int J Climatol 37(2):672–687

Tirkey AS, Ghosh M, Pandey AC, Shekhar S (2018) Assessment of climate extremes and its long term spatial variability over the Jharkhand state of India. Egypt J Remote Sens Space Sci 21(1):49–63

Ur MA, Tariq R, Wangchuk K, Muttil N (2021) A critical review of water resources and their management in Bhutan. mdpi.com. https://doi.org/10.3390/hydrology8010031

Vogel E, Donat MG, Alexander LV, Meinshausen M, Ray DK, Karoly D, Meinshausen N, Frieler K (2019) The effects of climate extremes on global agricultural yields. iopscience.iop.org. https://doi.org/10.1088/1748-9326/ab154b

Yuan Z, Yin J, Wei M, Yuan Y (2021) Spatio-temporal variations in the temperature and precipitation extremes in Yangtze River Basin, China during 1961–2020. Atmosphere 12(11):1423

Zhang X, Yang F (2004) RClimDex (1.0) user manual. Clim Res Branch Environ Canada 22:13–14

Zhang J, Felzer BS, Troy TJ (2016) Extreme precipitation drives groundwater recharge: the Northern High Plains Aquifer, central United States, 1950–2010. Hydrol Process 30(14):2533–2545. https://doi.org/10.1002/HYP.10809

Zhou B, Xu Y, Wu J, Dong S, Shi Y (2016) Changes in temperature and precipitation extreme indices over China: analysis of a high-resolution grid dataset. Int J Climatol 36(3):1051–1066

Chapter 7
Pathways to Accelerate Agricultural Production System Technologies in India: A Vision for New Generation Watersheds

Deep Banerjee, Krishna Reddy Kakumanu, and Swatantra Kumar Dubey

Abstract To achieve water and food security, ensure sustainable production and consumption, and reduce land degradation, watershed management is one of the key global strategies. Several initiatives have been introduced in the Indian subcontinent to enhance crop and water productivity while managing natural resources. While some of these initiatives initially demonstrated success, their long-term viability faced challenges due to various environmental and socio-economic factors. The Government of India has also launched several initiatives, underscoring the need for an extensive layout to guide the identification, planning, implementation, monitoring, and evaluation of watershed development activities. To address this need, our study adopts a holistic approach combining geospatial and agroecological methodologies to locate potential watershed development zones in India. We integrated appropriate agronomic, chemical, mechanical, and agrostological techniques customized for soil management and the conservation of green and blue water across various basins and agroecological zones. These proposed practices are further expanded toward value chain development, ensuring nutritional and economic benefits. Various stakeholders have been identified at the state, district, block, and village levels across regions to facilitate the promotion, implementation, and monitoring of these site-specific interventions. These stakeholders are intended to play a crucial role in maintaining the agricultural value chain and expediting the adoption of the proposed technologies and management strategies. The study finally adapts a system thinking approach to assemble these improved land use strategies into a project cycle. Such a critical combination of multi-disciplinary impact pathways provides the potential roadmap toward achieving new generation watersheds in India.

Keywords Watershed management · New generation watersheds · Production system technologies · Soil–water conservation

D. Banerjee (✉) · S. K. Dubey
International Rice Research Institute (IRRI) South Asia Regional Centre, Varanasi, Uttar Pradesh, India
e-mail: d.banerjee@irri.org

K. R. Kakumanu
National Institute of Agricultural Extension Management, Hyderabad, Telangana, India

A. Nanda et al. (eds.), *Navigating the Nexus*, Water Science and Technology Library 102, https://doi.org/10.1007/978-3-031-76532-2_7

7.1 Introduction

Watershed is not merely a geo-hydrological unit but rather has become a socio-political–ecological identity that ensures a proper supply of food, and socioeconomic security for the associated people (Flotemersch et al. 2016). Whereas those supplies are sourced from the natural resources existing in the watersheds. On the verge of ensuring livelihood, often those resources are exploited or wasted leading to watershed degradation. Watershed degradation primarily refers to the decline in physical, chemical, biological, and economic characteristics of a watershed, leading to a reduction in the productive capacity of its land and water resources. This process also induces significant changes in hydrological patterns at different scales, resulting in diminished water flow quality, quantity, and timing. Such degradation arises from the interaction of physiographic, climatic factors, and anthropogenic activities. Consequently, addressing the risks of ecological decline, diminished economic opportunities, and amplified social issues (Padbhushan et al. 2022; Vörösmarty et al. 2010; Molden et al. 2007; Allan 2004) requires a coupled approach of hydro-climatic, agricultural, and economic strategies and tools.

After 1970, the Indian government initiated several watershed development programs. However, it led to limited adoption due to their compartmentalized approach, inadequate planning, oversight, and technological stagnation. In recent years DoLR established a consortium approach guideline to encourage sustainable land use in India's various agro-ecologies (DoLR 2021; Raju et al. 2008; Rao 2000).

Recent studies have revealed a predominant focus on earth, agricultural, and environmental sciences within watershed management research, with less attention given to operations research for methodology development, Monitoring, Reporting, and Verification (MRV). Bibliometric assessments indicate that over the past decade, more than 4500 publications were developed on watershed management in India. Among these, environmental science ranked highest, comprising 6.94% of the top 1000 fields of study. The top three contributing disciplines overall were Geosciences, Hydrology, and Life Sciences. Within these fields, there has been a notable transformation in the definition of watershed, with a shift towards targeting specific watershed scales and prioritizing studies on geomorphologies (Shekar et al. 2023; Singh et al. 2021; Ghosh et al. 2021; Govindaraj and Lakshumanan 2019), hydraulic structures (Venkata Rao et al. 2024; Verma et al. 2023; Guihéneuf et al. 2014), and site-specific crop management strategies (Sahoo et al. 2024; Kar et al. 2023; Wani et al. 2014). However, there remains a gap at a macro level, where an integrated approach is needed to address multiple dimensions and provide policymakers with the necessary insights to formulate legislative initiatives for enhancing soil and water conservation, as well as fostering resilient and sustainable watersheds.

Building upon this rationale, enhanced production system technologies are integrated to formulate a blueprint for the development of Next-Generation watersheds. This framework (a) initially delineates the necessary elements for enhancing watershed potential across India, and (b) explores how these elements and management strategies can be incorporated into a project management cycle. The overarching aim is to combine agricultural and engineering facets to foster sustainable land use and environmental resilience, while also establishing connections with relevant institutions to ensure effective planning, implementation, and monitoring throughout the Indian subcontinent. To illustrate such a framework for new generation watershed management in the Indian context some specific objectives are followed: (a) To identify the components and prepare activity ensembles for integrated watershed management. (b) To devise a project cycle management framework encompassing all these activity ensembles.

7.2 Targeting and Monitoring Systems

For the implementation of a successful watershed program and to ensure sustainability, there must be a well-structured plan as well as a prodigious monitoring system. New generation watersheds require robust technological or instrumental measurements for planning and monitoring parameters at different scales.

7.2.1 Geospatial Applications

Geographic Information Systems (GIS) and Remote Sensing (RS) technologies rank among the top for planning at the initial level. Planning at the watershed level necessitates the generation and analysis of a plethora of interconnected data. Remote sensing data provides useful and current spatial information on natural resources and terrain parameters which is a valuable tool for planning and developing watershed monitoring systems that consider all natural and socioeconomic factors.

Satellite-based remote sensing constitutes an important source of data from which updated land use/land cover (LULC) changes can be proficiently extracted and examined. Furthermore, these land-use pattern changes are examined and simulated with digital soil mapping, crop-fallow mapping, drainage density mapping, inundation modeling, settlement pattern changes, etc. (Pettorelli et al. 2014; Hansen et al. 2013; Friedl et al. 2010; Giri et al. 2007). Different bio-physical aspects like vegetation, moisture, water turbidity, soil, fluvial and climatic phenomena, etc., and socio-economic aspects like built-up areas, demography, markets, etc. require different suitable sensors. Over the years satellite imageries have improved in terms of resolution and specificity. Currently, available sensors (Table 7.1) and recent advancements in microwave interferometry from satellites such as Kompstat-3, WorldView-3, and ALOS/Daichi, as well as laser altimetry from aerial platforms, have aided the generation of Digital Elevation Models to be faster and more precise. This high-resolution (spatial resolution of < 6 m) satellite imageries will be very useful for sub-watershed or micro-watershed level applications (Alqurashi and Kumar 2013; Lu and Weng 2007; Rosenqvist et al. 2007).

Table 7.1 Suitability of various remote sensing sensors in watershed studies

Level of study	Suitable spatial resolutions	Sensors	Application potential
Basin level (RF—1:250,000)	50 to 150 m	IRS-WiFS, AWiFS, LISS-I, Sentinel-2, WorldPop	Capturing comprehensive baseline information on land cover, natural resources, and socioeconomic factors driving; monitoring changes at large scale
Watershed Level (RF—1:50,000)	20 to 50 m	IRS-LISS-III, SPOT-MLA, TM, ETM, CBERS-2, Carbon Mapper	Collecting natural resources data for prioritization, planning, and monitoring of watershed projects
Sub/ Micro-Watershed Level (RF—1:10,000 or larger)	0.5 to 20 m	RS-LISS-IV, SPOT-PLA, SPOT-7, Cartosat-1/2, IKONOS, QuickBird, Worldview-4, Pelican, Pleiades Neo, Albedo, SuperView Neo, KOMPSAT-3A, Gaofen-2	Planning, implementation, and tracking watershed initiatives, as well as stereo data collection for DEM preparation

Adapted from Wani et al. (2011)

7.2.2 Crop and Water Modelling

Crop simulation models have been widely used to understand, observe, and experiment with different cropping systems components, and they've played an important role in scientific research and resource management due to their ability to (a) provide a system approach, (b) evaluate long-term impact of interventions, (c) analyze intervention risks, and (d) provide answers faster and less expensively than traditional experimentation (Tribouillois et al. 2022; Kisekka et al. 2017). Whereas the natural processes of the flow of water, sediment, chemicals, nutrients, and microbial organisms within watersheds can be simulated through the site and context-specific Watershed models Such models, are also useful to quantify the impact of human activities on these processes. Integrated simulation of these two model approaches can effectively address a range of hydrologic, agricultural, environmental, and economic issues. The recent crop models (DSSAT, APSIM, ORYZA, WOFOST, SWAP, Cropsyst, etc.) are quite diverse and vary significantly in sophistication and data and computational requirements. These models are typically understood as (a) Empirical (Historical data based), and (b) Dynamic (process based). In different spatial scales (farm, regional, global etc.) (Müller et al. 2021; Bassu et al. 2014; Elliott et al. 2014; Asseng et al. 2013) They can be effectively incorporated into contemporary watershed models (Table 7.2) by critically assessing their analysis parameters, input needs, output adaptability, scalability, processing capacity, computational efficiency, and capability to replicating changes in catchment.

Table 7.2 Suitable models for planning and monitoring interventions in different watersheds

	Model	Capability	Limitations	References
1	SHETRAN	Simulates various hydrological parameters like runoff, and peak runoff transport	Does not calculate the flow through the unsaturated zone	Lai et al. (2024), Birkinshaw et al. (2010), Ewen et al. (2000)
2	AGNPS	Can estimate soil loss and runoff	Depends on single storm event and is data intensive Sub-surface flow prediction is not available	Kasthuri Thilagam et al. (2023), Zhang et al. (2020a, b), Jena and Tiwari (2016)
3	KINEROS	Hydrologic variables like runoff, peak runoff rate as well as soil loss simulations, etc. are considered	Depends on single storm event and chemical elements and sub-surface flow cannot be simulated	Pambudi (2021)
4	RHEM	Capable of analyzing small agricultural catchments and rangelands hydrologic parameters	Estimations are dependent on single storm event	Hernandez et al. (2017), Al-Hamdan et al. (2015)
5	SWM	Useful for long-term simulations as it uses variable time steps (daily)	Parameter calibration is a laborious task	Mirchi et al. (2011); Crawford and Linsley (1966)
6	SWRRB	Hydrologic cycle and associated parameters can be simulated at daily time-step	Incorporates different assumptions and has extensive data requirement	Šípek and Tesař (2015), Wyseure (1991)
7	SEDIMOT III	The model can estimate flood and sediment during transitions from undistributed to distributed conditions	Limited to small field catchments and depends on single storm event	Barfield et al. (2010)
8	ANSWERS	This model can project runoff, peal runoff rate, nutrient, sediment yield and other hydrological parameters	Depends on a single storm event, requires intensive data, and is unable to estimate the sub-surface flow component	Parsons et al. (2004)

(continued)

Table 7.2 (continued)

	Model	Capability	Limitations	References
9	TOPMODEL	Surface, sub-surface flow, sediment yield, and solute transport can be estimated	Limited to catchments with shallow homogenous soils and very few dry periods	Park et al. (2019), Xue et al. (2018)
10	WESP	Runoff as well as soil loss can be predicted	Depends on single storm event Erosion and deposition parameter information are unavailable	da Silva et al. (2018)
11	SWIM	Key hydrological processes can be simulated utilizing daily time steps	The model is quite complicated Gully erosion cannot be simulated	Krysanova et al. (2015)
12	EUROSEM	Sediment yield, runoff, erosion, and deposition can be simulated	Suitability is limited to small catchments	Ikeda et al. (2019), Hamidpour et al. (2013)
13	LISEM	Runoff and soil erosion/ sediment yield can be estimated	Intensive data requirement and depends on single storm event	Vargas et al. (2021)
14	WEPP	Predicts soil erosion after any hillside disturbance on an event basis	Requires extensive input data	Saghafian et al. (2015)
15	SWAT	Capable of extensively analyzing hydrological and agricultural processes in large basins using daily or sub-daily input data	Source code is unavailable, approximately 400 variables are needed to parameterize, over-predicts sediment load at the outlet, lacks a robust mechanism to account for in-channel and floodplain storage, and underpredicts soil erosion over a 30% slope for interrill erosion	Janjić and Tadić (2023), Pang et al. (2020), Zhang et al. (2020a, b)
16	HEC-HMS/ HEC-RAS	Evaluating overland flow and soil loss	Critically check for representing grain size and settling depth	Thakur et al. (2017), Gibson et al. (2010)

(continued)

Table 7.2 (continued)

	Model	Capability	Limitations	References
17	GSSHA/ CASC2D	It can be used in a variety of environments; The fate of associated pollutants can also be tracked through the coupled system. Scalable	Highly capable and advanced computational support is required	Singh and Frevert (2010), Downer and Ogden (2004)
18	MIKE SHE	Directly incorporates satellite-derived precipitation data, along with various alternative model configurations or structures, which can be seamlessly connected to other DHI software and additional models	Unsuitable for simulating saltwater intrusion or heat flow Limited applications in Flood hydraulics, flood hazards, and dam break analysis	Ma et al. (2016)

Note SHETRAN System Hydrologique European Transport, *AGNPS* Agricultural Non-Point Source model, *KINEROS* Kinematic Runoff and Erosion Model, *RHEM* Range-land Hydrology and Erosion Model, *SWM* Stanford Watershed Model, *SWRRB* Simulator for Water Resources in Rural Basins, *SEDIMOT III* Sedimentology and Distributed Modeling Techniques, *ANSWERS* Areal Non-Point Source Watershed Environment Response Simulation, *TOPMODEL* Topography Based Hydrological Model, *WEPP* Watershed Erosion Prediction Project, *SWAT* Soil and Water Assessment Tool, *GSSHA* Gridded Surface/Subsurface Hydrologic Analysis, *HEC-1/HEC-HMS* Hydrologic Engineering Center-Hydrologic Modeling System, *HSFP* Hydrological Simulation Program—FORTRAN, *MIKE SHE* Système Hydrologique Europeén, *LISEM* Llmburg Soil Erosion Model, *EUROSEM* European Soil Erosion Model, *SWIM* Soil and Water Integrated Model, *WESP* Water Erosion Simulation Program
Adapted from Gull and Shah (2020), Daniel (2011)

To understand the technologies on which the models are generally developed can be classified into 3 major types—(a) Artificial Neural Networks (ANN), (b) Fuzzy Logic, and (c) Genetic Algorithms (Giusti and Marsili-Libelli 2015; Wani et al. 2011). However, the crop and watershed models might serve some distinctive purposes, our current propositions suggest an integrated approach for a robust planning and monitoring approach of watershed interventions.

7.2.3 Land Classification and Groundwater Potential Mapping

To understand the importance of managing arable and non-arable land in watersheds, we have to first understand land capabilities. The main aim of soil and water conservation in watersheds is to use the land according to its capability and treatment according to its needs. This land characteristic database is in utter need of planning, implementation, and evaluation of improved watershed development programs.

There are commonly three factors that determine land capabilities, i.e., (a) environmental factors (b) internal factors, and (d) external factors. Based on these factors, USDA has classified land capability (Fig. 7.1) into 8 'Classes', 4 'subclasses' and several 'Units' based on their distinct characteristics (Klingebiel and Montgomery 1961). With an increasing number of classes, land use limitations and vulnerability to different hazards increase. On that account for different spatial scales and different production interventions, land suitability needs to be mapped with their sub-classes (i.e., 'e', 'w', 'c', 's').

Simultaneously along with the land-use potential, understanding the water potential of different aquifers is equally important. According to the Aquifer Atlas of India, there are 14 Principal aquifer systems (Fig. 7.2) combined into three major groundwater systems that cover all the states and union territories.

(a) **Unconsolidated formations-alluvial**: These aquifers are spread over Indo-Gangetic Brahmaputra plains, coastal states, and deserts of Rajasthan and Gujrat. They have water holding potential up to 600 m but geographically recharge potential deviates due to differences in rainfall. They have a groundwater potential yield of 10–40 L s^{-1}.
(b) **Consolidated/semi-consolidated formations—sedimentary, basalt, and crystalline rocks**: These aquifers can be seen in the peninsular India. Water availability depends on secondary porosity developed due to weathering and fracturing. Scope for availability at shallow depths (20–40 m) in some areas and deeper depths (100–200 m) in other areas. They have a groundwater potential yield of 1–25 L s^{-1}.
(c) **Hilly**: These are mostly Himalayan crystalline aquifers and have low storage capacity due to the heavy runoff potential of landscapes. Under-confined and unconfined aquifers range around 300 m depth. They have a groundwater potential yield of < 1 L s^{-1}.

Central Ground Water Board, Ministry of Water Resource, Government of India provides further robust methodologies for real-time aquifer mapping applicable in different watershed scales. With the integration of land-use capability and aquifer mapping, critical zones of intervention can be identified that can enhance crop productivity as well as protect blue water levels. Additionally, ensemble groundwater modeling with proper quality filters (Yadav et al. 2020; Lal and Datta 2020) can also be integrated with the aquifer maps for predicting groundwater levels and generating suitability maps for different interventions.

7.2.4 Instrumental Approaches

Aside from the larger area approaches, for micro-level measurements instrumental methods are equally important. Table 7.3 highlights such instruments that can not only collect field-level data for watershed planning but also can work as intervention process control mechanisms.

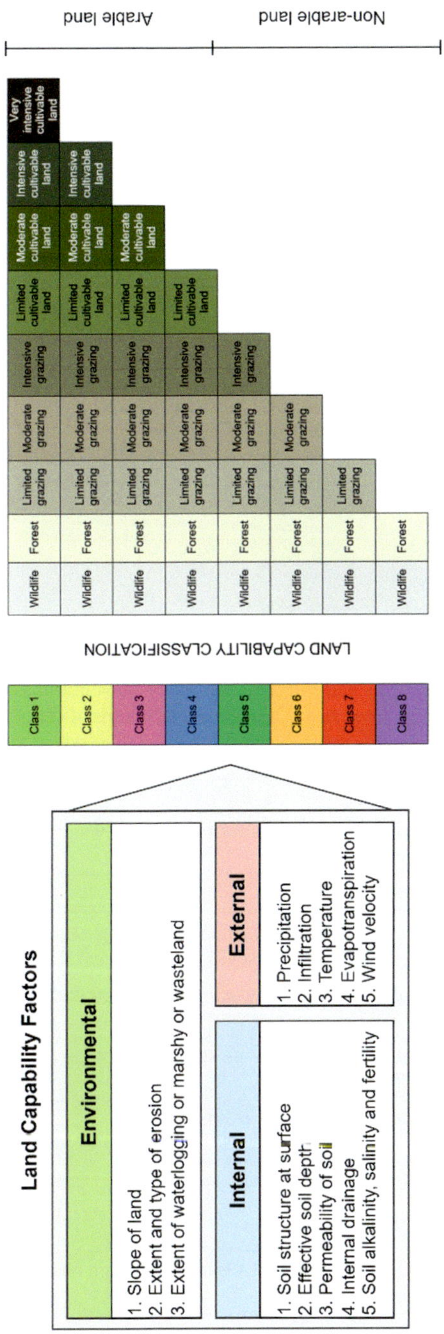

Fig. 7.1 USDA land capability classes and associated factors for classification

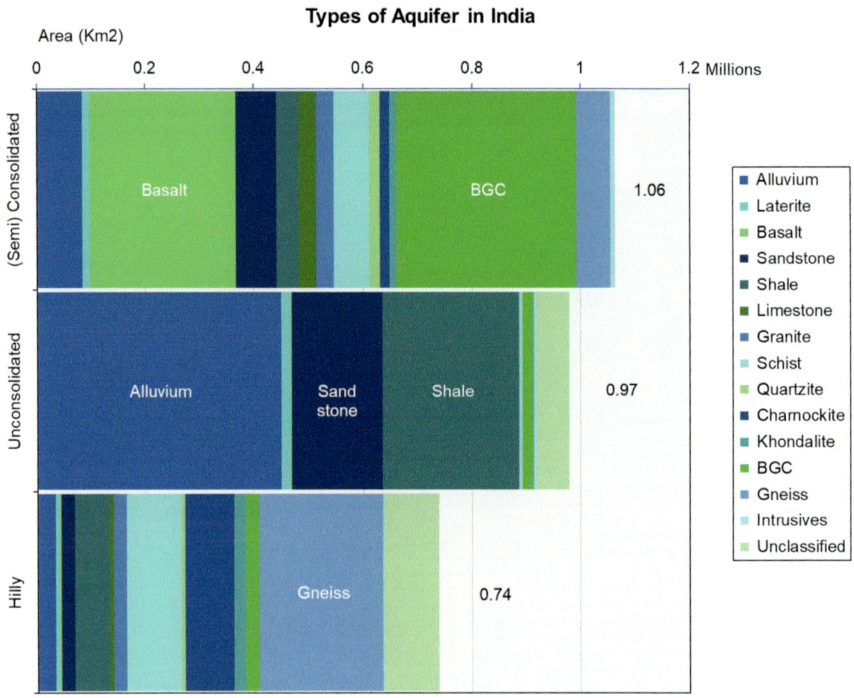

Fig. 7.2 Major groundwater systems and principal aquifer systems in India. *Note* The horizontal axis indicates the area covered by three major groundwater systems of India. Each block in three major groundwater systems highlights the area covered by the 14 principal aquifer systems. *Data Source* India-WRIS (n.d.), Dhiman et al. (2012)

Table 7.3 Field-level data collection instruments for micro-level watershed planning and monitoring

Sl. No.	Instruments	Main purpose
1	Total survey station	Coordinates identification, measuring angle and distance, and processing land surveying data, and aerial photogrammetry
2	Telemetry monitoring	Watershed boundary delineation, Longitudinal resource mapping
3	Digital rainfall recorder	Rainfall recording with standard program and user-friendly software
4	Rainfall event logger	Tracking rainfall over time
5	Digital evaporation recorder	Monitoring water evaporation
6	Digital water level recorder	Monitoring water levels of wells, open channels, dams, lakes, and aquifers

Adapted from Bankey et al. (2021)

7.2.5 *Agroecological Approaches*

Agroecology differs significantly from other approaches to sustainable development. It is founded on bottom-up and territorial processes, assisting in the delivery of contextualized solutions to local problems. The practice of agroecology relies on three major principles: (a) Common characteristics, (b) Context features, and (c) Enabling environment. India as a diverse country not only faces complexity in culture or tradition but also in biophysical characteristics. Based on three parameters, (a) Climatic factors (b) Soil properties, and (c) Physiological settings (topography and drainage) India is divided into 20 agroecological zones.[1] Within those 20 agroecological zones, we will be able to find 127 agro-climatic zones. Through these ecological variables, Indian landscapes can be classified. This landscape-based approach to planning contains these basic parameters:

(a) Geographical area (million ha.)
(b) Gross cropped area (million ha.)
(c) Physiography
(d) Precipitation (mm)
(e) Potential Evapotranspiration (PET) (mm)
(f) Length of Growing Period (days)
(g) Major crops.

This robust classification can play a remarkable role in planning the cropping pattern, irrigation requirement, and land-use structure. Suitable farming systems can also be identified through the inclusion of several Regional Research Stations, Krishi Vigyan Kendras, State Agricultural Universities, and Non-Governmental Organizations.

7.3 Sustainable Intensification and Conservation Measures

After proper planning and targeting, specific intervention methodologies need to be identified. Throughout the years, four types of watershed management measures were identified that can enhance soil and water conservation through in situ and ex-situ practices in different stressed ecologies. These measures (Mello et al. 2021; Mekonnen et al. 2021; Neupane and Guo 2019; Gupta et al. 2018; Fairfax and Russell 2014; Lal 2004) can be broadly understood as (a) Agronomic measures, (b) Mechanical measures, (c) Chemical measures, and (d) Agrostological measures. Furthermore, suitable Agroforestry models also can be included in different terrains and ecologies to control soil erosion and enhance system productivity.

[1] Agroecological zone can be explained as a land unit, derived from climatic zone, correlated with landforms, climate, and the length of growing period (LGP) (Balasubramanian 2013).

7.3.1 *Agronomic Measures*

Cropping system agronomy plays a significant role in soil and water conservation through various practices and techniques that promote sustainable agriculture and land management. These measures are applicable where the slope is gentle ($\leq 2\%$). Some key techniques include:

1. **Conservation Tillage**: Conservation tillage measures such as no-till or reduced tillage, to conserve soil moisture, reduce erosion, and improve soil structure (Busari et al. 2015). By minimizing soil disturbance, conservation tillage methods help support soil integrity and reduce the risk of water runoff and soil erosion.
2. **Water Management**: Agronomy involves the efficient use of green and blue water resources through techniques such as irrigation scheduling, crop water level monitoring, precision irrigation (surface and sub-surface), and water-saving technologies (Unger et al. 2006; Bongiovanni and Lowenberg-Deboer 2004). Besides irrigation technologies, it is important to prioritize the implementation of crop-specific methods like Alternating Wetting and Drying (AWD) tailored to rice cultivation for broader adoption and impact. (Lampayan et al. 2015). By optimizing water use efficiency, agronomy helps minimize runoff and leaching, thus reducing the risk of soil erosion and nutrient loss.
3. **Crop Selection**: For effective conservation, farmers need to select appropriate crop varieties and manage cropping systems to match local soil and climatic conditions. By choosing crops that are well-adapted to the environment, tailored agronomy can reduce soil erosion, improve water infiltration, and enhance overall soil and water conservation efforts. Proper crop selection has a direct influence on green water management and also help in reducing excessive extraction of blue water Furthermore, crops that have co-benefits and marketing flexibility can enhance the value chain.

   Additionally in India, several energy crops are suitable for soil and water conservation while also serving as renewable sources of energy (Blanco-Canqui 2010; Sims et al. 2006). Food grains (maize, sugarcane, sorghum, wheat barley, rice, etc.), fiber crops (agave fiber nes, bast fibers, coir, cotton, Jute, manila hemp fiber, ramie, and sisal, etc.), pulses (bambara beans, carobs, chickpeas, cowpeas, lentils, dry peas, green peas, pigeon peas and pulses nes etc.), oilseeds (castor, groundnut, linseed, oilseed nes, rapeseed, safflower, sesame, soybean and sunflower etc.), and some horticultural crops indirectly produce biofuels, particularly ethanol through processes like combustion, gasification, or anaerobic digestion (Usmani 2020). Additionally, they are well suited for adoption in the Indian landscape due to their long-standing farming tradition, thus they are a suitable trade-off in the water-energy-food nexus (Kholod et al. 2021).
4. **Contour Farming and Terracing**: Precision agriculture promotes the use of contour farming and terracing techniques to control soil erosion on sloping land (Montgomery 2007). By contouring fields along the natural contours of

the landscape and constructing terraces, agronomy helps reduce water runoff, slow down soil erosion, and retain soil moisture.

5. **Intercropping**: Intercropping helps enhance soil fertility, reduce erosion, and improve water retention by maximizing land use efficiency and biodiversity. Different crops with varied root structures and canopy cover can prevent soil erosion, increase soil organic matter, and reduce nutrient leaching (Yin et al. 2020).
6. **Strip cropping:** Strip cropping helps control soil erosion by reducing the runoff velocity, trapping sediment, and modifying soil structure (Unger et al. 1991). The alternating crop strips also promote biodiversity and habitat for beneficial insects and wildlife.
7. **Cover cropping**: Cover crops help control soil erosion by reducing water runoff, accelerating infiltration, and enhancing soil structure (Novara et al. 2021). They also suppress weed growth, reduce nutrient leaching, and promote beneficial microbial activity in the soil.
8. **Natural farming**: Natural farming practices promote soil health and water conservation by enhancing soil structure, nutrient cycling, and biological diversity (Sanathara and Vibhute 2020). By avoiding harmful chemicals, natural farming helps preserve soil fertility, protect water quality, and maintain ecosystem balance within watersheds.
9. **Regenerative agriculture**: Regenerative agriculture practices, such as no-till farming, agroforestry, rotational grazing, and diversified cropping systems, help build soil organic matter, increase infiltration, and mitigate soil erosion. By improving soil structure and water retention capacity, regenerative agriculture contributes to watershed health and sustainability (LaCanne and Lundgren 2018).
10. **Crop rotation**: Crop rotation enhances soil health and water conservation by diversifying plant root systems, nutrient requirements, and pest pressures (El-Bably et al. 2015). It helps maintain soil structure, reduces soil erosion, and improves water retention, leading to more resilient and sustainable agricultural systems within watersheds.
11. **Agroforestry and Windbreaks**: Integrating trees and shrubs into agricultural landscapes through agroforestry and windbreak planting can be one of many effective solutions. Trees and windbreaks help reduce wind erosion, stabilize soil, and conserve water by providing shade, reducing evaporation, and improving soil structure (Oribi and Abdulkareem 2020). Furthermore, agroforestry has a major co-benefit of carbon sequestration (Ghale et al. 2022).

Overall, these agronomic practices play a crucial role in managing soil and water resources through sustainable agriculture practices therefore mitigating the environmental impacts of agricultural activities and ensuring the long-term viability of cropping systems.

7.3.2 Mechanical Measures

In the context of watershed management, mechanical measures refer to engineering or physical interventions intended to control water flow, reduce erosion, preserve soil, and water resources, and enhance the overall health of the watershed. These measures usually entail installing structures or implementing practices that modify the hydrological dynamics of the terrain. These measures are applicable where the slope is steep (> 2%). Some major interventions include

1. **Contour Bunding**: Contour bunding is the process of constructing earthen embankments along the contour lines of sloping land. These bunds resist surface water runoff, allowing it to infiltrate the soil and reducing soil erosion. These bunds can effectively trap sediment and help retain moisture (Walia et al. 2024).
2. **Graded Bunding**: Graded bunding is similar to contour bunding but here bunds are constructed on a graded slope. This method controls runoff by directing it gradually along the slope, minimizing soil erosion and boosting infiltration. This is particularly effective in areas with steep slopes or high rainfall intensity (Walia et al. 2024; Wolka et al. 2018).
3. **Broad Base Terrace**: Broad base terraces are built along the contour lines of hilly terrain to create broad, level platforms. These terraces slow down runoff and allow it to spread across the terrace surface. Broad base terraces are suitable for areas with moderate slopes to not only retain soil moisture but also prevent nutrient loss (Pijl et al. 2022).
4. **Bench Terracing**: Bench terraces are a series of steps or benches along steep hillsides. This method effectively reduces the slope gradient (Tando et al. 2019; Finkel 2018). These terraces provide flat areas for cultivation, thereby increasing arable land and enhancing agricultural productivity in hilly regions. Different types of bench terracing (Table 7.4) can be seen based on slope, rainfall intensity, and soil physical properties (Sharda et al. 2013).
5. **Zingg Terracing**: Zingg terracing is a specialized form of bench terracing that incorporates vegetation strips or grassed waterways between terraces. These are usually implemented in the areas that have a slope of 3–10%. These strips help reduce surface water velocity, trap sediment, and promote infiltration. During medium rainfall years, water-intensive crops like paddy are cultivated in the

Table 7.4 Terrace typology for varied soil and rainfall conditions

Type	Suitability
Level and tabletop	Evenly distributed medium rainfall areas (750 mm) with highly permeable deep soil. E.g., irrigated condition
Sloping outward	Low rainfall areas (< 750 mm) with permeable soil of medium depth. E.g., orchard planting
Sloping inward	High rainfall areas (> 750 mm) with soil of poor infiltration rate. E.g., hilly areas

Adapted from Das and Saikia (2012), Mal (2005)

receiving area while dry crops are cultivated in the donor area (Walia et al. 2024; Finkel 2018).

6. **Trenching**: Trenching means digging small channels or trenches along the contour lines and across the slope. These trenches intercept surface water runoff, allowing it for infiltration and aquifer recharge (Mishra and Tripathi 2013). Commonly in steep slope areas (> 33%) trenching is commonly implemented.
7. **Vegetative Barrier**: Vegetative barriers consist of vegetative cover such as grass, shrubs, or trees planted along the contours of agricultural fields. These barriers help stabilize runoff and soil erosion but also enhance biodiversity and provide additional benefits such as wildlife habitat and carbon sequestration.
8. **Grassed Waterways**: Grassed waterways are natural or engineered channels planted with grass or other vegetation to channel water runoff safely across agricultural fields.
9. **Gully Control**: Gullies are deep erosional channels caused by concentrated water runoff. Gully plugs can be built with ground wooden poles, stones, sandbags, or cementing materials that can be employed to stabilize gully edges, control water flow, and restore degraded areas (Frankl et al. 2021).
10. **Check Dam**: Check dams, whether permanent or temporary, are small masonry overflow weirs built perpendicular to the water flow direction in natural or artificial drainage systems, and across seasonal streams or gullies to decelerate water flow. Besides reducing soil erosion, these dams recharge the groundwater storage by reviving dug wells on the downstream side (Gao et al. 2020). To what extent a check dam will contribute to groundwater recharge will depend upon the natural recharge potential of the stream.

7.3.3 Chemical Measures

Often, different amendments, conditioners, and stabilizers are used for improving soil physical parameters, and modulating soil compaction and water quality enhancement. For soil stabilization, Polyacrylamide (PAM) can be used temporarily as it binds soil particles together which reduces runoff (Georgees et al. 2018). To control sedimentation, amendments such as flocculants can be applied to sediment-laden water that accelerates particle aggregation and settling. Chemicals like lime or sulfur can be used to treat acidic or alkaline soils to adjust the pH levels. This can minimize soil erosion and create a favorable environment for plant growth and microbial activity. To improve water quality and the health of ecosystems, chemical treatments like algaecides and biocides can be administered to water bodies to regulate algal blooms or invasive aquatic species. To stop soil erosion, slopes, and embankments can be stabilized using asphalt and bitumen-based compounds (Andavan and Kumar 2020; Balbahaith 2013). These substances combine to create strong, erosion-resistant surfaces that shield watersheds' most susceptible spots from deterioration. Herbicides based on asphalt and bitumen are used to suppress invasive plant species that might cause soil erosion and sedimentation by managing the vegetation along roads and streams.

Although chemicals can be effective tools for conserving water and soil, their application needs to be carefully controlled to reduce any potential harm to the ecosystem or non-target organisms. Sustainable watershed management requires appropriate application methods, dosage regulation, and attention to regional laws and environmental concerns.

7.3.4 Agrostological Measures

Agrostology, the branch of systematic botany dealing with the study of grasses, plays a crucial role in watershed management. Grasses are foundational components of many ecosystems and are integral to maintaining soil stability, preventing erosion, and promoting water infiltration. Here are some agrostological measures relevant to watershed management:

1. **Grass Species Selection**: Choosing appropriate grass species is essential for watershed management. Native grass species are often preferred because they are adapted to local soil and climate conditions, and they typically have deep root systems that help prevent soil erosion and promote water infiltration.

   In India, several grassland species are recommended for watershed management, but their suitability may vary depending on factors such as climate, soil type, and local ecological conditions. In general species like native grasses (*Vetiveria zizanioides, Cynodon dactylon*), fodder grasses (*Pennisetum purpureum, Sorghum spp.*), leguminous grasses (*Desmanthus spp., Stylosanthes spp.*), bamboo species (*Bambusoideae*), and other mixed grass species.
2. **Grassland Restoration**: Restoring degraded grasslands within a watershed can significantly improve watershed health. Restorative measures may include reseeding native grass species, controlling invasive species, and implementing grazing management practices to promote healthy grass growth.
3. **Riparian Buffer Zones**: Establishing riparian buffer zones composed of grasses and other vegetation along water bodies within a watershed helps filter sediment and pollutants, stabilize stream banks, and provide habitat for wildlife (Cole et al. 2020). Agrostological measures in riparian buffer zones may involve planting native grass species with deep, fibrous root systems to enhance soil stability and water absorption.
4. **Grassland Conservation Practices**: Implementing conservation practices such as rotational grazing, prescribed burning, and managed haying can help in maintaining healthy grasslands within a watershed. These practices promote biodiversity, reduce soil erosion, and improve water quality by enhancing the resilience of grassland ecosystems.

These agronomic, chemical, mechanical, agrostological, and agroforestry measures are often coupled together for integrated watershed management which can mitigate land degradation, reduce sedimentation in water bodies, and enhance overall watershed resilience.

7.4 Measuring and Managing Water Usage

In ensuring the success and subsistence of a watershed area, the outputs and internal processes should be monitored. Consequently, water economics plays a crucial role in this concern. Majorly three classical mechanisms are used to ensure water productivity, efficient water use, and associated benefits i.e., Water Budgeting, Project Valuation, and Water Use Efficiency. A water budget indicates that the rate of change in the watershed storage is balanced by the rate at which water flows into and out of the area. Thus, in water budgeting five major components are used to understand the balance, i.e., Precipitation (P), Water in Flow (Q_{in}), Change in Storage (ΔS), Evapotranspiration (ET), and Water out Flow (Q_{out}). Several water budgeting equations were developed beforehand in hydrologic studies, but if large scale water budgeting (Healy et al. 2007) is conducted then the measurement technique can be expressed as:

$$P + Q_{in}^{sw} + Q_{in}^{gw} = ET^{sw} + ET^{gw} + ET^{uz} + \Delta S^{sw} + \Delta S^{snow} + \Delta S^{uz} + \Delta S^{gw} + Q_{out}^{gw} + RO + Q^{bf}$$

where the superscripts indicate Surface Water (sw), Ground Water (gw), Unsaturated Zone (uz); RO is Surface Runoff; Q_{out}^{gw} means both Ground Water Flow out of the site and any withdrawal by pumping; and Q^{bf} is Base Flow (ground-water discharge to streams). It is unusual that every component of the equation will be significant at every case; some will be insignificant and can be disregarded. Indeed, careful boundary selection can help streamline the accounting process when choosing an accounting unit for creating a reliable water budget. Furthermore, to conduct water budgeting in soil-specific scenarios, the equation provided by Brye et al. (2000) can be used. The equation is expressed as:

$$ET = P - RO - D - \Delta S$$

where, P = Precipitation, RO = Surface Runoff, D = Drainage Below the Root Zone and ΔS = Change in Soil and Surface Water Storage.

On the other hand, for smaller-scale budgeting Lysimeters[2] can be incorporated (Brown et al. 2021). Additionally in India participatory water budgeting is one of the major highlights that not only encompasses anthropogenic factors but further contributes towards the community empowerment and ownership development of watershed management activities.

[2] Lysimeters comprise containers filled with soil, that are hydrologically isolated from the surrounding environment but mimic the hydrologic behavior of that environment (Fank 2011).

Additionally, in water economics, monetization of project inputs and outputs is necessary to understand the welfare aspects of watershed development processes. These evaluation techniques use a cost benefit analysis followed by some Multi-Criteria Decision-Making models.

In majority of cases, either monsoon deficiency or monsoon surplus is eventually connected to all water use inefficiency issues. Mostly people tend to rely only on mechanical practices to conserve water. Consequently, it turns out not to be cost-effective and universally manageable. For enhancing Water Use Efficiency (WUE), agronomic practices should be the base area of intervention. Previously several organizations like ICRISAT, NABARD, etc. have worked on the rainwater usage in agriculture, water harvesting and water productivity. As regionally, water cover percentage, chances of flood, and drought exist at various levels, universal cropping pattern will fail immensely.

As a solution, some measurement mechanism is needed that can from about optimum water requirement at field level. In different scenarios water and crop-associated indices i.e., Irrigation Efficiency index (IE), Water Conveyance Efficiency index (WCE), Water Application Efficiency index (WAE), Crop Water Use Efficiency index (CWUE), etc. can be used widely, whereas to measure water usage in a macro perspective recently NITI Aayog (2018) has provided a Composite Water Management Index which can be used as an efficient water management tool. To do so, they have identified nine major thematic areas (Table 7.5) and provided respective weights based on their importance. Then they were subdivided into 28 detailed thematic areas for deeper analysis.

Table 7.5 Indicator themes and weights

No	Themes	Weights (W_i)
1	Source augmentation and restoration of waterbodies	5
2	Source augmentation (Groundwater)	15
3	Major and medium irrigation—Supply side management	15
4	Watershed development—Supply side management	10
5	Participatory irrigation practices—Demand side management	10
6	Sustainable on-farm water use practices—Demand side management	10
7	Rural drinking water	10
8	Urban water supply and sanitation	10
9	Policy and governance	15
	Total	100

Source NITI Aayog (2018)

It is important to emphasize the unique process of data gathering for preparing the index. Apart from collecting data on multiple indicators for the first time, it also required significant collaboration and coordination within states and between central and state authorities.

For calculating the index, indicators were categorized into *positive* and *negative indicators* which are supposed to be calculated using two different formulas, i.e.,

To standardize values for positive indicators (those where higher values indicate better performance), we applied the following formula. Once scaled, the values are ranged from 0 to 1, with the top-performing state assigned a value of 1 and the lowest-performing state assigned a value of 0.

$$Scale\ value\ of\ positive\ indicator(S_i) = \frac{X_i - minimum\ value}{maximum\ value - minimum\ value} \tag{7.1}$$

Likewise, for negative indicators (where lower values indicate better performance), the scaled values are determined using the following method.

$$Scale\ value\ of\ negative\ indicator(S_i) = \frac{Maximum\ value - X_i}{maximum\ value - minimum\ value} \tag{7.2}$$

Following scaling, values were distributed within a range of 0 to 1, with the highest-performing state assigned a value of 1 and the lowest-performing state assigned a value of 0. Additionally, for binary indicators, a 'Yes' was allocated a score of 1, while a 'No' received a score of 0.

Subsequently, based on the indicators' respective weights, a Composite Index was computed for each state for the base year (FY XX-XX) and FY XX-XX using the subsequent formula:

$$Composite\ Index = \frac{\sum (W_i \times S_i)}{\sum W} \tag{7.3}$$

To determine the weight of an indicator, the weight assigned to a theme was evenly distributed among its individual indicators. Calculating scores for the two years allowed for monitoring and comparing state-level performance over time.

Nevertheless, there are other indices which can be used for measuring water usage in a specific area. For calculating these indices, collection of water flow data is one of the base requisites. That data can be collected using localized methods, buckets, irrigation weirs, Parshall flumes, water flow meters, cut-throat flumes, etc. can be used for canal irrigation whereas, notches or clips can be used to measure water usage from drip or sprinkler irrigation.

7.5 Scaling and Value Chain Development

Enhancing market visibility and financial returns in India's watershed management relies on effective scaling and promotional methods. Emphasizing digital platforms and quality-driven promotional campaigns, especially highlighting sustainable practices, is crucial. Farmer Producer Organizations (FPOs) are pivotal in adopting modern marketing approaches, collaborating with e-commerce, participating in exhibitions, and establishing common facility centers. These strategies aim to uplift market presence, strengthen value chains, and contribute to the economic prosperity of farmers in watershed areas (Rafi et al. 2023; Kumari et al. 2021; Prasad 2019; Singh et al. 2018). This strategic paradigm transcends conventional agricultural practices and water resource management, encapsulating the entirety of the production-to-consumption continuum. Foundational to value chain development are two primary pillars: institutional engagement and infrastructure enhancement. Table 7.6 provides and outline of these two pillars in the Indian context, for propelling value chain dynamics within watershed management regions.

In the dynamic landscape of watershed management areas in India, scaling and promotional strategies for agricultural products are intricately linked to the widespread adoption of AGROMET advisories. The integration of weather forecasts and climate information into farming practices becomes a pivotal component of alternative farming strategies. Through mobile applications like Skymet and AgriMetSoft, farmers gain valuable insights into optimal planting times, irrigation scheduling, and crop selection based on climatic conditions (Singh et al. 2023; Sikka et al. 2022; Singh et al. 2020; Pandey and Singh 2019; Palsaniya et al. 2016).

Seed villages and multiplication programs are one of the effective scaling and promotional strategies in India which not only ensures the availability of quality seeds for stress-tolerant crop varieties but also promotes community engagement and empowerment. To strengthen different alternative agricultural practices geared towards conserving soil and water, harnessing seed villages emerges as a pivotal strategy. This approach not only alleviates the burden on the system by ensuring timely seed production and distribution but has also proven to be invaluable across diverse cropping systems (Srinivas et al. 2010; Reddy et al. 2007). FPOs are critical in popularizing and implementing these programs, fostering collective efforts in seed production and distribution. Collaborations with organizations like the National Seeds Corporation (NSC) further enhance the reach and impact of seed village initiatives (Donovan et al. 2021). Through targeted promotional campaigns and farmer training programs, the benefits of using locally adapted seeds are emphasized, contributing to increased crop productivity and resilience in watershed areas.

Government and non-government organizations working together has an impact on how agricultural commodities are marketed and scaled in India's watershed management areas. The National Rural Employment Guarantee Act (NREGA) and watershed improvement projects in Karnataka are a perfect example of how they work together. Government-driven convergence like this improves the resilience of

Table 7.6 Value chain interventions for effective watershed management in Indian agro-ecologies

Pillars		Functions
Institutional Involvement	Government agencies	Formulate and implement policies supporting sustainable farming practices to ensure efficient water use and value chain development in watershed areas
	Research and educational institutions	• Conduct research on crop varieties, farming techniques, and value-addition methods • Bridge the gap between research institutions and farmers, disseminating information on best practices and market trends
	Banks and financial institutions	Provide loans and financial support to farmers and entrepreneurs involved in value chain activities
	Non-governmental organizations (NGOs)	• Engage in capacity building and awareness programs for farmers, promoting sustainable agricultural practices • Facilitate collective action among local communities, promoting cooperative models for value addition
	Agri-Tech startups and incubators	Introduce innovative technologies for efficient farming and value addition
	Market intermediaries	Work as intermediaries between farmers and markets, negotiating better prices and ensuring fair returns
	National watershed development project (NWDP)	Focuses on sustainable development in watershed areas, incorporating agriculture and livelihood enhancement components
Infrastructure development	Road and transport	Enhances accessibility for farmers and producers to reach markets, reducing post-harvest losses
	Market infrastructure	Cold storage, warehousing facilities, and modern market yards to facilitate quality preservation and transparent transactions
	Processing units and technology hubs	Enable farmers to add value to their produce, creating employment opportunities and generating higher returns
	Water management infrastructure	Efficient irrigation systems and water harvesting structures ensure optimal water use and mitigate water scarcity
	Training and skill development centers	Equip farmers with modern farming techniques, enhancing their participation in value-chain activities
	Financial infrastructure	Accessible banking and credit facilities support investments in value addition activities and technology adoption
	Digital infrastructure	Market information systems enable farmers to make informed decisions about their produce

(continued)

Table 7.6 (continued)

Pillars		Functions
	Community infrastructure	Common facility centers promote collaboration among farmers, reducing costs and improving efficiency
	Quality certification and testing labs	Ensure that products meet required standards, enhancing consumer confidence and market acceptance

rural communities, integrates sustainable agriculture, and conserves land and water. Moreover, NGOs are essential to the development of these cooperative projects.

7.6 Project Cycle Management

For the new generation watershed management is a dynamic framework is required that changes with time to accommodate different demands and situations. Appraisal, planning, implementation, and evaluation are its four main stages. Although these stages are usually portrayed chronologically, they are convolutedly linked together, creating an ongoing cycle of actions. It can be illustrated more like a spiral staircase, where advancement and progress are continuous since one step informs the next. Stakeholders must be actively involved in this process from the beginning. Their participation is essential to the success and long-term viability of the management initiatives—it is not merely a formality. Through the integration of varied areas of knowledge, stakeholders play a crucial role in establishing agreement, obtaining backing, and cultivating a feeling of control over the results.

7.6.1 Critical Stakeholders

As the implementation of watershed programs holds severe importance, multilevel involvement of different stakeholders and nodal agencies is required. In India Department of Land Resources, the Government of India has already mobilized for an institutional arrangement at the National, state, district, and project levels (DoLR 2021).

7.6.1.1 National Level

At the national level, there are mainly ministry-level arrangements and National level data portals are the significant focal points that serve the purpose of project development, and supervision. Major ministry-level arrangements include the National Level

Nodal Department (NLND) to maintain budgetary and non-budgetary resources, the National Steering Committee (NSC) will work on appraisal, inter-state fund allocation, co-ordination, communicating Best Management Practices (BMPs), modification of physical targets and project review. National Level Nodal Agency (NLNA) will work on project Management Information Systems, interact with State Level Nodal Agencies, capacity building, building and sharing knowledge repositories, etc. Additionally, the National Data Center (NDC) is mandated to compile, store, and analyze project-related and land resource information.

7.6.1.2 State Level

At state level, major three focal points are serving for development, implementation, and monitoring of watershed projects in rainfed and irrigated geographies of India i.e., (a) State Level Sanctioning Committee (SLSC) for strategy development and supervision of watershed development projects within the States/UTs., (b) State Level Nodal Department (SLND) is responsible for coordinating with the SLSC, setting up and supporting SLNA, to clarify and implement guidelines as and when required, to ensure timely fund availability for the SLNA, as well as coordinating with NLNA., and (c) State Level Nodal Agency (SLNA) reviews programs and provide enabling mechanism to set up State level data cell and ensure regular reporting to the NLNA/NLND.

7.6.1.3 District Level

Watershed Cell cum Data Centers (WCDC) are situated at district level to manage watershed development projects during implementation. The data centers should have a team of trained professionals.

7.6.1.4 Project Level

At the grassroots level, several formal and informal institutions support is needed for effective implementation, i.e., (a) Project Implementing Agency (PIA) may include relevant line Departments, autonomous organizations under State/Central Governments, Government Institutes/Research Bodies, Panchayats, Voluntary Organizations (VOs), development agencies set up by Private Trusts and Industries to operate their Corporate Social Responsibility (CSR) activities and the like. (b) The Watershed Development Team (WDT) should be present closest to the target areas. They are intended to collaboratively work with a team of experts/professionals at the district and State levels. (c) Peoples' Institutions: These institutions include but are not limited to Gram Sabha (GS), Watershed Committee (WC), and Water User Groups (WUGs).

7.6.2 Watershed Project Cycle

In the domain of agriculture and hydrology, the intricate planning and implementation activities inherent in watershed management demand comprehensive monitoring and evaluation over specific periods. Assessments such as Ex-Ante, Ex-Post, Process Evaluation, and Impact Evaluation are indispensable to sustain the multifaceted layers of work plans. These evaluations serve as crucial tools to logically frame the technologies and activities aimed at enhancing production systems within agricultural watersheds. Figure 7.3 illustrates the overlapping and interdependent nature of these activities. It highlights the intricate relationships and dependencies among various components of watershed management.

By elucidating Fig. 7.3, we see that different activities within watershed management overlap and are mutually reliant. Their integration is vital for achieving the overarching goal of improving production systems in agricultural watersheds. Furthermore, a diverse array of stakeholders, including scientists, engineers, policymakers, administrators, corporations, NGOs, and village-level institutions, must actively engage in the planning and evaluation processes. This multi-stakeholder engagement fosters a holistic and practical pathway for the development of next-generation watersheds. Through collaborative efforts and inclusive decision-making processes, stakeholders can effectively address the complexities and challenges associated with watershed management. By leveraging their collective expertise and resources, they can devise sustainable solutions that optimize agricultural productivity, enhance water resource management, and promote ecosystem resilience within watershed areas.

7.6.2.1 Planning, Appraisal and Design

In this stage the preparatory work is mostly conducted that formulates the project blueprint. This stage commonly includes (a) baseline study where to understand the current state of agricultural practices, environmental conditions, and socio-economic factors within the target watersheds across the Indian subcontinent. (b) Stakeholder Assessments are further conducted to identify and analyze key stakeholders' interests, priorities, influence, and potential contributions to the project. (c) Problem Identification is conducted to identify the gap to be addressed. (d) Based on the problem statements, Specific, Measurable, Achievable, Relevant, and Time-Bound objectives are developed project through collaborative discussions with stakeholders. (e) Development of Theory of Change-Impact Pathways: following the objectives, a logical framework, activity monitoring indicators, theory of change is developed to clearly envision the impact pathways. (f) Feasibility Analysis and Appraisal: here stakeholders engage and co-analyze the technical, financial, environmental, and socio-economic feasibility of proposed interventions. (g) Finally, a detailed budget is prepared where allocation, stakeholder inputs, risks, and costs are clearly mentioned.

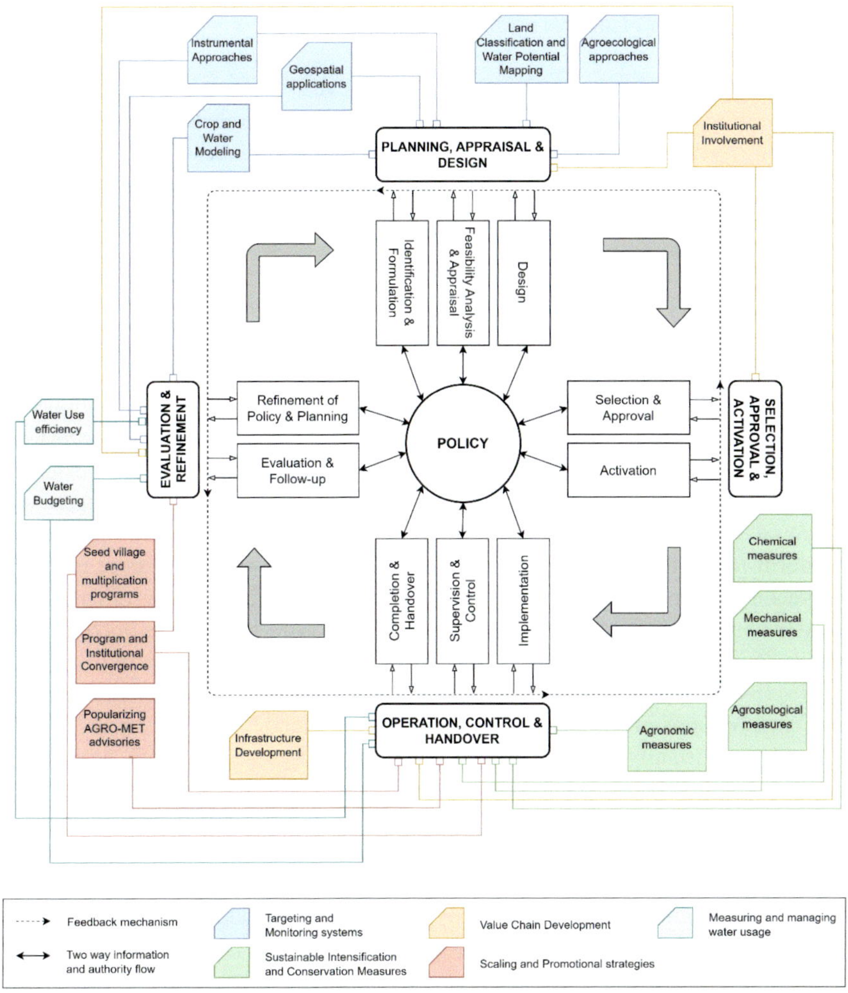

Fig. 7.3 Integration of improved production system technologies in project life cycle

7.6.2.2 Selection, Approval and Activation

In this stage, the stakeholders go through an iterative process of developing required documentation and revision of the project plans. After rigorous discussion with the stakeholders, the plan is finalized and approved for implementation.

7.6.2.3 Operation, Control and Handover

This is a substantial component of the project cycle itself, therefore it takes the longest time and resources. All the implementation activities based on targeting and monitoring systems, sustainable intensification and conservation measures, water measurement and management, value chain development and finally scaling and promotion is taken care of. Activities are properly tracked and monitored using SMART objectives and indicators developed during project design.

7.6.2.4 Evaluation and Refinement

This is the reflective stage of the project where comprehensive evaluations are conducted after the completion of watershed management activities to assess the overall effectiveness, efficiency, and sustainability of interventions. The extent to which project objectives were achieved and identify factors contributing to success or failure are assessed here which also includes gathering feedback from stakeholders to inform future decision-making processes. At distinct stages and time intervals, different evaluations are conducted but most commonly (a) processes are evaluated by assessing the efficiency of project activities, identifying bottlenecks or challenges, and implementing corrective measures as needed to ensure smooth progress. In the end, (b) Impact Assessment is conducted to measure the effect of watershed management interventions in the long run on key socio-economic and environmental indicators. Here normally changes in land use patterns, water availability, agricultural productivity, livelihoods, and ecosystem health are assessed over time.

In conclusion, the amalgamation of monitoring, evaluation, and stakeholder involvement is pivotal for the effective execution of watershed management activities. It establishes the foundation for nurturing innovation, sustainability, and adaptability within agricultural environments, thereby facilitating the enduring welfare of farming communities and the conservation of ecological resources.

7.7 Discussion

The proposed framework integrates agricultural, engineering, and economic components necessary for effectively execute and monitor watershed management projects throughout the Indian subcontinent. GIS and RS technologies are highlighted for their crucial role in initial planning at different spatial levels. These tools provide up-to-date spatial information on natural resources and terrain parameters, essential for developing comprehensive watershed monitoring systems. Satellite-based remote sensing offers valuable data on land use/land cover changes, aiding in planning and decision-making processes. Additionally, there is immense significance of crop simulation and watershed models in understanding different cropping and hydrologic systems and their long-term environmental and economic impacts. The integration of these models is intended to provide a systemic approach to evaluating interventions and assessing risks more efficiently than traditional methods. Agricultural and conservation measures are one of the key components for enhancing soil and water conservation across diverse ecological scenarios. Agronomic, Mechanical, Chemical, and Agrostological practices, along with agroforestry models, can effectively contribute to soil erosion control and system productivity enhancement. Efficient water management is one of the most critical precision agriculture indicators for watershed sustainability. Water budgeting, project valuation, and water use efficiency are found as classical mechanisms to optimize water productivity. Recent initiatives like the Composite Water Management Index provide comprehensive tools for macro-level water management. The Farm gate interventions with proper value addition measures and effective scaling and promotional strategies, market visibility, and adoption potential can be enhanced, and financial returns can be assured in watershed management. Digital platforms, FPOs, and collaborations with governmental as well as non-governmental organizations are identified as key strategies for promoting sustainable practices and strengthening value chains.

Multilevel involvement of stakeholders and nodal agencies is also emphasized for the successful implementation of watershed programs. Institutional arrangements at national, state, district, and grassroots levels are outlined, delineating the roles and responsibilities of different stakeholders. Furthermore, comprehensive monitoring and evaluation mechanisms are indispensable for sustaining watershed management efforts. Various assessment methodologies, including ex-ante, ex-post, process, and impact evaluations, play a crucial role in guiding future interventions and enhancing agricultural productivity within watersheds. The multilevel stakeholder involvement will provide a strong MRV mechanism through several exercises including stakeholder mapping, capacity building, developing data sharing collection and sharing protocols, and designing adaptive management as well as reporting systems, etc. will be required that can enrich the soil and water conservation activities by providing a transparent, precise, effective, and reliable project management system.

7.8 Conclusion

Our approach to watershed management is based on the concepts of sustainability, diversity, and replicability. By carefully combining engineering solutions with crop-based techniques, we provide a comprehensive strategy that tackles the various aspects of watershed management. Drawing from an extensive review our methodology underscores the paramount importance of technological advancements alongside multi-stakeholder strategies. Sustainable livelihoods within diverse farming systems hinge upon the constructive interaction of institutions, schemes, and stakeholders within a cohesive project-based ecosystem. By harmonizing biophysical, geoscientific, and socio-economic approaches, we foster an interdependent relationship that catalyzes the transformation of watershed projects from vision to reality.

References

Aayog N (2018) Composite water management index: a tool for water management. http://archive.nyu.edu/handle/2451/42272

Al-Hamdan OZ, Hernandez M, Pierson FB, Nearing MA, Williams CJ, Stone JJ, Boll J, Weltz MA (2015) Rangeland hydrology and erosion model (RHEM) enhancements for applications on disturbed rangelands. Hydrol Process 29(3):445–457. https://doi.org/10.1002/hyp.10167

Allan JD (2004) Landscapes and riverscapes: the influence of land use on stream ecosystems. Annu Rev Ecol Evol Syst 35(1):257–284. https://doi.org/10.1146/annurev.ecolsys.35.120202.110122

Alqurashi A, Kumar L (2013) Investigating the use of remote sensing and GIS techniques to detect land use and land cover change: a review. https://doi.org/10.4236/ars.2013.22022

Andavan S, Kumar BM (2020) Case study on soil stabilization by using bitumen emulsions–a review. Mater Today: Proc 22:1200–1202. https://www.sciencedirect.com/science/article/pii/S221478531934132X

Asseng S, Ewert F, Rosenzweig C, Jones JW, Hatfield JL, Ruane AC, Boote KJ, Thorburn PJ, Rötter RP, Cammarano D (2013) Uncertainty in simulating wheat yields under climate change. Nat Clim Change 3(9):827–832. https://www.nature.com/articles/nclimate1916

Balasubramanian A (2013) Agro-ecological zones of India

Balbahaith FAS (2013) Evaluation of sulfur modified foamed asphalt for local soil stabilization. PhD thesis, King Fahd University of Petroleum and Minerals (Saudi Arabia). https://search.proquest.com/openview/a4ce5d4987f4f68020dd8d33793dbfd2/1?pq-origsite=gscholar&cbl=2026366

Bankey B, Renuka R et al (2021) Integrated watershed management. National Institute of Agricultural Extension Management (MANAGE) and ICAR—Indian Institute of Soil and Water Conservation (ICAR-IISWAC)

Barfield BJ, Hayes JC, Stevens E, Harp SL, Fogle A (2010) SEDIMOT III model. In: Watershed models. CRC Press

Bassu S, Brisson N, Durand J, Boote K, Lizaso J, Jones JW, Rosenzweig C, Ruane AC, Adam M, Baron C, Basso B, Biernath C, Boogaard H, Conijn S, Corbeels M, Deryng D, De Sanctis G, Gayler S, Grassini P, Waha K (2014) How do various maize crop models vary in their responses to climate change factors? Glob Change Biol 20(7):2301–2320. https://doi.org/10.1111/gcb.12520

Birkinshaw SJ, James P, Ewen J (2010) Graphical user interface for rapid set-up of SHETRAN physically-based river catchment model. Environ Model Softw 25(4):609–610. https://doi.org/10.1016/j.envsoft.2009.11.011

Blanco-Canqui H (2010) Energy crops and their implications on soil and environment. Agron J 102(2):403–419. https://doi.org/10.2134/agronj2009.0333

Bongiovanni R, Lowenberg-Deboer J (2004) Precision agriculture and sustainability. Precision Agric 5(4):359–387. https://doi.org/10.1023/B:PRAG.0000040806.39604.aa

Brown S, Wagner-Riddle C, Debruyn Z, Jordan S, Berg A, Ambadan JT, Congreves KA, Machado PVF (2021) Assessing variability of soil water balance components measured at a new lysimeter facility dedicated to the study of soil ecosystem services. J Hydrol 603:127037. https://www.sciencedirect.com/science/article/pii/S0022169421010878

Brye KR, Norman JM, Bundy LG, Gower ST (2000) Water-budget evaluation of prairie and maize ecosystems. Soil Sci Soc Am J 64(2):715–724. https://doi.org/10.2136/sssaj2000.642715x

Busari MA, Kukal SS, Kaur A, Bhatt R, Dulazi AA (2015) Conservation tillage impacts on soil, crop and the environment. Int Soil Water Conserv Res 3(2):119–129. https://doi.org/10.1016/j.iswcr.2015.05.002

Cole LJ, Stockan J, Helliwell R (2020) Managing riparian buffer strips to optimise ecosystem services: a review. Agric Ecosyst Environ 296:106891. https://www.sciencedirect.com/science/article/pii/S0167880920300761

Crawford NH, Linsley RK (1966) Digital simulation in hydrology: Stanford watershed model IV. Department of Civil Engineering, Stanford University

Daniel EB (2011) Watershed modeling and its applications: a state-of-the-art review. Open Hydrol J 5(1):26–50. https://doi.org/10.2174/1874378101105010026
Das MM, Saikia MD (2012) Watershed management. PHI Learning Pvt. Ltd.
da Silva RM, Santos CAG, dos Santos JYG (2018) Evaluation and modeling of runoff and sediment yield for different land covers under simulated rain in a semiarid region of Brazil. Int J Sedim Res 33(2):117–125. https://doi.org/10.1016/j.ijsrc.2017.04.005
Dhiman SC, Kumar S, Sinha SK, Chandra R, Kujur A, Sahoo BK, Dhayamalar D, Rani VR, Swaroop SK, Verma JR (2012) Aquifer systems of India. Central Ground Water Board, Ministry of Jal Shakti, Department of Water Resources River Development and Ganga Rejuvenation, Government of India. https://cgwb.gov.in/cgwbpnm/public/uploads/documents/168741951268 0023437file.pdf
DoLR (2021) Guidelines for new generation watershed development projects (WDC-PMKSY 2.0) | Department of Land Resources | Ministry of Rural Development | Govt. of India. https://dolr.gov.in/en/document/guidelines-new-generation-watershed-development-projects-wdc-pmksy-20
Donovan J, Rutsaert P, Spielman D, Shikuku KM, Demont M (2021) Seed value chain development in the Global South: key issues and new directions for public breeding programs. Outlook Agric 50(4):366–377. https://doi.org/10.1177/00307270211059551
Downer CW, Ogden FL (2004) GSSHA: model to simulate diverse stream flow producing processes. J Hydrol Eng 9(3):161–174. https://doi.org/10.1061/(ASCE)1084-0699(2004)9:3(161)
El-Bably AZ, Abd El-Hafez SA, Mahmoud MA, Oud SA (2015) A new conceptual framework for water conservation based on addressing water balance, crop rotation and economics. Int J Water Resources Arid Environ 4(2):120–127. https://www.psipw.org/attachments/article/375/4e.pdf
Elliott J, Deryng D, Müller C, Frieler K, Konzmann M, Gerten D, Glotter M, Flörke M, Wada Y, Best N, Eisner S, Fekete BM, Folberth C, Foster I, Gosling SN, Haddeland I, Khabarov N, Ludwig F, Masaki Y, Wisser D (2014) Constraints and potentials of future irrigation water availability on agricultural production under climate change. Proc Natl Acad Sci 111(9):3239–3244. https://doi.org/10.1073/pnas.1222474110
Ewen J, Parkin G, O'Connell PE (2000) SHETRAN: distributed river basin flow and transport modeling system. J Hydrol Eng 5(3):250–258. https://doi.org/10.1061/(ASCE)1084-0699(2000)5:3(250)
Fairfax SK, Russell E (2014) Guide to U.S. environmental policy. CQ Press
Fank J (2011) Lysimeters: a tool for measurements of soil fluxes. In: Gliński J, Horabik J, Lipiec J (eds) Encyclopedia of agrophysics. Springer Netherlands, pp 428–431. https://doi.org/10.1007/978-90-481-3585-1_85
Finkel HJ (2018) Engineering measures for soil and water conservation: terracing and benching. In: Semiarid soil and water conservation. CRC Press, pp 85–91. https://doi.org/10.1201/9781351076586-7/engineering-measures-soil-water-conservation-terracing-benching-herman-finkel
Flotemersch JE, Leibowitz SG, Hill RA, Stoddard JL, Thoms MC, Tharme RE (2016) A watershed integrity definition and assessment approach to support strategic management of watersheds. River Res Appl 32(7):1654–1671. https://doi.org/10.1002/rra.2978
Frankl A, Nyssen J, Vanmaercke M, Poesen J (2021) Gully prevention and control: techniques, failures and effectiveness. Earth Surf Proc Land 46(1):220–238. https://doi.org/10.1002/esp.5033
Friedl MA, Sulla-Menashe D, Tan B, Schneider A, Ramankutty N, Sibley A, Huang X (2010) MODIS Collection 5 global land cover: algorithm refinements and characterization of new datasets. Remote Sens Environ 114(1):168–182. https://doi.org/10.1016/j.rse.2009.08.016
Gao Y, Zhang X, Zhang X, Li D, Yang M, Rongxiang H, Tian J (2020) Building check dams systems to achieve water resource efficiency: Modelling to maximize water and ecosystem conservation benefits. Hydrol Res 51(6):1409–1436. https://iwaponline.com/hr/article-abstract/51/6/1409/77938
Georgees RN, Hassan RA, Evans RP, Jegatheesan P (2018) Resilient response characterization of pavement foundation materials using a polyacrylamide-based stabilizer. J Mater Civ Eng 30(1):04017252. https://doi.org/10.1061/(ASCE)MT.1943-5533.0002109

Ghale B, Mitra E, Sodhi HS, Verma AK, Kumar S (2022) Carbon sequestration potential of agroforestry systems and its potential in climate change mitigation. Water Air Soil Pollut 233(7):228. https://doi.org/10.1007/s11270-022-05689-4

Ghosh PK, Mukhopadhyay R, Jana NC (2021) Quantitative analysis of drainage basin parameters towards better management of Damodar River, Eastern India. J Geol Soc India 97(7):711–734. https://doi.org/10.1007/s12594-021-1753-8

Gibson SA, Pak JH, Fleming MJ (2010) Modeling watershed and riverine sediment processes with HEC-HMS and HEC-RAS. Watershed Manage 2010:1340–1349. https://doi.org/10.1061/41143(394)120

Giri C, Pengra B, Zhu Z, Singh A, Tieszen LL (2007) Monitoring mangrove forest dynamics of the Sundarbans in Bangladesh and India using multi-temporal satellite data from 1973 to 2000. Estuar Coast Shelf Sci 73(1):91–100. https://doi.org/10.1016/j.ecss.2006.12.019

Giusti E, Marsili-Libelli S (2015) A fuzzy decision support system for irrigation and water conservation in agriculture. Environ Model Softw 63:73–86. https://doi.org/10.1016/j.envsoft.2014.09.020

Govindaraj V, Lakshumanan C (2019) Morphometric analysis and characterization for subwatershed delineation in Gomukhi river basin of Villupuram district, Tamilnadu using GIS. In: Rao PJ, Rao KN, Kubo S (eds) Proceedings of international conference on remote sensing for disaster management. Springer International Publishing, pp 805–815. https://doi.org/10.1007/978-3-319-77276-9_73

Guihéneuf N, Boisson A, Bour O, Dewandel B, Perrin J, Dausse A, Viossanges M, Chandra S, Ahmed S, Maréchal JC (2014) Groundwater flows in weathered crystalline rocks: impact of piezometric variations and depth-dependent fracture connectivity. J Hydrol 511:320–334. https://doi.org/10.1016/j.jhydrol.2014.01.061

Gull S, Shah SR (2020) Watershed models for assessment of hydrological behavior of the catchments: a comparative study. Water Pract Technol 15(2):261–281. https://doi.org/10.2166/wpt.2020.030

Gupta V, Kumar A, Gupta M, Sharma B, Singh M, Banotra M, Bhagat S, Sharma A (2018) Agrostological interventions and their utilization prospects for soil and water conservation in hill agriculture—a review. Int J Curr Microbiol Appl Sci 7:3514–3522. https://doi.org/10.20546/ijcmas.2018.701.413

Hamidpour M, Jalalian A, Afyuni M, Ghorbani B (2013) Sensitivity analysis, calibration and validation of EUROSEM model for prediction of runoff in the South Karoun Watershed. JWSS-Isfahan Univ Technol 16(62):49–60. https://jstnar.iut.ac.ir/article-1-2493-en.html

Hansen MC, Potapov PV, Moore R, Hancher M, Turubanova SA, Tyukavina A, Thau D, Stehman SV, Goetz SJ, Loveland TR, Kommareddy A, Egorov A, Chini L, Justice CO, Townshend JRG (2013) High-resolution global maps of 21st-century forest cover change. Science 342(6160):850–853. https://doi.org/10.1126/science.1244693

Healy RW, Winter TC, LaBaugh JW, Franke OL (2007) Water budgets: Foundations for effective water-resources and environmental management, vol 1308. US Geological Survey Reston, Virginia. https://pubs.usgs.gov/circ/2007/1308/

Hernandez M, Nearing MA, Al-Hamdan OZ, Pierson FB, Armendariz G, Weltz MA, Spaeth KE, Williams CJ, Nouwakpo SK, Goodrich DC, Unkrich CL, Nichols MH, Holifield Collins CD (2017) The rangeland hydrology and erosion model: a dynamic approach for predicting soil loss on rangelands. Water Resour Res 53(11):9368–9391. https://doi.org/10.1002/2017WR020651

Ikeda H, Wakamatsu T, Nakaya K, Ishii T (2019) Modeling of precipitation-infiltration rate using non-steady surface runoff model (EUROSEM). J Japan Soc Hydrol Water Resources 32(1):35–46. https://www.cabdirect.org/cabdirect/abstract/20219920382

India-WRIS (n.d.) Retrieved 18 April 2024 from https://indiawris.gov.in/wris/#/Aquifer

Janjić J, Tadić L (2023) Fields of application of SWAT hydrological model—a review. Earth 4(2):331–344. https://www.mdpi.com/2673-4834/4/2/18

Jena SK, Tiwari KN (2016) Hydrologic modeling of watersheds using remote sensing, GIS and AGNPS. In: Modeling methods and practices in soil and water engineering. Apple Academic Press

Kar G, Patra PK, Singh AK (2023) Spatial variability of soil organic carbon, pH and electrical conductivity and its influencing factors in a watershed of coastal region of Odisha, India. Commun Soil Sci Plant Anal. https://doi.org/10.1080/00103624.2023.2211114

Kasthuri Thilagam V, Manivannan S, Khola OPS (2023) Deriving land management practices for reduced nutrient movement from an agricultural watershed using the AGNPS model. Sustainability 15(5), Article 5. https://doi.org/10.3390/su15054001

Kholod N, Evans M, Khan Z, Hejazi M, Chaturvedi V (2021) Water-energy-food nexus in India: a critical review. Energy Clim Change 2:100060. https://doi.org/10.1016/j.egycc.2021.100060

Kisekka I, DeJonge K, Ma L, Paz J, Mankin K (2017) Crop modeling applications in agricultural water management. Trans ASABE 60:1959–1964. https://doi.org/10.13031/trans.12693

Klingebiel AA, Montgomery PH (1961) Land-capability classification. Soil Conservation Service, US Department of Agriculture. https://books.google.com/books?hl=en&lr=&id=c4kwAAAAYAAJ&oi=fnd&pg=PA3&dq=Land+Capability+Classification.+United+States+Department+of+Agriculture,+Soil+Conservation+Service.&ots=usMWDdL6o1&sig=aqVecqVdJBHaER2fHmw0j1DmHAM

Krysanova V, Hattermann F, Huang S, Hesse C, Vetter T, Liersch S, Koch H, Kundzewicz ZW (2015) Modelling climate and land-use change impacts with SWIM: lessons learnt from multiple applications. Hydrol Sci J 60(4):606–635. https://doi.org/10.1080/02626667.2014.925560

Kumari S, Bharti N, Tripathy KK (2021) Strengthening agriculture value chain through collectives: comparative case analysis. Int J Rural Manage 17(1_suppl):40S–68S. https://doi.org/10.1177/0973005221991438

LaCanne CE, Lundgren JG (2018) Regenerative agriculture: merging farming and natural resource conservation profitably. PeerJ 6:e4428. https://peerj.com/articles/4428/?td=tw

Lai YG, Abban B, Politano M (2024) A process-based mesh-distributed watershed model for water runoff and soil erosion simulation. Int J River Basin Manage 22(1):71–88. https://doi.org/10.1080/15715124.2022.2101465

Lal R (2004) Soil carbon sequestration impacts on global climate change and food security. Science 304(5677):1623–1627. https://doi.org/10.1126/science.1097396

Lal A, Datta B (2020) Performance evaluation of homogeneous and heterogeneous ensemble models for groundwater salinity predictions: a regional-scale comparison study. Water Air Soil Pollut 231(6):320. https://doi.org/10.1007/s11270-020-04693-w

Lampayan RM, Rejesus RM, Singleton GR, Bouman BAM (2015) Adoption and economics of alternate wetting and drying water management for irrigated lowland rice. Field Crop Res 170:95–108. https://doi.org/10.1016/j.fcr.2014.10.013

Lu D, Weng Q (2007) A survey of image classification methods and techniques for improving classification performance. Int J Remote Sens 28(5):823–870. https://doi.org/10.1080/01431160600746456

Ma L, He C, Bian H, Sheng L (2016) MIKE SHE modeling of ecohydrological processes: merits, applications, and challenges. Ecol Eng 96:137–149. https://doi.org/10.1016/j.ecoleng.2016.01.008

Mal BC (2005) Introduction to soil and water conservation engineering. Kalyani Publications (India)

Mekonnen M, Abeje T, Addisu S (2021) Integrated watershed management on soil quality, crop productivity and climate change adaptation, dry highland of Northeast Ethiopia. Agric Syst 186:102964. https://doi.org/10.1016/j.agsy.2020.102964

Mello I, Laurent F, Kassam A, Marques GF, Okawa CMP, Monte K (2021) Benefits of conservation agriculture in watershed management: participatory governance to improve the quality of no-till systems in the Paraná 3 watershed, Brazil. Agronomy 11(12), Article 12. https://doi.org/10.3390/agronomy11122455

Mirchi A, Watkins D, Madani K (2011) Modeling for watershed planning, management, and decision making. Watersheds: Manage Restor Environ Impact:221–244

Mishra PK, Tripathi KP (2013) Soil and water conservation research for land management in India. Indian J Dryland Agric Res Dev 28(1):1–18. https://www.indianjournals.com/ijor.aspx?target=ijor:ijdard&volume=28&issue=1&article=001&type=pdf

Molden D, Oweis TY, Pasquale S, Kijne JW, Hanjra MA, Bindraban PS, Bouman BA, Cook S, Erenstein O, Farahani H (2007) Pathways for increasing agricultural water productivity. https://ageconsearch.umn.edu/record/158005/files/H040200.pdf

Montgomery DR (2007) Soil erosion and agricultural sustainability. Proc Natl Acad Sci 104(33):13268–13272. https://doi.org/10.1073/pnas.0611508104

Müller C, Franke J, Jägermeyr J, Ruane AC, Elliott J, Moyer E, Heinke J, Falloon PD, Folberth C, Francois L (2021) Exploring uncertainties in global crop yield projections in a large ensemble of crop models and CMIP5 and CMIP6 climate scenarios. Environ Res Lett 16(3):034040. https://doi.org/10.1088/1748-9326/abd8fc/meta

Neupane J, Guo W (2019) Agronomic basis and strategies for precision water management: a review. Agronomy 9(2), Article 2. https://doi.org/10.3390/agronomy9020087

Novara A, Cerda A, Barone E, Gristina L (2021) Cover crop management and water conservation in vineyard and olive orchards. Soil Till Res 208:104896. https://www.sciencedirect.com/science/article/pii/S0167198720306784

Oribi MO, Abdulkareem AK (2020) Scenarios to reduce evaporation from class A evaporation pan by using windbreaks. Przegląd Naukowy. Inżynieria i Kszta\ltowanie Środowiska 29(3 [89]). https://yadda.icm.edu.pl/yadda/element/bwmeta1.element.agro-1e9f6173-1f55-4f13-a66a-d0cf7d1eb463/c/art7.pdf

Padbhushan R, Kumar U, Sharma S, Rana DS, Kumar R, Kohli A, Kumari P, Parmar B, Kaviraj M, Sinha AK, Annapurna K, Gupta VVSR (2022) Impact of land-use changes on soil properties and carbon pools in India: a meta-analysis. Front Environ Sci 9. https://doi.org/10.3389/fenvs.2021.794866

Palsaniya DR, Rai SK, Sharma P, Ghosh PK (2016) Natural resource conservation through weather-based agro-advisory. Curr Sci:256–258. https://ischolar.sscldl.in/index.php/CURS/article/view/120773

Pambudi AS (2021) Review of the Kineros model for predicting the effect of land use on the surface run-off. Indonesian J Appl Res (IJAR) 2(3), Article 3. https://doi.org/10.30997/ijar.v2i3.150

Pandey KC, Singh AK (2019) Weather and agro advisory services to farmers and its benefits. Clim Chang 5:116–123. http://www.discoveryjournals.org/climate_change/current_issue/v5/n18/A5.pdf

Pang S, Wang X, Melching CS, Feger K-H (2020) Development and testing of a modified SWAT model based on slope condition and precipitation intensity. J Hydrol 588:125098. https://www.sciencedirect.com/science/article/pii/S0022169420305588

Park D, Fan H-J, Zhu J-J, Oh S-E, Um M-J, Jung K (2019) Evaluation of reliable digital elevation model resolution for TOPMODEL in two mountainous watersheds, South Korea. Appl Sci 9(18), Article 18. https://doi.org/10.3390/app9183690

Parsons J, Thomas D, Huffman R (2004) Agricultural non-point source water quality models: their use and application. South Cooper Ser Bull 398

Pettorelli N, Laurance WF, O'Brien TG, Wegmann M, Nagendra H, Turner W (2014) Satellite remote sensing for applied ecologists: opportunities and challenges. J Appl Ecol 51(4):839–848. https://doi.org/10.1111/1365-2664.12261

Pijl A, Wang W, Straffelini E, Tarolli P (2022) Soil and water conservation in terraced and non-terraced cultivations: an extensive comparison of 50 vineyards. Land Degrad Dev 33(4):596–610. https://doi.org/10.1002/ldr.4170

Prasad CS (2019) Farming as an enterprise: ten years of FPO movement in India. State of India's Livelihood Report; Access Development Services, New Delhi, India, pp 37–48. https://www.indiaspend.com/wp-content/uploads/2020/10/Ten-Years-of-FPO-Movemet-in-India.pdf

Rafi D, Laxmi B, Charishma E, Babu KS (2023) Strengthening millet value chain through farmer producer organizations. Indian J Ext Educ 59(3):26–31. https://acspublisher.com/journals/index.php/ijee/article/view/8544

Raju K, Aziz A, Sundaram S, Sekher M, Wani S, Sreedevi T (2008) Guidelines for planning and implementation of watershed development program in India: a review. Global Theme on Agroecosystems Report No. 48

Rao C (2000) Watershed development in India: recent experience and emerging issues. Econ Pol Wkly 35:3943–3947. https://doi.org/10.2307/4409924

Reddy C, Tonapi V, Navi S, Bezkorowajnyj P, Navi S, Seetharama N (2007) Seed system innovations in the semi-arid tropics of Andhra Pradesh. https://dr.lib.iastate.edu/handle/20.500.12876/57645

Rosenqvist A, Shimada M, Ito N, Watanabe M (2007) ALOS PALSAR: a pathfinder mission for global-scale monitoring of the environment. IEEE Trans Geosci Remote Sens 45(11):3307–3316. https://doi.org/10.1109/TGRS.2007.901027

Saghafian B, Meghdadi AR, Sima S (2015) Application of the WEPP model to determine sources of run-off and sediment in a forested watershed. Hydrol Process 29(4):481–497. https://doi.org/10.1002/hyp.10168

Sahoo S, Singha C, Govind A (2024) Prediction of pulse suitability in rice fallow areas using fuzzy AHP-based machine learning methods in Eastern India. Paddy Water Environ. https://doi.org/10.1007/s10333-024-00970-0

Sanathara MM, Vibhute B (2020) A review of organic farming for soil conservation towards sustainable and natural farming in India. The Institution of Engineers (India) 9. https://www.researchgate.net/profile/Mahadeva-Madegowda/publication/369092303_17Land_Use_Land_Cover_Change_Detection_of_Yagachi_Watershed_Karnataka_4-2020_AEDB/links/64096aa0b1704f343fb68658/17Land-Use-Land-Cover-Change-Detection-of-Yagachi-Watershed-Karnataka-4-2020-AEDB.pdf#page=11

Sharda VN, Sena DR, Shrimali SS, Khola OPS (2013) Effects of an intercrop-based conservation bench terrace system on resource conservation and crop yields in a sub-humid climate in India. Trans ASABE 56(4):1411–1425. https://elibrary.asabe.org/abstract.asp?aid=43979

Shekar PR, Mathew A, Arun PS, Gopi VP (2023) Sub-watershed prioritization using morphometric analysis, principal component analysis, hypsometric analysis, land use/land cover analysis, and machine learning approaches in the Peddavagu River Basin, India. J Water Clim Change 14(7):2055–2084. https://doi.org/10.2166/wcc.2023.221

Sikka AK, Alam MF, Mandave V (2022) Agricultural water management practices to improve the climate resilience of irrigated agriculture in India. Irrig Drain 71(S1):7–26. https://doi.org/10.1002/ird.2696

Sims REH, Hastings A, Schlamadinger B, Taylor G, Smith P (2006) Energy crops: current status and future prospects. Glob Change Biol 12(11):2054–2076. https://doi.org/10.1111/j.1365-2486.2006.01163.x

Singh VP, Frevert DK (2010) Gridded surface/subsurface hydrologic analysis (GSSHA) model: a model for simulating diverse streamflow-producing processes. In: Watershed models. CRC Press, pp 155–182. https://doi.org/10.1201/9781420037432-15/gridded-surface-subsurface-hydrologic-analysis-gssha-model-model-simulating-diverse-streamflow-producing-processes

Singh G, Budhiraja P, Vatta K (2018) Sustainability of farmer producer organisations under agricultural value networks in India: a case of Punjab and Gujarat. Indian J Agric Econ 73(3):70–85. https://www.researchgate.net/profile/Gurpreet-Singh-98/publication/328497720_Sustainability_of_Farmer_Producer_Organisations_under_Agricultural_Value_Networks_in_India_A_Case_of_Punjab_and_Gujarat/links/5fe5c49c45851553a0eaf5c0/Sustainability-of-Farmer-Producer-Organisations-under-Agricultural-Value-Networks-in-India-A-Case-of-Punjab-and-Gujarat.pdf

Singh SP, Mishra SR, Kumar V, Saran B, Jaiswal P (2020) Economic impact and usefulness of agromet advisory services for wheat crop of Siddhartha Nagar district of Uttar Pradesh. Assessment (Gadgil, 1989 and Anon., 2002) 3:1. https://www.thepharmajournal.com/archives/2020/vol9issue12S/PartB/S-9-10-24-352.pdf

Singh WR, Barman S, Patidar N (2021) Prioritization of sub-watersheds based on morphometric parameters in Pare watershed, Arunachal Pradesh, India. Int J Scien Res Sci Eng Technol:53–64. https://doi.org/10.32628/IJSRSET21816

Singh KK, Ghosh K, Bhan SC, Singh P, Vishnoi L, Balasubramanian R, Attri SD, Goroshi S, Singh R (2023) Decision support system for digitally climate informed services to farmers in India. J Agrometeorol 25(2):205–214. http://journal.agrimetassociation.org/index.php/jam/article/view/2094

Šípek V, Tesař M (2015) Validation of a mesoscale hydrological model in a small-scale forested catchment. Hydrol Res 47(1):27–41. https://doi.org/10.2166/nh.2015.220

Srinivas T, Bishaw Z, Rizvi J, Niane AA, Manan AR, Amegbeto K (2010) ICARDA's approach in seed delivery: technical performance and sustainability of village-based seed enterprises in Afghanistan. J New Seeds 11(2):138–163. https://doi.org/10.1080/15228861003754156

Tando PKV, Susanto S, Suryatmojo H, Farizi FA (2019) Evaluation of terrace design as soil and water conservation technique in Karangkobar catchment, Banjarnegara, Indonesia. IOP Conf Ser: Earth Environ Sci 355(1):012022. https://doi.org/10.1088/1755-1315/355/1/012022/meta

Thakur B, Parajuli R, Kalra A, Ahmad S, Gupta R (2017) Coupling HEC-RAS and HEC-HMS in precipitation runoff modelling and evaluating flood plain inundation map. World Environ Water Resources Congress 2017:240–251. https://doi.org/10.1061/9780784480625.022

Tribouillois H, Constantin J, Murgue C, Villerd J, Therond O (2022) Integrated modeling of crop and water management at the watershed scale: optimizing irrigation and modifying crop succession. Eur J Agron 140:126592. https://doi.org/10.1016/j.eja.2022.126592

Unger PW, Stewart BA, Parr JF, Singh RP (1991) Crop residue management and tillage methods for conserving soil and water in semi-arid regions. Soil Tillage Res 20(2–4):219–240. https://www.sciencedirect.com/science/article/pii/016719879190041U

Unger PW, Payne WA, Peterson GA (2006) Water conservation and efficient use. In: Dryland agriculture. John Wiley and Sons, Ltd, pp 39–85. https://doi.org/10.2134/agronmonogr23.2ed.c3

Usmani RA (2020) Potential for energy and biofuel from biomass in India. Renew Energy 155:921–930. https://doi.org/10.1016/j.renene.2020.03.146

Vargas MM, Beskow S, de Mello CR, de Moura MM, Nunes MCM, Faria LC, Aquino LS (2021) Capability of LISEM to estimate flood hydrographs in a watershed with predominance of long-duration rainfall events. Nat Hazards 109(1):593–614. https://doi.org/10.1007/s11069-021-04850-2

Venkata Rao G, Nagireddy NR, Keesara VR, Sridhar V, Srinivasan R, Umamahesh NV, Pratap D (2024) Real-time flood forecasting using an integrated hydrologic and hydraulic model for the Vamsadhara and Nagavali basins, Eastern India. Nat Hazards. https://doi.org/10.1007/s11069-023-06366-3

Verma RK, Pandey A, Mishra SK, Singh VP (2023) A procedure for assessment of environmental flows incorporating inter- and intra-annual variability in dam-regulated watersheds. Water Resour Manage 37(8):3259–3297. https://doi.org/10.1007/s11269-023-03502-3

Vörösmarty CJ, McIntyre PB, Gessner MO, Dudgeon D, Prusevich A, Green P, Glidden S, Bunn SE, Sullivan CA, Liermann CR, Davies PM (2010) Global threats to human water security and river biodiversity. Nature 467(7315), Article 7315. https://doi.org/10.1038/nature09440

Walia SS, Kaur K, Kaur T (2024) Soil and water conservation techniques in rainfed areas. In: Walia SS, Kaur K, Kaur T (eds) Rainfed agriculture and watershed management. Springer Nature Singapore, pp 115–124. https://doi.org/10.1007/978-981-99-8425-1_15

Wani SP, Roy PS, Rao AVRK, Barron J, Ramachandran K, Balaji V (2011) Application of new science tools in integrated watershed management for enhancing impacts. CRC Press (Taylor and Francis), pp 159–204. https://oar.icrisat.org/3569/

Wani SP, Chander G, Sahrawat KL (2014) Science-led interventions in integrated watersheds to improve smallholders' livelihoods. NJAS: Wageningen J Life Sci 70–71(1):71–77. https://doi.org/10.1016/j.njas.2014.07.001

Wolka K, Mulder J, Biazin B (2018) Effects of soil and water conservation techniques on crop yield, runoff and soil loss in Sub-Saharan Africa: a review. Agric Water Manage 207:67–79. https://www.sciencedirect.com/science/article/pii/S0378377418306516

Wyseure GCL (1991) SWRBB: a basin scale simulation model for soil and water resources management. Agric Water Manage 20(1):82–83. https://ideas.repec.org//a/eee/agiwat/v20y1991i1p82-83.html

Xue L, Yang F, Yang C, Wei G, Li W, He X (2018) Hydrological simulation and uncertainty analysis using the improved TOPMODEL in the arid Manas River basin, China. Scien Reports 8(1):452. https://doi.org/10.1038/s41598-017-18982-8

Yadav B, Gupta PK, Patidar N, Himanshu SK (2020) Ensemble modelling framework for groundwater level prediction in urban areas of India. Sci Total Environ 712:135539. https://doi.org/10.1016/j.scitotenv.2019.135539

Yin W, Chai Q, Zhao C, Yu A, Fan Z, Hu F, Fan H, Guo Y, Coulter JA (2020) Water utilization in intercropping: a review. Agric Water Manage 241:106335. https://www.sciencedirect.com/science/article/pii/S0378377420303620

Zhang H, Wang B, Li Liu D, Zhang M, Leslie LM, Yu Q (2020a) Using an improved SWAT model to simulate hydrological responses to land use change: a case study of a catchment in tropical Australia. J Hydrol 585:124822. https://www.sciencedirect.com/science/article/pii/S0022169420302821

Zhang T, Yang Y, Ni J, Xie D (2020b) Best management practices for agricultural non-point source pollution in a small watershed based on the AnnAGNPS model. Soil Use Manag 36(1):45–57. https://doi.org/10.1111/sum.12535

Chapter 8
Analysing the Impact of Climate Change on Crop Yield and Its Consequences for the Water-Energy-Food Nexus in an Indian River Basin

Krishna Mondal, Vishal Gupta, Chandranath Chatterjee, and Rajendra Singh

Abstract Climate change-induced variations in temperature, precipitation, and related factors substantially threaten crop yield, thereby jeopardizing food security. In the context of the Kangsabati river basin in India, where rice cultivation is vital for farmers' livelihoods and food security, understanding the potential impacts of climate change on rice production becomes imperative. The interconnection of water, energy, and food (WEF Nexus) underscores the need to explore how alterations in food security, driven by climate change-induced crop yield changes, may reverberate across water and energy security. This study addresses a notable gap by utilizing the CMIP6-GCM to estimate climate change effects on crop yield, explicitly focusing on the Kangsabati River basin. Unlike previous research, this investigation integrates the influence of climate change-induced crop yield on the WEF Nexus at the basin level. The AquaCrop model is employed to simulate rice yields across historical (1981–2014), near-future (2021–2060), and far-future (2061–2100) periods under two climate scenarios (SSP245 and SSP585) using the EC-Earth3 GCM model. Findings indicate a projected decline in crop yield for both seasons (Kharif and Rabi) in both SSP245 and SSP585 scenarios. SSP245 anticipates a 10–15% decrease in Kharif paddy yield and a 15–20% reduction in Rabi season, while SSP585 predicts a more severe decline of 15–20% in Kharif and 20–25% in Rabi. Furthermore, the study reveals a diminishing trend in the Food Sub-index (FSI) and Water-Energy-Food Nexus Index (WEFNI), with SSP585 exhibiting a more pronounced decrease than SSP245. Notably, downstream areas in the basin demonstrate higher food and overall WEF security than other regions. These insights are instrumental for policymaking, enabling the formulation and implementation of sustainable water, energy, and food

K. Mondal (✉)
Irrigation and Drainage Engineering Department, College of Agricultural Engineering and Post Harvest Technology, Ranipool, Gangtok, India
e-mail: krishnamondal160@gmail.com

V. Gupta · C. Chatterjee · R. Singh
Agricultural and Food Engineering Department, Indian Institute of Technology Kharagpur, Kharagpur, India

A. Nanda et al. (eds.), *Navigating the Nexus*, Water Science and Technology Library 102, https://doi.org/10.1007/978-3-031-76532-2_8

security strategies that account for the anticipated impacts of climate change in the Kangsabati River basin.

Keywords Climate change · Rice · AquaCrop model · Food sub-index · Water-energy-food nexus · CMIP6

8.1 Introduction

Human activity increases atmospheric greenhouse gas levels, resulting in global warming and affecting rainfall, temperature, and solar radiation. As per the Sixth Assessment Report from the IPCC (Hennessy et al. 2022), recent catastrophic weather events and severe environmental and societal harm are direct results of human-caused climate change. Changes in global rainfall frequency, pattern, and severity could reduce low and medium rainfall events and increase heavy rainfall events in India (Pai et al. 2015). India will also have higher minimum and maximum temperatures (Chaturvedi et al. 2012). Climate change affects crop output by changing global surface temperature and rainfall (Lizumi and Ramankutty 2015). The growing variability in climatic variables due to climate change has harmed crop production worldwide (Chun et al. 2016). Kang et al. (2009) found that temperature increases and unpredictable rainfall can affect crop productivity, food quality, water usage efficiency, and water availability. Climate change-induced fluctuations in temperature, rainfall, and other variables can reduce crop productivity, threatening food security. According to the 2022 IPCC assessment, the globe will hit the 1.5 ºC threshold over the next two decades, creating the urgent need for significant carbon emission reductions to prevent an environmental disaster that could harm wheat and rice productivity. Many studies indicate that a 1–2 ^{0}C temperature increase during paddy harvesting can decrease rice yield by 10–20% (Zhang et al. 2015). To feed a projected population of 9.4 billion by 2050, the Food and Agriculture Organization anticipates a 60% increase in global food production, highlighting agriculture as one of the sectors most impacted by climate change (Food and Agriculture Organization (FAO) 2020). With 1.62 billion people, India will need 32% more food than in 2017 (Organisation for Economic Co-operation and Development (OECD) 2017).

The latest data and crop behavior under different climate change scenarios are needed to effectively estimate climate change's effects on agricultural productivity. Crop models help predict crop development under diverse climatic, soil, and management circumstances (Banerjee et al. 2016). In the last forty years, numerous crop simulation models have emerged, such as EPIC (Williams et al. 1989), CropSyst (Stockle and Nelson 1994), APSIM (Keating et al. 2003), and DSSAT (Jones et al. 2003). These models forecast rice production using cultivar, management, soil, and climate. Due to the global variety of crops and environmental circumstances, these models require various input variables and parameter values. FAO developed the AquaCrop model to address this concern, aiming to balance easiness, accuracy, and toughness (Steduto et al. 2009). The AquaCrop model mimics the yields of wheat, rice, maize, barley, soybean, and sorghum, correlating them with water consumption across different daily irrigation scenarios. It employs metrics like thermal time, growing degree days, or calendar days in its simulations. The model's growth module is driven by water, and transpiration is approximated before being converted into biomass using conservative and non-conservative parameters.

Shared socioeconomic pathways (SSPs) have recently emerged as an essential aspect of societal evolution, and they are included in the sixth phase of the IPCC's coupled model intercomparison project (CMIP6). Since 2016, numerous studies have examined the data and concluded that the new CMIP6 climate scenarios are more plausible than the Representative Concentration Pathways (Wyser et al. 2020). India's agriculture sector needs accurate CMIP6 climate estimates for crop output. Rising evapotranspiration owing to warming (Wilcox and Makowski 2014) and precipitation variability can reduce food production in India. Crop growth and yields are significantly impacted by the heightened frequency and severity of floods and droughts (Boonwichai et al. 2019). When looking at national and regional scales, it is noted that temperature fluctuations have a more pronounced effect on agricultural productivity than alterations in precipitation (Asseng et al. 2019). Society's prosperity and economic growth hinge on the crucial pillars of water, energy, and food. The twenty-first century, marked by climate change, demographic shifts, aging infrastructure, global trade networks, political interventions, and various other challenges, places substantial demands on these resources (Mondal et al. 2023a). The extensive links or nexus between these three sectors make these difficulties even more difficult. Any change in one sector would affect the other two. Mortada et al. (2018) understand that water, energy, and food research is essential for sustainable development and social and economic advancement.

The WEF nexus model optimises using food, energy, and water resources to increase resource efficiency (Mondal et al. 2023c). Several WEF nexus models have been created to study this complex relationship (Taguta et al. 2022). Most of these models have not incorporated all WEF connectivity characteristics, including climate change's impact on the WEF nexus. Mondal et al. (2023b) created a WEF Nexus model considering most WEF connectivity criteria. Grasping the WEF nexus and comprehending the impact of climate change is pivotal. Assessing how climate change might influence food security becomes particularly challenging in areas already grappling with food insecurity (Richardson et al. 2018). Although several simulation models have investigated how climate change may impact agricultural production, only a handful have incorporated the CMIP6-GCM. Furthermore, none of the studies has quantified how climate change-induced crop yield affects the water-energy-food nexus. Considering these difficulties, this study seeks the following goals: 1. To examine how climate change affects agriculture yield using CMIP6 data and AquaCrop in a basin; 2. To explore how climate change-induced crop production declines affect the WEF nexus in the basin using the WEF Nexus model. This study tries to understand WEF essential interconnections better and how climate change-induced crop production may affect them. This research may help policymakers and stakeholders make educated decisions to preserve these crucial resources for future generations.

8.2 Materials and Method

8.2.1 Study Area

The Kangsabati River, originating near Jhalda at Murguma in the Chota Nagpur plateau, flows for 465.23 km, traversing Purulia, Bankura, and Paschim Medinipur districts. In terms of geography, the Kangsabati River basin occupies the space between latitudes 86° 0′ E and 87° 30′ E and longitudes 22° 15′ N and 23° 30′ N. The elevations within this region span from 2 to 659 m above sea level (Fig. 8.1). The basin encompasses 24 administrative blocks, and rice cultivation dominates the region during both Kharif (monsoon) and Rabi (dry) seasons. The red laterite soil type prevails in the area, supporting the cultivation of predominantly paddy during the monsoon and dry seasons. The climate exhibits hot and humid summers (April–May), followed by a monsoon period (June–September). Autumn months are moderately hot and dry, while December and January are cool. Summer temperatures in the Kangsabati River basin fluctuate between 23 and 45 °C, while winter temperatures range from 3 to 20 °C. The area receives between 1100 and 1500 mm of rain annually. The soil, taxonomically classified as Haplustaff, is predominantly red laterite sandy loam.

8.2.2 Data Used

8.2.2.1 Climate Data

Data on daily rainfall, lowest temperature (Tmin), and highest temperature (Tmax) were provided by the India Meteorological Department (IMD) for the periods 1981–2014 and 2015–2020, enabling an examination of climate change trends. The CMIP6 dataset for India, sourced from the ZENODO website (https://zenodo.org/record/3873998#.Y3qlXnZBzIU), and empirical Quantile Mapping (EQM) was used to downscale historical (1951–2014) and predicted (2015–2100) climate scenarios, including SSP126, SSP245, SSP370, and SSP585 (Mishra et al. 2020). These bias-corrected future datasets are widely used (Das et al. 2023; Swain et al. 2023). For the transition period (2015–2020), near future (2021–2060), and far future (2061–2100), the study utilized the CMIP6 GCM (EC-Earth3) dataset. From the available 13 GCMs, the EC-EARTH3 model was selected for its reported suitability for the study area and nearby catchments (Mahato et al. 2022; Swain et al. 2023). Focused on SSP245 (538.9 ppm CO_2 equivalent) and SSP585 (936.4 ppm CO_2 equivalent), the analysis delved into the anticipated impact of these scenarios on the basin's rice yield and its implications for the water-energy-food nexus. SSP245 envisions sustainable growth efforts by most nations, while SSP585 presents a worst-case scenario involving unconventional development (O'Neill et al. 2016).

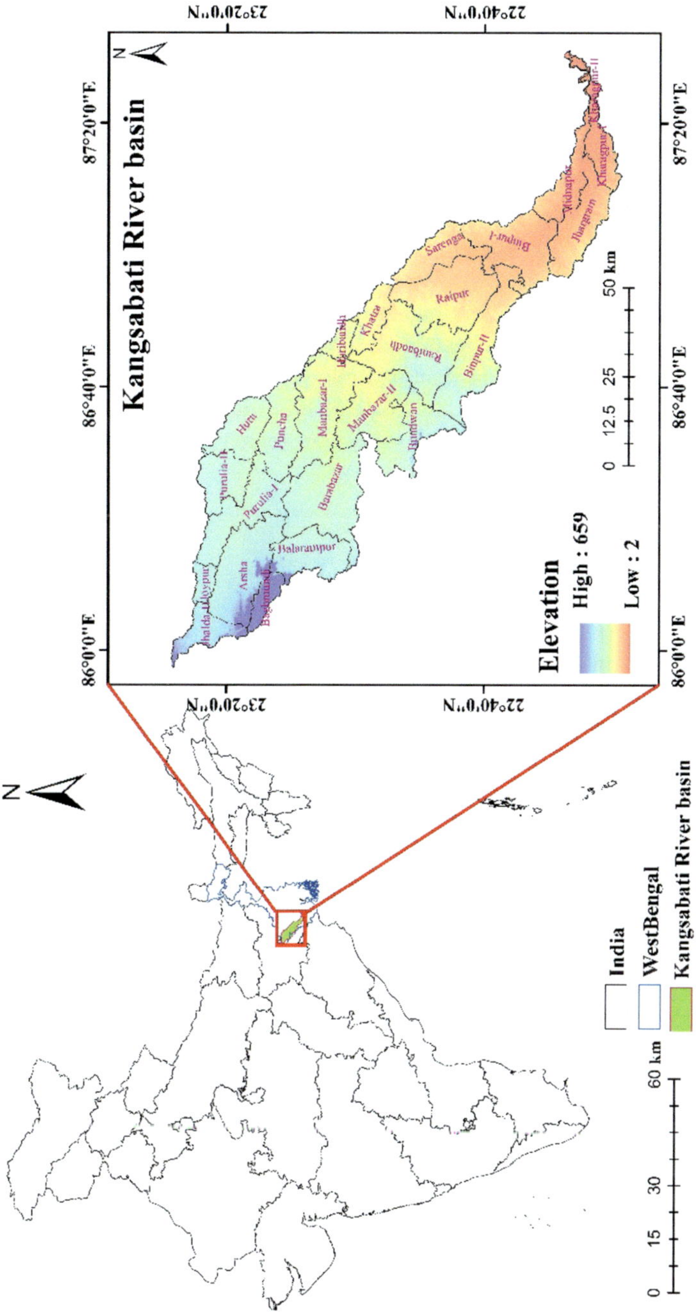

Fig. 8.1 Map of the Kangsabati river basin

8.2.2.2 Soil Data

The study extracted soil physical parameters, extending to a depth of 15 cm, and essential soil metrics, encompassing sand, silt, clay, organic carbon, and bulk density ratios, from literature sources (Debnath et al. 2021). This dataset originates from the experimental field affiliated with the Agricultural and Food Engineering Department at IIT Kharagpur, situated within the Kangsabati River basin.

8.2.2.3 Crop Data

To calibrate and validate the crop model, the field data gathered between 2015 and 2017 were referred from Debnath et al. (2021). The data consisted of sowing, transplantation, fertilization, and irrigation timings. The experiments used the IR36 cultivar of rice, and a fixed transplanting date of July 14th was assumed to coincide with the onset of monsoon season in the study area, thereby excluding any influence of the growing season on the rice yield.

8.2.2.4 Irrigation Data

Evapotranspiration (ET), a crucial aspect of the hydrological cycle, significantly influences runoff and water demand for irrigation. While various methods exist to compute potential evapotranspiration (PET), many are intricate models reliant on numerous parameters, limiting their applicability in data-scarce regions. The Hargreaves-Samani (HS) approach has been employed in this study, offering a simplified alternative that only necessitates the maximum and minimum temperature data for PET computation. Almorox and Grieser (2015) enhanced the HS method by formulating distinct equations adjusting constants based on climatic zones. The study applied the suitable equation (Eq. 8.1) from this modified approach to estimate PET, enabling effective irrigation scheduling in the study area with minimal data requirements.

$$ET_{HS} = k_1(T_m + 17.8) \times (T_x - T_n)^{k_2} \times R_a + k_3 \tag{8.1}$$

where, ET_{HS} = Evapotranspiration $\left(\frac{\text{mm}}{\text{day}}\right)$, T_n = Daily minimum temperature (°C), T_x = Daily maximum temperature (°C), T_m = Mean temperature (°C), $k_1 = 0.0023$, $k_2 = 0.5$, $k_3 = 0$, R_a = Extraterrestrial Radiation (MJ m^{-2} day^{-1}).

8.2.2.5 Input Data for the WEF Nexus Analyses

Analyzing the Water-Energy-Food Nexus using the WEF Nexus model developed by Mondal et al. (2023b) necessitates a wealth of secondary data (Table 8.1). All the input data for 2011 and 2021 were from different government reports and agencies and through field surveys. The decadal changes of each variable were analyzed for future data projection, except for demographic data. Data on demographics were gathered from the International Institute for Applied System Analysis-SSP Public Database, accessible at https://tntcat.iiasa.ac.at/SspDb. Data projections were made for 2031, 2041, 2051, 2061, 2071, 2081, and 2091.

8.2.3 *Model Description*

8.2.3.1 Aquacrop

The FAO developed the AquaCrop model (Steduto et al. 2009) to excel in simulating herbaceous crop production across various irrigation scenarios, encompassing rainfed, additional, shortage, and complete irrigation. Its growth module is water-centric, with transpiration playing a pivotal role in influencing biomass. A crop-specific parameter called biomass water productivity has been introduced to ensure adaptation to varying environments and seasons. This parameter is normalized for air CO_2 concentration and atmospheric evaporative demand. AquaCrop facilitates simulations at daily time steps, employing thermal or calendar time. The modeling process is designed to be user-friendly, leading to the culmination of crop yield simulation through a clear and comprehensible sequence of four steps, as outlined in work by Steduto et al. (2009).

8.2.4 *Water-Energy-Food Nexus Model*

The WEF Nexus model represents a cutting-edge distributed model designed for comprehensive multiscale analysis of the WEF nexus. Employing an indicator-based approach, specifically the Modified Pardee Rand WEF Nexus Index, the model's intricate governing equations are meticulously outlined in Table 8.2. Comprising three core modules—Water, Energy, and Food—each module meticulously computes the accessibility and availability indices for its respective resource, utilizing them to determine the security or sub-index specific to that resource. The outputs of these modules are seamlessly integrated into the primary equation of the water-energy-food nexus model, ultimately computing the overarching WEF Nexus Index or security.

Table 8.1 Input data for WEF Nexus model simulation (Mondal et al. 2023c)

Sl. No	Area	Data information	Unit	Sources
1	Demographic data	Total number of population and age group-wise population		India Census Reports (2011); Ministry of Statistics and Programme, Government of India (GoI); National Family Health Survey-5; https://www.indiastat.com/; https://www.ceicdata.com
2		Total number of households		
3		Total number of farmers		
4	Water security-related data	Farmers having access to sufficient water for irrigation		
5		Population having access to sufficient water for drinking purposes		
6		Population having access to sufficient water for sanitation purpose		
7		Precipitation	mm/year	https://mausam.imd.gov.in/
8		Evapotranspiration	mm/year	https://appeears.earthdatacloud.nasa.gov/
9		Geographical area	ha	India Census Reports (2011)
10		Water required for domestic purposes age group-wise	m^3	https://www.indiastat.com/; Ministry of Agriculture and Farmers Welfare, GoI
11		Water required for per unit meat, milk, egg, and crop production [for crop production crop water productivity of major food grains (paddy, wheat, maize) and vegetables (potato) have been considered]	kg/m^3	
12	Energy security-related data	Farmers having access to modern farm equipment for agriculture		India Census Reports (2011); National Family Health Survey-5
13		Farmers having access to sufficient electricity for irrigation		https://www.indiastat.com/; https://www.ceicdata.com
14		Population having access to sufficient electricity for household purposes		Ministry of Power, GoI; Ministry of Petroleum and Natural Gas, GoI; National Family Health Survey-5; https://www.indiastat.com/
15		Number of Household having access to modern fuel (like liquefied petroleum gas (LPG)) for cooking purposes		
16		Total electrify production (renewal and non-renewal)	kWh/year	

(continued)

Table 8.1 (continued)

Sl. No	Area	Data information	Unit	Sources
17		Total number of tractors		https://www.indiastat.com/
18		Agriculture worker		India Census Reports (2011); National Family Health Survey-5
19		Energy required for domestic purposes	kWh/year	
20		Energy required for crop production, meat production, milk production, egg production	kWh/kg	
21	Food security-related data	Egg production	number/ year	Animal Husbandry and Dairying, GoI, National Dairy Development Board; Ministry of Fisheries, GoI; https://www.indiastat.com/
22		Milk production	ton/year	
23		Meat production (poultry, sheep, buffalo, pig), fish production	ton/year	
24		Hunger index		Ministry of Statistics and Programme Implementation is a ministry, GoI; https://www.indiastat.com/
25		Food price level index		
26		Average crop productivity of major food grains and vegetables	kg/ha	Ministry of Agriculture and Farmers Welfare, GoI; https://www.indiastat.com/
27		Cropping intensity with respect to major food grains (paddy, wheat, maize) and vegetables (potato)		
28		Cultivated area	ha	
29		Age group-wise dietary requirement	ton/year	

8.2.5 Methodological Framework

The methodology employed in this study is visually outlined in Fig. 8.2 and comprises two primary components. The first process involves calibrating, validating, and simulating the Aquacrop model. The second component focuses on preparing the database and simulating the WEF Nexus model.

The Aquacrop model is configured for the IR36 rice cultivar, explicitly targeting the Kharif and Rabi seasons within the Kangsabati river basin at the block scale. Model inputs encompass daily climate data, soil information, irrigation, crop

Table 8.2 Indicators and equations regarding the WEF Nexus model (Mondal et al. 2023b)

WEF nexus indices	Equation	Indicators definition (unit)
Water availability (WAv)	$(\mathrm{WAv}) = \frac{\mathrm{wavi}}{\mathrm{wreq}} = \frac{\mathrm{A(P-ET)}}{\mathrm{WR_d+WR_{cp}+WR_m+WR_{ml}+WR_e}}$ (8.2)	wavi: wreq: water requirement (m^3) water available (m^3) A: geographical area (ha) P: precipitation (mm) ET: evapotranspiration WR_d, WR_{cp}, WR_m, WR_{ml}, WR_e: water required for domestic purposes, crop production, meat production, milk production, egg and electricity production, respectively (m^3)
Water accessibility (WAc)	$\mathrm{WAc} = \sqrt[2]{\mathrm{PDS*FSW_I}}$(8.3)	PDS: % population having access to sufficient water for drinking and sanitation FSW_I: % farmer having access to sufficient water for irrigation
Water sub-index (WSI)	$\mathrm{WSI} = \sqrt[2]{\mathrm{WAv * WAc}}$(8.4)	WAv : water availability (ratio) WAc: water accessibility (ratio)
Energy availability (EAv)	$\frac{\mathrm{Eavi}}{\mathrm{Ereq}} = \frac{\mathrm{TES +0.28\ (ME +HLE)}}{\mathrm{ER_d+ER_{cp}+ER_{ch}+ER_m+ER_{ml}+ER_e}}$ (8.5) $\mathrm{TES} = \mathrm{TE + SE + HE + WE}$ (8.6)	Eavi: energy available (kWh) Ereq: energy requirement (kWh) TES: total electricity supply (kWh) TE, SE, HE, WE: thermal, solar, hydropower, wind energy, respectively (kWh) ME: machinery energy (MJ) HLE: human labour energy (MJ) ER_d, ER_{cp}, ER_{ch}, ER_m, ER_{ml}, ER_e: energy required for domestic purposes, crop production, meat production, milk production, and egg production, respectively (kWh)
Energy accessibility (EAc)	$\mathrm{EAc} = \sqrt[4]{\mathrm{P_e*H_{mf}*F_{fe}*F_{sei}}}$(8.7)	P_e: % population having access to electricity for household purposes H_{mf}: % household having access to modern fuel F_{fe}: % farmers having access to modern farm equipment F_{sei}: % farmer having access to sufficient electricity for irrigation

(continued)

Table 8.2 (continued)

WEF nexus indices	Equation	Indicators definition (unit)
Energy sub-index (ESI)	$ESI = \sqrt[2]{EAv*EAc}$(8.8)	EAv: energy availability (ratio) EAc: energy accessibility (ratio)
Food availability (FAv)	$\frac{Favi}{Freq} = \frac{C_p+F_p+M_p+Ml_p+E_p}{\sum_i DR_i*Pop_i}$(8.9)	Favi: food availability (ton) Freq: food requirement (ton) C_p, F_p, M_p, Ml_p, E_p: crop production, fish production, meat production, egg production, respectively (ton) DR: dietary requirement (kg/year) Pop: population i: different age group
Food accessibility (FAc)	$FAc = \sqrt[2]{(1/FPLI)*(1\ -HI)}$(8.10)	FPLI: food price level index (ratio) HI: hunger index (ratio)
Food sub-index (FSI)	$FSI = \sqrt[2]{FAv * FAc}$(8.11)	FAv: food availability (ratio) FAc: food accessibility (ratio)
Water-energy-food nexus index (WEFNI)	$WEFNI = \sqrt[3]{WSI * ESI * FSI}$(8.12)	WSI: water sub-index ESI: energy sub-index FSI: food sub-index

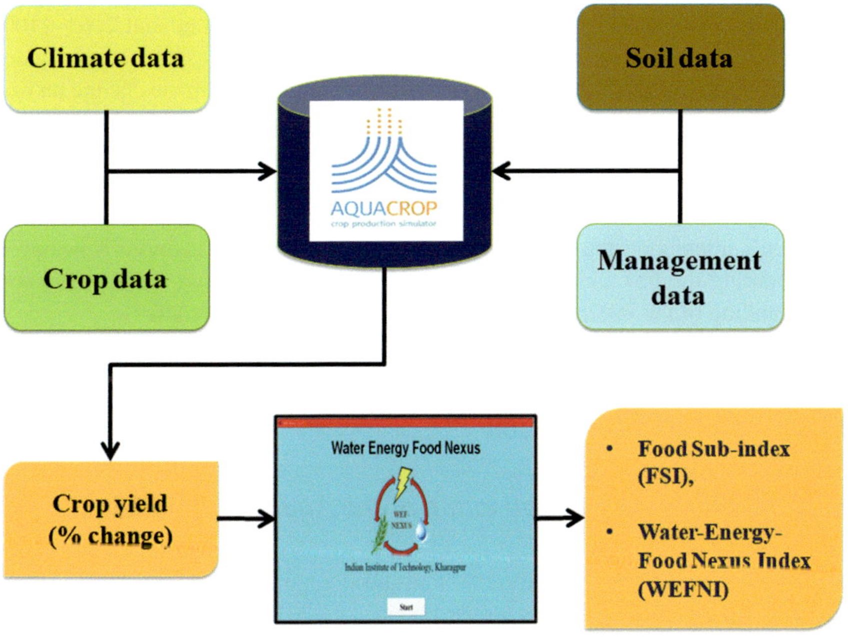

Fig. 8.2 Overall methodological framework

management data, and crop parameters. Calibration of the Aquacrop model is conducted using field observation data from the 2015 Kharif and Rabi seasons, followed by validation using data from the subsequent years, 2016 and 2017. Model performance is rigorously assessed using two key statistical indices: root mean square error (RMSE) and percentage bias (PBIAS).

Attaining low Root Mean Square Error (RMSE) and Percent Bias (PBIAS) values below 15% signifies a robust agreement between the simulated and observed data, as established by Kadiyal et al. in 2015. The RMSE and PBIAS are computed using the following equations.

$$RMSE = \left[\frac{\sum_{i=1}^{n} (P_i - O_i)^2}{n} \right]^{0.5} \tag{8.13}$$

$$PBIAS = \frac{P_i - O_i}{O_i} \times 100 \tag{8.14}$$

where P_i = Simulated yield, O_i = observed yield, and n = observations' number.

Following the calibration and validation of the Aquacrop model, simulations were conducted for various future years. The historical rice yields spanning 1981–2014 were simulated using observed weather data from the Indian Meteorological Department (IMD). Additionally, predictions were made for 2021–2060 and 2061–2100, considering climatic data forecasted for each block within the research region under the SSP245 and SSP585 scenarios. To evaluate the impact of climate change on rice production, future yields have been compared with those from the historical period, and the percentage change in crop yield has been accurately calculated.

Along with other data from Table 8.1, this percentage change in crop yield is used as input data for the WEF Nexus model. The WEF Nexus model then examines the basin's interconnected dynamics of WEF resources. This analysis is achieved by computing the FSI and the basin's WEFNI values, providing a comprehensive understanding of the interrelationships between these crucial resources.

8.3 Results and Discussion

8.3.1 Calibration and Validation of the AquaCrop Model

The AquaCrop model underwent calibration and validation processes utilizing experimental field data for the IR36 cultivar obtained in 2015, 2016, and 2017, respectively. The AquaCrop model's output yield was generated by adjusting the main menu's crop, management, and soil parameters. The simulated yield closely aligned with the observed yield during the calibration year 2015 for both the Kharif and Rabi seasons. Specifically, the model projected rice yields of 3.302 tons/ha and 5.088 tons/ha during the Kharif and Rabi seasons, respectively, compared to observed values of 3.550 tons/ha and 5.315 tons/ha. Table 8.3 shows the calibrated values of the AquaCrop parameters for the IR36 rice variety for the Kharif and Rabi seasons, respectively. Following model calibration, the model has been validated for 2016 and 2017 for both the Kharif and Rabi seasons. Table 8.4 present the actual and simulated rice yields and the statistical indices, which are getting from the model calibrartion. As evident from Table 8.4, the RMSE values are 0.48 and 0.51 tons/ha for the Kharif and Rabi seasons, respectively, which fall within the permissible range of 0.2–0.5 tons/ha (Sutedy and Qomariyah 2022). The PBIAS values are – 11.46% and – 10.68% for the Kharif and Rabi seasons, respectively i.e., within the permissible limit of 15% (Cardoso de Salis et al. 2019). Thus, the AquaCrop model is considered well-calibrated and validated. However, the negative PBIAS in both years shows that the model underestimates the rice yield.

Table 8.3 Calibrated the AquaCrop model parameter values for the Kharif and Rabi seasons

Parameter description	Unit	Calibrated values (Kharif)	Calibrated values (Rabi)
Time to start of canopy senescence	day	81	93
Maximum canopy cover (CC_x)	%	81	81
Time to maximum canopy cover	day	59	79
Water productivity (WP)	kg/m^3	0.83	0.89
Canopy growth coefficients (CGC)	% per day	10.91	12.73
Canopy size of transplanted seedling	cm^2/plant	6.25	6.25
Canopy decline coefficient (CDC)	% per day	0.01	0.03
Time to recover transplant	days	7	9
Average root expansion	cm/day	0.8	1
Reference harvest index	%	40	43

Table 8.4 Validation results of the Aquacrop model for the Kharif and Rabi seasons

Season	Year	Observed yield (tons/ha)	Simulated yield (tons/ha)	PBIAS (%)	RMSE (tons/ha)
Kharif	2016	3.365	3.181	– 11.46	0.48
	2017	3.998	3.337		
Rabi	2016	4.745	4.258	– 10.68	0.51
	2017	4.742	4.215		

8.3.2 Analysing the Impact of Climate Change on Rice Yield

The analysis of climate change impact on rice yield encompasses both near and far future periods for the Kharif and Rabi seasons, employing the SSP245 and SSP585 scenarios from the EC-Earth3 GCM of CMIP6 across all 24 blocks in the Kangsabati river basin. Across all future projection simulations, there was a consistent decline in rice yield compared to the historical period. Figures 8.3 and 8.5 illustrate the percentage reduction in rice yields under SSP245 and SSP585 scenarios for the near future (2021–2060) and far future (2061–2100) for the Kharif and Rabi seasons, respectively. As evident from Fig. 8.3, in the near-future scenario under SSP245, most blocks experience a yield reduction of 5–10%. However, Arsha and Baghmundi blocks are likely to experience a 10–15% reduction in rice yield compared to the historical period. On the contrary, under the SSP245 far-future scenario, a large number of blocks (Arsha, Baghmundi, Jhalda-II, Joypur, Purulia-Balarampur, Barabazar, Bundwan, Binpur-II, and Ranibandh) show a decline of 10–15% in rice yield, while others show a decline of 5–10%. In the SSP585 near-future scenario (Fig. 8.3), most blocks show a decline of 10–15% in rice yield, while a few blocks (Arsha, Baghmundi, Balarampur, Purulia-I, Purulia-II, Hura, Barabazar, Puncha, Haribandh, and Bundwan) show a reduction of 15–20%. For the SSP585 far-future scenario, all blocks in the study area show a 15–20% decline in rice yield. Dubey et al. (2021) also reported up to a 15% reduction in rice yield in West Bengal by the end of the twenty-first century under future RCP scenarios. Debnath et al. (2021) similarly documented a substantial 30–60% reduction in rice yield within the study area attributed to the influence of climate change.

In the near-future scenario of SSP245 in the Rabi season (Fig. 8.4), most of the blocks show a reduction of 10–15% in rice yield, except a few blocks (Midnapur, Jhargram, Kharagpur-I, and Kharagpur-II), which experience a reduction of 5–10% compared to the historical period. Under the SSP245 far-future scenario, a few blocks (Binpur-I, Midnapur, Jhargram, Kharagpur-I, and Kharagpur-II) show a decline of 10–15%, while other blocks have a reduction of 15–20%. In the SSP585 near-future scenario (Fig. 8.4), most blocks show a reduction of 15–20%, while a few blocks (Arsha, Baghmundi, Jhalda-II, Joypur, Purulia-I, Purulia-II, Balarampur, and Barabazar) exhibit a reduction of 20–25%. In the SSP585 far-future scenario, most blocks in the study area show a reduction of 20–25%, while a few blocks (Arsha, Baghmundi, Jhalda-II, Joypur, Purulia-I, Purulia-II, Balarampur, Barabazar, Hura, and Puncha) experience a 20–30% reduction.

In summary, rice yield is expected to decline in the future due to the adverse impact of climate change and related water and temperature stresses affecting the physiological characteristics of the crop. The SSP585 scenario considered the worst-case, assumes higher greenhouse gas emissions and more severe climate change impacts than SSP245, leading to more significant yield reductions. Additionally, compared to the Kharif season, the Rabi season will likely experience a higher yield reduction due to high-temperature stress. Figures 8.3 and 8.4 also demonstrate that the upstream portion of the basin experiences higher crop reduction compared to the

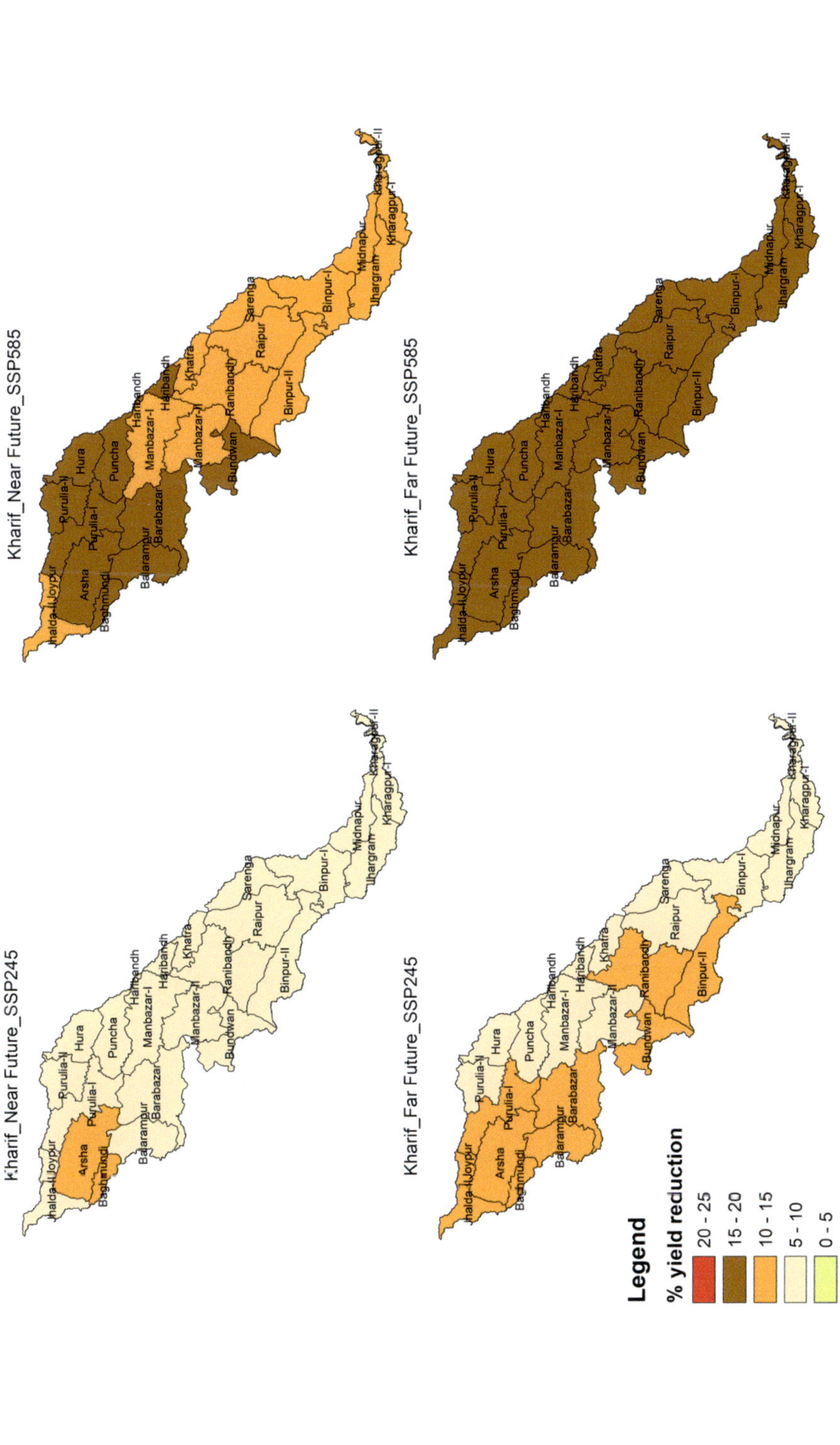

Fig. 8.3 Percent (%) reduction in the rice yield in the Kharif season under SSP245 and SSP585 scenarios for near-future and far-future periods in the Kangsabati River basin

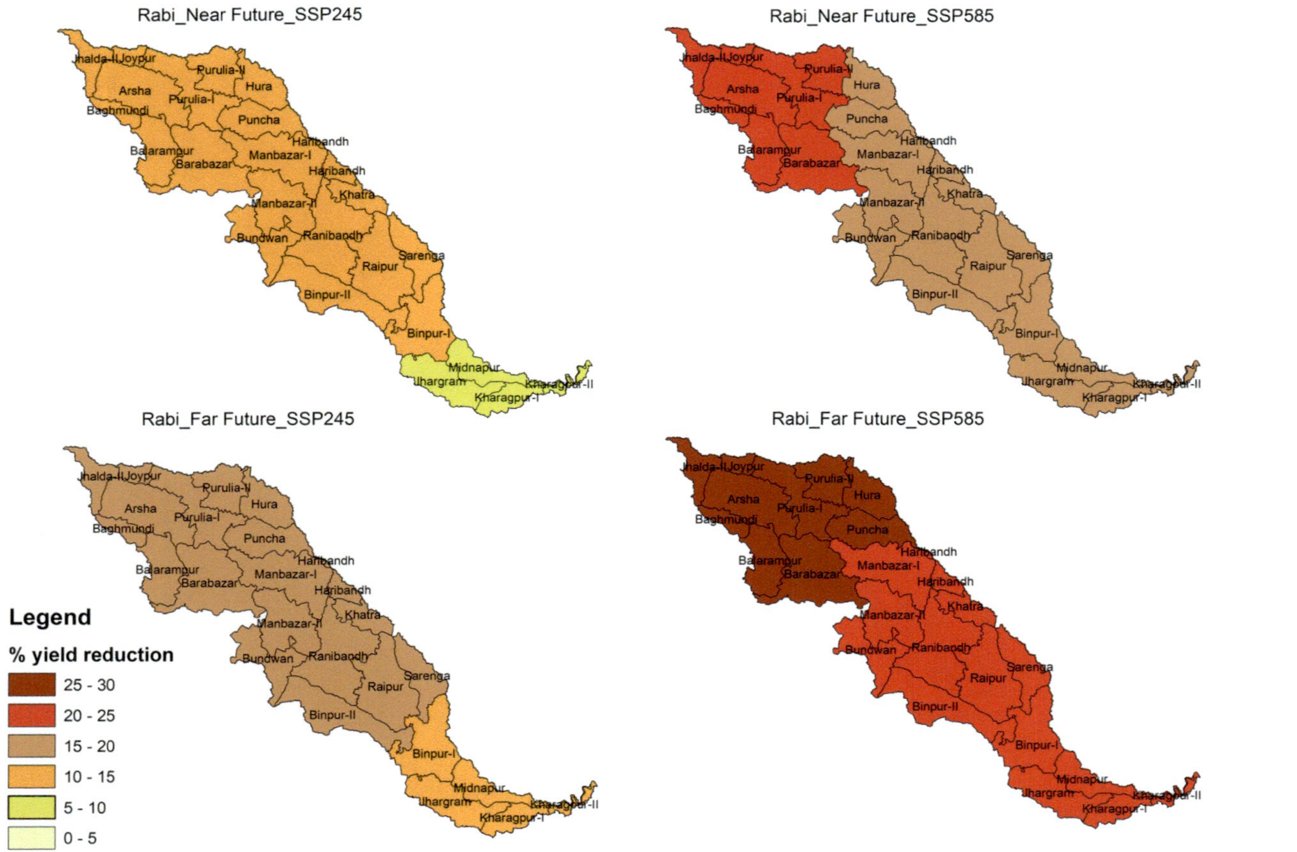

Fig. 8.4 Percent (%) reduction in the rice yield in the Rabi season under SSP245 and SSP585 scenarios for near-future and far-future periods in the Kangsabati river basin

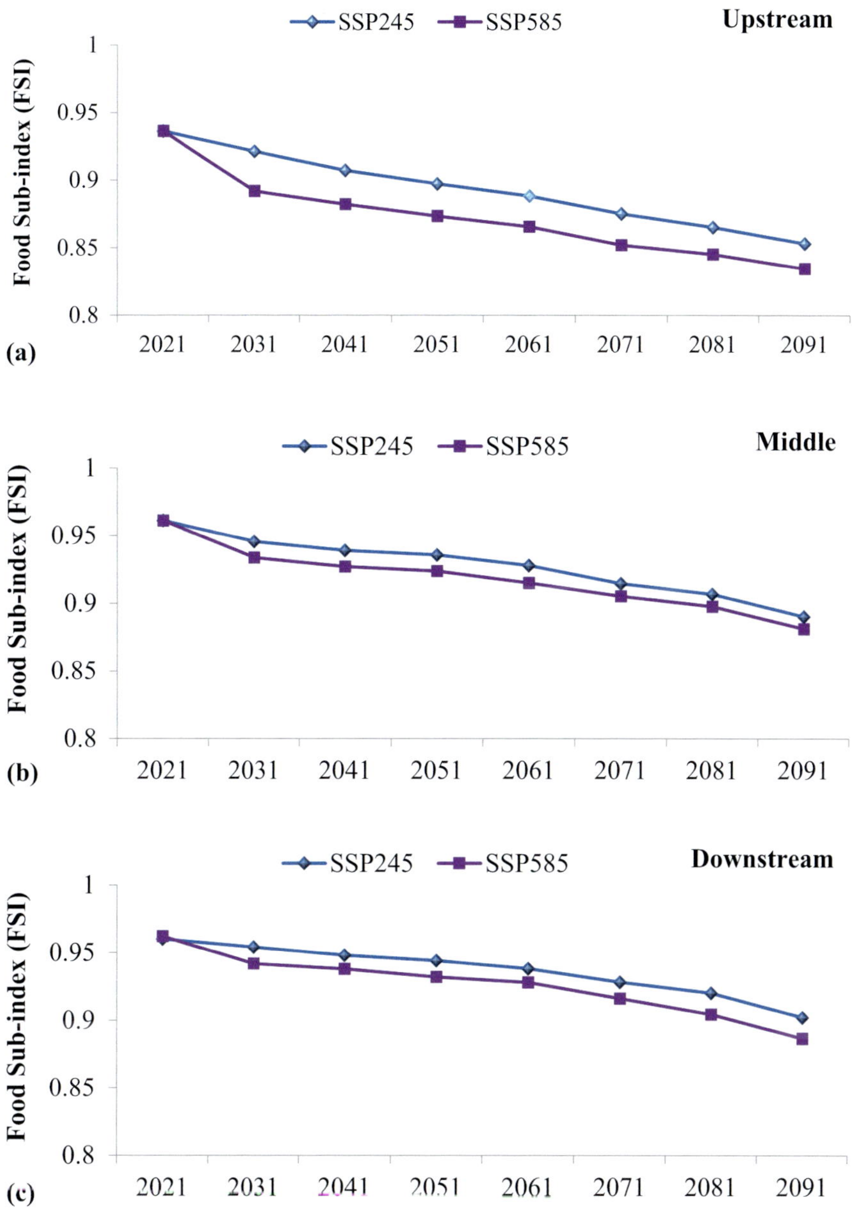

Fig. 8.5 Food Sub-Index (FSI) of **a** upstream, **b** middle, and **c** downstream blocks of the Kangsabati river basin under SSP245 and SSP585 scenarios

middle or downstream blocks due to higher temperatures and lower rainfall in the upstream portion.

8.3.3 Analysis of Climate Change-Induced Rice Yield Impact on FSI and WEFNI

The evaluation of climate change's impact on rice yield, influencing the Food Sub-Index (FSI) and Water-Energy-Food Nexus Index (WEFNI) in the future, is carried out utilizing the Water-Energy-Food Nexus model under SSP245 and SSP585 scenarios across all 24 blocks of the Kangsabati river basin. Figures 8.5 and 8.6 visually represent the outcomes of the model simulations. Based on the similarity in results, the study area is categorized into three sections: the upstream, middle, and downstream portions of the Kangsabati River basin blocks.

Figure 8.5 presents the FSI, reflecting food security in the upstream, middle, and downstream blocks of the Kangsabati River basin in the years 2021, 2031, 2041, 2051, 2061, 2071, 2081, and 2091 under SSP245 and SSP585. The FSI values range from 0 to 1, with higher values indicating better food security. In the upstream blocks (Fig. 8.5a), in both scenarios, FSI decreases over time. In SSP245, FSI starts at 0.936 in 2021 and decreases to 0.853 in 2091, while in SSP585, FSI starts at 0.936 in 2021 and decreases more rapidly to 0.834 in 2091. In the middle blocks (Fig. 8.5b), FSI is generally higher than in the upstream blocks but decreases over time in both scenarios. SSP245's index starts at 0.961 in 2021 and decreases to 0.89 in 2091. In SSP585, the index starts at the same value of 0.961 in 2021 and decreases to 0.881 in 2091. In the downstream blocks (Fig. 8.5c), FSI values are the highest among the three blocks and decrease over time in both scenarios. In SSP245, the index starts at 0.96 in 2021 and decreases to 0.90 in 2091. SSP585's index starts at 0.96 in 2021 and decreases to 0.89 in 2091. Overall, it may be concluded that food security decreases throughout the basin under both scenarios. The middle and downstream blocks generally have better food security than the upstream blocks under both scenarios. It may be noted that these projections are subject to uncertainties and assumptions made in the modeling process. The forecasts can inform policy decisions and initiatives to improve food security by minimizing the effects of climate change.

Figure 8.6 presents the WEFNI values, reflecting the overall WEF security of the area from 2021 to 2091 under SSP245 and SSP585. The figure shows that WEFNI under SSP245 is consistently higher than SSP585 for all years except 2081 and 2091. However, WEFNI exhibits a fluctuating trend under both SSP scenarios. These fluctuations may be attributed to changes in the water, food sub-indices, and amount of rainfall each year, which may directly impact the availability of water, energy, and food resources in the region.

In the upstream blocks of the Kangsabati River basin (Fig. 8.5a), WEFNI varies from 0.845 in 2021 to 0.753 in 2091 under the SSP245 scenario. In SSP585, WEFNI varies from 0.845 in 2021 to 0.737 in 2041 and then increases to 0.8 in 2091. Under

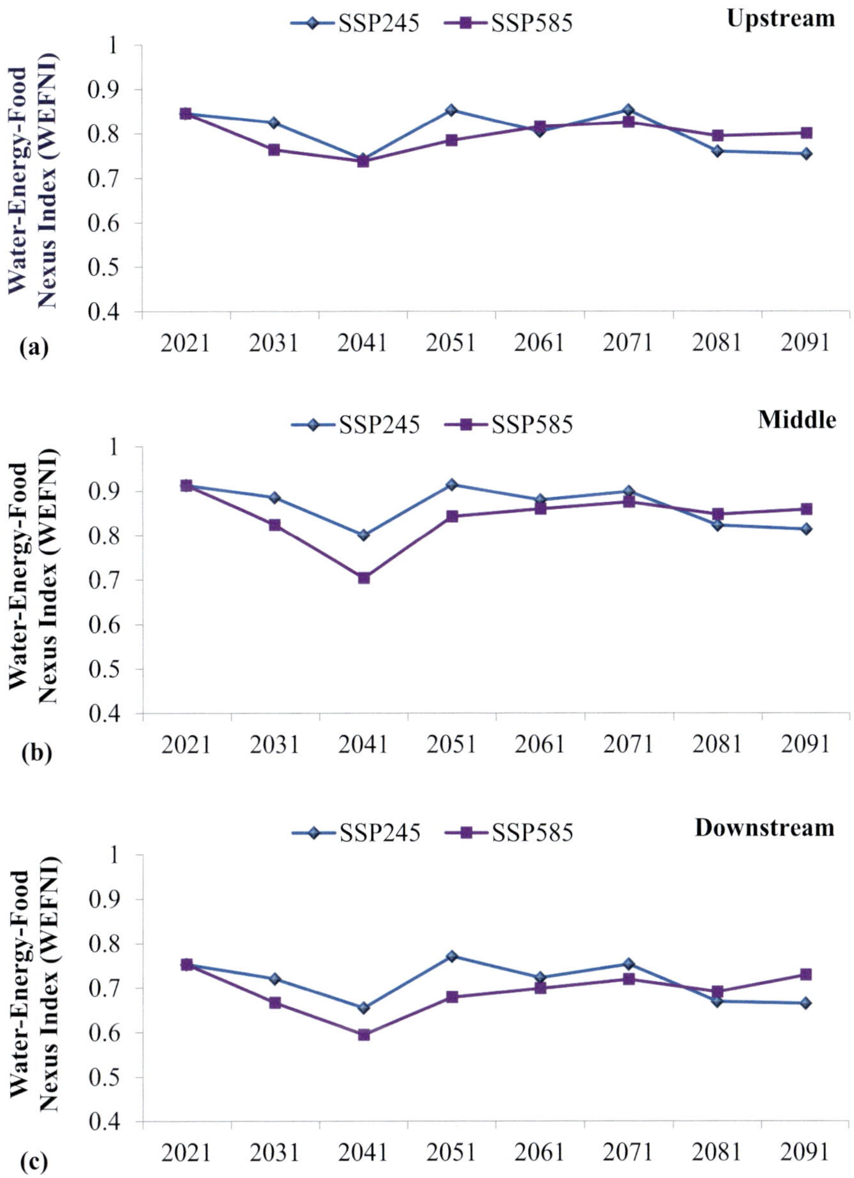

Fig. 8.6 Water-energy-food nexus index (WEFNI) of **a** upstream, **b** middle, and **c** downstream blocks of the Kangsabati river basin under SSP245 and SSP585 scenarios

the SSP245 scenario, WEFNI ranges in the middle blocks (Fig. 8.5b) from 0.912 in 2021 to 0.813 in 2091. WEFNI decreases from 0.912 in 2021 to 0.703 in 2041 and then rises to 0.858 in 2091 under the SSP585 scenario. For blocks in the downstream end of the Kangsabati River basin (Fig. 8.5c), WEFNI varies from 0.752 in 2021 to 0.664 in 2091. However, under SSP585, WEFNI values decline from 0.752 in 2021 to 0.594 in 2041 and then increase to 0.728 by 2091. The middle blocks of the basin generally show higher WEFNI than the upstream and downstream blocks. Also, WEFNI values under SSP245 are generally higher than those under SSP585. The changes in WEFNI do not solely depend on the reduction in paddy yield. This study shows that the decreasing trend in WEFNI is more pronounced due to changes in the amount of rainfall and the Water Scarcity Index (WSI). The results suggest that future WEF security in the Kangsabati basin may face significant challenges, particularly under the SSP585 scenario. The results may guide policy and investment decisions and monitor and evaluate progress toward achieving resource sustainability in the basin.

8.4 Conclusions

Climate change substantially impacts world food production, reducing crop growth due to weather variability. This study primarily examines how climate change affects rice yield and what it means for the water-energy-food nexus in the Kangsabati River basin. The main findings are as follows:

1. Both future scenarios (SSP245 and SSP585) project a decrease in rice yield for both Kharif and Rabi seasons.
2. SSP245 anticipates a 10–15% decline in rice yield during the Kharif season, while SSP585 predicts a 15–20% reduction. During the Rabi season, SSP245 forecasts a 15–20% decrease, and SSP585 indicates a 20–25% decline.
3. The Food Security Index (FSI) shows a continuous decline in food security across all study area blocks, signalling a future decrease. Notably, the decline is more significant in the SSP585 scenario than SSP245.
4. Downstream blocks exhibit higher food security than other basin areas, with FSI values ranging from 0.96 to 0.90 in SSP245 and 0.96 to 0.89 in SSP585.
5. Regarding overall Water-Energy-Food (WEF) security, SSP245 demonstrates higher WEF Nexus Index (WEFNI) values than SSP585. The middle portion blocks of the basin are identified as the most secure. For SSP245, WEFNI values range from 0.912 in 2021 to 0.813 in 2091. In contrast, SSP585's WEFNI values vary from 0.912 in 2021 to 0.703 in 2041, increasing to 0.858 in 2091.

References

Almorox J, Grieser J (2015) Calibration of the Hargreaves-Samani method for the calculation of reference evapotranspiration in different Köppen climate classes. Hydrol Res 47(2):521–531

Asseng S, Martre P, Maiorano A, Rötter RP, O'Leary GJ, Fitzgerald GJ, Ewert F (2019) Climate change impact and adaptation for wheat protein. Glob Change Biol 25(1):155–173

Banerjee S, Das S, Mukherjee A, Mukherjee A, Saikia B (2016) Adaptation strategies to combat climate change effect on rice and mustard in Eastern India. Mitig Adapt Strat Glob Change 21:249–261

Boonwichai S, Shrestha S, Babel MS, Weesakul S, Datta A (2019) Evaluation of climate change impacts and adaptation strategies on rainfed rice production in Songkhram River Basin, Thailand. Sci Total Environ 652:189–201

Cardoso de Salis HH, Monteiro da Costa A, Moreira Vianna JH, Azeneth Schuler M, Künne A, Sanches Fernandes LF, Leal Pacheco FA (2019) Hydrologic modeling for sustainable water resources management in urbanized karst areas. Int J Environ Res Public Health 16(14):2542

Chaturvedi RK, Joshi J, Jayaraman M, Bala G, Ravindranath NH (2012) Multi-model climate change projections for India under representative concentration pathways. Curr Sci 791–802

Chun JA, Li S, Wang Q, Lee WS, Lee EJ, Horstmann N, Park H, Veasna T, Vanndy L, Pros K, Vang S (2016) Assessing rice productivity and adaptation strategies for Southeast Asia under climate change through multiscale crop modeling. Agric Syst 143:14–21

Das J, Das S, Umamahesh NV (2023) Population exposure to drought severities under shared socioeconomic pathways scenarios in India. Sci Total Environ 867:161566

Debnath S, Mishra A, Mailapalli DR, Raghuwanshi NS, Sridhar V (2021) Assessment of rice yield gap under a changing climate in India. J Water Clim Change 12(4):1245–1267

Dubey M, Mishra A, Singh R (2021) Climate change impact analysis using bias-corrected multiple global climate models on rice and wheat yield. J Water Clim Change 12(4):1282–1296

FAO (2020) Global agriculture towards 2050. https://www.fao.org/fileadmin/templates/wsfs/docs/Issues_papers/HLEF2050_Global_Agriculture.pdf

Hennessy K, Lawrence J, Mackey B (2022) IPCC sixth assessment report (AR6): climate change 2022-impacts, adaptation and vulnerability: regional factsheet Australasia. https://policycommons.net/artifacts/2264302/ipcc_ar6_wgii_factsheet_australasia/3023355/

Jones JW, Hoogenboom G, Porter CH, Boote KJ, Batchelor WD, Hunt et al (2003) The DSSAT cropping system model. Eur J Agron 18(3–4):235–265

Kadiyala MDM, Jones JW, Mylavarapu RS, Li YC, Reddy MD (2015) Identifying irrigation and nitrogen best management practices for aerobic rice–maize cropping system for semi-arid tropics using CERES-rice and maize models. Agric Water Manage 149:23–32

Kang Y, Khan S, Ma X (2009) Climate change impacts on crop yield, crop water productivity and food security–a review. Prog Nat Sci 19(12):1665–1674

Keating BA, Carberry PS, Hammer GL, Probert ME, Robertson MJ, Holzworth et al (2003) An overview of APSIM, a model designed for farming systems simulation. Eur J Agron 18(3–4):267–288

Lizumi T, Ramankutty N (2015) How do weather and climate influence cropping area and intensity? Glob Food Sec 4:46–50

Mahato PK, Singh D, Bharati B, Gagnon AS, Singh BB, Brema J (2022) Assessing the impacts of human interventions and climate change on fluvial flooding using CMIP6 data and GIS-based hydrologic and hydraulic models. Geocarto Int 37(26):11483–11508

Mishra V, Bhatia U, Tiwari AD (2020) Bias-corrected climate projections for South Asia from coupled model intercomparison project-6. Sci Data 7(1):1–13

Mondal K, Chatterjee C, Singh R (2023a) An analytical framework for state level water-energy-food nexus analysis in India: Insight from implemented policies. Environ Sci Policy 141:33–49

Mondal K, Tantuway RB, Chatterjee C, Singh R (2023b) Development of a water-energy-food nexus model for multiscale studies. Irrig Drainage, 5–8. https://doi.org/10.1002/ird.2800

Mondal K, Chatterjee C, Singh R (2023c) Examining the coupling and coordination of water-energy-food nexus at a sub-national scale in India-Insights from the perspective of Sustainable Development Goals. Sustain Prod Consumption 43:140–154

Mortada S, Abou Najm M, Yassine A, El Fadel M, Alamiddine I (2018) Towards sustainable water-food nexus: an optimization approach. J Clean Prod 178:408–418

O'Neill BC, Tebaldi C, Van Vuuren DP, Eyring V, Friedlingstein P, Hurtt G et al (2016) The scenario model intercomparison project (ScenarioMIP) for CMIP6. Geoscientific Model Dev 9(9):3461–3482

Organisation for Economic Co-operation and Development (OECD) (2017) Building food security and managing risk in Southeast Asia. OECD Publishing, Paris, France. https://doi.org/10.1787/9789264272392-en

Pai DS, Sridhar L, Badwaik MR, Rajeevan M (2015) Analysis of the daily rainfall events over India using a new long period (1901–2010) high resolution (0.25× 0.25) gridded rainfall data set. Clim Dyn 45(3):755–776

Richardson KJ, Lewis KH, Krishnamurthy PK, Kent C, Wiltshire AJ, Hanlon HM (2018) Food security outcomes under a changing climate: impacts of mitigation and adaptation on vulnerability to food insecurity. Clim Change 147:327–341

Steduto P, Hsiao TC, Raes D, Fereres E (2009) AquaCrop—the FAO crop model to simulate yield response to water: i. Concepts and underlying principles. Agron J 101:426–437

Stockle CO, Nelson R (1994) Cropping systems simulation: model users manual (Version 1. 02. 00). Biological Systems Engineering Department, Washington State University, p 167

Sutedy MF, Qomariyah NN (2022) Text to image latent diffusion model with dreambooth fine tuning for automobile image generation. In: 5th international seminar on research of information technology and intelligent systems (ISRITI). IEEE, pp 440–445

Swain SS, Kumar SB, Mishra A, Chatterjee C (2023) Sensitive or resilient catchment?: A Budyko-based modeling approach for climate change and anthropogenic stress under historical to CMIP6 future scenarios. J Hydrol 622:129651

Taguta C, Senzanje A, Kiala Z, Malota M, Mabhaudhi T (2022) Water-energy-food nexus tools in theory and practice: a systematic review. Front Water 4:40

Wilcox J, Makowski D (2014) A meta-analysis of the predicted effects of climate change on wheat yields using simulation studies. Field Crop Res 156:180–190

Williams JR, Jones CA, Kiniry JR, Spanel DA (1989) The EPIC crop growth model. Trans ASAE 32(2):497–0511

Wyser K, Kjellström E, Koenigk T, Martins H, Döscher R (2020) Warmer climate projections in EC-Earth3-Veg: the role of changes in the greenhouse gas concentrations from CMIP5 to CMIP6. Environ Res Lett 15(5):054020

Zhang Y, Zhao Y, Chen S, Guo J, Wang E (2015) Prediction of maize yield response to climate change with climate and crop model uncertainties. J Appl Meteorol Climatol 54(4):785–794

Part III
Hydro-chemical Environment under Climate Change

Chapter 9
Exploring the Effects of Contemporary Agricultural Techniques on Soil and Water Pollution in Rice Fields

Ankita Paul and Arti Bhatia

Abstract Rice field agroecosystems are pivotal for global food security, providing rice, a staple for over half the world's population, while sustaining farmers' livelihoods and cultural heritage. However, conventional rice cultivation faces challenges jeopardizing environmental sustainability and agricultural efficiency, including soil degradation, water scarcity, and pest management. Urgent action is needed as unsustainable practices worsen environmental degradation. Addressing these challenges requires integrating sustainable practices, innovative technologies, and policy interventions. Precision agriculture, organic farming, and agroecology can boost productivity while reducing environmental impact. Leveraging emerging technologies like gene editing and smart irrigation offers promising solutions. Aligned with UN Sustainable Development Goals, collaborative efforts are essential to reshape rice cultivation sustainably. Collaboration among stakeholders is crucial to harmonize agricultural practices with environmental and societal needs, ensuring that rice cultivation contributes to global food security while safeguarding ecosystems and communities for generations to come.

Keywords Agroecosystems · Environmental sustainability · Gene editing · Global food security · Precision agriculture · Stakeholder collaboration

9.1 Introduction

Rice field agroecosystems (RAEs) are important human-made habitats providing rice (*Oryza sativa*), a staple crop for more than half of the world's population, and providing revenue, subsistence, and cultural services to farmers (Hollaus et al.

A. Paul (✉) · A. Bhatia
Division of Environment Science, ICAR-Indian Agricultural Research Institute, New Delhi 110012, India
e-mail: ankita06paul@gmail.com

A. Bhatia
e-mail: arti.bhatia@icar.gov.in

A. Nanda et al. (eds.), *Navigating the Nexus*, Water Science and Technology Library 102, https://doi.org/10.1007/978-3-031-76532-2_9

2022). Rice is cultivated in various agroecosystems, adopting agricultural methods and social and environmental factors. Rainfed upland rice, irrigated rice, lowland rice, and flood-prone rice systems are the primary categories the International Rice Research Institute (IRRI) divides RAEs into. However, alternate land uses, switching to other input-intensive crops, decreased yield, disease and pest incidence, labor shortages, and climate variability are just a few of the issues that conventional rice field agroecosystems must address (Bhatt et al. 2023).

Modern rice cultivation methods, involving intensive wet tillage and repeated puddling operations, form an impermeable layer at 15–20 cm depth, hampering water infiltration and root growth (Kukal and Aggarwal 2003). Subsurface compaction negatively affects subsequent crops' establishment, seed emergence, and yield (Kukal and Aggarwal 2003). Long-term intensive rice–wheat cropping in north-western India causes soil salinization, nutrient deficiencies, and reduced nitrogen (N) supply capacity, potentially leading to yield reduction and paddy field abandonment (Killebrew and Wolf 2010). Soil degradation, declining water table, inadequate crop residue recycling, and lack of organic fertilization contribute to micro-nutrient deficiencies and yield stagnation (Khedwal et al. 2023).

Declining crop response to fertilizers poses a significant challenge, compelling farmers to apply fertilizers above recommended doses. In the rice–wheat cropping system in the Indo-Gangetic Plain, the net negative balance of nitrogen, phosphorus, and potassium (NPK) is 2.22 metric tonne (MT) per annum (Biswakarma et al. 2023). This trend jeopardizes future crop productivity improvements. Decreasing water productivity, especially for less water efficient C3 crops like rice, is another major concern (Sharma et al. 2015). Despite high land productivity in Punjab and Haryana (4 tonnes per hectare for rice), water productivity remains low at 0.22–0.60 kg m^{-3}, highlighting inappropriate irrigation practices (Kumar et al. 2022).

On the other hand, unsustainable irrigation practices have led to widespread over-exploitation, causing a rapid decline in groundwater levels (Rodell et al. 2009). This results in water-stressed conditions, impacting per-capita water availability. The excessive cultivation of water-intensive crops like rice exacerbates the problem, contributing to groundwater depletion and environmental concerns. The situation is expected to worsen, threatening water sustainability (Kumar et al. 2022). Excessive N fertilizer uses in intensive rice–wheat cropping systems, particularly in coarse-textured soils, leads to groundwater pollution through nitrate leaching (Bhatt et al. 2016).

Arsenic-contaminated soils in the middle Indo-Gangetic plain pose a serious threat, with arsenic levels in groundwater under rice cultivation exceeding safe limits. This contamination, reaching 312 $\mu g\ L^{-1}$, may affect soil and grain toxicity (Srivastava et al. 2015). Arsenic content in soil surpassing permissible limits, up to 35 mg kg^{-1}, poses risks to human health, with unsafe arsenic levels (0.179–0.932 mg kg^{-1}) found in 8 of 17 rice varieties (Kumar et al. 2022).

The decline in total factor productivity in agriculture, particularly in the rice–wheat cropping system of north-western India, poses a threat to sustainable farming and food security (Kumar and Mittal 2006). Factors such as excessive tillage leading to land degradation and the development of herbicide-resistant weeds contribute to stagnant or declining crop productivity (Aggarwal et al. 2000). Weed management is a critical issue in rice cultivation, with diverse weed flora causing significant yield reductions. It is worth noting that the overreliance on herbicides poses risks of evolution and spread of herbicide-resistant weeds (Kumar et al. 2022). Weedy rice, exacerbated by the shift to direct-seeded rice, demands effective, integrated weed management strategies that balance herbicides with non-chemical methods for long-term success, considering the harmful implications of excessive herbicide uses (Chhokar et al. 2014).

In India, annual crop residue generation exceeds 869 MT, with a 288.14 MT surplus. Cereal crops contribute 478.74 MT, notably rice (169.14 MT) and wheat (149.08 MT) (Deka et al. 2023). Managing rice residue is challenging due to silica content, narrow time window, high handling costs, labor-intensive removal, and lacking storage systems (Rathour et al. 2023). Farmers resort to burning and degrading air quality with CO, CO_2, CH_4, and particulate matter. Incorporating residue in soil improves long-term soil health but faces challenges like energy requirements and N immobilization. Direct seeding with resource-conserving machines is a promising residue management option, offering benefits like reduced soil erosion, improved organic carbon, and climate-resilient crop production systems (Kumar et al. 2023). However, widespread adoption faces hurdles, requiring infrastructure development and policy interventions. These challenges highlight the urgency of addressing soil health and water management for sustainable rice cultivation.

As we navigate the complexities of modern agriculture, a concerted effort towards sustainable practices is essential to safeguard the long-term viability of rice cultivation and the well-being of the communities dependent on these agroecosystems. This book chapter aims to investigate the impacts of contemporary Agricultural practices on soil and water Pollution in rice Fields in line with the United Nations' Agenda 2030 Sustainable Development Goals (SDGs) Goal 2 (Zero Hunger), Goal 3 (good health and well-being), Goal 6 (clean water and sanitation), Goal 15 (Life on Land). To achieve these sustainable development goals, it is crucial to consider innovative approaches and technologies in rice cultivation. Precision agriculture, organic farming, and agroecological practices are among the strategies gaining attention for their potential to enhance productivity while minimizing environmental impact (Duff et al. 2021). Organic farming emphasizes natural processes, avoiding synthetic chemicals, and promoting biodiversity, which can contribute to healthier soil and water ecosystems. Agroecological practices integrate ecological principles into agricultural systems, fostering resilience and sustainability. Initiatives like integrated pest management, crop rotation, and water conservation practices can contribute to the resilience of rice field agroecosystems (HLPE 2019). Moreover, exploring the potential of emerging technologies, such as gene editing for disease-resistant crops or smart irrigation systems, could offer innovative solutions to some of the existing challenges. Additionally, addressing conventional rice field agroecosystems' challenges requires a multidisciplinary approach. Collaborative efforts from researchers, policymakers, farmers, and local communities are essential to develop and implement strategies that balance agricultural productivity with environmental and social considerations. By aligning our agrarian practices with the United Nations' Sustainable Development

Goals, we can work towards a future where rice cultivation meets the nutritional needs of a growing population and contributes to a more resilient and sustainable global food system.

9.2 An Overview of Soil and Water Pollution

9.2.1 Factors Contributing to Soil and Water Pollution

Soil and water pollution result from the introduction of harmful substances into the environment, affecting soil quality and water resources. Various human activities and natural processes contribute to these forms of pollution. Understanding these contributing factors is crucial for developing effective strategies to prevent and mitigate soil and water pollution. Some key factors contributing to soil and water pollution are:

A. **Industrial Activities**: Activities involving furnaces or other manufacturing processes disperse contaminants into the environment. Copper (Cu), lead (Pb) released from industrial activities of enterprises are the main sources of heavy metals (HMs) in the atmosphere (Liu et al. 2023a, b). Untreated textile mill effluents are a major concern, associated with various hazardous problems for ecosystems and human health. Complex effluent with several dyes, metal contaminations, and other organic agents used for softening, printing, and heat stabilizing accumulate in natural sources, soil, and water if disposed of without or after complete treatment (Mohan and Shukla 2022). Industries release pollutants, including chemicals, heavy metals, and toxins, into water bodies through improper disposal and inadequate treatment facilities (Talema 2023).
B. **Agricultural Practices**: Using herbicides, pesticides, insecticides, and fertilizers in agricultural activities contributes significantly to soil pollution (Upadhyay et al. 2020). The plants grown in such soil can uptake harmful components and pass these through various physiological pathways within the food chain. These soil contaminations ultimately affect an area's vegetation and pollute our future (FAO and UNEP 2021). Excessive use of fertilizers contributes to nutrient runoff, causing eutrophication in water bodies (Fig. 9.1), and elevated nutrient levels lead to algal blooms, oxygen depletion, and harm to aquatic life health (Ongley 1996).

C. **Waste Disposal**: Both accidental and deliberate chemical waste dumping, including illegal practices, introduce hazardous substances into the soil (Kumar et al. 2019). Hazardous substances from illegally dumped waste infiltrate the soil, threatening ecosystems and human health. Landfills, designed for waste storage, can become sources of soil pollution when contaminants leach into the surrounding soil and groundwater. The decomposition of organic waste generates leachate, a liquid containing pollutants such as heavy metals, organic

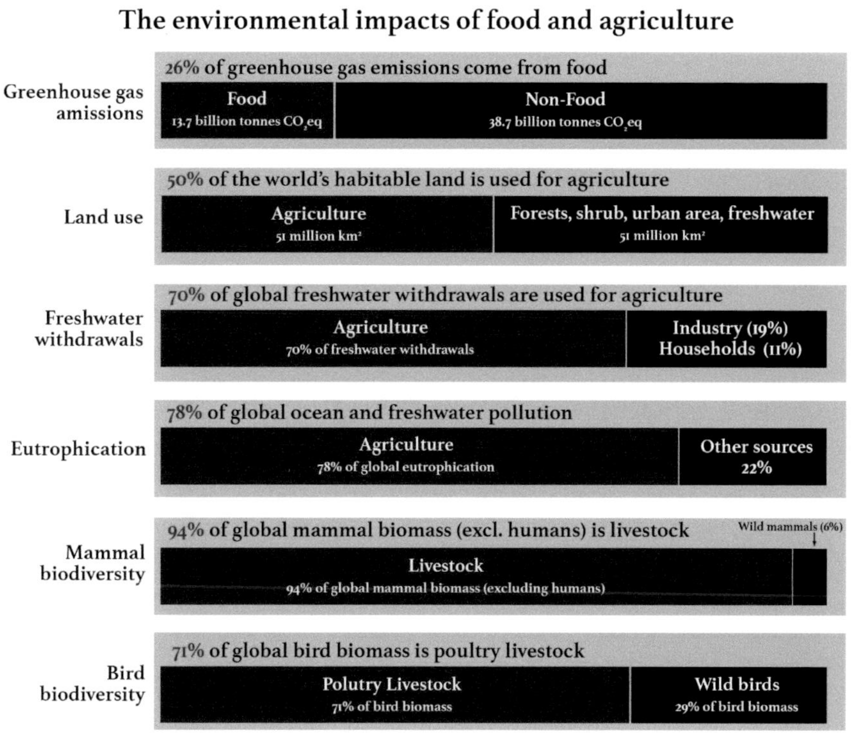

Fig. 9.1 The environmental impacts of food and agriculture. *Data source* Poore and Nemecek (2018); FAO (2017, 2022); Bar-On et al. (2018). Licensed under CC-BY by the author Hannah Ritchie

compounds, and harmful chemicals (Bazienė et al. 2020). Improper disposal of solid waste, including hazardous materials, into water bodies contributes to pollution. Leachate from landfills can contaminate nearby water sources (Bangani et al. 2023). Effective waste management strategies, including proper landfill design, liners, and monitoring systems, are crucial to prevent soil pollution from landfills. Improper disposal of plastic waste results in the accumulation of non-biodegradable materials in water bodies, causing harm to aquatic life and ecosystems (Zhou et al. 2021).

D. **Urbanization and Construction**: The release of construction dust containing substances like asbestos during demolition highlights the importance of adopting secure and regulated construction practices. Managing various chemicals at construction sites, especially those capable of dispersing particulate matter threatens soil health (Sandil and Kumar 2021). Urban areas characterized by impermeable surfaces generate stormwater runoff that transports pollutants such as oil, heavy metals, and chemicals into water bodies, significantly contributing to water pollution. The emission of toxic vehicle pollutants during transportation activities in urban areas introduces contaminants to the soil (Kumar et al. 2019).

E. **Mining Activities and Oil Spills**: The crushing and processing of raw materials in mining activities lead to the emission of toxic substances, especially heavy metals (Mulenga 2022). These pollutants can persist and accumulate, posing risks to aquatic ecosystems and downstream communities (Talema 2023). Unintentional spills and leaks during the storage, transportation, or utilization of chemicals, such as gasoline and diesel at gas stations, are critical contributors to soil pollution (Adeniran et al. 2023). Implementing appropriate handling and containment measures is crucial to avoid such occurrences. Whether accidental or intentional, releasing oil into water bodies, frequently resulting from shipping and industrial activities, can cause significant harm to marine ecosystems. Oil spills harm aquatic life and result in enduring environmental consequences (Talema 2023).

F. **Natural Accumulation and Atmospheric Deposition**: Imbalances between atmospheric deposition and precipitation water cause the concentration and accumulation of specific compounds and heavy metals in arid environments (Liu et al. 2023a, b). This natural process highlights the intricate dynamics between atmospheric interactions and soil pollution. Airborne pollutants, including industrial emissions and atmospheric deposition, settle into water bodies (Shi et al. 2021). This deposition introduces contaminants like heavy metals and chemicals into aquatic ecosystems.

G. **Sewage Discharge**: Sewer systems are crucial in managing urban wastewater, but leaks and overflows can lead to soil pollution (Ashraf et al. 2014). Inadequate sanitation systems and untreated sewage discharges introduce pathogens, nutrients, and organic matter into water bodies, posing risks to public health and ecosystems (Ducci et al. 2022). The elevated nutrient levels, mainly N and phosphorus, contribute to water pollution by triggering eutrophication, leading to imbalances in aquatic ecosystems. Furthermore, the chemical pollutants in sewage, including heavy metals and pharmaceuticals, can accumulate in water, affecting water quality and aquatic life. In agriculture, applying sewage sludge as fertilizer may lead to nutrient imbalances and the potential transfer of pathogens to the soil, impacting crops and groundwater quality.

H. **Climate Change**: Climate change is intricately intertwined with the issues of soil and water pollution, creating a complex interplay of environmental challenges. The impacts of climate change significantly contribute to the worsening of soil and water pollution. Factors such as extreme weather events, temperature variations, rising sea levels, changes in precipitation patterns, and melting glaciers and permafrost all influence the quality of soil and water systems (Ahmed et al. 2020); melting glaciers and permafrost release stored contaminants, affecting water and soil environments (Langer et al. 2023). Temperature changes influence the persistence and mobility of contaminants in soil and water, while rising sea levels contribute to saltwater intrusion, impacting freshwater sources. Altered precipitation patterns concentrate pollutants, posing risks to ecosystems and human health.

I. **Lack of Environmental Policies and Regulation**: Insufficient or inadequately enforced environmental regulations lead to irresponsible waste disposal and pollution (Sawhney and Sawhney 2023). The absence of strict policies encourages industries and individuals to neglect proper environmental stewardship. In Austria, the Environmental Impact Assessment (EIA) is integrated into the Environmental Impact Assessment Act (UVP-G 2000) and various lawsrelated to soil reform like Lower Austria Soil Protection Law and Burgenland Soil Protection Law, and it is well implemented (Pröbstl-Haider 2022). In India, it is recommended that independent environmental courts be established to ensure the efficient and effective enforcement of environmental laws (Verma 2021). This is essential because the decisions made by the National Green Tribunal are ineffective, as they can be contested in the Supreme Court, leading to a time-consuming process since Supreme Court decisions often take years to be delivered.

9.2.1.1 Impacts of Soil Pollution

The level of soil pollution is continuously increasing, and the consequences of soil pollution are wide-ranging (Table 9.1) and can significantly impact the environment, human health, and socio-economic well-being (Naidu et al. 2021).

9.2.1.2 Prevention and Remediation

To prevent soil contamination, implement best practices to reduce or eliminate the release of pollutants into the soil. These may include using cleaner production techniques, applying integrated pest management, adopting organic farming methods, recycling or treating waste, preventing leaks or spills, or using protective barriers. Also, educate and train workers, farmers, and communities on handling and disposing of hazardous substances safely and responsibly (Liu et al. 2023a, b). Further, monitoring soil quality regularly and systematically is necessary to prevent contamination. Use various tools and indicators to measure the soil's physical, chemical, and biological properties and the presence and concentration of pollutants, sensors, drones, or satellites to collect and analyze data remotely and efficiently.

Depending on the type and extent of contamination, we can use various methods such as excavation, encapsulation, bioremediation, phytoremediation, or thermal treatment to remove, isolate, or degrade pollutants in the soil. Using natural attenuation or enhanced recovery techniques to accelerate the natural processes of dilution, dispersion, or degradation of pollutants in the soil (Mangkoedihardjo and Samudro 2023). Various strategies are used to restore soil functions and ecosystem services to improve soil health, fertility, and productivity, such as adding organic matter, applying fertilizers, planting cover crops, or implementing soil conservation practices. The use of ecological engineering or restoration ecology methods to enhance

Table 9.1 Comprehensive overview of impacts of soil pollution

Sr. No.	Impacts	Description	References
1.	Damage to health	Soil pollutants entering the food chain lead to various illnesses. Additionally, the spread of antibiotics in the environment contributes to developing drug-resistant pathogens, posing a threat to public health	Steffan et al. (2017)
2.	Poorer harvests	Soil pollution agents jeopardize global food security by reducing both the quantity and quality of harvests, which also leads to shortages and affects the nutritional content of crops	Gavrilescu (2021)
3.	Climate change	Globally, agriculture is reported to account for 34% of land area usage, contributes circa 24% of global greenhouse gas (GHG) emissions, and accounts for 70% of freshwater usage. Land use change and land degradation contribute to GHG emissions, with an estimated 17% of current annual emissions coming from such activities	UN Water (2015)
4.	Population displacement	Soil degradation and climate change are projected to drive significant population displacement, potentially leading to the migration of millions of people by 2050	Clement (2021)
5.	Species extinction	Soil contamination is identified as an important factor that could contribute to the sixth mass extinction event in history. The decline in wildlife populations, as indicated by reports like WWF's living planet report, underscores the severity of the issue	WWF's living planet report
6.	Desertification	The impact of soil pollution contributes to desertification, and projections suggest that arid areas could host 45% of the world's population by 2050. Additionally, global wetland areas have significantly decreased over the past three centuries	UNCCD (n.d.)
7.	Economic impact	Global economic losses from soil degradation are expected to be substantial, accounting for more than half of the world's annual gross domestic product (GDP). This reflects the broad economic implications of deteriorating soil quality in various sectors	Sartori et al. (2019)

soil biodiversity, resilience, and functionality, such as reintroducing native species, creating habitats, or facilitating succession, is recommended (Wang et al. 2022a, b).

9.2.1.3 Impacts of Water Pollution

Water pollution has profound and far-reaching impacts on the environment and human health (Ridzuan 2021). The consequences of water pollution are diverse, impacting various ecosystems and sectors in detrimental ways (Table 9.2). This contamination can lead to adverse effects such as the depletion of aquatic life, the degradation of water quality, and the potential introduction of harmful substances into the food chain. Therefore, addressing water pollution is crucial for safeguarding ecosystems and promoting public health.

9.2.1.4 Prevention and Remediation

The indiscriminate disposal of treated and untreated solid and liquid wastes into water leads to the contamination of receiving river water with nutrients, organic chemicals, metals, and nanomaterials. Various physical, chemical, biological, ecological, and engineering methods are available for treating polluted river water. Widely employed engineering and physical processes encompass aeration, water transfer, mechanical algae removal, construction of hydraulic structures, and dredging river sediment. Aeration, a sustainable and versatile technique, significantly enhances microbial community diversity and breaks down organic chemicals in river water. However, some engineering methods may threaten the river ecosystem while effectively remediating polluted river water. Consequently, it is advisable to combine physical engineering techniques with biological and ecological engineering processes to facilitate the environmental restoration of rivers (Adejumoke et al. 2018). Riverbank filtration, a natural, gradual, and self-sustainable process, efficiently removes organic and inorganic contaminants from river water without adverse effects.

Microbial agents, ecological floating beds, constructed wetlands, and biofilm reactor techniques utilize microorganisms and plant-based bioremediation processes to decompose organic chemicals and eliminate nutrients and metals from river water. Despite variations in their purification capabilities, these methods demonstrate high economic, environmental, and ecological benefits, requiring minimal maintenance and avoiding secondary pollution. Biofilm reactors prove highly efficient in remediating polluted river water through the growth of microbial communities in biofilms. Directly mixing microbial agents with river water moderately removes Ammoniacal N (NH_3-N), Chemical oxygen demand (COD), and Total phosphorus (TP), while engineering-based applications significantly degrade organic matter and remove COD, Biochemical oxygen demand (BOD), and nutrients. Ecological floating bed techniques are widely applicable for water treatment unaffected by water fluctuation, river waves, and inundation (Saini et al. 2023). The key influential parameter for their effectiveness lies in carefully selecting plant species. Incorporating high-capacity adsorbent materials into the floating mat matrix enhances contaminant removal efficiency. Hybrid, integrated, sequential, and engineering-based floating

Table 9.2 Comprehensive overview of impacts of water pollution

Sr. no	Impacts	Description
1.	Ecological impact	**Biodiversity loss**: water pollution can lead to declining aquatic biodiversity, affecting fish, insects, plants, and other organisms. Contaminated water bodies may no longer support diverse ecosystems **Habitat destruction**: polluted water can alter or destroy aquatic habitats, disrupting many species' breeding and feeding grounds
2.	Human health impact	**Waterborne diseases**: contaminated water is a breeding ground for harmful microorganisms, leading to waterborne diseases such as cholera, dysentery, and gastrointestinal infections **Toxic contaminants**: exposure to pollutants in water like heavy metals, pesticides, and industrial chemicals can have adverse health effects (Liu et al. 2022)
3.	Economic impact	**Impact on fisheries**: water pollution can harm fish populations, affecting commercial and recreational fisheries. This, in turn, has economic implications for communities dependent on fishing industries **Cost of water treatment**: treating polluted water for safe consumption becomes a significant financial burden for governments and communities, often requiring advanced technologies and substantial investments **Impact on tourism**: polluted water bodies are less attractive for tourism and recreational activities. This can negatively impact local economies dependent on tourism-related revenue
4.	Agricultural impact	**Soil contamination**: runoff from water pollution can contaminate soil, affecting agricultural productivity and the quality of crops. Using polluted water for irrigation may lead to the accumulation of harmful substances in food crops (Water pollution from and to agriculture—International Decade for Action 2018) **Livestock health**: animals that rely on contaminated water sources can suffer from health issues, impacting livestock production
5.	Eutrophication	**Algal blooms**: excessive nutrient runoff, often from agricultural activities, can lead to algal blooms in water bodies. These blooms deplete oxygen levels, leading to "dead zones" where aquatic life cannot survive
6.	Climate impact	**Disruption of hydrological cycle**: water pollution can disrupt the natural hydrological cycle, affecting precipitation patterns and contributing to climate change. This impact extends beyond local ecosystems (Adejumoke et al. 2018)
7.	Long-term consequences	**Persistent contaminants**: some pollutants, such as certain chemicals and heavy metals, may persist in the environment for extended periods, causing long-term and irreversible damage (Liu et al. 2022)
8.	Social impact	**Displacement and migration**: communities relying on polluted water sources may face displacement, migration, or social unrest due to the challenges posed by water scarcity and contamination

bed wetlands demonstrate maximum water purification efficiency, overcoming drawbacks in single-constructed floating beds or wetlands. Despite showing contradictory and moderate performances, aquatic animals are not recommended for restoring polluted river water due to potential limitations in their effectiveness (Anawar and Chowdhury 2020).

9.2.2 Concerns on a Global Scale

Water and soil pollution are pressing global challenges that transcend borders, impacting ecosystems, human health, and socio-economic stability on an unprecedented scale. These concerns underscore the urgent need for coordinated international efforts to address the multifaceted consequences of pollution (Tozer 2023). Water pollution poses a dire threat to public health, manifesting as a global crisis with far-reaching consequences. In regions lacking access to clean water, contaminated sources become breeding grounds for waterborne diseases, perpetuating health disparities and affecting millions. Simultaneously, marine, and freshwater ecosystems suffer degradation from pollutants like plastics, chemicals, and nutrient runoff. It jeopardizes biodiversity, disrupting ecological balance and threatening aquatic life and food chains. Moreover, water pollution intensifies the already critical issue of global water scarcity, impacting drinking water supplies, agriculture, industry, and water-dependent economies. The transboundary nature of water pollution necessitates international cooperation to develop and enforce effective management strategies. Interactions between water pollution and climate change create a concerning feedback loop, with extreme weather events exacerbating pollution and vice versa (Mdee et al. 2022).

Similarly, soil pollution poses a significant threat to global food security, particularly in agricultural practices involving pesticides, herbicides, and chemical fertilizers. Rural communities relying on agriculture bear the brunt, as contaminated soils compromise crop production, exacerbating poverty and food insecurity (Lü et al. 2015). The persistence of organic pollutants and heavy metals in soils poses a long-term threat to ecosystems. At the same time, emerging contaminants like pharmaceuticals and microplastics raise concerns about their impact on biodiversity and human health. Limited regulatory frameworks compound the challenge of managing these novel pollutants, and the loss of essential ecosystem services further undermines soil fertility and ecosystem resilience in the face of environmental changes. Both water and soil pollution demand urgent attention and concerted global efforts to mitigate their far-reaching and interconnected impacts.

9.3 Innovations in Rice Farming: Current Trends and Technological Progress

9.3.1 Varieties with High Yield Potential

Japonica varieties are typically grown in the temperate regions of Asia, Africa, and Latin America. While yields are generally high in these areas, the rice types must be further improved for yield potential and resistance to insects and diseases and for environmental stress tolerance, especially cold. High grain quality is a crucial requirement. Indica, including aus type and tropical japonica rice varieties, are good sources for desirable traits with improved yield and resilience (Wang et al. 2022a, b). Traits are targeted for disease resistance, drought tolerance, and nutrient efficiency. However, resulting progeny often suffer from spikelet sterility, poor quality, and other undesirable characteristics. New quantitative breeding strategies are employed to systematically improve the mean value of desirable traits in temperate japonica backgrounds (Wang et al. 2022a, b). Pusa 2090, developed by Indian Agricultural Research Institute (IARI), New Delhi, is an enhanced iteration derived from a cross between CB-501, an early-maturing Japonica rice line, and Pusa-44. It combines the quicker maturation cycle of CB-501 and high yield attributes of Pusa-44, presenting a compelling alternative. The shorter duration of 120–125 days to mature, while maintaining a comparable yield of 3.5 t ha^{-1}, addresses the core issue of stubble burning in the states of Punjab, Haryana, Rajasthan, and Uttar Pradesh in India (AICRIP 2023). A list of varieties is furnished below (Table 9.3).

9.3.2 Mechanization

The advent of Autonomous Farming Machinery heralds a transformative era in agriculture, particularly rice farming. By seamlessly integrating robotics and AI-driven machines, these technological marvels bring efficiency and reduced labor dependency to various aspects of the farming process, from planting to harvesting and overall field maintenance. In tandem with the rise of autonomous machinery, the incorporation of Data Analytics for Decision Support has become a game-changer in farm management. This shift towards automation optimizes resource utilization and addresses challenges associated with GHG emissions, offering a promising solution for the future of rice cultivation (Castelein et al. 2022). By leveraging big data analytics, farmers can employ predictive modeling and data-driven insights to make informed decisions. This innovative approach enhances productivity by comprehensively understanding various factors influencing crop growth, resource allocation, and farm operations (Wolfert et al. 2017). As rice farming embraces these cutting-edge technologies, the synergy between autonomous machinery and data analytics propels agriculture into heightened efficiency, sustainability, and resilience.

Table 9.3 High-yielding rice varieties with potential reaction to pests and diseases and adverse conditions

Sr. no	Variety	Ecology	Year of release	Duration	Grain type	Released by	Average grain yield (t/ha)	Reaction to diseases and pests	Reference
1	CR Dhan 413 (Reeta-Panidhan)	Rainfed shallow lowland	2021	145	SB	SVRC (Odisha)	6.5	Resistant to BPH, WBPH, stem borer (dead heart) and moderate resistance to white ear head, leaf, plant hopper, leaf folder and case worm	ICAR-NRRI (1968–2019)
2	CR Dhan 803 (Trilochan)	Rainfed shallow lowland	2021	150	SB	SVRC (Odisha)	7	Resistant to stem borer (dead heart) and BPH and moderately resistant to white ear head attack, WBPH, leaf folder, plant hopper, and case worm. Moderately resistant to neck blast and rice tungro virus	ICAR-NRRI (1968–2019)
3	CR Dhan 316	Irrigated	2021	130–135	LB	SVRC (Odisha)	6.53	Resistance against stem borer, moderately resistant to leaf folder	ICAR-NRRI (1968–2019)
4	CR Dhan 411	Shallow low land	2021	140	SB	SVRC (Odisha)	5.6	Moderate tolerance to leaf blast, neck blast, brown spot, RTD, and BLB, moderately resistant against stem borer	ICAR-NRRI (1968–2019)
5	CR Dhan 702	Irrigated and shallow lowland	2021	140–145	LS	SVRC (Odisha)	7.5–8	Moderately resistant to leaf blast, neck blast, and resistant to false smut; and moderately resistant to Gal midge	ICAR-NRRI (1968–2019)
6	CR Dhan 703	Boro and irrigated shallow lowland	2021	165 (Boro) and 140–145 (irrigated)	LS	SVRC (Odisha)	7.5–8	Moderate resistance to leaf BLB, RTV and resistant to false smut; and moderate resistance to Gal midge	ICAR-NRRI (1968–2019)
7	Pusa Basmati 1847	Irrigated	2021	120	LS	IARI (Pusa, New Delhi)	5.7	Bacterial blight resistance, blast resistance	ICAR-IARI (2021)

(continued)

Table 9.3 (continued)

Sr. no	Variety	Ecology	Year of release	Duration	Grain type	Released by	Average grain yield (t/ha)	Reaction to diseases and pests	Reference
8	Pusa Basmati 1885	Irrigated	2021	135–140	LS	IARI (Pusa, New Delhi)	4.68	Bacterial blight resistance, blast resistance	ICAR-IARI (2021)
9	Pusa Basmati 1886	Irrigated	2021	135–140	LS	IARI (Pusa, New Delhi)	4.49	Bacterial blight resistance, blast resistance	ICAR-IARI (2021)
10	Pusa Basmati 1882	Irrigated	2022	134	LS	IARI (Pusa, New Delhi)	4.69	Tolerant to reproductive stage drought stress	ICAR-IARI (2022)

SVRC: State variety release committee; grain: LS—long slender, LB—long bold, SB—short bold; resistance: BLB—bacterial leaf blight, BPH—brown plant hopper, RTV—rice Tungro virus, WBPH—white back plant hopper

9.3.3 Irrigation Practices

Water-saving technologies are pivotal in rice farming practices, addressing water scarcity concerns and promoting sustainable agricultural practices. Among these technologies, drip irrigation and controlled flooding are innovative approaches that optimize water usage while maintaining crop productivity (Arouna et al. 2023). Drip irrigation, characterized by the targeted water delivery directly to the plant's root zone, minimizes water wastage and fosters efficient nutrient absorption (Fig. 9.2). Controlled flooding, on the other hand, involves

(a) Strategic water management.
(b) Enabling farmers to maintain optimal water levels in rice fields.
(c) Conserving water resources without compromising crop health.

In addition to these methods, integrating efficient water usage through sensor-driven irrigation systems further refines water management in rice farming. These advanced systems leverage real-time sensors to monitor soil moisture levels, weather conditions, and crop water requirements. By providing precise and timely data, sensor-driven irrigation systems enable farmers to make informed decisions, ensuring that water is applied where and when it is needed most (Millán et al. 2020). This conserves water and contributes to improved crop yields and resource efficiency. Adopting these water-saving technologies underscores a commitment to sustainable agriculture, aligning rice farming practices with the imperative of responsible water stewardship for a resilient and productive future.

Fig. 9.2 Drip irrigation in rice field: targeted water delivery directly to the plant's root zone

9.3.4 Exploring the Realm of Biotechnology and Genetic Modification

Clustered Regularly Interspaced Short Palindromic Repeats (CRISPR) technology has emerged as a revolutionary tool in rice breeding, presenting a breakthrough in the quest to improve the nutritional content of this staple crop, especially in developing nations where combating malnutrition is a critical challenge (Zafar et al. 2020). In the collaborative efforts between rice researchers and nutritionists, biotechnological and breeding methods are being harnessed to enhance the nutritional profile of rice (Kobayashi et al. 2023). CRISPR technology allows precise gene editing, creating customized rice varieties with desired nutrition, yield, and resilience traits (Fig. 9.3). Recently, a CRISPR/Cas12a promoter editing (CAPE) system has been developed to improve rice agronomic traits by editing specific gene promoters. A high-yield rice with better lodging resistance than the Green Revolution *OsSD1* mutant was generated by editing the *OsD18* promoter (Zhou et al. 2023). Meanwhile, CRISPR/Cas12i3, a newer addition to the genomic editing arsenal, presents promising advancements over its predecessor, offering potentially heightened specificity and efficiency. Although less extensively explored than CRISPR/Cas9, CRISPR/Cas12i3 demonstrates its potential in rice enhancement efforts (Lv et al. 2023). This accelerates the breeding process for faster crop improvement and opens avenues to address specific nutritional deficiencies by precisely editing genes associated with nutritional content. This personalized approach not only ensures optimal nutrition but also contributes to addressing region-specific dietary needs (Sbai 2024).

Furthermore, integrating GIS maps generated from remote sensing data in Variable Rate Application strategies allows for the customized implementation of inputs such as water, fertilizers, and pesticides. This precision-driven approach optimizes resource utilization, minimizes environmental impact, and aligns with sustainable agricultural practices.

9.3.5 Integrated Pest Management

Integrated Pest Management (IPM) in rice fields is a holistic and sustainable approach that combines various strategies to manage pest populations effectively while minimizing environmental impact and ensuring economic viability for farmers (Hùng et al. 2023).

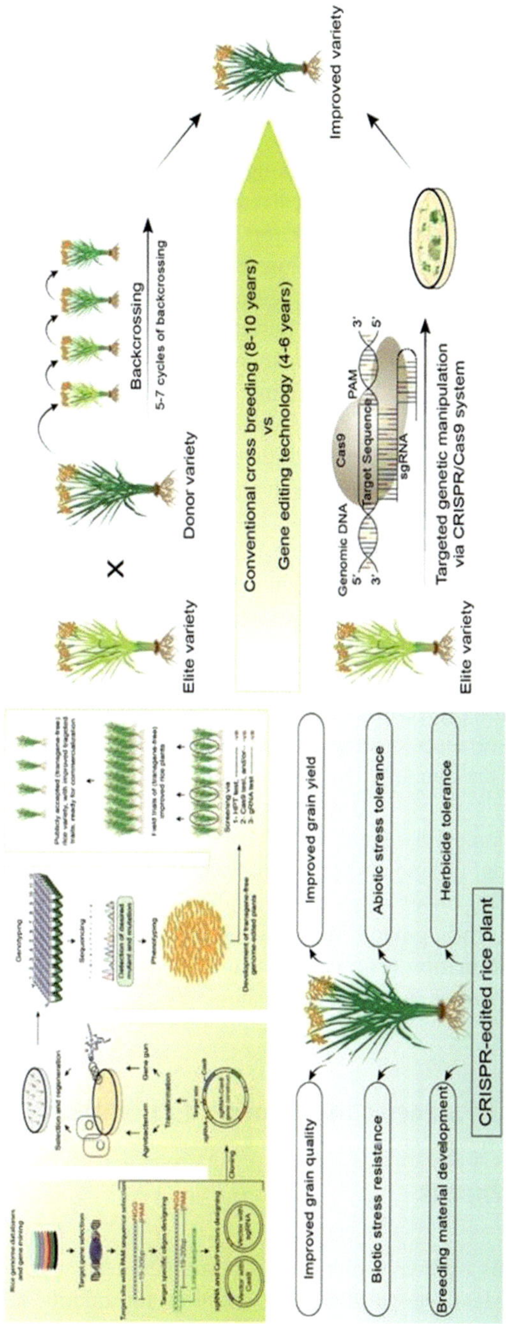

Fig. 9.3 CRISPR-edited rice plant: recreated from source (Yadav and Ramakrishna 2023)

9.3.6 *Exploring Diversification and Crop Rotation*

With its diverse agro-climatic zones, India is navigating the path of agricultural sustainability by strategically implementing diversification and crop rotation practices. As India grapples with the complexities of modern agriculture, the exploration of diversification and crop rotation emerges as a crucial paradigm for sustainable and resilient farming practices Understanding diversification in Indian agriculture involves comprehensively examining critical dimensions that shape agricultural practices (Gupta et al. 2021). The focus on Crop Diversity and Economic Resilience explores the economic benefits of cultivating various crops and underscores diversification's essential role in mitigating monoculture risks. Crop rotation with species diversification contributes to controlling pests, diseases, and weeds and improves soil fertility and conservation, leading to increased profitability in grain production systems (Volsi et al. 2022). Transitioning to Horticultural and Non-traditional Crops, the increasing prominence of these alternatives' sheds light on associated market opportunities and challenges tied to diversifying crop portfolios. Crop Rotation Practices for Soil Health emphasizes the pivotal role of crop rotation in maintaining soil fertility, addressing strategies for effective nutrient management, and preventing soil degradation. Pest and Disease Control underscores the utilization of crop rotation as a natural tool for managing pests and diseases, supporting these insights with case studies highlighting successful pest control through strategic crop rotation. Lastly, Government Initiatives and Policy Implications focus on the National Mission on Sustainable Agriculture (NMSA), promoting diversification and sustainable practices. Evaluating government incentives and support mechanisms designed to encourage farmers to adopt diversified cropping systems provides a comprehensive understanding of the dynamic landscape of diversification in Indian agriculture (Kishore et al. 2018).

Rice farming embraces these cutting-edge technologies and sustainable practices like implementing organic and eco-friendly methods to reduce environmental impact and incorporating cover crops and natural pest control for enhanced sustainability. The ongoing developments in precision agriculture, genetic engineering, sustainability, and automation promise to ensure food security while minimizing the environmental impact of rice cultivation.

9.4 Agricultural Factors Contributing to Pollution

9.4.1 *Chemical Inputs*

Contamination of soils by heavy metals and pesticides results in the degradation of fertile agricultural land, jeopardizing the quantity and quality of food production. Although no worldwide assessment specifies the extent of arable land losses due to chemical pollution, regional accounts highlight substantial actual or potential declines (Naidu et al. 2021). The persistence of pesticide residues in soil and water threatens soil health and aquatic ecosystems, impacting both target and non-target species. Nitrogen and P runoff from fertilized fields can lead to water pollution, causing eutrophication and endangering aquatic life. Moreover, the development of herbicide-resistant weeds and the overreliance on chemical herbicides pose challenges to sustainable weed management (Gage et al. 2019). By acknowledging these factors, we can develop strategies to mitigate the environmental impact of chemical inputs by exploring alternative, eco-friendly practices, such as integrated pest management and organic farming.

9.4.1.1 Pesticides and Herbicides

Soil harbours a wealth of diverse biological organisms that play essential roles, including nutrient cycling, soil structure upkeep, carbon transformation, and the control of pests and diseases. Penoxsulam 21.7% w/w SC is a selective herbicide that controls grasses, sedges, and broad-leaf weeds in transplanted rice (Sansa et al. 2016). Fenoxaprop-P-ethyl (FPPE) is a selective ACCase herbicide extensively used in paddy fields to inhibit barnyard grass. However, prolonged use of FPPE herbicide elicits phytotoxicity in cultivated rice and herbicide resistance in weeds (Antony and Ramanathan 2022). In recent years, numerous terrestrial invertebrates have experienced a decline, with key factors identified as habitat loss and agricultural intensification leading to agrichemical pollution (Gunstone et al. 2021).

9.4.1.2 Impact of Fertilizer

Since the mid-1920s, human activities have doubled the natural N deposition rate onto land by employing N fertilizers (inorganic and manure), burning fossil fuels, and replacing natural vegetation with N-fixing crops like soybeans. The primary human-induced source of N in the environment is N fertilizer, with a significant portion applied to agricultural land globally since 1980. Approximately half of the applied N leaches from agricultural fields, contaminating surface and groundwater, consequently increasing nitrate concentrations in water resources (Park et al. 2023). In agricultural systems, extensive fertilizer use enhances plant production, but the surplus N, typically nitrate, may leach into streams and rivers and eventually reach drinking water. The Haber–Bosch process, involving the reaction of N_2 with H_2 to produce synthetic fertilizers for agriculture, has notably increased in recent decades, contributing to nearly 80% of the N in human tissues (Howarth 2008). A substantial amount of N applied to agricultural and urban areas enters rivers and nearshore coastal systems, leading to anoxia or hypoxia in nearshore marine environments, ultimately reducing biodiversity. Elevated N levels in aquatic systems contribute to freshwater ecosystem acidification, and the surplus N, when converted into nitrates, can contaminate neighbouring water bodies and air. Nitrate air emissions pose risks to groundwater and drinking water and contribute to the eutrophication of surface water bodies and terrestrial ecosystems. Nitrogen surpluses can lead to adverse environmental impacts, including landscape degradation, biodiversity loss, and the formation of GHG. Furthermore, there are health concerns, including colorectal cancer, bladder and breast cancer, thyroid disease, methemoglobinemia, and neural tube defects associated with N exposure (Münzel et al. 2023). Ammonia particle emissions from agriculture may also lead to inhalation-related health effects due to fine and ultrafine particles (Fig. 9.4).

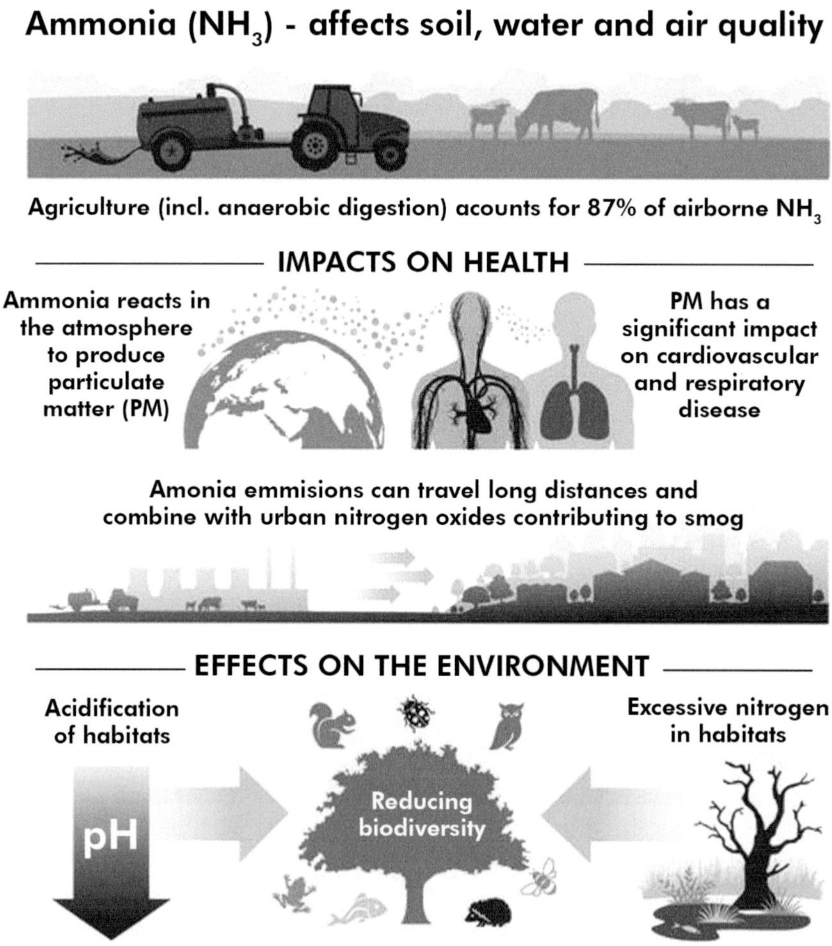

Fig. 9.4 Impact of fertilizers on health and environment

9.4.2 Soil Erosion

Soil erosion resulting from rice cultivation poses a significant environmental challenge. The unique practices involved in rice farming, including flooding fields and constructing terraces, contribute to the erosion of valuable topsoil. The water runoff from flooded rice fields, especially on sloped terrain, carries away fertile soil, compromising agricultural sustainability. Additionally, improper construction and maintenance of terraces lead to erosion, further exacerbating the issue (Deng et al. 2021). The temporary reduction in vegetative cover during various phases of rice cultivation and prolonged flooding weakens the soil structure, making it more susceptible to erosion. To address this concern, adopting sustainable farming practices such as contour plowing, cover cropping, and efficient water management is crucial in preserving soil health and mitigating the adverse effects of erosion associated with rice cultivation (Cárceles Rodríguez et al. 2022).

9.4.3 Water Management

9.4.3.1 Inadequate Irrigation Methods

Water plays a crucial role in the growth and development of rice plants. However, continuous flooded irrigation leads to significant water loss through evaporation, seepage, runoff, and percolation, resulting in nutrient depletion and groundwater pollution (Mboyerwa et al. 2021). The restricted expansion of rice fields is attributed to various factors, including the absence of irrigation facilities, the crop's high-water demand, declining water tables, labour-intensive farming practices, poor-quality by-products like straw, and soil degradation, aggravated by erratic rainfall patterns (Kumar et al. 2022). The water productivity of rice is also lower (0.60–1.60 kg m^{-3}) than other crops. On average, it takes 2500 L of water to produce one kg of rice, with variations ranging from 800 L to over 5000 L (Najmuddin et al. 2018). Approximately 50% of global rice production is impacted by water shortages. Projections suggest that by 2025, between 15 and 20 million hectares of irrigated rice could face water scarcity issues. The profound impact of extreme weather events, like droughts and floods, has already disrupted rice production (Mboyerwa et al. 2021). With climate change exacerbating these challenges and the growing water demands due to rapid economic growth and urbanization, the imperative to boost food production and enhance agricultural water efficiency within constrained water resources has emerged as a pressing concern for the agricultural sector.

9.4.3.2 Issues with Water Logging and Salinization

Challenges associated with waterlogging and salinization in agricultural areas are significant concerns. Waterlogging, where excess water accumulates in the soil, can lead to oxygen deprivation for plant roots, negatively impacting crop growth and overall productivity. Salinization, conversely, refers to the accumulation of salts in the soil, posing a threat to crop health by disrupting water uptake and nutrient absorption (Nath et al. 2021). These issues are exacerbated by factors such as poor drainage systems, inappropriate irrigation practices, and changes in land use. Addressing waterlogging and salinization requires comprehensive strategies, including implementing effective drainage systems, sustainable irrigation practices, and cultivating salt-tolerant crop varieties (Wichelns and Qadir 2015). Additionally, awareness campaigns and farmer education initiatives are crucial for promoting practices that mitigate these challenges and ensure the long-term viability of agricultural landscapes.

9.5 Strategies for Mitigation

9.5.1 Precision Farming

In modern agriculture, satellite imagery has emerged as a powerful tool for crop monitoring and yield prediction. The integration of satellite technology enables farmers to acquire real-time insights into the health and development of their crops, facilitating proactive decision-making. Furthermore, Internet of Things (IoT) and sensor technologies have revolutionized precision agriculture (Singh and Sharma 2024). Intelligent sensors integrated into the agricultural landscape continuously monitor crucial parameters such as soil conditions, water usage, and plant health. The real-time data collected from these sensors gives farmers a comprehensive understanding of their fields, allowing for precise and informed decision-making (Tripathi et al. 2022) to enhance resource optimization and facilitate the implementation of precision applications of fertilizers and pesticides. Farmers can tailor their approach by leveraging real-time data and applying inputs precisely where and when needed, thus minimizing environmental impact and maximizing agricultural efficiency. The synergy between satellite imagery and IoT-enabled sensor technologies marks a significant stride toward sustainable and data-driven agricultural practices in the contemporary farming landscape.

9.5.2 Organic Farming Practices

Organic rice farming practices involve cultivating rice without synthetic pesticides, herbicides, and chemical fertilizers, emphasizing environmentally friendly and sustainable methods. Instead, organic farmers employ natural alternatives such as compost, manure, cover cropping, and crop rotation to enhance soil fertility and manage pests and diseases. Organic rice fields often integrate diversified cropping systems, promoting biodiversity and reducing the reliance on external inputs. Implementing terracing and contour farming helps slow water runoff, minimizing soil erosion on sloped terrain. Emphasis is placed on maintaining soil health, fostering a balanced ecosystem, and minimizing the environmental impact of agricultural activities (Hazra et al. 2016). Integrating trees and shrubs within or around rice fields helps stabilize soil, reducing erosion and enhancing biodiversity. Organic certification standards set clear guidelines for organic rice production, ensuring adherence to organic principles throughout cultivation (Johannes et al. 2019). By adopting organic rice farming practices, farmers contribute to healthier ecosystems, improved soil quality, and the production of free rice from synthetic residues, meeting the growing demand for sustainable and environmentally conscious agricultural products.

9.5.3 Integrated Pest Management

Integrated pest management (IPM) is a promising technology with environmental benefits. In rice production systems, insect pests and weeds have historically caused substantial output losses. While traditional methods of controlling insects, pests, and weeds have proven effective, environmental concerns have led scientists to explore alternative pest management approaches. In response to the awareness of sustainable conventional agriculture, integrated pesticide management (IPM) has been widely adopted (Table 9.4). This multimodal pesticide management method is designed to mitigate adverse environmental impacts, acknowledging the need for a more environmentally friendly approach to plant protection.

9.5.4 Cover Cropping and Crop Rotation

Introducing cover crops into agricultural practices represents a proactive strategy that offers numerous benefits for soil health and sustainability. Cover crops provide essential ground cover, acting as a protective shield against the impact of raindrops on bare soil. This shielding effect is crucial in preventing soil erosion, as the cover crops create a physical barrier that mitigates the erosive force of rainwater. Cover crops also improve soil structure by promoting root growth and microbial activity. Their extensive root systems help bind soil particles together, enhancing soil aggregation

Table 9.4 Key components of IPM in rice cultivation (Hajjar et al. 2023)

Sr. No	Components	Description
1.	Biological control	**Predators and parasitoids**: introduce natural enemies of pests, such as predatory insects and parasitic wasps, to control pest populations biologically. These natural predators help maintain a balance in the ecosystem **Microbial agents**: use microbial organisms like fungi, bacteria, or viruses that specifically target pests. Biopesticides derived from these microorganisms can be applied to control pest infestations
2.	Cultural practices	**Crop rotation**: rotate rice crops with non-host crops to disrupt the life cycle of pests to reduce the buildup of pest populations and minimize the need for chemical interventions **Adjusting planting dates**: optimize planting schedules to avoid peak periods of pest activity. Timely planting can help escape certain pests or synchronize the crop with natural enemies **Resistant varieties**: plant rice varieties that are resistant or tolerant to specific pests. Resistant varieties can withstand pest pressure without significant yield losses
3.	Mechanical and physical control	**Traps and barriers**: use physical barriers, such as sticky traps, pheromone traps, or light traps, to capture and reduce pest populations. Barriers can also prevent the movement of pests between fields **Mechanical weed control**: proper weed management reduces pest habitats and disrupts their life cycles. Mechanical weeders or manual weeding can be employed to control weed growth
4.	Chemical control	**Selective pesticides**: use pesticides judiciously and selectively if necessary. Choose pesticides that target specific pests while minimizing harm to beneficial organisms. Rotate pesticides to avoid resistance buildup **Threshold-based applications**: monitor pest populations regularly and apply pesticides only when pest levels exceed predetermined thresholds, prevent unnecessary pesticide use, and preserve natural predator–prey relationships
5.	Monitoring and surveillance	**Regular field inspections**: implement routine field inspections to monitor pest populations, crop health, and the presence of natural enemies. Early detection allows for timely and targeted intervention **Pheromone traps and monitoring devices**: use pheromone traps and other monitoring devices to track the population dynamics of pests. This information helps in making informed decisions about pest management strategies

(continued)

Table 9.4 (continued)

Sr. No	Components	Description
6.	Community involvement and education	**Farmer training**: provide training and educational programs to farmers on the principles and practices of IPM. Empowering farmers with knowledge helps in the proper implementation of IPM techniques **Community-based approaches**: encourage collaboration among farmers, researchers, and extension services to share experiences and information. Community involvement fosters collective efforts in implementing sustainable pest management practices
7.	Record keeping and data analysis	**Farm records**: maintain detailed records of pest occurrences, interventions, and crop performance. Analyzing this data helps in making informed decisions for future pest management strategies **Decision support systems**: implement digital tools and decision support systems that utilize weather conditions, pest life cycles, and crop health data to recommend optimal pest management practices

and stability (Sharma et al. 2018). Furthermore, incorporating reduced or no-till farming methods complements the positive impact of cover crops on soil structure and erosion control. By minimizing soil disturbance during planting and cultivation, these practices preserve the integrity of the soil, preventing erosion risk (Song et al. 2019). Reduced-till and no-till methods help maintain the natural arrangement of soil particles, reduce compaction, and enhance water infiltration, fostering a healthier and more resilient soil environment.

The synergistic approach of introducing cover crops and adopting reduced or no-till farming methods contributes to sustainable soil management. This combined strategy safeguards against erosion and promotes soil structure, water retention, and health. As agriculture continues evolving, these practices are integral to environmentally conscious and resilient farming.

9.5.5 *Conservation Tillage*

Conservation tillage has emerged as a transformative approach in rice farming, signifying a departure from traditional cultivation methods. Conservation tillage improves soil health by fostering a conducive environment for beneficial microorganisms (Cherlinka 2024). The benefits of conservation tillage in rice farming are manifold, including erosion control, efficient water use, enhanced soil fertility, and weed suppression. Conservation tillage aligns with environmentally friendly approaches by mitigating the risk of soil erosion and promoting sustainable water management. Challenges such as farmer adaptation and equipment compatibility must be addressed, emphasizing the importance of education and suitable machinery (Nandan et al. 2019). Conservation tillage represents a cornerstone of sustainable agricultural practices, offering a pathway to resilient and environmentally friendly rice farming. As the global agricultural landscape evolves, adopting conservation tillage is a promising strategy for achieving a harmonious balance between productivity, environmental stewardship, and long-term food security.

9.5.6 Effective Irrigation Management

Well-managed irrigation is pivotal in increasing yields compared to dryland farming, ensuring a positive return on investment. This irrigation method allows the incorporation of fertilizers and pesticides, enabling the application of fertilizers as needed by the crop. This approach proves to be environmentally sustainable while yielding economic benefits. Various tools, such as irrigation scheduling programs and soil moisture sensors, can enhance irrigation water use efficiency (Arouna et al. 2023). Numerous water-efficient technologies for Rice cultivation have been introduced, including direct-seeded rice (DSR), the system of rice intensification (SRI), aerobic rice, and drip-irrigated rice (Vijayakumar 2023). These innovations aim to reduce water consumption and enhance productivity and profitability in paddy cultivation. However, adopting these technologies depends on regional factors such as climate suitability, soil type, resource availability, and expertise. DSR, the SRI, and alternate wetting and drying (AWD) are cost-effective choices suitable for field implementation and should be actively promoted for widespread adoption (Mallareddy et al. 2023). Despite the demonstrated effectiveness of these improved irrigation methods, their adoption among farmers could be more active. Farmers need training in new irrigation technologies like smart irrigation employing sensors and IoT to address this. Tailored programs with attractive incentives should be developed and implemented to encourage the adoption of these technologies in water-scarce areas (Koech and Langat 2018). Furthermore, optimizing the irrigation system's design, maintenance, and operation can enhance application efficiency. Given agricultural producers' typically narrow profit margins, the additional investment cost of irrigation systems warrants careful consideration. Evaluating the design, maintenance expenses, and energy usage of the equipment is essential to maximize profitability in agriculture.

9.6 Research Opportunities and Prospective Future

9.6.1 Evaluating the Effects of Agrochemicals

Agrochemicals, which include fertilizers, pesticides, and herbicides, are commonly used in rice farming to enhance productivity and protect crops. However, their indiscriminate or excessive use can lead to ecological and health concerns (Nacorda et al. 2019). Some vital aspects to consider when evaluating the effects of agrochemicals in rice fields are described below (Table 9.5).

9.6.2 Studying the Interactions of Microorganisms in Soil Health

Microorganisms, when associated with rice plants, can either utilize the plants as a habitat or colonize the roots or surfaces of the rice plant. Once established, these microorganisms emit bioactive signals that receptors on rice plant cells can detect. Subsequently, they contribute to the augmentation of plant growth, provide resilience to abiotic stress, boost resistance against diseases, induce beneficial alterations in gene expression patterns, enhance various physiological and biochemical traits, and improve the nutrient absorption of rice plants (Doni et al. 2022). To comprehensively understand microbial functions and activities in rice field soils, omics technologies, including metagenomics, metatranscriptomics, and metaproteomics, must be utilized.

9.6.3 Investigating the Availability of Contaminants in the Environment

While instrumental in pest control, pesticides, and herbicides pose concerns related to environmental impact and potential health risks. Similarly, using fertilizers, though essential for crop productivity, demands careful management to prevent nutrient imbalances and soil degradation. Contaminants in irrigation water further compound these challenges, affecting water quality and crops. Soil health is intricately linked to fertilizer usage, with the potential for nutrient imbalances and adverse effects on microbial ecosystems. Additionally, heavy metals in the soil pose environmental and health risks. Exposure to elevated levels of HMs induces various physiological and metabolic changes in rice. However, the visual evidence of HM toxicity varies depending on the specific metal and its target sites within the plant. The most common visual indicators include diminished plant growth, leaf chlorosis, necrosis, turgor loss, reduced seed germination, and impaired photosynthetic machinery (Fig. 9.5).

Table 9.5 Holistic assessment of agrochemical effects on rice field (Nacorda et al. 2019)

Sr. No	Parameters	Description
1.	Soil health	**Nutrient imbalance**: excessive use of fertilizers can lead to nutrient imbalances in the soil, affecting overall health and fertility. Monitoring soil nutrient levels is essential to prevent over-fertilization **Soil structure**: agrochemicals may impact soil structure, affecting water retention and drainage. Assessing soil texture, compaction, and moisture content helps evaluate these effects
2.	Water quality	**Runoff and leaching**: agrochemicals can contribute to water pollution through runoff and leaching into nearby water bodies. Regular water quality testing helps identify the presence of pesticides, herbicides, and excess nutrients **Aquatic ecosystems**: evaluate the impact of agrochemical runoff on marine ecosystems, including potential harm to fish, invertebrates, and overall biodiversity in nearby water bodies
3.	Crop health	**Pesticide residues**: monitor pesticide residues on rice crops to assess the effectiveness of pest management practices and ensure that residue levels comply with safety standards (Onorati et al. 2022) **Herbicide impact**: assess the impact of herbicides on weed control and crop growth. Consider the development of herbicide-resistant weeds and their implications for long-term crop management
4.	Biodiversity	**Impact on beneficial organisms**: evaluate the effects of pesticides on non-target organisms, including beneficial insects and soil microorganisms. Protecting natural predators contributes to sustainable pest management **Crop diversity**: consider the impact of agrochemicals on the diversity of crops grown in rice fields. Monoculture and the excessive use of certain chemicals may lead to a decline in crop diversity
5.	Human health	**Residue levels in rice**: regularly test rice crops for pesticide residues to ensure they comply with safety standards for human consumption **Health monitoring**: conduct health assessments of farmers and communities exposed to agrochemicals. Monitor for potential long-term health effects associated with pesticide exposure
6.	Socioeconomic aspects	**Farmers' practices**: understand farmers' practices regarding agrochemical use and their awareness of sustainable alternatives. Promote education and training programs for responsible agrochemical applications **Economic impact**: evaluate the economic impact of agrochemical use, considering both short-term gains and long-term sustainability. Explore alternative practices that balance productivity with environmental and health considerations
7.	Regulatory compliance	**Compliance with regulations**: ensure that agrochemical use meets local and international regulations. Monitor changes in laws and adapt practices accordingly to stay within legal limits

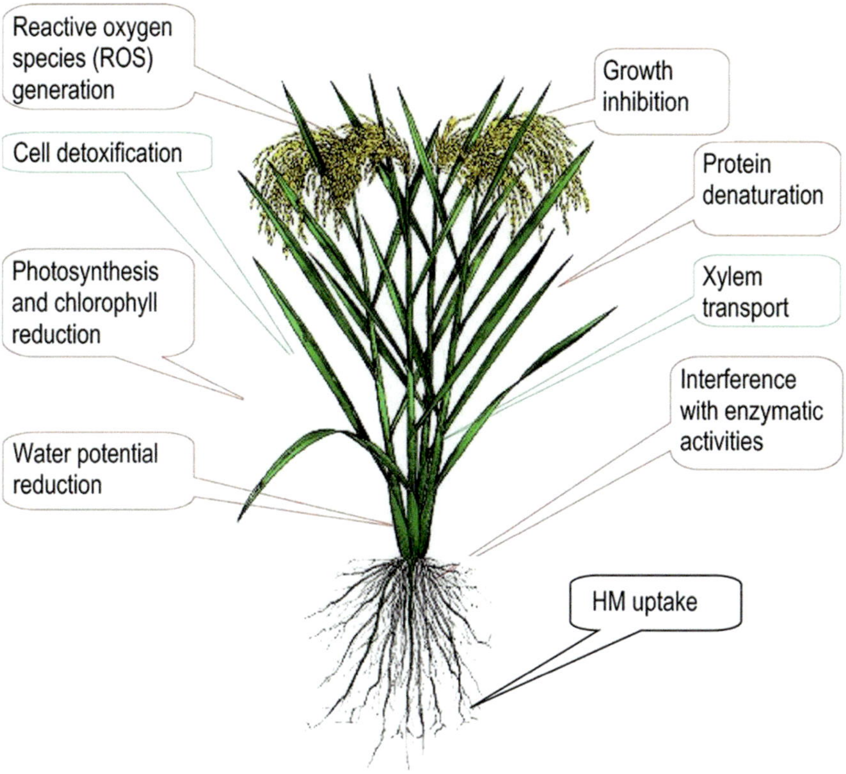

Fig. 9.5 Uptake mechanism of heavy metals in paddy plants: toxic element transfer to the food chain in paddy rice system

These effects are closely linked to biochemical, ultrastructural and molecular alterations in plant tissues and cells caused by the presence of HMs (Hossain et al. 2012). While offering opportunities for improved traits, genetic modifications in rice farming raise ethical and environmental considerations that warrant careful examination (Rajmohan et al. 2020). Addressing these interconnected issues requires a holistic and sustainable approach, integrating best practices, technological innovations, and regulatory measures to ensure the long-term viability of rice farming while minimizing adverse impacts on the environment and human health.

9.6.4 Utilizing Remote Sensing and GIS for Various Applications

Remote Sensing (RS) and Geographic Information Systems (GIS) play a significant role in modern agriculture, including rice farming. Figure 9.6 illustrates the diverse

Fig. 9.6 Applications of precision techniques in rice cultivation: GIS integrates various layers of spatial data to make informed decisions

advantages of utilizing RS and GIS in agriculture, such as enhanced crop yield estimation, monitoring, soil moisture mapping, and risk assessment (Rane et al. 2023). GIS integrates various layers of spatial data, enabling farmers and policymakers to make informed decisions based on a comprehensive understanding of the landscape and environmental conditions. These technologies provide various benefits, as mentioned in Table 9.6.

9.6.5 *Eco-friendly Alternatives*

Promoting eco-friendly alternatives in rice farming is imperative for fostering sustainable agriculture, minimizing environmental impact, and ensuring long-term productivity. Embracing organic farming practices stands out as a foundational strategy, advocating substituting synthetic fertilizers with organic alternatives like compost,

Table 9.6 Remote sensing and GIS integration for enhanced agricultural management and environmental monitoring

Sr. No	Benefits	Applications
1.	Crop monitoring and management	**NDVI analysis**: remote sensing allows for the calculation of the normalized difference vegetation index (NDVI), indicating the health and vigour of crops. Monitoring NDVI over time helps farmers assess crop growth stages, detect stress, and optimize irrigation and fertilizer application (Huang et al. 2020)
2.	Land use and land cover mapping	**Crop classification**: remote sensing imagery can classify different land cover types, distinguishing between rice fields, other crops, and non-agricultural areas. GIS tools aid in creating accurate land use maps and facilitating land management decisions (Dash et al. 2023)
3.	Water management	**Water stress detection**: remote sensing helps identify water stress in rice fields by monitoring changes in surface temperature and vegetation indices. GIS can then map and analyze water stress patterns, enabling efficient water management strategies (Sishodia et al. 2020)
4.	Disease and pest monitoring	**Early detection**: remote sensing can capture subtle changes in crop reflectance caused by diseases or pest infestations. GIS can process and analyse this data, providing early warning systems for farmers to take prompt action (Sishodia et al. 2020)
5.	Yield prediction	**Vegetation indices for yield estimation**: remote sensing data, especially from satellite or UAV platforms, can derive vegetation indices correlating with crop yield. GIS models can then predict potential yields across different areas of a rice field (Patra et al. 2022)
6.	Weather monitoring and forecasting and flood monitoring and risk assessment	**Climate data integration**: GIS can integrate remote sensing data with climate information, creating spatially explicit weather maps. This aids in predicting weather patterns and planning agricultural activities accordingly (Farhadi and Najafzadeh 2021) **Flood inundation mapping**: GIS tools can integrate remote sensing data to create flood inundation maps, helping farmers assess flood risks and plan for appropriate mitigation measures (Avand and Moradi 2021)
7.	Soil health assessment	**Soil property mapping**: remote sensing can be combined with GIS to map soil properties and variations. This information is crucial for making informed decisions regarding soil amendments and improving soil health (Yuzugullu et al. 2020)
8.	Carbon sequestration and environmental impact assessment	**Biomass estimation**: remote sensing can be used to estimate biomass in rice fields, providing insights into carbon sequestration. GIS helps in spatially analyzing the environmental impact of agricultural practices (Issa et al. 2020)

manure, and green manure to enhance soil fertility and microbial activity. Biological pest control methods emphasize introducing and promoting natural predators and parasites, such as ladybugs or ducks, to manage pest populations without chemical interventions. Crop rotation and diversification, a cornerstone of sustainable agriculture, break pest and disease cycles, improve soil structure, and reduce reliance on chemical inputs (Eyni-Nargeseh et al. 2023). The System of Rice Intensification (SRI) prioritizes water and resource efficiency, advocating for careful management of rice plants to achieve increased yields with reduced water, seed, and fertilizer requirements. From mechanical weed control to agroforestry practices, biodegradable mulching, natural additives like biochar, and water-saving techniques such as Alternate Wetting and Drying (AWD), these practices collectively contribute to rice cultivation in an environmentally responsible manner. Implementing cover crops, biological N fixation through Azolla cultivation and exploring aerobic rice production further emphasize the diverse eco-friendly alternatives available to rice farmers (Hester et al. 2022). Conservation tillage, community-based management, agroecological farming systems, and eco-labeling initiatives underscore a holistic approach towards sustainable rice farming, encouraging biodiversity, nutrient cycling, and resilience in the entire farming ecosystem. Active participation in certification programs ensures rice production adheres to stringent, environmentally friendly, and sustainable standards, promoting responsible agricultural practices for a greener future.

9.6.6 Socioeconomic Impacts

Sustainable development necessitates resources, with a focus on enhancing energy efficiency to mitigate GHG emissions. Urgent adoption of low carbon technologies in agriculture can provide economic and social benefits while safeguarding the climate for future generations. Low carbon technologies such as zero tillage, use of nitrification inhibitors, site specific nutrient management and integrated nutrient management are not only environmentally friendly but also economically feasible. Integrated nutrient management and zero tillage can increase net return by 9–10% in the Indo-Gangetic Plains (IGP) of India (Pathak and Aggarwal 2012). However, *t*he intensity and probability of low carbon technology adoptions are influenced by factors such as farmers' education level, climate change awareness, machinery ownership, technical support, and subsidies (Chen and Chen 2022). While organic rice farming offers numerous socioeconomic benefits, it's essential to recognize that the outcomes can vary based on local context, market dynamics, and government policies (Panpakdee 2023). Government support through subsidies, incentives, and favorable policies can further encourage farmers to adopt organic practices, positively impacting their socioeconomic conditions. Additionally, successful adoption often requires a supportive infrastructure, including extension services, research, and market access, to ensure the long-term viability of organic farming practices.

9.6.7 Interactions of Climate Change

Climate change results in a reduction in both the frequency and volume of precipitation across different regions and periods and ultimately leads to an increased occurrence of droughts and floods, directly impacting the production of rice (Bharambe et al. 2023). Comprehending the complex connection between hydrological processes and rice cultivation is crucial. The limited availability of reliable and comprehensive long-term weather and streamflow data poses obstacles in accurately capturing the local-scale processes and critical conditions for robust modelling. More efforts are needed to improve data availability, collection method, and data quality to address these challenges and enhance the reliability of climate-hydrological-crop models (Ansari et al. 2023). This involves assessing water availability in each irrigation zone, the capability of surface and groundwater reservoirs to retain water during drought periods and establishing a crop calendar based on the available water resources.

9.6.8 Exploring Cutting-Edge Solutions for Pollution Remediation

Addressing pollution in rice farming necessitates innovative and environmentally sustainable solutions. Several approaches can be employed for pollution remediation in rice farming. Microbial bioremediation involves utilizing specific microbes to break down and detoxify soil, water, or air pollutants, particularly effective for heavy metals or organic pollutants (Sharma et al. 2022). Phytoremediation employs certain plant species with the capacity to absorb, accumulate, or transform pollutants in their tissues, aiding in remediating soil and water contaminants. Nanotechnology offers nanoparticle-based solutions engineered to absorb, degrade, or immobilize pollutants, with applications including nano-fertilizers and nanoparticles for contaminant removal. Applying biochar to rice fields has emerged as a promising strategy to enhance crop productivity and soil health. Biochar, a carbon-rich material produced from the pyrolysis of organic matter, offers several benefits when incorporated into rice cultivation practices. By improving soil structure and water retention, biochar promotes better root development and nutrient uptake in rice plants. Its high surface area and porous structure also facilitate nutrient retention in the soil, reducing nutrient leaching and enhancing fertilizer efficiency, as shown in Fig. 9.7. Studies have shown that biochar application can increase rice yields, particularly in soils with poor fertility or water-holding capacity. Moreover, biochar contributes to carbon sequestration and reduces GHG emissions, aligning with sustainable agriculture practices. As research continues to uncover the optimal application rates and timing for biochar, integrating this innovative approach into rice farming holds significant potential for enhancing agricultural productivity and environmental sustainability (Liu et al. 2022). Floating treatment wetlands, constructed as floating islands with vegetation, provide an eco-friendly approach to water purification. Ozone treatment, electrokinetic soil remediation, hydroponic and aeroponic systems, smart irrigation systems, satellite imaging, and remote sensing, along with the development of pollution-tolerant rice varieties, green infrastructure implementation, community-based approaches, and policy enforcement, collectively contribute to pollution remediation in rice farming. Integrating these cutting-edge solutions with sustainable agricultural practices promotes environmental health and long-term sustainability.

9.7 Concluding Remarks

This comprehensive exploration of research opportunities in rice farming underscores the multifaceted nature of sustainable agricultural practices. Evaluating agrochemical effects highlights the need to carefully consider soil health, water quality, crop health, biodiversity, human health, socioeconomic aspects, and regulatory compliance. Efforts in this domain must prioritize continuous monitoring, collaborative research, and adherence to regulations to ensure the long-term sustainability of rice field agroecosystems. The study of microorganisms in soil health, utilizing advanced omics technologies, presents an exciting avenue for enhancing our understanding of the intricate relationships that contribute to plant growth, disease resistance, and overall crop health. By harnessing the potential of microorganisms, future research can pave the way for innovative and sustainable approaches to optimize rice production. Investigating the availability of contaminants in the environment, utilizing remote sensing and GIS applications, and promoting eco-friendly alternatives collectively contribute to a holistic and responsible approach to rice farming. These avenues offer valuable insights into mitigating environmental impact, enhancing resource efficiency, and fostering resilience in the face of climate change. The socioeconomic impacts of low-carbon technologies underline the significance of understanding and addressing the factors influencing their adoption. Government support, education, and awareness play pivotal roles in encouraging sustainable practices among farmers, ultimately contributing to the overall well-being of communities. Recognizing the profound interactions of climate change on rice production emphasizes the need for adaptive strategies. A thorough assessment of water resources and the implementation of crop calendars based on climate considerations are essential steps toward mitigating the adverse effects of climate change on rice farming. Lastly, exploring cutting-edge solutions for pollution remediation, such as microbial bioremediation, phytoremediation, nanotechnology, and biochar application, presents innovative approaches to address environmental challenges. Integrating these solutions with intelligent irrigation systems, satellite imaging, and community-based approaches further strengthens the potential for sustainable and eco-friendly rice farming. It is crucial for stakeholders, including farmers, policymakers, and researchers, to collaborate to balance agricultural productivity and environmental stewardship. Through

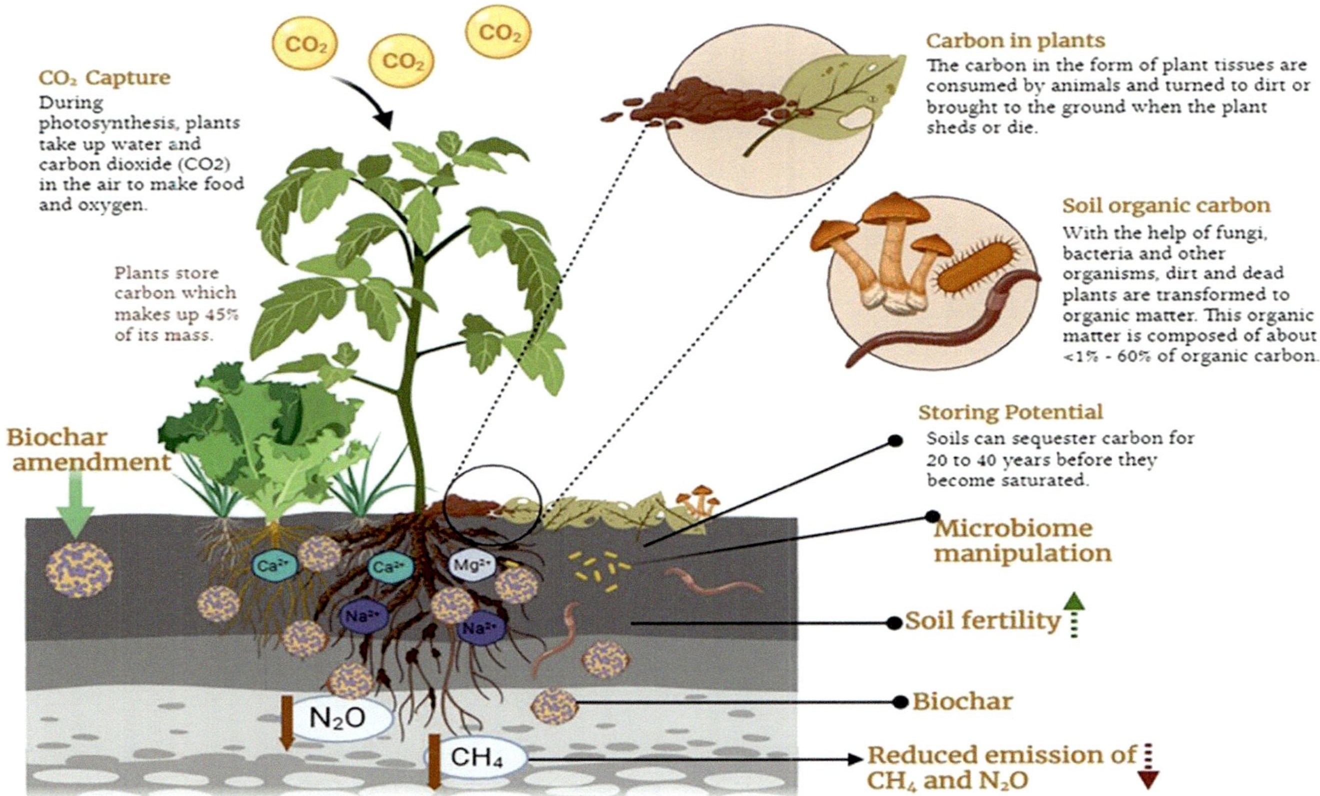

Fig. 9.7 Processes that occur in soil amended with biochar: biochar binds with minerals and fertilizers and helps in its solubilization, facilitating their accelerated plant uptake

informed decision-making and adopting sustainable practices, we can contribute to a healthier and more resilient future for rice cultivation.

References

Adejumoke AI, Babatunde OA, AbimbolaP O, Tabitha AAA, Dada AO, ToyinA O (2018) Water pollution: effects, prevention, and climatic impact. InTech eBooks 33:33–47. https://doi.org/10.5772/intechopen.72018

Adeniran M, Oladunjoye MA, Doro KO (2023) Soil and groundwater contamination by crude oil spillage: a review and implications for remediation projects in Nigeria. Front Environ Sci 11:495. https://doi.org/10.3389/fenvs.2023.1137496

Aggarwal PK, Bandyopadhyay SK, Pathak H, Kalra N, Chander S, Kumar S (2000) Analysis of yield trends of the rice-wheat systems in north-western India. Outlook Agric 29(4):259–268. https://doi.org/10.5367/000000000101293329

Ahmed T, Zounemat-Kermani M, Scholz M (2020) Climate change, water quality and water-related challenges: a review with focus on Pakistan. Int J Environ Res Public Health 17(22):8518. https://doi.org/10.3390/ijerph17228518

All India Coordinated Rice Improvement Project (AICRIP) (2023) Draft proceedings: 58th annual rice research group meetings, 10–11th April & 4–5th May, 2023 (hybrid mode). ICAR–Indian Institute of Rice Research, Indian Council of Agricultural Research

Anawar HM, Chowdhury RK (2020) Remediation of polluted river water by biological, chemical, ecological and engineering processes. Sustainability 12(17):7017. https://doi.org/10.3390/su12177017

Ansari A, Pranesti A, Telaumbanua M, Alam T, Taryono T, Wulandari RA, Nugroho BDA (2023) Evaluating the effect of climate change on rice production in Indonesia using multi modelling approach. Heliyon 9(9):1–15. https://doi.org/10.1016/j.heliyon.2023.e19639

Antony A, Ramanathan K (2022) Searching of novel herbicides for paddy field weed management—a case study with acetyl-CoA carboxylase. Agronomy 12(7):1635. https://doi.org/10.3390/agronomy12071635

Arouna A, Dzomeku IK, Abdul-Ganiyu S, Rahman NA (2023) Water management for sustainable irrigation in rice (Oryza sativa L.) production: a review. Agronomy 13(6):1522. https://doi.org/10.3390/agronomy13061522

Ashraf MA, Maah MJ, Yusoff I (2014) Soil contamination, risk assessment and remediation. Environ Risk Assess Soil Contam 1:3–56. https://doi.org/10.5772/57287

Avand M, Moradi H (2021) Using machine learning models, remote sensing, and GIS to investigate the effects of changing climates and land uses on flood probability. J Hydrol 595:125663. https://doi.org/10.1016/j.jhydrol.2020.125663

Bangani L, Kabiti HM, Amoo OT, Nakin MDV, Magayiyana Z (2023) Impacts of illegal solid waste dumping on the water quality of the Mthatha River. Water Pract Technol 18(5):1011–1021. https://doi.org/10.2166/wpt.2023.053

Bar-On YM, Phillips R, Milo R (2018) The biomass distribution on earth. PNAS 115(25):6506–6511. https://doi.org/10.1073/pnas.1711842115

Bazienė K, Tetsman I, Albrektienė R (2020) Level of pollution on surrounding environment from landfill aftercare. Int J Environ Res Public Health 17(6):2007. https://doi.org/10.3390/ijerph17062007

Bharambe KP, Shimizu Y, Kantoush SA, Sumi T, Saber M (2023) Impacts of climate change on drought and its consequences on the agricultural crop under worst-case scenario over the Godavari River Basin, India. Clim Serv 32:100415. https://doi.org/10.1016/j.cliser.2023.100415

Bhatt R, Kukal SS, Busari MA, Arora A, Yadav M (2016) Sustainability issues on rice–wheat cropping system. Int Soil Water Conserv Res 4:64–74. https://doi.org/10.1016/j.iswcr.2015.12.001

Bhatt A, Kumar NA, Dhanya CS, Vipindas P (2023) Restoring rice paddies and rice agro-ecosystem services through a participatory seed conservation and exchange programme. Maiko Nishi 137. https://doi.org/10.1007/978-981-99-1292-6_8

Biswakarma N, Pooniya V, Zhiipao RR, Kumar D, Shivay YS, Das TK, Behara B et al (2023) Identification of a resource-efficient integrated crop management practice for the rice-wheat rotations in south Asian Indo-Gangetic Plains. Agric Ecosyst Environ 357:108675. https://doi.org/10.1016/j.agee.2023.108675

Cárceles Rodríguez B, Durán-Zuazo VH, Soriano Rodríguez M, García-Tejero IF, Gálvez Ruiz B, Cuadros Tavira S (2022) Conservation agriculture as a sustainable system for soil health: a review. Soil Syst 6(4):87. https://doi.org/10.3390/soilsystems6040087

Castelein R, Broeze JK, Axmann H, Guo X, Soethoudt J (2022) Mechanization in rice farming reduces greenhouse gas emissions, food losses, and constitutes a positive business case for smallholder farmers—results from a controlled experiment in Nigeria. Cleaner Eng Technol 8:100487. https://doi.org/10.1016/j.clet.2022.100487

Chen Z, Chen F (2022) Socio-economic factors influencing the adoption of low carbon technologies under rice production systems in China. Carbon Balance Manage 17(1):19. https://doi.org/10.1186/s13021-022-00218-6

Cherlinka V (2024) Conservation tillage farming: practices to protect soil. EOS data analytics. https://eos.com/blog/conservation-tillage/

Chhokar RS, Sharma RK, Gathala MK, Pundir AK (2014) Efect of crop establishment techniques on weeds and rice yield. Crop Prot 64:7–12. https://doi.org/10.1016/j.cropro.2014.05.016

Clement V (2021) Groundswell part 2: acting on internal climate migration. https://openknowledge.worldbank.org/entities/publication/2c9150df-52c3-58ed-9075-d78ea56c3267

Dash P, Sanders SL, Parajuli P, Ouyang Y (2023) Improving the accuracy of land use and land cover classification of landsat data in an agricultural watershed. Remote Sens 15(16):4020. https://doi.org/10.3390/rs15164020

Deka TJ, Osman AI, Baruah DC, Rooney DW (2023) Assessment of bioenergy and syngas generation in India based on estimation of agricultural residues. Energy Rep 9:3771–3786. https://doi.org/10.1016/j.egyr.2023.02.0542352-4847

Deng C, Zhang G, Liu Y, Nie X, Li Z, Liu J, Zhu D (2021) Advantages and disadvantages of terracing: a comprehensive review. Int Soil Water Conserv Res 9(3):344–359. https://doi.org/10.1016/j.iswcr.2021.03.002

Doni F, Suhaimi NSM, Mispan MS, Fathurrahman F, Marzuki BM, Kusmoro J, Uphoff N (2022) Microbial contributions for rice production: from conventional crop management to the use of 'Omics' technologies. Int J Mol Sci 23(2):737. https://doi.org/10.3390/ijms23020737

Ducci L, Rizzo P, Pinardi R, Solfrini A, Maggiali A, Pizzati M, Balsamo F, Celico F (2022) What is the impact of leaky sewers on groundwater contamination in urban semi-confined aquifers? A test study related to fecal matter and personal care products (PCPs). Hydrology 10(1):3. https://doi.org/10.3390/hydrology10010003

Duff H, Hegedus PB, Loewen S, Bass T, Maxwell BD (2021) Precision agroecology. Sustainability 14(1):106. https://doi.org/10.3390/su14010106

Eyni-Nargeseh H, Asgharipour MR, Rahimi-Moghaddam S, Gilani A, Damghani AM, Azizi K (2023) Which rice farming system is more environmentally friendly in Khuzestan province, Iran? A study based on energy analysis. Ecol Model 481:110373. https://doi.org/10.1016/j.ecolmodel.2023.110373

FAO (2017) The future of food and agriculture – Trends and challenges. Retrieved from https://www.dochas.ie/assets/Files/FAO_2017.pdf

FAO (2022) AQUASTAT - FAO's global information system on water and agriculture. https://www.fao.org/aquastat/en/. Accessed 15 Feb 2024

Farhadi H, Najafzadeh M (2021) Flood risk mapping by remote sensing data and random forest technique. Water 13(21):3115

Food and Agriculture Organization (FAO), United Nations Environment Programme (UNEP) (2021) Global assessment of soil pollution: report. Rome. https://doi.org/10.4060/cb4894en

Gage KL, Krausz RF, Walters SA (2019) Emerging challenges for weed management in herbicide-resistant crops. Agriculture 9(8):180. https://doi.org/10.3390/agriculture9080180

Gavrilescu M (2021) Water, soil, and plants interactions in a threatened environment. Water 13(19):2746. https://doi.org/10.3390/w13192746

Gunstone T, Cornelisse TM, Klein K, Dubey A, Donley N (2021) Pesticides and soil invertebrates: a hazard assessment. Front Environ Sci 9:122. https://doi.org/10.3389/fenvs.2021.643847

Gupta N, Pradhan S, Jain A, Patel N (2021) Sustainable agriculture in India 2021: what we know and how to scale up. Council on Energy, Environment and Water, New Delhi. https://www.ceew.in/sites/default/files/CEEW-FOLU-Sustainable-Agriculture-in-India-2021-20Apr21.pdf

Hajjar MJ, Ahmed N, Alhudaib K, Ullah H (2023) Integrated insect pest management techniques for rice. Sustainability 15(5):4499. https://doi.org/10.3390/su15054499

Hazra KK, Swain DK, Bohra A, Singh S, Kumar N, Nath CP (2016) Organic rice: potential production strategies, challenges and prospects. Org Agric 8:39–56. https://doi.org/10.1007/s13165-016-0172-4

Hester ER, Vaksmaa A, Valè G, Monaco S, Jetten MSM, Lüke C (2022) Effect of water management on microbial diversity and composition in an Italian rice field system. FEMS Microbiol Ecol 98(3):fiac018. https://doi.org/10.1093/femsec/fiac018

HLPE (2019) Agroecological and other innovative approaches for sustainable agriculture and food systems that enhance food security and nutrition. A report by the high-level panel of experts on Food Security and Nutrition of the Committee on World Food Security, Rome

Hollaus A, Schunko C, Weisshaidinger R, Bala P, Vogl CR (2022) Indigenous farmers' perceptions of problems in the rice field agroecosystems in the upper Baram, Malaysia. J Ethnobiol Ethnomed 18(1):26. https://doi.org/10.1186/s13002-022-00511-1

Hossain MA, Piyatida P, da Silva JAT, Fujita M (2012) Molecular mechanism of heavy metal toxicity and tolerance in plants: central role of glutathione in detoxification of reactive oxygen species and methylglyoxal and in heavy metal chelation. J Bot 2012. https://doi.org/10.1155/2012/872875

Howarth RW (2008) Coastal nitrogen pollution: a review of sources and trends globally and regionally. Harmful Algae 8(1):14–20. https://doi.org/10.1016/j.hal.2008.08.015

Huang S, Tang L, Hupy JP, Wang Y, Shao G (2020) A commentary review on the use of normalized difference vegetation index (NDVI) in the era of popular remote sensing. J Forest Res 32(1):1–6. https://doi.org/10.1007/s11676-020-01155-1

ICAR-Indian Agricultural Research Institute (2021) IARI annual report 2022. New Delhi—110012, India

ICAR-Indian Agricultural Research Institute (2022) IARI annual report 2021. New Delhi—110012, India

ICAR-NRRI (n.d.) Rice varieties developed by ICAR-NRRI, Cuttack during 1968–2019. Retrieved from https://icar-nrri.in/released-varieties/

Issa S, Dahy B, Ksiksi T, Saleous N (2020) A review of terrestrial carbon assessment methods using geo-spatial technologies with emphasis on arid lands. Remote Sens 12(12):2008. https://doi.org/10.3390/rs12122008

Johannes HP et al (2019) IOP Conf Ser Mater Sci Eng 546:022008

Khedwal RS, Chaudhary A, Sindhu VK, Yadav DB, Kumar N, Chhokar RS, Dahiya S et al (2023) Challenges and technological interventions in rice–wheat system for resilient food–water–energy-environment nexus in North-western Indo-Gangetic Plains: a review. Cereal Res Commun 1-23. https://doi.org/10.1007/s42976-023-00355-9

Killebrew K, Wolf H (2010) Environmental impacts of agricultural technologies. EPAR Brief No. 65, University of Washington. https://econ.washington.edu/sites/econ/fles/old-site-uploads/2014/06/2010-Environmental-Impacts-of-Ag-Technologies.pdf. Accessed 14 Feb 2024

Kishore A, Pal BD, Joshi K, Aggarwal PK (2018) Unfolding government policies towards the development of climate smart agriculture in India. Agric Econ Res Rev 31(conf):123–137. https://doi.org/10.5958/0974-0279.2018.00028.9

Kobayashi K, Wang X, Wang W (2023) Genetically modified rice is associated with hunger, health, and climate resilience. Foods 12(14):2776. https://doi.org/10.3390/foods12142776

Koech R, Langat P (2018) Improving irrigation water use efficiency: a review of advances, challenges and opportunities in the Australian context. Water 10(12):1771. https://doi.org/10.3390/w10121771

Kukal SS, Aggarwal GC (2003) Puddling depth and intensity effects in rice–wheat system on a sandy loam soil: I. development of subsurface compaction. Soil Tillage Res 72:1–8. https://doi.org/10.1016/S0167-1987(03)00093-X

Kumar P, Mittal S (2006) Agricultural productivity trends in India sustainability issues. Agric Econ Res Rev 19:71–88

Kumar N, Chaudhary A, Ahlawat OP, Naorem A, Upadhyay G, Chhokar RS, Singh GP et al (2023) Crop residue management challenges, opportunities and way forward for sustainable food-energy security in India: a review. Soil Tillage Res 228:105641. https://doi.org/10.1016/j.still.2023.105641

Kumar R, Kumar V, Sharma A, Singh N, Kumar R, Katnoria JK, Rodrigo-Comino J (2019) Assessment of pollution in roadside soils by using multivariate statistical techniques and contamination indices. SN Appl Sci 1:1–11. https://doi.org/10.1007/s42452-019-0888-3

Kumar N, Chhokar RS, Meena RP et al (2022) Challenges and opportunities in productivity and sustainability of rice cultivation system: a critical review in Indian perspective. Cereal Res Commun 50:573–601. https://doi.org/10.1007/s42976-021-00214-5

Langer M, von Deimling TS, Westermann S, Rolph R, Rutte R, Antonova S, Grosse G et al (2023) Thawing permafrost poses environmental threat to thousands of sites with legacy industrial contamination. Nat Commun 14(1):1721. https://doi.org/10.1038/s41467-023-37276-4

Liu Y, Li H, Hu T, Mahmoud A, Li J, Zhu R, Jing P et al (2022) A quantitative review of the effects of biochar application on rice yield and nitrogen use efficiency in paddy fields: a meta-analysis. Sci Total Environ 830:154792. https://doi.org/10.1016/j.scitotenv.2022.154792

Liu P, Wu Q, Hu W, Tian K, Huang B, Zhao Y (2023a) Effects of atmospheric deposition on heavy metals accumulation in agricultural soils: evidence from field monitoring and Pb isotope analysis. Environ Pollut 330:121740. https://doi.org/10.1016/j.envpol.2023.121740

Liu Z, Yin Y, Zhang Y, Shi S (2023b) Legal system of soil pollution remediation in China and its regulation and guidance to soil pollution remediation. Sustainability 15(15):11504. https://doi.org/10.3390/su151511504

Lü Y, Song S, Wang R, Liu Z, Meng J, Sweetman AJ, Wang T et al (2015) Impacts of soil and water pollution on food safety and health risks in China. Environ Int 77:5–15.https://doi.org/10.1016/j.envint.2014.12.010

Lv P, Su F, Chen F, Yan C, Xia D, Sun H, Li S, Duan Z, Ma C, Zhang H, Wang M, Niu X, Zhu J, Zhang J (2023) Genome editing in rice using CRISPR/Cas12i3. Plant Biotechnol J 22(2):379–385. https://doi.org/10.1111/pbi.14192

Mallareddy M, Thirumalaikumar R, Balasubramanian P, Naseeruddin R, Nithya N, Mariadoss A, Vijayakumar S et al (2023) Maximizing water use efficiency in rice farming: a comprehensive review of innovative irrigation management technologies. Water 15(10):1802. https://doi.org/10.3390/w15101802

Mangkoedihardjo S, Samudro H (2023) Preventive remediation methods minimize soil pollution. Int J Adv Appl Sci 12(1):60–65. https://ijaas.iaescore.com/index.php/IJAAS/article/view/20712/13022

Mboyerwa PA, Kibret K, Mtakwa PW, Aschalew A (2021) Evaluation of growth, yield, and water productivity of paddy rice with water-saving irrigation and optimization of nitrogen fertilization. Agronomy 11(8):1629. https://doi.org/10.3390/agronomy11081629

Mdee A, Ofori A, Lopez-Gonzalez G, Stringer L, Martin-Ortega J, Ahrari S, Camargo-Valero MA (2022) The top 100 global water questions: results of a scoping exercise. One Earth 5(5):563–573. https://doi.org/10.1016/j.oneear.2022.04.009

Millán S, Campillo C, Casadesús J, Pérez-Rodríguez JM, Prieto M (2020) Automatic irrigation scheduling on a hedgerow olive orchard using an algorithm of water balance readjusted with soil moisture sensors. Sensors 20(9):2526. https://doi.org/10.3390/s20092526

Mohan D, Shukla SP (2022) Hazardous consequences of textile mill effluents on soil and their remediation approaches. Cleaner Eng Technol 7:100434. https://doi.org/10.1016/j.clet.2022.100434

Mulenga C (2022) Soil governance and the control of mining pollution in Zambia. Soil Secur 6:100039. https://doi.org/10.1016/j.soisec.2022.100039

Münzel T, Hahad O, Daiber A, Landrigan PJ (2023) Soil and water pollution and human health: what should cardiologists worry about? Cardiovasc Res 119(2):440–449. https://doi.org/10.1093/cvr/cvac082

Nacorda ER, Vasquez PPG, Lador RP, MoraGarcia C (2019) Assessment of agrochemical use among rice field farmers in municipalities surrounding Lake Mainit, Philippines. Int J Dev Sustain 8(9):633–644. https://www.researchgate.net/publication/338660101_Assessment_of_agrochemical_use_among_rice_field_farmers_in_municipalities_surrounding_Lake_Mainit

Naidu R, Biswas B, Willett IR, Cribb J, Singh BK, Nathanail CP, Aitken RJ (2021) Chemical pollution: a growing peril and potential catastrophic risk to humanity. Environ Int 156:106616. https://doi.org/10.1016/j.envint.2021.106616

Najmuddin O, Rasul G, Hussain A, Molden D, Wahid S, Debnath B (2018) Low water productivity for rice in Bihar, India—a critical analysis. Water 10(8):1082. https://doi.org/10.3390/w10081082

Nandan R, Singh V, Singh SS, Kumar V, Hazra KK, Nath CP, McDonald A et al (2019) Impact of conservation tillage in rice–based cropping systems on soil aggregation, carbon pools and nutrients. Geoderma 340:104–114.https://doi.org/10.1016/j.geoderma.2019.01.001

Nath A, Samanta S, Banerjee S, Danda AA, Hazra S (2021) Threat of arsenic contamination, salinity and water pollution in agricultural practices of Sundarban Delta, India, and mitigation strategies. SN Appl Sci 3:1–15. https://doi.org/10.1007/s42452-021-04544-1

Ongley ED (1996) Control of water pollution from agriculture, vol 55. Food & Agriculture Organization. https://www.fao.org/3/w2598e/w2598e04.htm

Onorati F, Tornambé A, Paina A, Maggi C, Sesta G, Berducci MT, D'Antoni S et al (2022) Ecotoxicological and chemical approach to assessing environmental effects from pesticide use in organic and conventional rice paddies. Water 14(24):4136. https://doi.org/10.3390/w14244136

Panpakdee C (2023) The social-ecological resilience indicators of organic rice production in Northeastern Thailand. Org Agric 13(3):483–501. https://doi.org/10.1007/s13165-023-00441-6

Park JR, Jang YH, Kim EG, Lee GS, Kim KM (2023) Nitrogen fertilization causes changes in agricultural characteristics and gas emissions in rice field. Sustainability 15(4):3336. https://doi.org/10.3390/su15043336

Pathak H, Aggarwal PK (2012) Low carbon technologies for agriculture: a study on rice and wheat systems in the Indo-Gangetic Plains. Indian Agric Res Inst 78:xvii

Patra SS, Suneetha P, Rout S, Dwivedi DK (2022) Estimation of rice crop acreage using remote sensing and GIS. ResearchGate. https://www.researchgate.net/publication/358424223_Estimation_of_rice_crop_acreage_using_Remote_Sensing_and_GIS

Poore J, Nemecek T (2018) Reducing food's environmental impacts through producers and consumers. Science 360(6392):987–992. https://doi.org/10.1126/science.aaq0216

Pröbstl-Haider U (2022) EIA effectiveness in sensitive alpine areas: a comparison of winter tourism infrastructure development in Germany and Austria. Sustainability 14(15):9775. https://doi.org/10.3390/su14159775

Rajmohan KS, Chandrasekaran R, Varjani S (2020) A review on occurrence of pesticides in environment and current technologies for their remediation and management. Indian J Microbiol 60:125–138. https://doi.org/10.1007/s12088-019-00841-x

Rane N, Giduturi M, Choudhary SP, Pande CB (2023) Remote sensing (RS) and geographical information system (GIS) as a powerful tool for agriculture applications: efficiency and capability in agricultural crop management. Zenodo (CERN European Organization for Nuclear Research). https://doi.org/10.5281/zenodo.7845187

Rathour RK, Devi M, Dahiya P, Sharma N, Kaushik N, Kumari D, Bhatia RK et al (2023) Recent trends, opportunities and challenges in sustainable management of rice straw waste biomass for green biorefinery. Energies 16(3):1429. https://doi.org/10.3390/en16031429

Ridzuan S (2021) Inequality and water pollution in India. Water Policy 23(4):985–999. https://doi.org/10.2166/wp.2021.057

Rodell M, Velicogna I, Famiglietti JS (2009) Satellite-based estimates of groundwater depletion in India. Nature 460:999–1002. https://doi.org/10.1038/nature08238

Saini S, Tewari S, Dwivedi J, Sharma V (2023) Biofilm-mediated wastewater treatment: a comprehensive review. Mater Adv. https://doi.org/10.1039/d2ma00945e

Sandil S, Kumar R (2021) Soil contamination from construction projects. Ecol Health Eff Build Mater 205–244. https://doi.org/10.1007/978-3-030-76073-1_12

Sansa S, Syriac KE, Raj SK (2016) Penoxsulam as post-emergence herbicide for weed control in transplanted rice. Indian J Weed Sci 48(2):215. https://doi.org/10.5958/0974-8164.2016.00052.6

Sartori M, Philippidis G, Ferrari E, Borrelli P, Lugato E, Montanarella L, Panagos P (2019) A linkage between the biophysical and the economic: assessing the global market impacts of soil erosion. Land Use Policy 86:299–312. https://doi.org/10.1016/j.landusepol.2019.05.014

Sawhney A, Sawhney A (2023) Compliance dilemmas in Indian environmental policy | East Asia Forum. East Asia Forum. https://eastasiaforum.org/2018/10/10/compliance-dilemmas-in-indian-environmental-policy/

Sbai A (2024) Precision agriculture: accuracy promotes success. Infomineo. https://infomineo.com/agriculture/precision-agriculture-accuracy-promotes-success/

Sharma B, Molden D, Cook S (2015) Water use efficiency in agriculture: measurement, current situation and trends. In: Drechsel P, Hefer P, Magen H, Mikkelsen R, Wichelns D (eds) Managing water and fertilizer for sustainable agricultural intensification. International Fertilizer Industry Association (IFA), International Water Management Institute (IWMI), International Plant Nutrition Institute (IPNI), and International Potash Institute (IPI), France, Paris, pp 39–64

Sharma P, Singh A, Kahlon CS, Brar AS, Grover KK, Dia M, Steiner RL (2018) The role of cover crops towards sustainable soil health and agriculture—a review paper. Am J Plant Sci 9(9):1935–1951. https://doi.org/10.4236/ajps.2018.99140

Sharma P, Bano A, Singh S, Dubey NK, Chandra R, Iqbal HM (2022) Recent advancements in microbial-assisted remediation strategies for toxic contaminants. Cleaner Chem Eng 2:100020. https://doi.org/10.1016/j.clce.2022.100020

Shi M, Geng B, Zhao T, Wang F (2021) Influence of atmospheric deposition on surface water quality and DBP formation potential as well as control technology of rainwater DBPs: a review. Environ Sci Water Res Technol 7(12):2156–2165. https://doi.org/10.1039/d1ew00520k

Singh G, Sharma S (2024) A comprehensive review on the Internet of Things in precision agriculture. Multimed Tools Appl. https://doi.org/10.1007/s11042-024-19656-0

Sishodia RP, Ray RL, Singh SK (2020) Applications of remote sensing in precision agriculture: a review. Remote Sens 12(19):3136. https://doi.org/10.3390/rs12193136

Song Z, Seitz S, Li J, Goebes P, Schmidt K, Kühn P, Shi X, Scholten T (2019) Tree diversity reduced soil erosion by affecting tree canopy and biological soil crust development in a subtropical forest experiment. For Ecol Manage 444:69–77. https://doi.org/10.1016/j.foreco.2019.04.015

Srivastava PK, Singh M, Gupta M, Singh N, Kharwar RN, Tripathi RD, Nautiyal CS (2015) Mapping of arsenic pollution with reference to paddy cultivation in the middle Indo-Gangetic Plains. Environ Monit Assess 187(4):e198. https://doi.org/10.1007/s10661-015-4418-5

Steffan JJ, Brevik EC, Burgess LC, Cerdà A (2017) The effect of soil on human health: an overview. Eur J Soil Sci 69(1):159–171. https://doi.org/10.1111/ejss.12451

Talema A (2023) Causes, negative effects, and preventive methods of water pollution in Ethiopia. Qual Assur Saf Crops Foods 15(2):129–139. https://doi.org/10.15586/qas.v15i2.1271

Tozer L (2023) Water pollution 'timebomb' threatens global health. Nature. https://doi.org/10.1038/d41586-023-02337-7

Tripathi R, Guru PK, Mohanty S, Goud BR (2022) Advanced techniques for precision farming in rice. ResearchGate. https://www.researchgate.net/publication/360485061_Advanced_Techniques_for_Precision_Farming_in_Rice#:~:text=Precision%20rice%20farming%20is%20implemented,models%2C%20and%20decision%20support%20systems

UN Water (2015) UN world water development report 2015. Retrieved from https://www.unwater.org/publications/un-world-water-development-report-2015

UNCCD (n.d.) Global land outlook 2nd edition. Retrieved from https://www.unccd.int/resources/global-land-outlook/glo2#:~:text=Global%20Land%20Outlook%202nd%20edition,reduce%20and%20reverse%20land%20degradation

Upadhyay P, Vaishampayan A, Jaiswal SK (2020) Soil pollution caused by agricultural practices and strategies to manage it. Plant Responses Soil Pollut 119–132. https://doi.org/10.1007/978-981-15-4964-9_7

Verma A (2021) Law of environment in India: problems and challenges in its enforcement. Res Ambition Int Multi e-J 6(II):17–26. https://doi.org/10.53724/ambition/v6n2.04

Vijayakumar S (2023) Revolutionizing rice farming: maximizing yield with minimal water to sustain the Hungry Planet. IntechOpen. https://doi.org/10.5772/intechopen.112167

Volsi B, Higashi GE, Bordin I, Telles TS (2022) The diversification of species in crop rotation increases the profitability of grain production systems. Sci Rep 12(1):19849. https://doi.org/10.1038/s41598-022-23718-4

Wang HH, Yang J, Hao N (2022a) Consumers' willingness to pay for rice from remediated soil: potential from the public in sustainable soil pollution treatment. Int J Environ Res Public Health 19(15):8946. https://doi.org/10.3390/ijerph19158946

Wang H, Xiong R, Zhou Y, Tan X, Pan X, Zeng Y, Zeng Y et al (2022) Grain yield improvement in high-quality rice varieties released in southern China from 2007 to 2017. Front Sustain Food Syst 6:986655. https://doi.org/10.3389/fsufs.2022.986655

Water Pollution from and to Agriculture—International Decade for Action (2018) International decade for action. https://wateractiondecade.org/2017/12/09/water-pollution-from-and-to-agriculture/

Wichelns D, Qadir M (2015) Achieving sustainable irrigation requires effective management of salts, soil salinity, and shallow groundwater. Agric Water Manage. https://doi.org/10.1016/j.agwat.2014.08.016

Wolfert S, Ge L, Verdouw C, Bogaardt M (2017) Big data in smart farming—a review. Agricult Syst 153:69–80. https://doi.org/10.1016/j.agsy.2017.01.023

Yadav R, Ramakrishna W (2023) Biochar as an environment-friendly alternative for multiple applications. Sustainability 15(18):13421. https://doi.org/10.3390/su151813421

Yuzugullu O, Lorenz F, Fröhlich P, Liebisch F (2020) Understanding fields by remote sensing: soil zoning and property mapping. Remote Sens 12(7):1116. https://doi.org/10.3390/rs12071116

Zafar K, Sedeek KE, Rao GS, Khan MZ, Amin I, Kamel R, Mahfouz MM et al (2020) Genome editing technologies for rice improvement: progress, prospects, and safety concerns. Front Genome Editing 2:5. https://doi.org/10.3389/fgeed.2020.00005

Zhou J, Wen Y, Marshall MR, Zhao J, Gui H, Yang Y, Zang H et al (2021) Microplastics as an emerging threat to plant and soil health in agroecosystems. Sci Total Environ 787:147444. https://doi.org/10.1016/j.scitotenv.2021.147444

Zhou J, Liu G, Zhao Y, Zhang R, Tang X, Li L, Zhang Y et al (2023) An efficient CRISPR–Cas12a promoter editing system for crop improvement. Nat Plants 9(4):588–604. https://doi.org/10.1038/s41477-023-01384-2

Chapter 10
Engineered Bioremediation for Decontamination of Aquifers

Bhavna Nigam, Dheeraj Rathore, Anuj Kumar Yadav, Ratna Chauhan, and Indra Jeet Chaudhary

Abstract Globally land resources and groundwater are naturally available for human utilization. But after globalization and industrialization the use of hazardous chemicals increased due to unmanageable conditions. The chemical disposed on the ground surface and many other anthropogenic activities by humans like use of pesticides and other contaminate the soil and groundwater. Underground storage tanks and pipelines frequently damage the soils and groundwater. The different type of contaminant by percolation through the ground reach to the aquifer and get affected which causes serious problem. Therefore countries spend lot of money and use many technologies to extract and remediate the groundwater contamination. For many years, these polluted sites have been cleaned up using in situ and ex situ bioremediation techniques. These techniques have potential to environmental friendly contamination from soil and groundwater. Therefore the aim of the chapter is provide knowledge about groundwater contamination and its remediation technology. This chapter also provides critical review about in situ and ex situ bioremediation technologies and its applicability. The longer-term success of bioremediation may well depend upon developing environmental friendly and cost-effective treatments techniques for degradation of contaminants, especially in groundwater. Therefore these effort can change country economy and sustainable uses of fresh water.

Keywords Groundwater pollution · Engineered bioremediation · In situ and ex situ techniques · Monitoring and control

B. Nigam · D. Rathore
School of Environment and Sustainable Development, Central University of Gujarat, Gandhinagar, Gujarat 382030, India

A. K. Yadav
Department of Physics, Dr. Hari Singh Gour Vishwavidyalaya, Sagar, Madhya Pradesh 470003, India

R. Chauhan · I. J. Chaudhary (✉)
Department of Environmental Science, Savitribai Phule Pune University, Pune, Maharashtra 411007, India
e-mail: indrajeet.cug@gmail.com

A. Nanda et al. (eds.), *Navigating the Nexus*, Water Science and Technology Library 102, https://doi.org/10.1007/978-3-031-76532-2_10

10.1 Introduction

Water is most important factor for both human survival and other living things on Earth. On Earth, it exists as solids (ice and glaciers), gases (vapour in the atmosphere), and predominantly liquids (oceans, lakes, rivers, and groundwater, among other places). Groundwater is the primary source of drinking water in the world because of its accessibility, abundance, and purity. For irrigation proposes about 60% of agricultural area and drinking water around 85% of the population used groundwater in India (Worldbank 2012). The amount of rainfall, the topography, and aquifer recharge from water catchments all affect groundwater availability. Rapid urbanization, industry, and other activities have made urban groundwater scarcity problem (La Vigna 2022). Population growth raises the need for fresh water supplies, necessitating the construction of hundpums and the digging of bore wells. Groundwater levels and its physico-chemical characteristics could decline day by day. Groundwater is defined as water that lies below the surface and fills the pores (Arora et al. 2019). A vast number of people worldwide depend on groundwater for their daily needs. Highly mineralized groundwater makes up 1.69% of all groundwater and doesn't need significant processing before usage (Makubalo 2020). Their mobility in the subsurface causes them to get filtered and mineralized. However, numerous industrial effluents infiltrate and reach beneath water table due to industrialization and an increase in anthropogenic activity (Arif et al. 2021). Approx 33% of groundwater is not safe for drinking purposes in India (Vig et al. 2023). The major groundwater pollutants are either geogenic which includes dissolved inorganic and organic substances, or anthropogenic substances like pesticide, metals and industrial by products. Organic and inorganic toxic substances enter into aquifers through leachates of industrial effluents (Nemčić-Jurec et al. 2022). According to Papazotos (2021), the majority of heavy metals (HMs) are released into groundwater sources by both natural and man-made processes. These releases include arsenic (As), zinc (Zn), lead (Pb), mercury (Hg), and manganese (Mn), iron (Fe), copper (Cu), and nickel (Ni) (Iqbal et al. 2023a, b). Toxic metal exposure can have both acute and long-term consequences on human health. It is known to cause cancer, hyperkeratosis, melanosis, lung diseases, hypertension, and peripheral vascular diseases, among other health outcomes (Ricolfi et al. 2020; Iqbal et al. 2023a, b).

The globe is investing billions of dollars and employing a variety of strategies to extract or remediate these contaminations. One method that makes it simple and practical to remove or treat this kind of contamination is bioremediation (Sharma 2020). For bioremediation to be effective, organic wastes must be biologically degraded under controlled conditions to a mild state or to concentrations below limits set by regulatory authorities. This, microorganisms must attack the pollutants enzymatically and transform them into harmless products. Only in environments that support microbial growth and activity can bioremediation be effective; this is why it frequently involves modifying environmental parameters to hasten microbial growth and breakdown. In comparison to other approaches like cremation, bioremediation techniques are often more cost-effective. Additionally, certain pollutants can be treated on-site, lowering the danger of exposure for clean-up personnel or wider exposure in the event of transportation accidents (King et al. 2023).

The process's methodology is not technically challenging, but it may take a great deal of experience and knowledge to put it into practice. This involves carefully examining the location and determining the necessary conditions. The typical remediation method involves either covering or capping the polluted area or pumping up contaminated soil and removing it to the designated location. There are a few shortcomings. Excavation, processing, and transportation of hazardous materials constitute the first method's simple means of moving the contaminated materials, which poses significant dangers (Tripathi and Ram 2018). The search for additional disposal landfills is particularly challenging. While contaminants stay on site and necessitate long-term monitoring and maintenance of the designed barrier, increasing costs and liabilities, the cover or cap method only offers a temporary solution. The more effective and sophisticated techniques compared to the earlier ones, which either totally remove pollutants or at the very least convert them into harmless compounds. High-temperature incineration and several forms of chemical decomposition (such as UV oxidation and base-catalyzed de-chlorination) are a few of the technologies that have been employed (Tripathi and Ram 2018). Although they can be very effective at lowering levels of a variety of contaminants, they have a number of disadvantages, the main ones being their technological complexity, the expense of applying them on a small scale, and the lack of public acceptance, particularly in the case of incineration, which may expose nearby residents and workers to higher levels of contaminants.

Geological processes like ore production, rock weathering, and leaching, as well as human population growth, urbanization, and industrial activity, can all lead to groundwater contamination and also water scarcity. This contaminant causes various human health effects. Therefore the chapter aims to focus on groundwater pollution and its effect and engineered bioremediation techniques for decontamination of aquifers. The study will be helpful for the management of groundwater contamination and its pollution control technology.

10.2 Factors of Bio-remediation

Bio-remediation is a complicated process, and for it to be optimized, a number of environmental conditions must be satisfied, including pH, temperature, soil type, oxygen content, and so forth (Kebede et al. 2021). These circumstances are necessary for the biological development of microorganisms that control the pollution. Table 10.1 lists a few of the requirements and suitable conditions for bioremediation work. One of the main factors influencing microbial activity is the microorganisms' capacity to break down organic resources into energy sources. An aerobic heterotrophic organism can effectively obtain energy from a contaminant's average oxidation state of C (Lin et al. 2022). Therefore, reduced energy yields associated with higher oxidation states translate into decreased energy output for microbial breakdown. The outcome of any biodegradation procedure depends on a variety of environmental factors (pH, moisture content, temperature, EC, availability of electron acceptors and carbon, and energy sources), microbiological factors (population diversity, enzyme activities, and biomass concentration), and substrate factors (physical and chemical characteristics, molecular concentration, and structure). These factors affect how long it takes for the bacteria to become accustomed to the substrate. It is well recognized that the concentration and molecular structure of pollutants have a significant impact on the viability of bioremediation, the kind of microbial transformation that takes place, and whether the molecule will function as a primary, secondary, or co-metabolic substrate (Saravanan et al. 2023).

Nearly every type of environmental setting can support the growth of microorganisms. Microbes can proliferate in the presence of any hazardous compound or waste stream, and they can accumulate and grow in a variety of environments, including sub-zero and extremely high temperatures, aqueous media, and oxygen and non-oxygen environments (Bala et al. 2022). Microbial growth requires the presence of a carbon source, which provides the energy needed for enzymes that can be utilized to remove or break down pollutants. These microorganisms can be divided into the following groups: Aerobic bacteria include Mycobacterium, Rhodococcus, Sphingomonas, Alcaligenes, and Pseudomonas. These microorganisms remove pesticides and hydrocarbons, including chemicals and alkanes (Bhattacharjee et al. 2020). The primary source of energy and carbon accumulation in the presence of oxygen is a large number of these bacteria. Anaerobic microorganisms are not as useful as aerobic ones. It can be applied to dechlorinate the solvent trichloroethylene (TCE), bioremediate river sediments of polychlorinated biphenyls (PCBs), and remove chromium. Without oxygen, it functions (Tripathi and Ram 2018). According to Arya et al. (2018), ligninolytic fungi are utilized to break down a very rare variety of harmful or important environmental contaminants. Corn cobs, sawdust, and straw are examples of common substrates. For instance, white rot fungi Methylotrophic bacteria are aerobic bacteria that thrive by using methane as a source of carbon and energy. Methane mono-oxygenase, the first enzyme in the aerobic degradation pathway, is active against a large variety of substances, including 1,2-dichloromethane and chlorinated aliphatic trichloroethylene. It also has a broad substrate range.

Table 10.1 Factor affecting in bioremediation (Azubuike et al. 2016)

S. no	Factor and process in bioremediation	
1.	Micro-organism	Aerobic and anaerobic
2.	Biological process	Catabolism and anabolism
3.	Environmental factor	Temperature, pH, oxygen content, electron acceptor or donor
4.	Nutrient	Carbon, nitrogen, oxygen etc.
5.	Soil moisture	25–28% of water holding capacity
6.	Type of soil	Clay and silty soil

10.2.1 Nutrient

The most crucial element that is needed in great quantities to support the growth of microorganisms is carbon. Other than this, around 95% of the weight is required to be made up of oxygen, hydrogen, and nitrogen (Godswill et al. 2020). The type of bioremediation needed depends on the type of soil and the level of contamination it contains. Sulfur and phosphorus concentrations can help remove 70% of the toxins (Hort et al. 2009). The ratio of carbon to nitrogen and carbon to phosphorus in food is 10:1 and 30:1, respectively. Microbial cell development and division require nutrients (Prasad et al. 2021). In order to promote bioremediation, nutrients can be provided in a form that is useful or through an organic substrate amendment (Patel et al. 2022), which also acts as an electron donor. Sufficient amounts of trace nutrients are typically available for microbial growth can promote bioremediation capabilities.

10.2.2 Soil

Water with a pollutant concentration of 5% or more can be treated by directing it to various soil stratum interfaces (Ossai et al. 2020). Active agents in the interface help to partially remove contaminants from the oils. We can install bioremediation to clean up the soil after moving on to this (Sales da Silva et al. 2020). This procedure is the only one that, at this experimental level, converts the polluted soil into dirt that is good for sites and landscaping. To improve the effectiveness of this procedure, numerous other changes are being considered. Bioremediation can be used to directly treat low concentrations of pollutants. It is a laborious procedure that takes six months to a year or longer to purify soil with 2% or less oil; however, if there is only 0.8% or less heavy oil present, the soil can be cleaned in one to two months. With a little work, these techniques aid in the recycling and reuse of soil, which is a green process.

10.2.3 pH

pH plays important role in ex-situ as well as in-situ bioremediation processes. pH of the media affects the nature of pollutant, and availability of proton ions which would bind with the metals, dyes and other pollutants. Microorganisms like bacteria and fungi required optimum pH conditions. The favourable bioremediation pH for bacterial growth lies between 6.5 and 8.5 (Kharangate-Lad and D'Souza 2021). The change in pH may alter the pollutant mobility and binding capacity. The process of degrading contaminants is sometimes hampered by changes in groundwater pH, which prevents the growth of dechlorination bacteria (O'Connor et al. 2018). Although dechlorination bacteria are highly sensitive to an acidic pH, hydrochloric acid can be produced throughout the process (Chen et al. 2020). According to previous research, the ideal pH range for groundwater is between 6.8 and 7.8, and dechlorination is rarely seen when the pH is below 6 (Ortiz-Medina et al. 2023). Thus, some research indicates that in order to maximize reductive dechlorination and avoid a large pH drop, pH buffering would be necessary (Liu et al. 2021). The solubility strength of DNAPL in groundwater may be limited by the impact of acidification on biological dechlorination (Lin et al. 2021). By using non-acidified electron donors or pH buffering, this groundwater acidification issue can be resolved (Lu et al. 2023).

10.2.4 Moisture Content

The availability of moisture content affects the solubility of pollutants like metals, dyes & pollutants in soil. Bioremediation of organic pollutants is highly dependent on moisture for microbial growth. The biodegradability is reduced on presence of higher moisture content present in soil (Ayilara and Babalola 2023). Soil moisture content is a key factor in soil bioremediation. Soil has to have access to enough water to sustain microbiological activity. While too much water can block soil pores and prevent oxygen from diffusing toward microbes, too little water can also suppress microbial activity. Therefore, it is crucial to provide soil with an ideal water content for bioremediation. Previous studies have demonstrated that when the soil's water content reaches 60% of its water-holding capacity, microbial activity in the soil reaches its peak. During the bioremediation process, soil water evaporation cannot be prevented.

10.2.5 Electrical Conductivity

Electrical conductivity is measurements of anions & cation present in environmental media. Salimi et al. (2012) studied that high EC levels of irrigation water increased Cd uptake and decreased Zn concentration in the shoots. The salinity of groundwater, which differs among locations as a result of various geological and environmental factors, significantly affects the bioremediation effect. Determining the structure of a pollutant plume requires precise mapping of the groundwater's electrical conductivity and redox potential. An aquifer's redox potential map shows the concentrations of redox-active components and the biodegradation of organic matter, whereas an electrical conductivity map shows the groundwater's mineralization.

10.3 Groundwater Pollution

Pollutants that are released onto the ground seep through various ground layer interfaces and mix or come into contact with the groundwater table (Liu et al. 2021). These substances undergo more than acceptable changes in temperature, pH, color, total dissolve solids (TDS), dissolve oxygen (DO), and other physical and chemical properties (Nayar 2020). These changes are brought on by human activity, chemical releases from industries, and the use of various chemicals, such as pesticides, in agricultural practices. Unwanted substances or contaminants in groundwater can also lead to pollution of the water supply. It has an impact on an aquifer's plume. A substantial portion of the ground is contaminated by the movement and dispersion of water inside the aquifer. Its advancing edge, also known as the plume edge, makes surface water poisonous for both people and wildlife when it comes into contact with groundwater wells, springs, rivers, etc. Groundwater models can be used to analyze plume front migration (He et al. 2022). The hydrology, geology, soil qualities, and kind of contaminants all contribute to the analysis of groundwater pollution (Li et al. 2021).

The presence of foreign material in groundwater that poses a health concern to humans or wildlife is one way to depict groundwater toxicity. These can be classified into two categories: acute toxicity and chronic toxicity. They can also be caused by psychological issues, other health problems, or persistent neurological disorders. The toxicity caused by a brief exposure to a toxicant, measured by LD50, is known as acute toxicity (Zhang et al. 2022). The deadly concentration needed to kill 50% of the population is known as an LC50, while the LD50 is the fatal dose at which 50% of the population is killed in a specific amount of time (Tripathi and Ram 2018). It is important to consider that various populations of a species may exhibit varying degrees of tolerance to hazardous substances. On the other hand, chronic toxicity is the kind of toxicity brought on by repeated exposure to a toxicant that can result in major problems including mutagenicity, carcinogenicity, and teratogenicity, which can lead to bioaccumulation-induced death (Lee et al. 2024). The buildup of chemicals or other compounds, such pesticides, within an organism is known as bioaccumulation. When an organism absorbs a material more quickly than it is eliminated by excretion and catabolism, this is known as bioaccumulation. Examples of this include Lowest Observed Effect Level (LOEL), No Observed Effect Level (NOEL), and others (Yang et al. 2020). The greatest dose or exposure level of a compound that has no discernible (observable) harmful impact on test animals is known as the NOEL, or no observable effect level. The lowest dosage level (LOAEL) at which tested anımal's exhibit negative effects after repeated exposure to a chemical is known.

Because groundwater moves through hidden aquifers across a wide area, it is more difficult to halt than surface water. Non-porous clayey aquifers filter water by filtration (adsorption and absorption), dilution, and occasionally chemical reactions and biological activity; certain pollutants transform into soil contaminants (Zeidan 2017). It is impossible to filter contaminants that seep via fissures and come into contact with surface water. We can get rid of the toxins or pollutants by employing several methods. Groundwater remediation is the process of changing the characteristics of contaminated groundwater by applying physical, biological, chemical, or a combination of treatments. Among the biological treatment methods are phytoremediation, bio-augumentation, bioventing, biosparging, and bioslurping (Tripathi and Ram 2018). Injecting ozone and oxygen gases, chemical precipitation, membrane separation, ion exchange, carbon absorption, aqueous chemical oxidation, and surfactant improved recovery are a few chemical treatment methods. Nanomaterials can be used to implement some chemical methods.

Groundwater, generally we extract from unconfined aquifers which percolates through gelogenic pathways (Singh et al. 2023). Evaluating groundwater pollution is important from an aesthetic and financial standpoint. Because of industrial activity, groundwater in metropolitan areas whether they are in the center of them or not has been contaminated. Groundwater pollution can result from the disposal of industrial waste, overuse of fertilizers, plastic waste, compost pit leachates, and sewage seepage (Ghosh et al. 2022). Water-consuming industries such as textile, paint, pulp and paper, cement, iron, and steel produce industrial effluents that are either improperly treated or released into open water bodies like rivers, eventually reaching the water table. Industrial regions' water quality was found to be low because of excessive extraction and aquifer channel contamination. Many parts of India are home to numerous harmful contaminants, including pesticides, nitrate, metals (such Cr, As), and fluoride. Groundwater contains fluoride in the forms of fluorosilicic acid (H_2SiF_6), sodium fluoride (NaF), and hydrogen fluoride (HF). Hydrogen fluoride (HF) is one of these that dissolves easily in water (Shaji et al. 2023). The World Health Organization states that fluoride exposure levels exceeding 1.5 mg/l are hazardous. At lower quantities, it is necessary for the development of teeth, bones, and other body components; but, at higher concentrations, it poses health risks, such as fluorosis (Prabhu and Meenakshi 2014). A common type of nitrogen in the biosphere that is typically necessary for plant growth is nitrate. Nitrate is transformed by microbiological and chemical activities into ammonia, nitrogen (gaseous), and nitrogen oxides in the biogeochemical cycles (Huno et al. 2018). Nitrate from anthropogenic causes, such as the use of pesticides and fertilizers in the agricultural sector, adds to groundwater pollution (Fig. 10.1). In addition to human activity, nitrate pollution is caused by manure pools, untreated industrial discharges, and aquifer geological processes. The WHO establishes 50 mg/L as the safe limit for drinking water; concentrations above this range can cause harmful health effects such as methemoglobinemia, also referred to as "blue baby syndrome" (Wang et al. 2023). Li et al. (2021) explained that in northern hemisphere countries major pollutants are arsenic, fluoride metal, redon and organic contaminates. Pesticides are used in the agriculture industry to boost crop, fruit, and vegetable production. Insecticides, fungicides, and herbicides are among its forms found throughout the globe. Pesticides are not biodegradable in an environment and are xenobiotics by nature. Approximately 62% of pesticides used in the agriculture sector worldwide are herbicides (Bexfield et al. 2021).

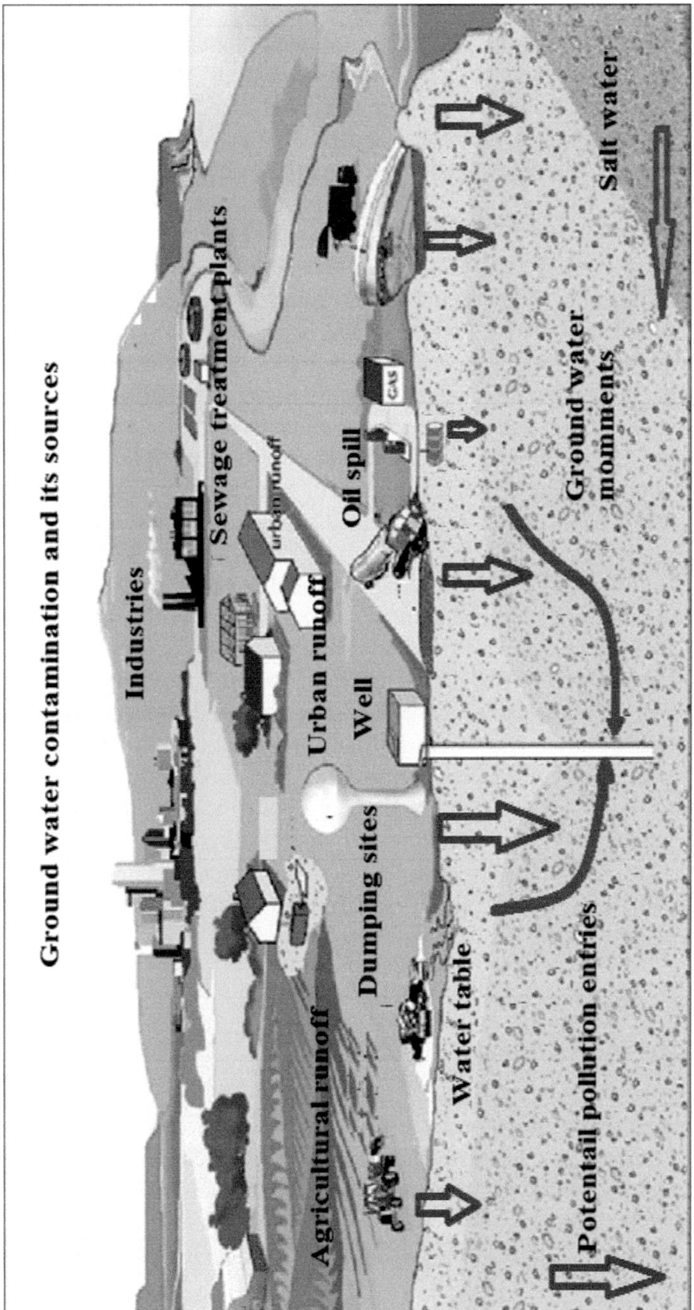

Fig. 10.1 Groundwater pollution and its sources

10.4 Groundwater Pollution Remediation Technologies

Groundwater remediation is the process of removing pollutants from water using physical, chemical, biological, or a mix of these methods. Ex-situ and in-situ are the two subcategories into which it can be separated. Injecting ozone and oxygen gases, chemical precipitation, membrane separation, ion exchange, carbon absorption, aqueous chemical oxidation, and surfactant improved recovery are a few chemical treatment methods. Nanomaterials may be used in the application of certain chemical processes (Yuhong et al. 2022). Moreover, the PRB technology has recently included electrokinetic (EK) ideas to enhance its capabilities in groundwater cleanup. Because of its ability to eliminate impurities from low-permeable media, the EK process can be utilized to solve PRB clogging issues brought on by precipitation reactions (Faisal et al. 2018). Furthermore, in their 2007 study, Chung and Lee examined the possible use of atomizing slag, a reactive material for RPB, in conjunction with EK to remove either organic or inorganic pollutants from contaminated groundwater (Faisal et al. 2018). The findings showed that the electro migration mechanism of the applied configuration enhanced the rate of cadmium removal. In a different study, He et al. (2018) treated Cr-contaminated soil using a coupled EK with PRB of ZVI. They came to the conclusion that this technique is workable for the cleanup process because it limits the precipitate portions that occur between the anode and cathode reservoirs on one side and the PRB porous volume on the other. With the recent global proliferation of nanotechnology, numerous researchers have begun to use nanotechnology principles to the treatment of passively contaminated groundwater. For in situ cleanup, using nanotechnology can be thought of as a quicker and more affordable option (Hussain et al. 2022). Nanoscale zeolites, nanoscale ZVI particles, carbon nanotubes, metal oxides, titanium dioxide, noble metals, nanoclays, magnetic nanoparticles, and nanomembranes are among the nanomaterials that have been investigated for application in nanoremediation (Kale et al. 2024). This methodology offers a general decrease in the levels of contaminants compared to other remediation strategies; nevertheless, it is still in the research stage with limited field applications (Song et al. 2017).

10.4.1 Ex-Situ Technology

Ex-Situ technique treats groundwater by pumping out the contaminated aquifer, dewatering it, treating the surface water using physical, chemical, or biological methods, and then reintroducing the cleaned water into the aquifer (Sethi et al. 2019). Taking groundwater out of the aquifer, treating it above ground, and then putting it back into the aquifer. Pumping groundwater out of the trench or well and treating it is the extraction process.

10.4.2 In-Situ Technology

In-Situ Technology includes treatment of groundwater within the aquifer (in the subsurface) by using thermal, chemical and biological treatment technology. Include treating groundwater (in situ) as opposed to pulling water out of an aquifer for treatment. Various pollutants, such as metals, bacteria, viruses, ammonia, cyanide, and fluoride, dense and light non-aqueous phase liquids (DNAPL and LNAPL), and inorganic chemicals (Tripathi and Ram 2018).

10.5 Phyto-Remediation Technology

Aquifer contamination is a serious environmental problem that has far-reaching effects on ecosystems and public health (Azam and Rafi 2023). When it comes to efficiently tackling the complex and diverse contaminants present in aquifers, traditional remediation strategies frequently fall short. The potential of phytoremediation, a promising green technology, to provide an affordable and sustainable solution for aquifer decontamination has drawn attention (Nwogwu et al. 2022). The purpose of this chapter is to present a thorough analysis of phytoremediation with an emphasis on its use as an engineering solution for aquifer decontamination. Understanding the fundamentals of phytoremediation, evaluating how well it removes different toxins, and talking about the technical features that improve its performance are among the goals (Shamsi et al. 2023).

Utilizing plants to remove pollutants from the environment is known as phytoremediation (Jeevanantham et al. 2019). It can be divided into four main categories: rhizofiltration, phytoextraction, phytostabilization, and phytodegradation (Fig. 10.2). Each variety targets and reduces distinct kinds of pollutants by taking advantage of particular plant traits. Because of their distinct physiological and biochemical processes, plants are able to absorb, store, and change pollutants. Whereas phytodegradation depends on the plant's capacity to decompose pollutants, phytoextraction entails the uptake of pollutants from the soil into the plant tissues (Anerao et al. 2022). Comprehending these mechanisms is essential to creating phytoremediation plans that work.

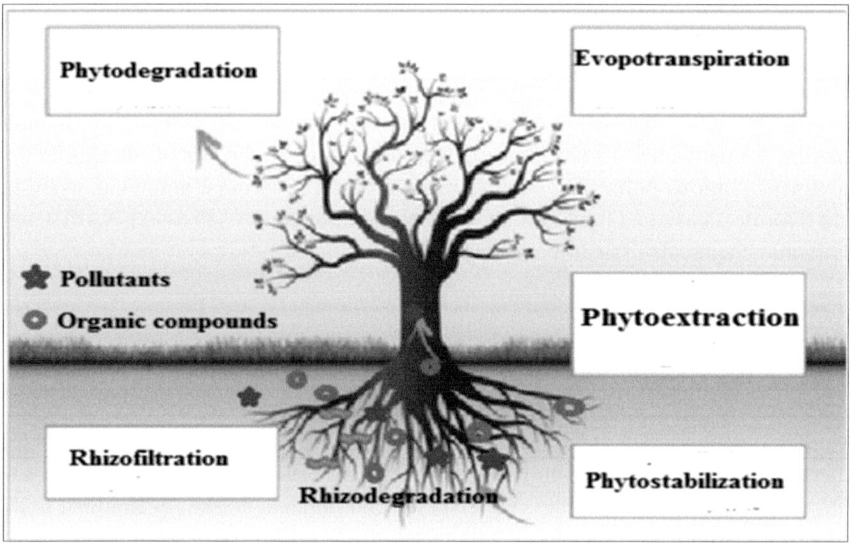

Fig. 10.2 Phyto-remediation of contaminant from soil and aquafer

10.6 Engineering Phytoremediation for Aquifer Decontamination

10.6.1 Plant Selection and Genetic Engineering

Careful plant selection is essential to improving phytoremediation for aquifer decontamination. Selecting hyper accumulator plants that have a high tolerance to the particular toxins found in the aquifer requires a thorough study of those pollutants (Jeevanantham et al. 2019). Genetic engineering is a novel way to modify the natural characteristics of particular plants so that they can be more effective in cleanup operations. Plant genes involved in contaminant absorption, translocation, and transformation pathways can be altered through genetic alteration (Koźmińska et al. 2018). Scientists can increase a plant's capacity for phytoextraction the process of drawing toxins from the soil by focusing on specific genes. This genetic fine-tuning enhances the plant's inherent resistance to toxins and makes it easier to maximize its remediation potential for particular chemicals found in aquifers.

10.6.2 Rhizosphere Management

The rhizosphere, which is the interface between the soil and plant roots, is essential to the effectiveness of phytoremediation (Antoniadis et al. 2017). To improve pollutant breakdown, engineering the rhizosphere entails carefully adjusting microbial communities and biochemical processes. One way to facilitate microbial activity and the consequent breakdown of contaminants is by introducing particular bacteria or modifications (Sulaeva et al. 2015). Plant exudates function as a carbon source for microbial development in a synergistic interaction that underpins microbial-assisted phytodegradation in the rhizosphere. Through a variety of enzymatic activities, this contact encourages the breakdown of pollutants (Gaur et al. 2018). A crucial component of phytoremediation techniques is the engineering of rhizosphere microbial populations, which has a direct impact on the total effectiveness of pollutant removal.

10.6.3 Hydraulic Control by Pumping

Hydrolic control technology widely used for decontamination of groundwater by pumping from wells (Padhye et al. 2023). Groundwater extracted and treated once contamination falls under permissible limits. Hydraulic control system may needs treatment more than decades. Water pumps from well is subjected to involve further treatment processes like air stripping, adsorption and micro-biological methods. Energy consumption and carbon footprint are major factor which leads major drawbacks. Energy cost may reduced by using renewable energy sources which minimize potential environmental side effects.

10.6.4 Permeable Reactive Barrier (PRB)

Permeable Reactive Barrier (PRB) generally used as alternative of pump and treat method for groundwater treatment. The advantages of this techniques are high efficiency, low operation cost & low energy requirements over conventional methods (Padhye et al. 2023). In this method, contaminated groundwater passes through permeable reactive zone. PRB is one of the advantages in-situ treatment techniques for aquifer remediation (Faisal et al. 2018). Jayasundara (2023) studied that PRB techniques may implemented for in-situ ground water remediation for open solid waste dumpsites. The dimensions of PRB techniques are genrally of < 5 m in width (parallel to flow), 10 m in depth, and 50 m in length (transverse to flow) with different reactive media (Faisal et al. 2018). Zeolite filled reactive media used in PRB for contaminated groundwater in column experiments (Medvidović et al. 2018).

10.7 Monitoring and Control

Robust monitoring and control techniques are essential to the performance of designed phytoremediation systems. Technologies for real-time monitoring are essential for monitoring microbial activity, tracking pollutant concentrations, and evaluating plant health (Roper et al. 2021). Precise data gathering is made possible by sophisticated sensors and analytical tools, which offer insights into the effectiveness of phytoremediation systems. Large-scale phytoremediation project management is further enhanced by automation and remote control technologies (Prasad et al. 2021). With the use of these technologies, environmental factors like fertilizer levels and irrigation may be changed in response to real-time data. Researchers and environmental engineers can optimize phytoremediation procedures by putting automated systems into place (Naz et al. 2022). This will make sure that remediation efforts are tailored to the unique requirements and difficulties of the aquifer under consideration. Thus, integrating plant biology, genetics, and environmental engineering is a multidisciplinary approach to the engineering of phytoremediation for aquifer decontamination. The efficacy and practicality of phytoremediation solutions can be improved by rhizosphere management, genetic alteration, meticulous plant selection, and sophisticated monitoring methods (Iqbal et al. 2023a, b). As long as these technical techniques are developed further, they could be a viable and environmentally responsible way to tackle the complicated problems associated with aquifer contamination and environmental rehabilitation.

10.8 Conclusion

Globally, people and plant depend on water for several purposes, mainly for drinking. Due to the overexploitation and improper management the groundwater depleted day by day. The groundwater contamination recognised due to human activities. Through ground penetration, many contaminants reach the aquifer and have an impact, which might result in serious problems. Consequently, countries allocate significant financial resources and human capital towards the advancement of diverse technologies aimed at eliminating and purifying pollutants. The elimination of groundwater pollution is accomplished via a variety of ex situ and in situ bioremediation techniques. The long-term viability of bioremediation may depend on developing in situ treatments that may dramatically accelerate the rates of pollutant degradation, especially in groundwater, in a predictable and cost-effective manner. Therefore, these initiatives may have an impact on the sustainable use of freshwater and the country's economy. Because groundwater degradation is a complex process that requires expensive treatment, it is imperative that it be prevented and controlled for effective management. Therefore development of ground water purification technology will be useful tool for sustainable water quality.

References

Anerao P, Kaware R, Kumar Khedikar A, Kumar M, Singh L (2022) Phytoremediation of persistent organic pollutants: concept challenges and perspectives. Phytoremediation technology for the removal of heavy metals and other contaminants from soil and water, pp 375–404

Antoniadis V, Levizou E, Shaheen SM, Ok YS, Sebastian A, Baum C, Prasad MN, Wenzel WW, Rinklebe J (2017) Trace elements in the soil-plant interface: phytoavailability, translocation, and phytoremediation–a review. Earth Sci Rev 171:621–645

Arif Z, Sethy NK, Mishra PK, Verma B (2021) Impact on groundwater quality resources due to industrial effluent. Groundwater geochemistry: pollution and remediation methods, pp 212–231

Arora B, Dwivedi D, Faybishenko B, Jana RB, Wainwright HM (2019) Understanding and predicting vadose zone processes. Rev Miner Geochem 85(1):303–328

Arya A, Ojha S, Singh S (2018) Arbuscular mycorrhizal fungi as phosphate fertilizer for crop plants and their role in bioremediation of heavy metals. Fungi and their role in sustainable development: current perspectives, pp 255–265

Ayilara MS, Babalola OO (2023) Bioremediation of environmental wastes: the role of microorganisms. Front Agron. https://doi.org/10.3389/fagro.2023.1183691

Azam A, Rafi NU (2023) Assessing the occurrence, causes, and health effects of arsenic contamination in groundwater: a comprehensive review. Int J Innovative Res Multi Field 9(7):107–111

Azubuike CC, Chikere CB, Okpokwasili GC (2016) Bioremediation techniques–classification based on site of application: principles, advantages, limitations and prospects. World J Microbiol Biotechnol 32:1–18

Bala S, Garg D, Thirumalesh BV, Sharma M, Sridhar K, Inbaraj BS, Tripathi M (2022) Recent strategies for bioremediation of emerging pollutants: a review for a green and sustainable environment. Toxics 10(8):484

Bexfield LM, Belitz K, Lindsey BD, Toccalino PL, Nowell LH (2021) Pesticides and pesticide degradates in groundwater used for public supply across the United States: occurrence and human-health context. Environ Sci Technol 55:362–372

Bhattacharjee G, Gohil N, Vaidh S, Joshi K, Vishwakarma GS, Singh V (2020) Microbial bioremediation of industrial effluents and pesticides. In: Bioremediation of pollutants. Elsevier, pp 287–302

Chen M, Tong H, Qiao J, Lv Y, Jiang Q, Gao Y, Liu C (2020) Microbial community response to the toxic effect of pentachlorophenol in paddy soil amended with an electron donor and shuttle. Ecotoxicol Environ Safety 205:111328.

Faisal AAH, Sulaymon AH, Khaliefa QM (2018) A review of permeable reactive barrier as passive sustainable technology for groundwater remediation. Int J Environ Sci Technol 15:1123–1138

Gaur N, Narasimhulu K, PydiSetty Y (2018) Recent advances in the bio-remediation of persistent organic pollutants and its effect on environment. J Clean Prod 198:1602–1631

Ghosh A, Manna MC, Jha S, Singh AK, Misra S, Srivastava RC, Srivastava PP, Laik R, Bhattacharyya R, Prasad SS, Singh SP (2022) Impact of soil-water contaminants on tropical agriculture, animal and societal environment. Adv Agron 176:209–274

Godswill AG, Somtochukwu IV, Ikechukwu AO, Kate EC (2020) Health benefits of micronutrients (vitamins and minerals) and their associated deficiency diseases: a systematic review. Int J Food Sci 3(1):1–32

He J, He C, Chen X, Liang X, Huang T, Yang X, Shang H (2018) Comparative study of remediation of Cr (VI)-contaminated soil using electrokinetics combined with bioremediation. Environ Sci Pollut Res 25:17682–17689

He Z, Liang F, Meng J, Li N (2022) Effects of groundwater fluctuation on migration characteristics and representative elementary volume of entrapped LNAPL. J Hydrol 610:127833

Hort C, Gracy S, Platel V, Moynault L (2009) Evaluation of sewage sludge and yard waste compost as a biofilter media for the removal of ammonia and volatile organic sulfur compounds (VOSCs). Chem Eng J 152(1):44–53

Huno SKM, Rene ER, van Hullebusch ED, Annachhatre AP (2018) Nitrate removal from groundwater: a review of natural and engineered processes. J Water Supply Res T 67(8SI):885e902

Hussain A, Rehman F, Rafeeq H, Waqas M, Asghar A, Afsheen N, Rahdar A, Bilal M, Iqbal HM (2022) In-situ, ex-situ, and nano-remediation strategies to treat polluted soil, water, and air–a review. Chemosphere 289:133252

Iqbal B, Li G, Alabbosh KF, Hussain H, Khan I, Tariq M, Javed Q, Naeem M, Ahmad N (2023) Advancing environmental sustainability through microbial reprogramming in growth improvement, stress alleviation, and phytoremediation. Plant Stress 100283

Jayasundara RBCD (2023) Permeable reactive barriers as an in situ groundwater remediation technique for open solid waste dumpsites: a review and prospect. Water Air Soil Pollut 234:1

Jeevanantham S, Saravanan A, Hemavathy RV, Kumar PS, Yaashikaa PR, Yuvaraj D (2019) Removal of toxic pollutants from water environment by phytoremediation: a survey on application and future prospects. Environ Technol Innov 13:264–276

Kale SS, Chauhan R, Nigam B, Gosavi S, Chaudhary IJ (2024) Effectiveness of nanoparticles in improving soil fertility and eco-friendly crop resistance: a comprehensive review. Biocatal Agric Biotechnol 103066

Kebede G, Tafese T, Abda EM, Kamaraj M, Assefa F (2021) Factors influencing the bacterial bioremediation of hydrocarbon contaminants in the soil: mechanisms and impacts. J Chem 2021:1–17

Kharangate-Lad A, D'Souza NC (2021) Current approaches in bioremediation of toxic contaminants by application of microbial cells; biosurfactants and bioemulsifiers of microbial origin. Rhizobiont Bioremediat Hazard Waste pp 217–263

King RB, Sheldon JK, Long GM (2023) Practical environmental bioremediation: the field guide. CRC Press

Koźmińska A, Wiszniewska A, Hanus-Fajerska E, Muszyńska E (2018) Recent strategies of increasing metal tolerance and phytoremediation potential using genetic transformation of plants. Plant Biotechnol Rep 12:1–14

La Vigna F (2022) Urban groundwater issues and resource management, and their roles in the resilience of cities. Hydrogeol J 30(6):1657–1683

Lee JS, Lee JS, Kim HS (2024) Toxic effects of triclosan in aquatic organisms: a review focusing on single and combined exposure to environmental conditions and pollutants. Sci Total Environ 170902

Li P, Karunanidhi D, Subramani T, Srinivasamoorthy K (2021) Sources and consequences of groundwater contamination. Arch Environ Contam Toxicol 80:1–10

Lin WH, Chien CC, Lu CW, Hou D, Sheu YT, Chen SC, Kao CM (2021) Growth inhibition of methanogens for the enhancement of TCE dechlorination. Sci Total Environ 787:147648

Lin L, Huang H, Zhang X, Dong L, Chen Y (2022) Hydrogen-oxidizing bacteria and their applications in resource recovery and pollutant removal. Sci Total Environ 835:155559

Liu JW, Wei KH, Xu SW, Cui J, Ma J, Xiao XL, Xi BD, He XS (2021) Surfactant-enhanced remediation of oil-contaminated soil and groundwater: a review. Sci Total Environ 756:144142

Lu X, Zhang X, Zhan N, Wang Z, Li S (2023) Factors contributing to soil acidification in the past two decades in China. Environ Earth Sci 82(3):74

Makubalo SS (2020) Hydrogeochemistry of the groundwater in the Namaqualand region, South Africa: implications for surficial uranium mineralization. Doctoral dissertation, University of Pretoria

Medvidović NV, Nuić I, Ugrina M, Trgo M (2018) Evaluation of natural zeolite as a material for permeable reactive barrier for remediation of zinc-contaminated groundwater based on column study. Water Air Soil Pollut 229:367

Iqbal KJ, Khan N, Fatima M, Lashari MH, Asad M, Ashraf A, Fatima H, Majeed H, Baboo I, Iqbal KJ, Asghar M (2023b) Determination of heavy metals (Pb, Cr, As, Hg, and Cd) into the body organs of selected fish, water, sediment, and soil samples from Head Punjnad and Head Taunsa, Punjab, Pakistan. PLoS ONE 18(9):e0288163

Nayar R (2020) Assessment of water quality index and monitoring of pollutants by physico-chemical analysis in water bodies: a review. Int J Eng Res Technol 9(01)
Naz M, Benavides-Mendoza A, Tariq M, Zhou J, Wang J, Qi S, Dai Z, Du D (2022) CRISPR/Cas9 technology as an innovative approach to enhancing the phytoremediation: concepts and implications. J Environ Manage 323:116296
Nemčić-Jurec J, Ruk D, Oreščanin V, Kovač I, Ujević Bošnjak M, Kinsela AS (2022) Groundwater contamination in public water supply wells: risk assessment, evaluation of trends and impact of rainfall on groundwater quality. Appl Water Sci 12(7):172
Nwogwu NA, Ajala OA, Ajibade FO, Ajibade TF, Adelodun B, Lasisi KH, Ugya AY, Kumar P, Omotade IF, Babalola TE, Adewumi JR (2022) Phytoremediation mechanisms of heavy metal removal: a step towards a green and sustainable environment. In: Innovative bio-based technologies for environmental remediation. CRC Press, pp 207–236
O'Connor D, Hou D, Ok YS, Song Y, Sarmah AK, Li X, Tack FM (2018) Sustainable in situ remediation of recalcitrant organic pollutants in groundwater with controlled release materials: a review. J Control Release 283:200–213
Ortiz-Medina JF, Yuncu B, Ross L, Elkins B (2023) The importance of proper pH adjustment and control to achieve complete in situ enhanced reductive dechlorination. Integr Environ Assess Manage 19(4):943–948
Ossai IC, Ahmed A, Hassan A, Hamid FS (2020) Remediation of soil and water contaminated with petroleum hydrocarbon: a review. Environ Technol Innov 17:100526
Padhye LP, Srivastava P, Jasemizad T, Bolan S, Hou D, Connor DO, Lamb D, Wang H (2023) Contaminant containment for sustainable remediation of persistent contaminants in soil and groundwater. J Hazard Mater 455:131575
Papazotos P (2021) Potentially toxic elements in groundwater: a hotspot research topic in environmental science and pollution research. Environ Sci Pollut Res 28(35):47825–47837
Patel AK, Singhania RR, Albarico FPJB, Pandey A, Chen CW, Dong CD (2022) Organic wastes bioremediation and its changing prospects. Sci Total Environ 824:153889
Prabhu SM, Meenakshi S (2014) Enriched fluoride sorption using chitosan supported mixed metal oxides beads: synthesis, characterization and mechanism. J Water Process Eng 2:96–104
Prasad MNV, de Jesus Alves L, Nunes FC (2021) Global remediation industry and trends. Handbook of assisted and amendment: enhanced sustainable remediation technology, pp 1–28
Ricolfi L, Barbieri M, Muteto PV, Nigro A, Sappa G, Vitale S (2020) Potential toxic elements in groundwater and their health risk assessment in drinking water of Limpopo National Park, Gaza Province, Southern Mozambique. Environ Geochem Health 42(9):2733–2745
Roper JM, Garcia JF, Tsutsui H (2021) Emerging technologies for monitoring plant health in vivo. ACS Omega 6(8):5101–5107
Sales da Silva IG, Gomes de Almeida FC, Padilha da Rocha e Silva NM, Casazza AA, Converti A, Asfora Sarubbo L (2020) Soil bioremediation: overview of technologies and trends. Energies 13(18):4664
Salimi M, Amin MM, Ebrahimi A, Ghazifard A, Najafi P (2012) Influence of electrical conductivity on the phytoremediation of contaminated soils to Cd2+ and Zn2+. Int J Environ Health Eng 1:11
Saravanan A, Kumar PS, Duc PA, Rangasamy G (2023) Strategies for microbial bioremediation of environmental pollutants from industrial wastewater: a sustainable approach. Chemosphere 313:137323
Sethi R, Di Molfetta A, Sethi R, Di Molfetta A (2019) Remediation of contaminated groundwater. Groundwater engineering: a technical approach to hydrogeology, contaminant transport and groundwater remediation, pp 331–409
Shaji E, Sarath KV, Santosh M, Prasad PK, Arya BK, Babu MS (2023) Fluoride contamination in groundwater: a global review of the status, processes, challenges, and remedial measures. Geosci Front 15:101734

Shamsi IH, Jin X, Zhang X, Feng Q, Ibrahim Z, Adil MF, Karim MF, Ullah N (2023) The dynamics of phytoremediation of heavy metals: recent progress and future perspective. Heavy metal toxicity and tolerance in plants: a biological, omics, and genetic engineering approach, pp 553–572

Sharma I (2020) Bioremediation techniques for polluted environment: concept, advantages, limitations, and prospects. In: Trace metals in the environment-new approaches and recent advances. IntechOpen

Singh KK, Tewari G, Kumar S et al (2023) Understanding urban groundwater pollution in the upper Gangetic Alluvial plains of northern India with multiple industries and their impact on drinking water quality and associated health risks. Groundw Sustain Dev 21:100902

Song B, Zeng G, Gong J, Liang J, Xu P, Liu Z, Zhang Y, Zhang C, Cheng M, Liu Y, Ye S (2017) Evaluation methods for assessing effectiveness of in situ remediation of soil and sediment contaminated with organic pollutants and heavy metals. Environ Int 105:43–55

Sulaeva I, Henniges U, Rosenau T, Potthast A (2015) Bacterial cellulose as a material for wound treatment: properties and modifications. A review. Biotechnol Adv 33(8):1547–1571

Tripathi SM, Ram S (2018) Bioremediation of groundwater: an overview. Int J Appl Eng Res 13(24):16825–16832

Vig N, Ravindra K, Mor S (2023) Evaluation of groundwater for drinking and irrigation applications concerning physicochemical and ionic parameters through multiple indexing approach: a case study around the industrial zone, Punjab, India. Environ Geochem Health 45(5):2077–2097

Wang J, Liu X, Beusen AHW, Middelburg JJ (2023) (2023) Surface-water nitrate exposure to world populations has expanded and intensified during 1970–2010. Environ Sci Technol 57(48):19395–19406

Worldbank (2012) https://www.worldbank.org/en/news/feature/2012/03/06/india-groundwater-critical-diminishing. Access 16 Feb 2024

Yang C, Rathman JF, Magdziarz T, Mostrag A, Kulkarni S, Barton-Maclaren TS (2020) Do similar structures have similar no observed adverse effect level (NOAEL) values? Exploring chemoinformatics approaches for estimating NOAEL bounds and uncertainties. Chem Res Toxicol 34(2):616–633

Yuhong F, Yaci L, Yasong L, Xilin B, Pengwei Z (2022) Prospect of groundwater pollution remediation methods and technologies in China. Geol China 49(2):420–434

Zeidan BA (2017) Groundwater degradation and remediation in the Nile Delta aquifer. The Nile Delta, pp 159–232

Zhang YY, Huang YF, Liang J, Zhou H (2022) Improved up-and-down procedure for acute toxicity measurement with reliable LD50 verified by typical toxic alkaloids and modified Karber method. BMC Pharmacol Toxicol 23(1):1–11

Chapter 11
Groundwater Remediation Using Meshless Local Petrov Galerkin (MLPG) Simulation Coupled with Differential Evolution (DE)

Sanjukta Das and T. I. Eldho

Abstract The Pump and Treat (PAT) is an effective remediation technique for the alleviation of pollutants such as industrial effluents and chemicals from the groundwater. This technique involves the extraction of groundwater from one or more pumping wells for the treatment above the ground surface. Remediation of contaminated aquifers is a tedious and resource consuming process. Thus, a proper planning and knowledge of physical processes is extremely essential, which can be achieved by using Simulation–Optimization (SO) models. For PAT, the SO models can be applied for the estimation of fate of contaminants, identification of optimal locations and rate of pumping and injection and analysis of the time and cost required to obtain the desired level of remediation. In this study, the Meshless Local Petrov Galerkin (MLPG) method is used as the simulation tool for solving the contaminant transport equation. The merits of this meshless method over the classic mesh based Finite Difference Method (FDM) and Finite Element Method (FEM), is the exploration of areas within the grids for optimal solutions along with the reduction of preprocessing efforts arising due to the mesh. MLPG being a local weak form meshless method, is robust due to the local integration of the governing partial differential equation. The MLPG model is coupled with Differential Evolution (DE), which is a simple yet efficient optimization technique. The MLPG-DE model is applied for the study of design of PAT remediation plans using a hypothetical heterogeneous aquifer, contaminated with Total Dissolved Solids (TDS), with a maximum value of 1356 mg/l in the aquifer. Since the allowable limit of TDS is 750 mg/l which cannot be achieved through natural attenuation in the specified three years period, the remediation design is presented for this aquifer. Using PAT remediation design by MLPG-DE model, the optimal pumping rates are estimated, and the concentration of TDS is observed to reduce to 734 mg/l in 3 years with a remediation cost of INR 27,55,953. This study highlights the effectiveness of the MLPG-DE model and its capability for the analysis of field problems.

S. Das (✉) · T. I. Eldho
Department of Civil Engineering, IIT Bombay, Mumbai 400076, India
e-mail: sanjukta.das1221@gmail.com

A. Nanda et al. (eds.), *Navigating the Nexus*, Water Science and Technology Library 102, https://doi.org/10.1007/978-3-031-76532-2_11

Keywords Groundwater remediation · Simulation optimization · Pump and treat · Meshless local Petrov Galerkin · Differential evolution

11.1 Introduction

The alarming concentrations of chemical pollutants in groundwater are associated with the over exploitation, waste disposal, mismanagement and unscientific development of groundwater resources (Kumar and Shah 2006). The restoration of a contaminated aquifer is an expensive and long-term process. The remediation techniques are based on the hydrogeological properties of the aquifer, the type and quantity of contaminant and the potential for future contamination (Aquifer Restoration 2023). A reliable prediction of future and fate of the contaminant taking into account the factors influencing the remediation process, coupled with the goal of minimization of the remediation cost for a desired remediation duration forms an effective tool for the design of management and remediation plans. The role of a hydrogeologist is the design of such efficient plans. Simulation–Optimization (SO) has been widely used for the aquifer restoration and management since the past few decades. An SO model typically consists of a numerical simulation tool for solving the groundwater contaminant transport equation and an optimization algorithm for choosing the best values of the decision variable for a given remediation objective of cost or remediation duration minimization. The simulation model is called during every fitness evaluation of the optimization model.

A groundwater contaminant transport simulation model is designed for solving the governing transport equation with an aim of obtaining predictions of concentration distribution in a future period. Numerous models have been developed for the transport simulation. The analytical models are exact solutions involving simplifying assumptions and cannot be applied for practical cases (Bear and Cheng 2010). Many numerical models are proven to be successful for transport study, and analytical models are required for the validation of these numerical models. The classic methods such as the Finite Difference Method (FDM), Finite Element Method (FEM), Finite Volume Method (FVM) rely on a background mesh which complicates the automation and refinement process for improving accuracy (Liu and Gu 2005). The usage of the mesh is limited in the meshless methods, thereby significantly improving the discretization of the aquifer and adaptive analysis. Various meshless models are proposed for the contaminant transport study. Kumar and Dodagoudar (2008) presented a weak form Element Free Galerkin Method (EFGM) for the transport simulation. Herrera et al. (2009) proposed a Monte Carlo Smoothed Particle Hydrodynamics (MC-SPH) which was observed to be extremely stable for advection dominant flows subjected to the usage of appropriate model parameters. A simple and straightforward Point Collocation Method (PCM) based transport model was developed by Meenal and Eldho (2012b) which is very accurate and fast for aquifers with Dirichlet boundaries. Ciftci et al. (2012) simulated the advection–dispersion phenomenon using the Radial Basis Function Collocation Method (RBFCM). Singh et al. (2016) presented another direct collocation meshless method, the Radial PCM (RPCM) for transport modelling. Nourani et al. (2018) coupled the Multi-quadrics Radial Basis Function (MQRBF) with Artificial Neural Network (ANN) to reduce the uncertainties in the transport modelling. Pathania and Eldho (2020) presented a stable EFGM based coupled flow and transport model. Swetha et al. (2023) proposed a Local Radial Point Interpolation Method (LRPIM) transport simulation tool. A hybrid Meshless Weak Strong (MWS) model was proposed by the authors of the present study (Das and Eldho 2022a), for obtaining reliable concentration prediction in less simulation time. The MLPG model used in this study was proposed by Atluri and Zhu (1998) and first applied for solving the transport equation by Swathi and Eldho (2017). Further, MLPG has been proven to be successful for groundwater problems by various researchers (Ilati and Dehghan 2016; Mohtashami et al. 2020; Das and Eldho 2022b, to mention a few). Thus, this reliable meshless technique is the proposed simulation tool in this study.

Differential Evolution (DE) is a metaheuristic evolutionary algorithm based on Darwin's theory of evolution, proposed by Storn and Price (1997). DE is applied for various groundwater problems. Zhang et al. (2007) coupled DE with FDM based MODFLOW for the Pump and Treat (PAT) remediation. Qu et al. (2010) analyzed the prediction of groundwater table by using DE with back propagation neural network. Considering the aquifer vulnerability, an optimal design for location production wells was developed by Elçi and Ayvaz (2014) using MODFLOW with hybrid DE. An SO model for the identification of source of contamination was developed by Gurarslan and Karahan (2015) using FDM based MODFLOW-MT3D and DE. Kisi et al. (2017) coupled Artificial Neural Network (ANN) with DE and Particle Swarm Optimization (PSO) for evaluating quality parameters such as Sodium Adsorption Ratio (SAR). PSO and DE were compared for the optimal well placement problem by Redoloza and Li (2021). Bai et al. (2022) proposed an adaptive mutation DE Markov Chain (AM-DEMC) for the contaminant source identification. Chang et al. (2021) combined an Ensemble Smoother (ES) and DEMC for contaminant source characteristics study. In this study, the MLPG simulation model is coupled with DE optimization.

In this study, an SO model is presented for the Pump and Treat (PAT) remediation design and this SO model is based on Meshless Local Petrov Galerkin (MLPG) simulation coupled with Differential Evolution (DE) optimization. PAT is effective for the restoration of aquifers contaminated with Total Dissolved Solids (TDS) which is the contaminant used in this study. The proposed model is hereby referred to as MLPG-DE-PAT in this paper. The objective of this study is to develop a robust SO model for optimal design of PAT remediation plans. MLPG is a local weak form meshless method and has the advantages of high accuracy and stability, easier adaptive analysis and lesser preprocessing than mesh based Finite Difference and Finite Element methods. Also, DE being a reliable metaheuristic tool, the MLPG-DE-PAT can be a robust SO model for effective PAT remediation. However, due to the higher simulation time required in MLPG owing to the background integration, the run time of MLPG-DE-PAT model is higher than the meshless strong form and other PAT models. The Improved Interpolating Moving Least Squares (IIMLS)

(Wang et al. 2012) is the approximation technique in MLPG simulation, as it satisfies the Kronecker delta property and eliminates the issue of ill-conditioned matrix. The MLPG5 method (Atluri 2004) is considered due to its high stability and thus the Heaviside step test function is used. Gauss Quadrature is used for integration, due to its faster performance and accurate results. Crank Nicolson method (Rastogi 2012) is applied for time discretization due to its implicit nature which results in stable solutions. The MLPG model is developed for coupled flow in confined aquifer and transport phenomena. The objective function is considered to be the minimization of the cost of remediation. The decision variables in DE are the pumping or injection rates at specified well locations to obtain a desired level of remediation in a given duration. With these criteria, the MLPG-DE-PAT model is developed in MATLAB R2020a. The presented SO model is demonstrated using a hypothetical rectangular aquifer. The MLPG-DE-PAT model aims at minimizing the remediation cost for aquifer restoration in a specified duration, thereby providing an estimate and improving the usage of financial resources in the remediation procedure. With the data inputs of aquifer geometry, boundary conditions and hydrogeological parameters, the presented model can be easily extended to field aquifers. With the advantages of better accuracy, efficient remediation and simpler applications, the MLPG-DE-PAT model can enhance the field remediation scenarios and can therefore be considered as a better alternative.

11.2 Model Development

11.2.1 Governing Equations for Groundwater Flow and Transport

The governing equation for groundwater flow in a confined aquifer in two-dimensions is given by (Bear 1979):

$$\nabla(T\nabla h) = \left(S\frac{\partial h}{\partial t}\right) \pm Q_w\delta(x - x_w)(y - y_w) - q \tag{11.1}$$

Here, h is the hydraulic head in m, T represents the transmissivity of confined aquifer, S is the storage coefficient, Q refers to the rate of pumping or recharge, the subscript w is the notation of the pumping well with location (x_w, y_w), q is the areal recharge in m/d and (x, y, t) are the cartesian coordinates and time. The Dirichlet and Neumann boundary conditions are given by the following equations:

$$h = h_1, \quad \textit{for nodes on Dirichlet boundary} \tag{11.2}$$

$$T_x\frac{dh}{dx} = g \ (or)\ T_y\frac{dh}{dy} = g, \quad \textit{for nodes on Neumann boundary} \tag{11.3}$$

where, h_l is the constant head and g is the constant inflow at boundaries. The head values obtained by solving the flow model can be used for the determination of velocities V_x and V_y using Darcy's law and in turn the dispersion coefficients D_x and D_y using the following equations (Rastogi 2012):

$$V_x = \frac{T_x}{B}\left(\frac{\partial h}{\partial x}\right) \textit{ and } V_y = \frac{T_y}{B}\left(\frac{\partial h}{\partial y}\right) \tag{11.4}$$

$$D_x = \frac{\alpha_l V_x^2 + \alpha_t V_y^2}{|\sqrt{V_x^2 + V_y^2}|} \textit{ and } D_y = \frac{\alpha_l V_y^2 + \alpha_t V_x^2}{|\sqrt{V_x^2 + V_y^2}|} \tag{11.5}$$

Here, α_l and α_t are the longitudinal and transverse dispersivities respectively. The velocities and dispersion coefficients computed using Eqs. (11.4) and (11.5) are the input to the transport model given by the equation (Bear 1972):

$$\frac{\partial}{\partial x}\left(D_x \frac{\partial C}{\partial x}\right) + \frac{\partial}{\partial y}\left(D_y \frac{\partial C}{\partial y}\right) - \frac{\partial (V_x C)}{\partial x} - \frac{\partial \left(V_y C\right)}{\partial y} = \frac{\partial C}{\partial t} \tag{11.6}$$

Here, the contaminant is considered to be in dissolved state and its concentration at (x, y, t) is indicated by C. The Dirichlet and Neumann boundary conditions are given by:

$$C = C_1 \quad \textit{for nodes on Dirichlet boundary} \tag{11.7}$$

$$D_x \frac{\partial C}{\partial x} n_x + D_y \frac{\partial C}{\partial y} n_y = g_1 \quad \textit{for nodes on Neumann boundary} \tag{11.8}$$

Here, C_1 and g_1 are constant concentration and flux at the respective boundaries.

11.2.2 MLPG Formulation

The MLPG formulations for the above governing equations are derived in our previous studies. The final equations for the formulation for flow equation can be written as (Das and Eldho 2022b):

$$K\left\{\hat{h}^{t+1}\right\} = F \tag{11.9}$$

In Eq. (11.9),

$$K_j = \left(\left(\int_{\partial\Omega_{s,j}} -\left(T_x \sum_{i=1}^{N} \frac{\partial \emptyset_i}{\partial x} n_x + T_y \sum_{i=1}^{N} \frac{\partial \emptyset_i}{\partial y} n_y\right) d\Omega\right)\theta + \int_{\Omega_s} \frac{S}{\Delta t} d\Omega\right) \tag{11.10}$$

$$F_j = Q_j - f_j A_{\Omega_{s,j}} + \left(\left(\int_{\partial\Omega_{s,j}} \left(T_x \sum_{i=1}^{N} \frac{\partial \emptyset_i}{\partial x} n_x + T_y \sum_{i=1}^{N} \frac{\partial \emptyset_i}{\partial y} n_y\right) d\Omega\right)(1-\theta) + \int_{\Omega_s} \frac{S}{\Delta t} d\Omega\right)\left(\hat{h}^t\right) \tag{11.11}$$

The MLPG formulation for the contaminant transport equation is given by (Das and Eldho 2022a):

$$K\left\{\hat{C}^{t+1}\right\} = F \tag{11.12}$$

In Eq. (11.12),

$$K_j = \left(\int_{\Omega_s} \frac{1}{\Delta t} d\Omega - \left(\begin{array}{l} \int_{\partial\Omega_{s,j}} \left(D_x \sum_{i=1}^{N} \frac{\partial \emptyset_i}{\partial x} n_x + D_y \sum_{i=1}^{N} \frac{\partial \emptyset_i}{\partial y} n_y\right) d\Omega \\ -\left(\int_{\Omega_{s,j}} V_x \sum_{i=1}^{N} \frac{\partial \emptyset_i}{\partial x} + \int_{\Omega_s} V_y \sum_{i=1}^{N} \frac{\partial \emptyset_i}{\partial y}\right)\theta d\Omega \end{array}\right)\right) \tag{11.13}$$

$$F_j = \left(\int_{\Omega_s} \frac{1}{\Delta t} d\Omega + \left(\begin{array}{l} D_j \int_{\partial\Omega_{s,j}} \left(D_x \sum_{i=1}^{N} \frac{\partial \emptyset_i}{\partial x} n_x + D_y \sum_{i=1}^{N} \frac{\partial \emptyset_i}{\partial y} n_y\right) d\Omega \\ -\left(\int_{\Omega_s} V_x \sum_{i=1}^{N} \frac{\partial \emptyset_i}{\partial x} + \int_{\Omega_s} V_y \sum_{i=1}^{N} \frac{\partial \emptyset_i}{\partial y}\right)(1-\theta) d\Omega \end{array}\right)\right)\hat{C}^t \tag{11.14}$$

The coupled MLPG flow and transport model is developed using Eqs. (11.9) to (11.14). In these equations, the subscript j denotes the node of interest, N is the number of support domain nodes of the node j, $\emptyset_i$ is the shape function of the ith support domain node of the jth node, Ω is the aquifer domain, Ω_s is the sub domain, $\partial\Omega_{s,j}$ is the boundary of the jth sub domain node, $\theta = 0.5$ is the Crank Nicolson coefficient and superscripts t and t + 1 represent present and next time steps. The MLPG simulation model is well validated with available analytical and numerical solutions and real field problems, before extending the model for remediation problem. The validation is available in our previous study, Das and Eldho (2022a, b). Further, in this study, the MLPG model is coupled with Differential Evolution for optimal PAT remediation design.

11.2.3 Differential Evolution

In this study, the optimization is a minimization problem. There are four major steps in DE (Storn and Price 1997), namely, initialization, mutation, crossover and selection. The initial population of candidates can be generated as:

$$P^{0}_{a,b} = P_{min,b} + r1_{a,b}\left(P_{max,b} - P_{min,b}\right) \tag{11.15}$$

a = 1, 2, 3 .... N_P and b = 1, 2, 3 .... N_D. Thus, the initial population $P^{0}_{a,b}$ has a dimension $N_P \times N_D$ where, N_P is the population size and N_D is the number of decision variables. Here, the population is bounded by the limits $P_{min,b}$ and $P_{max,b}$.

In the mutation, in an iteration *itr*, the donor vector $U^{itr}_{a,b}$ is obtained by adding the scaled difference of two randomly selected candidates P^{itr}_{r1} and P^{itr}_{r2} of the population to the third randomly selected candidate P^{itr}_{r3}:

$$U^{itr}_{a,b} = P^{itr}_{r3} + F\left(P^{itr}_{r1} - P^{itr}_{r2}\right) \tag{11.16}$$

Here, F is the positive scaling factor and $F \in (0, 1)$. In this study, the value of F is taken between 0.3 and 0.5. Further, the trial vector $W^{itr}_{a,b}$ is generated by the crossover operation, with a cross over probability Cr value taken between 0.7 and 1.

$$W^{itr}_{a,b} = \begin{cases} U^{itr}_{a,b} \ \textit{if} \ r2^{itr}_{a,b} \leq Cr \\ P^{itr}_{a,b} \ \textit{otherwise} \end{cases} \tag{11.17}$$

Finally, the candidate with best fitness (*fit*) among the trial and target vector is selected for the next generation. Therefore, for a minimization problem, the population for the next generation or iteration itr + 1 $P^{itr+1}_{a,b}$ can be written as:

$$P^{itr+1}_{a,b} = \begin{cases} W^{itr}_{a,b} \ \textit{if} \ \textit{fit}\left(W^{itr}_{a,b}\right) \leq \textit{fit}\left(P^{itr}_{a,b}\right) \\ P^{itr}_{a,b} \ \textit{otherwise} \end{cases} \tag{11.18}$$

11.2.4 SO Model

Based on the formulation in Sect. 2.2, the MLPG flow and transport model is verified in our previous studies (Das and Eldho 2022a, b) and found to be similar to the analytical solutions. Here, MLPG is coupled with DE to form MLPG-DE-PAT model, which is designed as a single objective SO model for remediation. Here, the pumping rates at specified locations of wells are required to be optimized to obtain the best combination of these pumping rates to obtain the design reduction of concentration within a specified duration of remediation. The objective function is considered as the minimization of the cost function, which can be written as (Sharief et al. 2008; Meenal and Eldho 2012a):

$$Minimize\ fit = cc_1 \sum_{rw=1}^{RW} \frac{Q_{rw}\gamma(t_{rem})}{\eta_p}\left(h_{rw}^{gl} - h_{rw}\right) + cc_2 \sum_{rw}^{RW} Q_{rw}(t_{rem}) + FixedC \tag{11.19}$$

In Eq. (11.19), the FixedC is the fixed cost and is not considered in optimization, as this is a fixed cost such as construction of borewell which cannot be reduced and hence optimizing is not required. RW is the total number of remediation wells, h_{rw}^{gl} and h_{rw} are the ground level and head at remediation well rw, γ is the unit weight of water, t_{rem} is the total remediation time and η_p is the efficiency of the pumping system. $\sum_{rw=1}^{RW} \frac{Q_{rw}\gamma(t_{rem})}{\eta_p}\left(h_{rw}^{gl} - h_{rw}\right)$ results in the energy consumption in Joule for lifting the water and cc_1 is the cost coefficient for energy consumption and is assumed as INR 5/kW h (INR stands for Indian National Rupees). Here $\sum_{rw}^{RW} Q_{rw}(t_{rem})$ represents the cost for treatment of water and cc_2 is cost coefficient for treatment and is assumed to be INR 6/m^3 of water treated. The coefficients cc_1 and cc_2 are considered based on the rates of electricity and wastewater treatment in India in 2023.

This SO model is subjected to the constraints on the heads, concentrations and pumping rates as follows:

$$h_{min} \le h_j \le h_{max};\ C_j \le C_{specified} \tag{11.20}$$

$$Q_{rw,min} \le Q_{rw} \le Q_{rw,max} \tag{11.21}$$

From Eq. (11.20), the head at each node j (h_j) must be within specified upper and lower limits, given by h_{min} and h_{max}. Also, the concentration at every node j must be less than a specified concentration $C_{specified}$. The pumping rates at every remediation well must be within the specified range $\left[Q_{rw,min}, Q_{rw,max}\right]$ as indicated by Eq. (11.21).

11.2.5 Modelling Procedure

The MLPG-DE-PAT modelling procedure can be summarized as follows:

1. The aquifer details such as the type and dimensions of the aquifer, the zonation structure, transmissivity, dispersivity, storage coefficient, initial and boundary conditions, pumping and recharge details are entered into the model. Also, the model parameters such as the sub domain and support domain details, penalty coefficient, discretization and required type of nodal arrangement are inputted into the model. Similarly, the DE parameters and the number of iterations and cost coefficients are entered.

2. The initial population or target vector of DE is generated. The fitness is evaluated. The MLPG flow and transport model is evoked during the fitness evaluation. Here, the population is the pumping rates of the remediation wells which is an input to the MLPG simulation.
3. The mutation and crossover operations are performed. The fitness of the trial vector is evaluated and again calling the MLPG model.
4. The population with the best fitness among the trial vector and target vector is selected for the next generation. This completes one iteration.
5. Steps 3 and 4 are repeated for all the iterations till the termination criterion is satisfied. In this study, the termination criterion is considered as the number of iterations.

The MLPG-DE-PAT modelling procedure is depicted in Fig. 11.1.

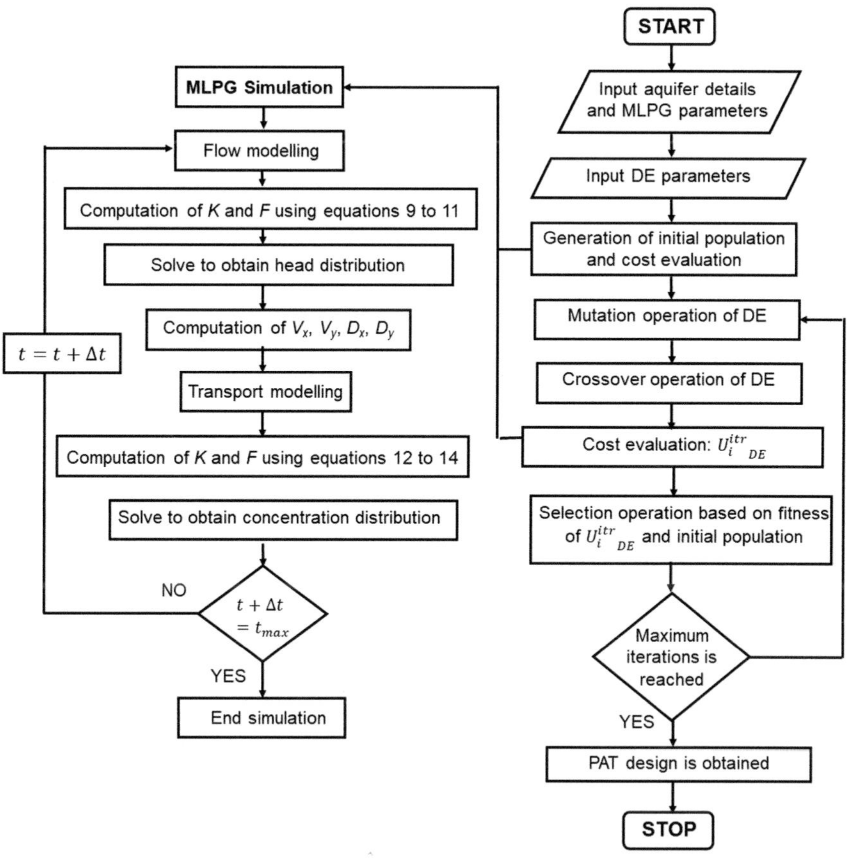

Fig. 11.1 MLPG-DE-PAT modelling procedure

11.3 Case Study

A two-dimensional rectangular hypothetical heterogeneous and anisotropic confined aquifer described in Fig. 11.2 is used for the demonstration of the MLPG-DE-PAT model. The dimensions and properties of this case study are taken from Meenal and Eldho (2012a). The aquifer has dimensions 1500 m × 1400 m and thickness 25 m. The aquifer is considered to have three zones of heterogeneity with hydraulic conductivities of 40, 50 and 60 m/day in x direction and 35, 40 and 50 m/day in y direction in the zones 1, 2 and 3 respectively. The transmissivities are obtained by multiplying the corresponding hydraulic conductivities with the thickness of the aquifer. The porosity and specific storage are considered as 0.3 and 0.004 respectively. As shown in Fig. 11.2, a pond (300 m × 200 m) located in the aquifer releases 2500 mg/l of Total Dissolved Solids (TDS) at 0.02 m/day. The longitudinal and transverse dispersivity are taken as 50 m and 10 m. Zone 2 is recharged at 0.0012 m/day. The West and East boundaries are Dirichlet boundaries with constant heads of 35 m and 30 m respectively. For transport modelling, these boundaries are fixed at zero concentration. The North and South boundaries have zero inflow. The aquifer is discretized into 240 equidistant nodes as shown in Fig. 11.3. The radii of sub and support domain are considered as (0.3 * Δx) and (3 * Δx) respectively. The ground level of the aquifer is considered to be 40 m. First, the MLPG simulation model is executed for 5 years with a 5 day time step to obtain the concentration data. This concentration is considered as the initial concentration for the MLPG-DE-PAT model. Therefore, the simulation is described in this section and the remediation using MLPG-DE-PAT is discussed in Sect. 11.4.

The aquifer is initially free from pumping and pollution and the steady state head over the aquifer, obtained from MLPG model is illustrated in Fig. 11.4a. During the period of 5 years when the contaminant is being disposed into the aquifer, a continuous pumping of 1000 m^3/day is considered at each of the three wells located at (200 m, 300 m), (200 m, 1100 m) and (1000 m, 700 m). The pond located in the aquifer that releases 2500 mg/l of Total Dissolved Solids (TDS) at 0.02 m/day is the source of pollution. After 5 years, at the end of simulation time, the release of TDS is considered to have stopped. The groundwater head after 5 years is illustrated in Fig. 11.4b. The distribution of TDS after 5 years is given in Fig. 11.4c. The FDM based MODFLOW-MT3DMS model (Aquaveo 2017) is developed for comparing the results obtained from MLPG model. The results of MLPG are observed to match with the MODFLOW-MT3DMS solution as shown in Fig. 11.5, thereby proving the model reliability. Moreover, MLPG is truly meshless, and the discretization and fine tuning is automatic and simple in MLPG when compared to MODFLOW-MT3DMS, which makes it an efficient alternative to mesh-based methods.

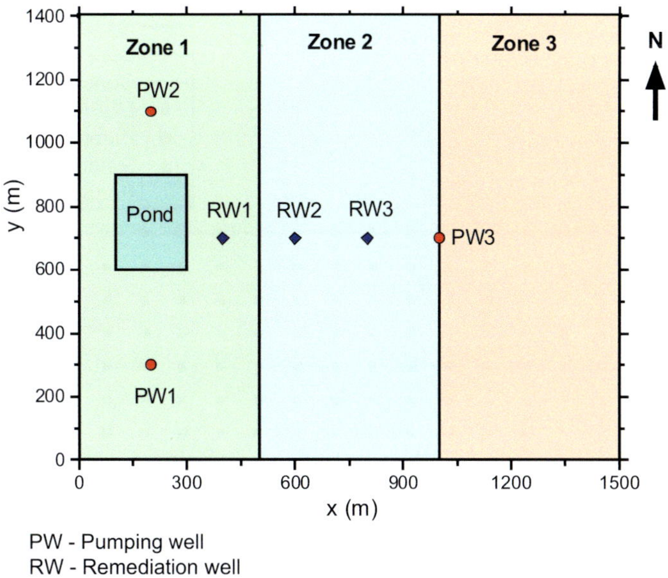

Fig. 11.2 Aquifer domain

11.4 Results and Discussion

The concentration described in Fig. 11.4c is the initial concentration in the MLPG-DE-PAT model. No further contaminant is released into the aquifer. The maximum TDS concentration is observed to be 1356 mg/l which is much greater than the acceptable TDS limit of 750 mg/l in India (Meenal and Eldho 2012a). Through natural attenuation, the TDS is observed to reduce 1135 mg/l after 3 years as illustrated in Fig. 11.6, which is greater than the acceptable limit. Thus, PAT remediation is applied to this aquifer, to convert the groundwater to be potable within a short duration of 3 years. The DE parameters are considered as: $N_P = 25$, $F = 0.45$, $Cr = 0.81$.

These parameter values are finalised by trial and error. The maximum number of iterations is taken as 15. The total time for remediation is considered as 3 years, with a time step of 5 days. Three wells are considered at RW1 (400 m, 700 m), RW2 (600 m, 700 m) and RW3 (800 m, 700 m) which serve as the pumping wells for the remediation process and the pumping rates of each of these wells is to be optimised to obtain the least remediation cost. Thus, $N_D = 3$. The lower and upper bounds of the pumping rates are considered as 0 and 2000 m^3/day. With the optimal rates at the three remediation wells, the concentration distribution at the end of 3 years of remediation is as shown in Fig. 11.7. It can be observed that the maximum TDS concentration has reduced within the permissible limit.

The details of the results of the MLPG-DE-PAT model are given in Table 11.1. The obtained optimal pumping rates are suggested for the least cost of remediation at the given locations and duration. It can be observed that lower pumping rates are

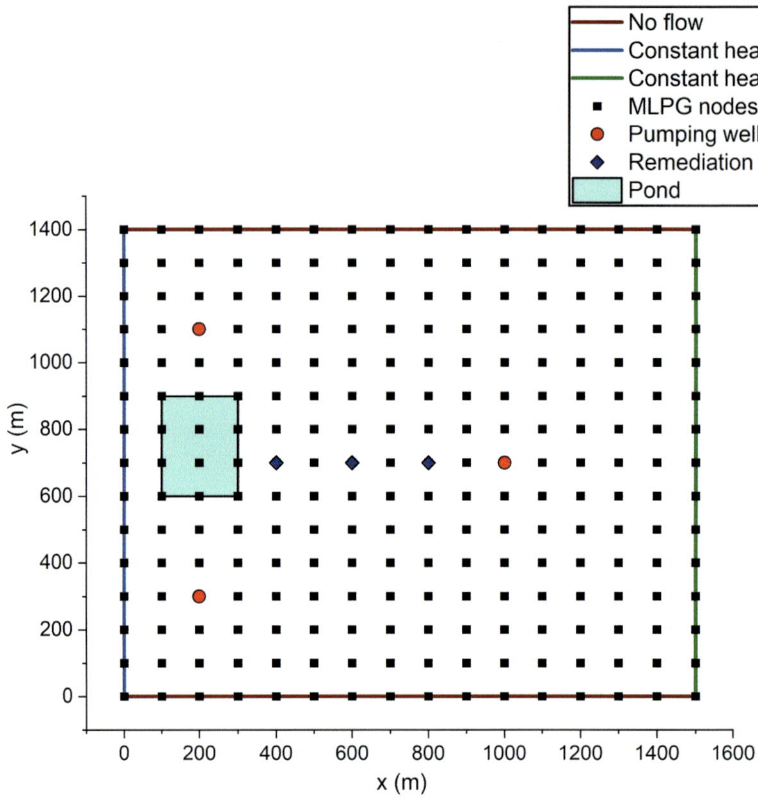

Fig. 11.3 Meshless nodal arrangement

sufficient for aquifer restoration in this study. The computational time is determined using a computer with 16 GB RAM and Intel Core i7 processor. Further, the variation of the objective function value is observed to reduce with the number of iterations as depicted in Fig. 11.8, thereby showing the convergence of DE. Also, from Fig. 11.8, it can be observed that the convergence is faster, and a better solution is obtained with an increase in the population size. The computational time for population size 10, 25 and 50 are 6.51, 11.30 and 22.62 min respectively. Since the increase in population size directly impacts the computational time, a population size of 25 is sufficient for this case study as a further increase has not much influence on the objective function value and the global optima could be attained in less than 15 iterations for $N_P = 25$.

Thus, from the observed results of the MLPG-DE-PAT model, it can be highlighted that the considered case study could be restored in a required remediation time, which is a very short duration. Thus, the PAT is an enhancement to reduction of the TDS, which would take a longer time to be reduced to permissible limits by natural attenuation. Also, with the best set of optimal pumping wells, the cost of remediation has been reduced, which proves the importance of the optimization through DE in this

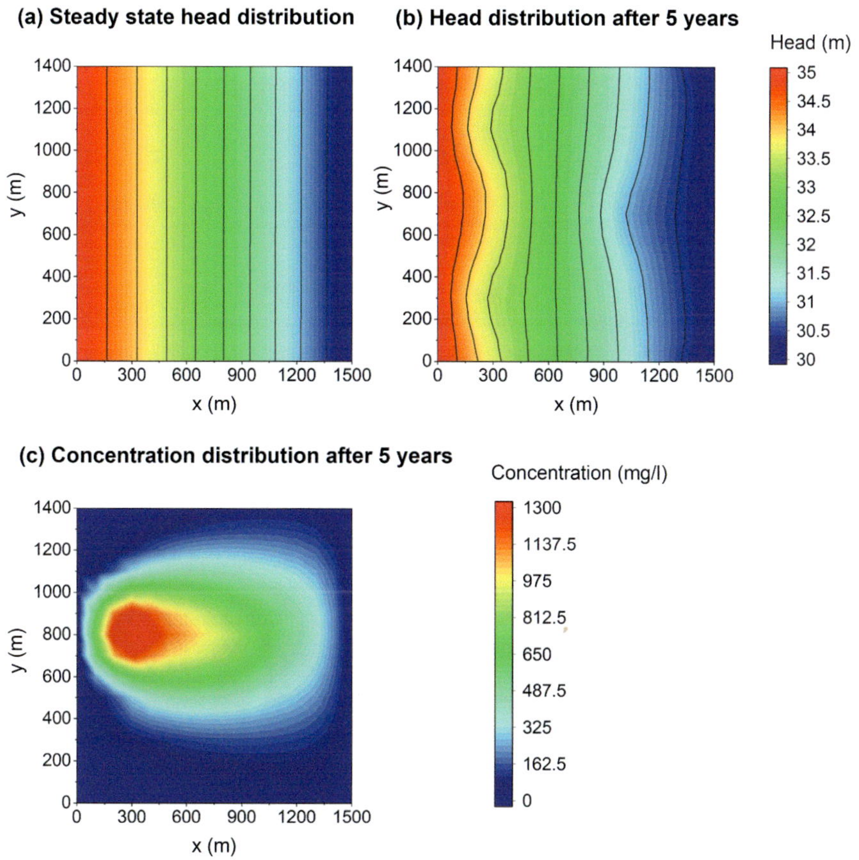

Fig. 11.4 Results of the MLPG simulation model

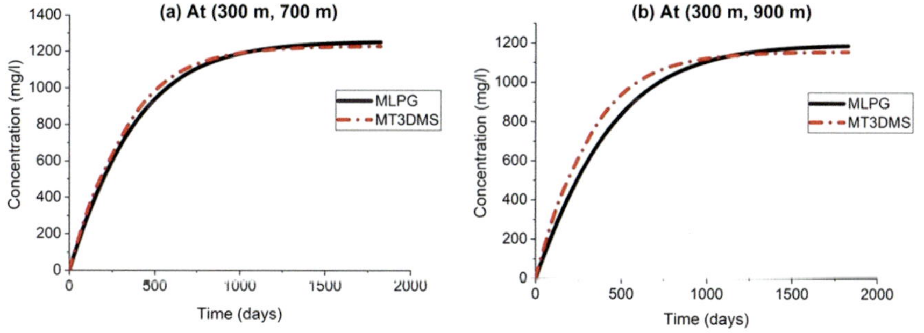

Fig. 11.5 Comparison of MLPG simulation results with MODFLOW-MT3DMS

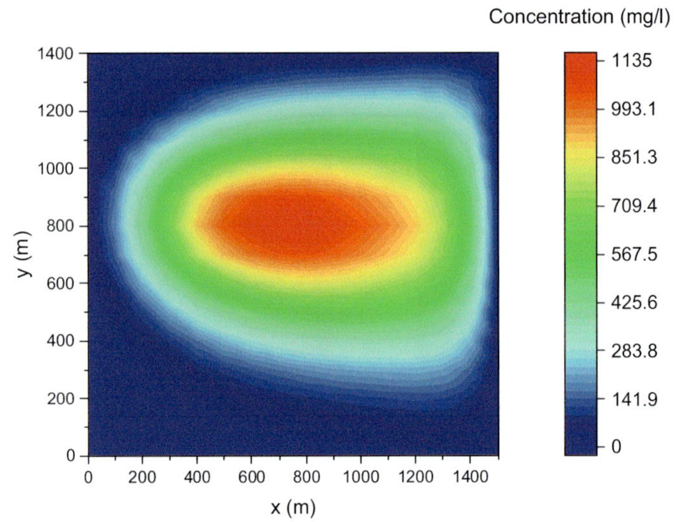

Fig. 11.6 Concentration distribution of TDS after 3 years without remediation

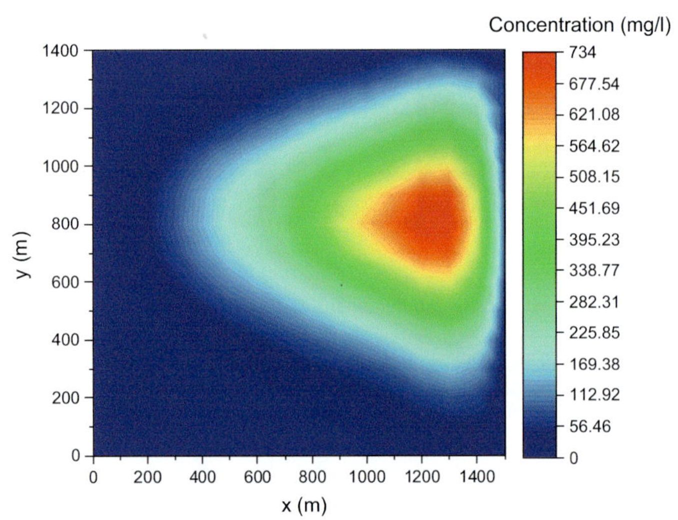

Fig. 11.7 Concentration distribution of TDS after 3 years with PAT remediation

Table 11.1 Results of the MLPG-DE-PAT model

Optimal pumping rates at RW1, RW2 and RW3 (m^3/day)	11.73, 10, 817.13
Remediation cost: extraction and treatment (INR)	27,55,953.20
Computational time with $N_P = 25$, $N_D = 3$ and 15 iterations	11.30 min

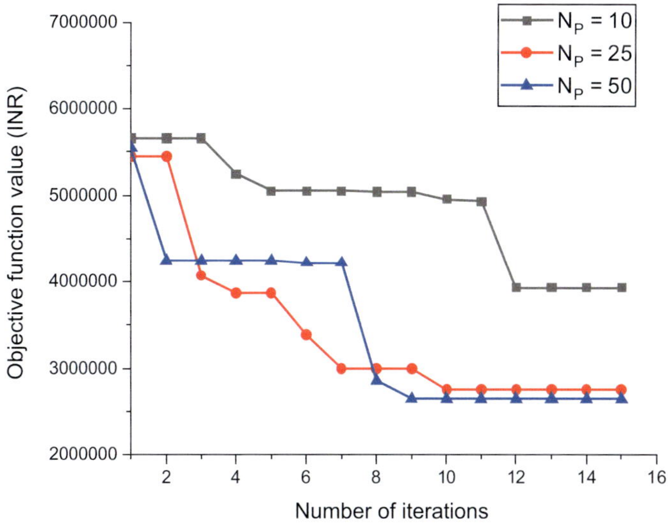

Fig. 11.8 Variation of the objective function value or the remediation cost with respect to the number of iterations and population size

study. In addition to the total remediation cost represented in Table 11.1, there is also a well installation cost. The well installation cost is a fixed value and depends on the number of active wells required for remediation. In the MLPG-DE-PAT modelling, the well installation cost is excluded as it is a fixed cost and does not require or improve by optimization. Also, the fixed well installation costs are not required to be considered if the remediation is performed through pre-installed wells in the aquifer. Thus, based on the number of well installations required, these fixed costs can be added to remediation cost estimated in Table 11.1 to obtain the combined installation, operation and maintenance cost. Thus, the MLPG-DE-PAT can be efficiently used for the cases where the contaminants have to be brought to the permissible limits in a relatively faster period or large amounts of concentrations are to be reduced for a safer usage of groundwater resources.

11.5 Conclusions

In this study, a Simulation–Optimization (SO) model based on weak form MLPG and DE is developed for the optimal design of Pump and Treat (PAT) remediation plans for restoration of contaminated confined aquifers. PAT, being an effective remediation technique, is proposed for the removal of TDS in groundwater. MLPG model for the coupled confined aquifer flow and transport is developed. The DE optimization is performed for obtaining the best pumping rates at a fixed number and locations of pumping wells. The minimal possible cost for the remediation is estimated for a given remediation duration. The objective function is considered as the minimization of the cost of pumping and treatment of extracted water in a given duration.

The application of the developed model is demonstrated using a heterogeneous hypothetical aquifer contaminated with TDS. The steady state and head distribution after five years with pumping and concentration distribution are determined and are observed to be similar to widely used MODFLOW-MT3DMS solutions. Further, this concentration at the end of release period is considered as the initial condition for the remediation problem. Since the concentration of TDS is greater than the allowable limit of 750 mg/l and the natural attenuation is not possible within a required remediation duration of 3 years, the PAT remediation considering pumping at 3 wells for 3 years is applied. It is observed that a total of 840 m^3/day pumping rate is sufficient for aquifer restoration and the minimum remediation cost is obtained, which excludes the fixed costs of installation of remediation wells and their maintenance.

The proposed model is thus proven to be efficient. MLPG being a weak form method, this model can be extended for even complex aquifers. Compared to the existing mesh-based methods, the discretization is simpler in MLPG, and the solutions are valid for high advective flows. However, due to the background integration, the computational time required in MLPG is relatively high when compared to mesh-based and strong form meshless methods. DE has excellent exploration capability and requires fewer parameters but has a disadvantage of susceptibility to local optimal solutions. Also, the model can be analyzed by using other optimization techniques, which can improve the convergence and performance of DE. In spite of the mentioned limitations, the presented model is robust as demonstrated in the case study. Therefore, with the present study, the proposed MLPG-DE-PAT model can be considered an alternative to existing methods and can be applied to field studies.

Acknowledgements The first author is thankful to the Ministry of Education, Government of India, for supporting this work through the Prime Minister Research Fellowship (PMRF) (PMRF ID: 1300144).

References

Aquaveo (2017) Groundwater modelling system version 10.2.6. Build date 22 September, 2017, Utah, USA

Aquifer Restoration (2023) Environmental encyclopaedia. <https://www.encyclopedia.com/environment/encyclopedias-almanacs-transcripts-and-maps/aquifer-restoration> (Retrieved on September 19, 2023 from Encyclopedia.com)

Atluri SN (2004) The meshless method (MLPG) for domain and BIE discretizations. Tech Science Press

Atluri SN, Zhu T (1998) A new meshless local Petrov-Galerkin (MLPG) approach in computational mechanics. Comput Mech 22(2):117–127. https://doi.org/10.1007/s004660050346

Bai Y, Lu W, Li J, Chang Z, Wang H (2022) Groundwater contamination source identification using improved differential evolution Markov chain algorithm. Environ Sci Pollut Res 1–14. https://doi.org/10.1007/s11356-021-17120-2

Bear J (1972) Dynamics of fluids in porous media. Elsevier, New York

Bear J (1979) Hydraulics of groundwater. McGraw Hill Publishing, New York
Bear J, Cheng AHD (2010) Modeling groundwater flow and contaminant transport, vol 23. Springer, Dordrecht
Chang Z, Lu W, Wang Z (2021) A differential evolutionary Markov chain algorithm with ensemble smoother initial point selection for the identification of groundwater contaminant sources. J Hydrol 603:126918. https://doi.org/10.1016/j.jhydrol.2021.126918
Ciftci E, Avci CB, Borekci OS, Sahin AU (2012) Assessment of advective–dispersive contaminant transport in heterogeneous aquifers using a meshless method. Environ Earth Sci 67:2399–2409. https://doi.org/10.1007/s12665-012-1686-z
Das S, Eldho TI (2022a) A meshless weak-strong form method for the simulation of coupled flow and contaminant transport in an unconfined aquifer. Transp Porous Media 143(3):703–737. https://doi.org/10.1007/s11242-022-01805-4
Das S, Eldho TI (2022b) Groundwater flow simulation using MLPG meshless method. J Indian Water Resour Soc 42(2). https://iwrs.org.in/journal/apr2022/6apr.pdf
Elçi A, Ayvaz MT (2014) Differential-evolution algorithm based optimization for the site selection of groundwater production wells with the consideration of the vulnerability concept. J Hydrol 511:736–749. https://doi.org/10.1016/j.jhydrol.2014.01.071
Gurarslan G, Karahan H (2015) Solving inverse problems of groundwater-pollution-source identification using a differential evolution algorithm. Hydrogeol J 23(6):1109. https://doi.org/10.1007/s10040-015-1256-z
Herrera PA, Massabó M, Beckie RD (2009) A meshless method to simulate solute transport in heterogeneous porous media. Adv Water Resour 32(3):413–429. https://doi.org/10.1016/j.advwatres.2008.12.005
Ilati M, Dehghan M (2016) Remediation of contaminated groundwater by meshless local weak forms. Comput Math Appl 72(9):2408–2416. https://doi.org/10.1016/j.camwa.2016.09.009
Kisi O, Keshavarzi A, Shiri J, Zounemat-Kermani M, Omran ESE (2017) Groundwater quality modeling using neuro-particle swarm optimization and neuro-differential evolution techniques. Hydrol Res 48(6):1508–1519. https://doi.org/10.2166/nh.2017.206
Kumar MD, Shah T (2006) Groundwater pollution and contamination in India: the emerging challenge. IWMI-TATA Water Policy Program Draft Paper 2006/1. https://www.iwmi.cgiar.org/iwmi-tata/files/pdf/ground-pollute4_FULL_.pdf
Kumar RP, Dodagoudar GR (2008) Two-dimensional contaminant transport modelling using meshless method. In: Proceedings of GeoCongress 2008: characterization, monitoring, and modeling of geosystems, pp 774–781. https://doi.org/10.1061/40972(311)9
Liu GR, Gu YT (2005) An introduction to meshfree methods and their programming. Springer Science and Business Media, Taylor and Francis Group, LLC
Meenal M, Eldho TI (2012a) Simulation–optimization model for groundwater contamination remediation using meshfree point collocation method and particle swarm optimization. Sadhana 37:351–369. https://doi.org/10.1007/s12046-012-0086-0
Meenal M, Eldho TI (2012b) Two-dimensional contaminant transport modeling using meshfree point collocation method (PCM). Eng Anal Boundary Elem 36(4):551–561. https://doi.org/10.1016/j.enganabound.2011.11.001
Mohtashami A, Hashemi Monfared SA, Azizyan G, Akbarpour A (2020) Computation of groundwater balance using numerical MLPG method (case study: birjand unconfined aquifer). Iran J Irrig Drainage 14(4):1460–1474. https://dorl.net/dor/20.1001.1.20087942.1399.14.4.28.7
Nourani V, Mousavi S, Sadikoglu F (2018) Conjunction of artificial intelligence-meshless methods for contaminant transport modeling in porous media: an experimental case study. J Hydroinf 20(5):1163–1179. https://doi.org/10.2166/hydro.2017.172
Pathania T, Eldho TI (2020) A moving least squares based meshless element-free Galerkin method for the coupled simulation of groundwater flow and contaminant transport in an aquifer. Water Resour Manage 34:4773–4794. https://doi.org/10.1007/s11269-020-02689-z

Qu J, Li Y, Zhou J (2010) Improved differential evolution based BP neural network for prediction of groundwater table. In: 2010 third international symposium on knowledge acquisition and modeling. IEEE, pp 36–39. https://doi.org/10.1109/KAM.2010.5646232

Rastogi AK (2012) Numerical groundwater hydrology. Penram International Publishing (India) Pvt. Ltd.

Redoloza F, Li L (2021) A comparison of extremal optimization, differential evolution and particle swarm optimization methods for well placement design in groundwater management. Math Geosci 53:711–735. https://doi.org/10.1007/s11004-020-09864-3

Sharief SMV, Eldho TI, Rastogi AK (2008) Optimal pumping policy for aquifer decontamination by pump and treat method using genetic algorithm. ISH J Hydraul Eng 14(2):1–17. https://doi.org/10.1080/09715010.2008.10514901

Singh LG, Eldho TI, Kumar AV (2016) Coupled groundwater flow and contaminant transport simulation in a confined aquifer using meshfree radial point collocation method (RPCM). Eng Anal Boundary Elem 66:20–33. https://doi.org/10.1016/j.enganabound.2016.02.001

Storn R, Price K (1997) Differential evolution–a simple and efficient heuristic for global optimization over continuous spaces. J Global Optim 11:341–359. https://doi.org/10.1023/A:1008202821328

Swathi B, Eldho TI (2017) A moving least squares based meshless local Petrov-Galerkin method for the simulation of contaminant transport in porous media. Eng Anal Boundary Elem 78:8–19. https://doi.org/10.1016/j.enganabound.2017.02.003

Swetha K, Eldho TI, Singh LG, Kumar AV (2023) Simulation of coupled flow and contaminant transport in an unconfined aquifer using the local radial point interpolation meshless method. Hydrogeol J 31(2):485–500. https://doi.org/10.1007/s10040-022-02558-6

Wang JF, Sun FX, Cheng YM (2012) An improved interpolating element-free Galerkin method with a non-singular weight function for two-dimensional potential problems. Chin Phys B 21(9):090204. https://doi.org/10.1088/1674-1056/21/9/090204

Zhang M, Mok CM, Polenghi-Gross I (2007) Optimization of groundwater extraction-and-treatment systems using differential evolution (DE) algorithm. In: Proceedings of world environmental and water resources congress 2007: restoring our natural habitat, pp 1–9. https://doi.org/10.1061/40927(243)337

Chapter 12
Transport of Nitrate, Phosphate and Potassium in Vadose Zone with Varying Rainfall Scenarios for a Semi-arid Cotton Growing Region of South India

Deepak Kumar, P. Vijay Kumar, Pankaj Kumar Gupta, and Suri L. Naidu

Abstract In terms of farm fertilizer Consumption, India is the world's second largest country after China. Reducing fertilizer use cannot be considered a viable option, since optimizing food output was the highest priority for the population's food and nutritional protection. Nitrogen, Phosphorous and Potassium fertilizers are farm fertilizers and are in great demand in Indian agriculture. This study investigated the effect of rainfall dynamics on farm fertilizer transport for rainfed Cotton Crop in Semi-arid Region in the Raichur district of South India. The solute transport and reaction parameters were assessed at the depths of 50, 100 and 150 cm of soil column. Hydrus-1D and Hydrus-2D software has been used for modelling transport of fertilizer in vadose zone. Rainfall data from 2015 to 2020, Soil data, Crop data, Soil hydraulic parameters, and Solute data has been used for this study. Results suggested that solute transport for average actual precipitation, 20% decreased precipitation and 20% increased precipitation at 150 cm depth of soil column was 0.000109, 0.00000000424, 0.000371 mmol cm^{-3} for nitrogen fertilizer, 0, 0, 0.00000014 $mmol^{-3}$ for phosphorous fertilizer and 0, 0, 0.0000000329 $mmol^{-3}$ respectively for potassium fertilizer. From this, it can be inferred that increased precipitation caused solutes to move more quickly than usual. Therefore, it can be concluded that due to climate change, changes in rainfall patterns and increased rainfall intensity might lead to excessive leaching of agricultural fertilizers into the vadoze zone and groundwater.

D. Kumar (✉) · P. Vijay Kumar
Department of Soil and Water Conservation Engineering, GBPUAT Pantnagar, Pantnagar, India
e-mail: deepak.swce.cot.gbpuat@gmail.com

P. K. Gupta
Centre for Rural Development and Technology, IIT Delhi, New Delhi, India

S. L. Naidu
National Institute of Hydrology, Roorkee, India

A. Nanda et al. (eds.), *Navigating the Nexus*, Water Science and Technology Library 102, https://doi.org/10.1007/978-3-031-76532-2_12

Keywords Vadoze zone · Nitrate · Phosphate · Potassium · Rainfed cotton crop · Raichur · Hydrus

12.1 Introduction

Cotton is India's most valuable fibre and cash crop, and it plays a significant role in the country's manufacturing and agricultural economies. It supplies the cotton textile industry with the basic raw material (cotton fibre). Cotton was a direct source of income for 6 million farmers in India, and the cotton trade and manufacturing employs 40–50 million people. There are ten major cotton-growing states in India. It is divided into three zones: the north zone, the central zone, and the south zone. Punjab, Haryana, and Rajasthan are included in the north region. Madhya Pradesh, Maharashtra, and Gujarat are included in the central region. Andhra Pradesh, Telangana, Karnataka, and Tamil Nadu are all part of the South region. The average yields of cotton in these regions were 1.8 quintal per hectare, which had almost the national average. Crop yields vary from year to year due to changes in rainfall and nutrient supply in the soil caused by climate change. According to intergovernmental panel on climate change (IPCC), climate change was postulated as state of change in the climatic conditions that was identified (e.g. using statistical tests) as changes in the mean or the variability of its properties, that persists for an extended period, typically decades or longer. UNFCCC (United Nations Framework Convention on Climate Change) defines the term climate change as change of climate that was attributed directly or indirectly to human activity that varies the composition of the global atmosphere and that was in addition to natural climate variability observed over significant time period. Climate change has a significant effect on rainfall patterns, which directly or indirectly affect groundwater recharge and the transport of solutes, which are important nutrients for plant development.

India was the world's second largest fertilizer producer after China, and it also ranks first in fertilizer imports. Complete N, P, and K fertilizer use in India rose from 78 thousand tons in 1965–66 to 26.5 million tonnes in 2009–2010. Reducing fertilizer use cannot be considered a viable option in India since optimizing food output was the highest priority for the population's food and nutritional protection. Improving fertilizer usage quality by maintaining steps such as balanced nutrient distribution, good planning, and water conservation will significantly reduce fertilizer leaching beyond the root zone of crops (Chen et al. 2010). Vadoze zone models and groundwater models plays important role in studying the dynamics of contaminants movement (Ch et al. 2013; Kumar et al. 2013). Nitrogen, phosphorous and potassium leaching can be modelled using various models like HYDRUS-1, HYDRUS-2 (Chowdary et al. 2004; Wang et al. 2010; Liu et al. 2013; Tafteh and Sepaskhah 2012).

Kumari et al. (2019) investigated Nitrate–N load and vulnerability of groundwater resources using in-situ observed hydrogeological data for Madhubani area of Bihar. The eastern part of the study area showed high vulnerability than northern part followed by western part. In-situ observed nitrate concentrations were well matched with simulated results. The high vulnerability in eastern and northern part was due to alluvial sandy lithologs and very shallow groundwater table. Gupta et al. (2020) evaluated the vulnerability of nitrate in groundwater resources. The study area selected was three districts of Bihar, India namely, Samastipur, Darbhanga and Madhubani. HYDRUS 1D was used to model nitrate transport. Results have shown high risk of nitrate-leaching vulnerability in southern part of study area which is dominated by Gangetic kankar in subsurface. Also, high pollution risk was reported in eastern north part of study area having alluvial deposition in subsurface.

A computer simulation model Hydrus2D/3D was used to simulate the NPK distribution for various depths of soil column. It was used to determine the important factors that influence the nutrient distribution under different rainfall dynamics. A detailed explanation of both water and solute flow through the vadoze zone was expected to reliably forecast environmental impacts associated with human activities such as overuse of fertilizer and irrigation. It is critical to account for the two-dimensional existence of the system when growing rain-fed crops. As a result, the Hydrus-2D mathematical modelling model has been used to simulate solute transport and has been shown to be a reliable predictor of both water and solute dynamics (Mmolawa and Or 2003; Skaggs et al. 2004; Gärdenäs et al. 2005). The Hydrus-2D program solves the Richards' equation for saturated–unsaturated water flow and the convection–dispersion equation for solute transport numerically. A sink word was often used in the flow equation to account for water uptake by plant roots (Simunek et al. 1999). Due to climate change, the rainfall intensity pattern are changing. Due to high rainfall, the leaching may affect the transport of fertilizers in the vadose zone as well as in aquifers. To study the effect of high rainfall on leaching of fertilizers, few hypothetical situations above and below the average rainfall of study area has been created. The leaching of fertilizers in vadose zone in cotton growing regions of Raichur is the northern part region of Karnataka was still unexplored. Keeping in view, the above scenerio, the main objective of this study was to evaluate the effect of rainfall dynamics on Nitrogen, Phosphorus and Potassium (NPK) concentration at various depths of soil column for cotton growing regions of South India.

12.2 Materials and Methods

12.2.1 Study Area

India occupies 2.4% area of world's land which has the geographical area of 329 M ha. Among that the Karnataka occupies geographical area of 191,791 km^2 which is about 5.83% of the total area of the country. Out of the total geographical area of Karnataka, about 12.31 M ha was under cultivation. Raichur is the northern part region of Karnataka, which covers an area of about 8386 km^2. The average annual rainfall varies from 400 to 800 mm. The study area is located in the semiarid region of Gopalapur in the Raichur district. The location map of study area is shown in

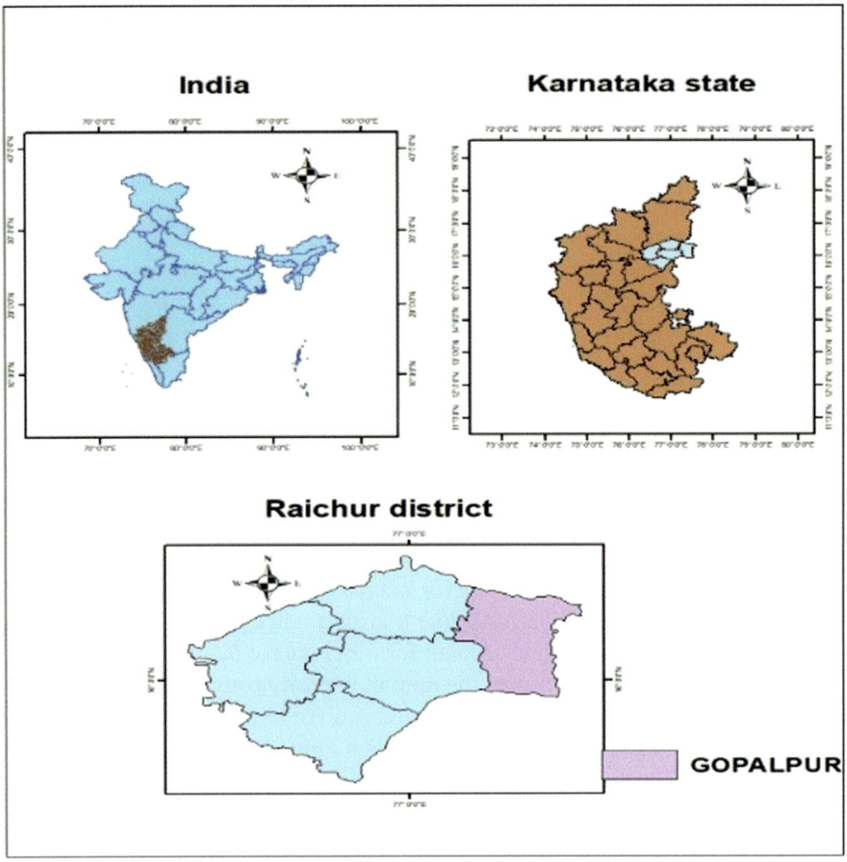

Fig. 12.1 Location map of the study area

Fig. 3.1. The latitude and longitude of the area are 16.46755° North and 76.89853° South respectively. This sub-watershed covers an area of 673.18 ha and drains into Krishna River. Major crop grown in this region was cotton which was under rain-fed condition. Rain-fed agriculture is a type of farming that relies on rainfall for water. The location map of study area is shown in Fig. 12.1.

12.2.2 Governing Equations and HYDRUS Model

The solute transport and reaction parameters were assessed using HYDRUS-2D, while soil water content were assessed using the HYDRUS-1D. The governing equations are discussed in following paragraphs.

HYDRUS-2D is a finite element model that numerically solves Richard's equation (as shown in Eq. 12.1) for soil water flow given below:

$$\frac{\partial \theta}{\partial t} = \frac{\partial}{\partial z}\left(D(\theta)\frac{\partial \theta}{\partial z} - K(\theta)\right) + S_r \tag{12.1}$$

where θ is water content, D is hydraulic diffusivity and is a function of θ. S_r is sink term which represents the volume of water removed per unit time from a unit volume of soil due to plant water uptake. In other words, S_r is the root water uptake that can be computed using Feddes RWU model (Feddes et al. 1978; Šimůnek et al. 2011). K(θ) is empirical unsaturated hydraulic conductivity function.

Governing flow equation in one dimension is given by the following modified form of the Richards equation (Šimůnek et al. 2006)

$$\frac{\partial \theta(h)}{\partial t} = \frac{\partial}{\partial z}\left[k\frac{\partial h}{\partial z} + 1\right] - s(h, t) \tag{12.2}$$

where, θ is volumetric water content [L^3L^{-3}]; h is pressure head [L]; S is sink term [T^{-1}]; z is spatial coordinates [L]; t is time [T]; k is unsaturated hydraulic conductivity [LT^{-1}]

Root water uptake (RWU), a sink term of Richard's equation was computed using Feddes' model (Feddes et al. 1978).

$$S(\psi, x, z) = \alpha(\psi, x, z) * b(x, z) * T_p * L \tag{12.3}$$

where water stress response function α (ψ, x, z) is a dimensionless function of soil matric potential head (ψ); Normalized water uptake distribution function b (x, z) is (cm^{-2}); T_p is potential transpiration rate ($cm\ d^{-1}$); and L (cm) is width of soil surface associated with transpiration (Deb et al. 2013).

HYDRUS-1D is a widely used software tool for simulating water, heat, and solute movement in one-dimensional variably saturated media. It is particularly effective for studying vertical solute transport in soil and other porous media. In this study, HYDRUS-1D has been used to estimate the vertical movement of solute in the vadose zone. Further, HYDRUS-2D has been used to get the lateral movement of solute beneath the surface of earth.

12.2.3 Soil Data

Soil data of study area is shown in Table 12.1. Figure 12.2 shows the columns indicating hundred nodes with 150 cm of soil depth, the soil column can be discredited by the help of nodal points. This column was to characterize the flow domain and parameters of spatial distribution like- hydraulic scaling factors, spatial distribution of soil materials, and parameters of root-water uptake.

Table 12.1 Particle size distribution and bulk density of soil

S. no	Soil depth (cm)	Textural class	Sand (%)	Silt (%)	Clay(%)	Bulk density (g/cm^3)
1	0–10	Clay loam	33	27	40	1.41
2	10–20	Clay loam	41	23	36	1.42
3	20–40	Clay loam	32	26	42	1.35
4	40–70	Sandy clay loam	53	17	30	1.37
5	70–100	Sandy clay loam	45	21	34	1.46
6	100–150	Sandy clay loam	44	21	32	1.57

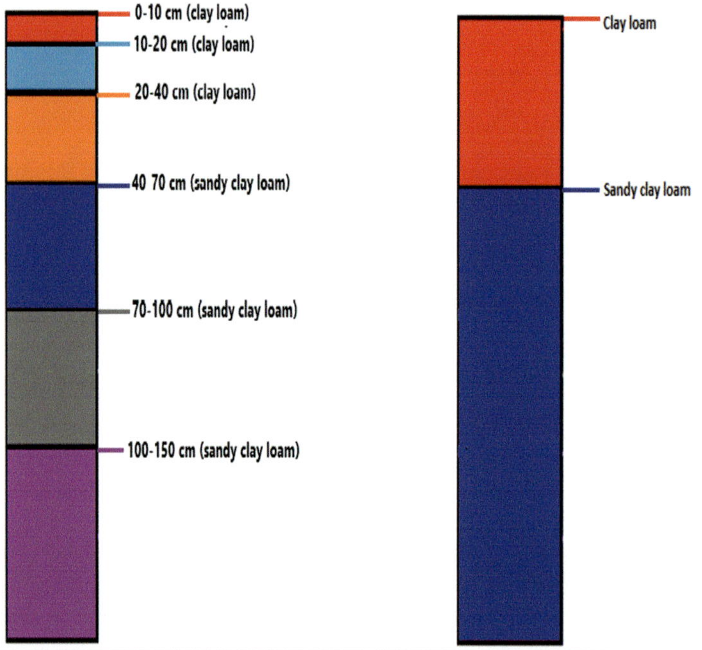

Fig. 12.2 Variation of soil type along 150 cm depth of soil column

12.2.4 Metrological Data

Data repositories and archives play a critical role as the source for the observational data used in the study of climate. The observations in the last 5 years have been collected from the metrological station. In order to observe the effect of the rainfall dynamics on solute transport and groundwater recharge, following scenarios has been considered.

(i) The 5 year metrological data was averaged to annual data.
(ii) The average annual data increased to 20% and
(iii) The average annual data decreased to 20%.

12.2.5 Crop Data

Cotton is a tropical and subtropical crop. For the successful germination of its seeds, a minimum temperature of 15 °C was required. Cotton was grown on a variety of soils. Soil should be black medium to deep about 90 cm with good drainage availability. Cotton does not tolerate water-logging condition. In South India the major portion of the rain-fed crops were planted in the month of June-July. The agronomical parameters of cotton are shown in Table 12.2. The fertilizer requirement for rain-fed cotton was 80:40:40 kg nitrogen, phosphorus and potassium per hectare were recommended. The application of 50% nitrogen and whole phosphorus and potash should be given at the time of sowing and the remaining 50% nitrogen should be applied after 4 weeks of sowing and the initial rooting depth of cotton plant was 0.01 cm for the root growth time of 3 days and the maximum rooting depth of cotton plant was 100 cm. For cotton crop, root water uptake parameters are shown in Table 12.3.

Table 12.2 Agronomical parameters of cotton crop

Name of the crop	Cotton
Scientific name	Gossypium
Planting time	July
Duration of crop	10 months
Dose of fertilizer	80:40:40 NPK kg ha^{-1}
Day after sowing (DAS)	
At the time of sowing	40:20:20 NPK kg ha^{-1}
30 DAS	20:10:10
80 DAS	20:10:10

Table 12.3 Root water uptake parameters

S. no	Feddes' parameters	Value
1	P0 (cm)	– 10
2	P0pt (cm)	– 25
3	P2H (cm)	– 200
4	P2L (cm)	– 600
5	P3 (cm)	– 14,000
6	r2H (cm/day)	0.5
7	r2L (cm/day)	0.1

Table 12.4 Water flow parameters

S. no	Soil depth (cm)	Q_r	Q_s	Alpha (cm^{-1})	n	Ks (cm day^{-1})	I
1	0–10	0.0888	0.45	0.0155	1.3543	11.16	0.5
2	10–20	0.0838	0.4413	0.0166	1.354	11.98	0.5
3	20–40	0.0926	0.47	0.0161	1.3528	16.6	0.5
4	40–70	0.0778	0.4484	0.019	1.3765	25.01	0.5
5	70–100	0.0797	0.481	0.0177	1.3398	11.15	0.5
6	100–150	0.0756	0.3989	0.0192	1.2813	6.31	0.5

12.2.6 Soil Hydraulic Properties

The knowledge of saturated hydraulic conductivity (k_s) is essential as an input parameter for the solution of the general transport equation of the water flow. The required water flow parameters are residual water content (θ_r), saturated water content (θ_s), saturated hydraulic conductivity (k_s), pore connectivity parameter (l) in the hydraulic conductivity function and was taken as 0.5 (Mualem 1976). Two other parameters Alpha (α) and n are empirical coefficients affecting the shape of hydraulic functions. Alpha (α) is a fitting parameter in soil retention curve (h-θ curve) and is inverse of air entry value. The parameter n is also a fitting parameter in the soil retention curve (dimensionless) (Kukal and Aggarwal 2002; Singh et al 2001). Water flow parameters has been tabulated in Table 12.4. Residual water content was obtained using gravimetric method, while saturated water content has been obtained using the porosity tests. Cylindrical infiltrometer was used to measure saturated hydraulic conductivity of soil. As mentioned above, value of alpha and pore connectivity parameter has been taken from standard literatures cited above.

12.2.7 Solute Parameters

The solute parameters are shown in Table 12.5. Input Parameters for calibration were longitudinal dispersivity (D_L), transverse dispersivity (D_T), diffusivity of water (cm^2 day^{-1}), bulk density, no. of solute used and reaction parameters. Initial estimates for optimisation: D_L was kept as 1/10 of soil column depth, D_T was kept as 1/10 of D_L (Deb et al. 2016). 9 data points were used as training data set for each treatment for optimization of parameters. For each solute (N, P and K) optimization was done separately in the order—nitrogen followed by phosphorus followed by potassium for period of 365 days. Defining initial conditions: 40 kg/ha of urea, 40 kg/ha of phosphorus and 40 kg/ha of potassium applied after few days of rainfall at the beginning of simulation was used as initial time independent solute concentration at soil surface. Optimised values were used for prediction of concentration of N, P,

Table 12.5 Solute transport parameters

Solute transport	Time weighting scheme	Crank Nicholson scheme
	Space weighting scheme	Galerkin finite elements
	Number of solutes	3
	Mass unit	mmol
	Initial conditions	Total concentration (mass_solute/volume_soil)
Specific parameters	**Material 1**	
	Bulk density	1.206
	Transverse dispersivity (cm)	1.5
	Longitudinal dispersivity (cm)	15
	Material 2	
	Bulk density	1.591
	Transverse dispersivity (cm)	1.1
	Longitudinal dispersivity (cm)	11

and K were compared with one another in order to determine the effect of rainfall dynamics. Core experiment was performed to obtain bulk density of soil.

12.3 Results and Discussions

This section presents effect of rainfall dynamics on NPK transport along the depth of soil column using HYRUS-2D model. The HYDRUS-2D model was calibrated for the soil hydraulic parameters (saturated moisture content, residual moisture content, saturated conductivity), soil physical parameter (texture) and solute transport parameters (longitudinal dispersivity, molecular diffusion coefficient, solute rate constants) with the values of soil water and NPK content pertaining to soil depths of 50, 100 and 150 cm, observed in the vadose zone of the cotton crop using the average of five years annual meteorological data from 2015 to 2020. In the present study, three scenarios has been studied for analyzing the impact of rainfall dynamics on the solute transport of NPK.

12.3.1 Scenario-1: Impact of Average Annual Precipitation (AAP) on NPK Transport

A fertilizer dosage of 80:40:40 kg ha^{-1} of NPK was applied after onset of monsoon that is at the beginning of July month because the chosen study area was located in the semiarid region. So for irrigation purpose it entirely depends on rainfall. The calibrated NPK content generally followed the linear relationship with respect to rainfall/irrigation applied along the soil depth. Model calibration results pertaining to solute transport up to 150 cm soil depth for annual rainfall of 659 mm has been presented. Soil Nitrogen content peaked at the depth of 50 cm about 0.03851 mmol cm^{-3}, then concentration decreased from 100 to 150 cm soil depth about 0.0024778 and 0.0000321 mmol cm^{-3} respectively, it continues to decrease thereafter as shown in Fig. 12.3. The higher Nitrogen content on the top soils (0–50 cm) was apparently attributable to fertilization on soil surface. After 50 cm depth of soil, the decrease in the concentration of Nitrogen content is due to the Nitrogen displacement through the unsaturated zone down to the ground surface which is forced by infiltrated rainfall.

Similarly, Soil Phosphorous content peaked at the depth of 50 cm about 0.003179 mmol cm^{-3}, then concentration decreased from 100 to 150 cm soil depth about 0.0000165 and 0 mmol cm^{-3} respectively, it continues to decrease thereafter as shown in Fig. 12.4. The higher phosphorus content on the top soils (0–50 cm) was apparently attributable to fertilization on soil surface. After 50 cm depth of soil, the decrease in the concentration of phosphorus content is due to the phosphorus

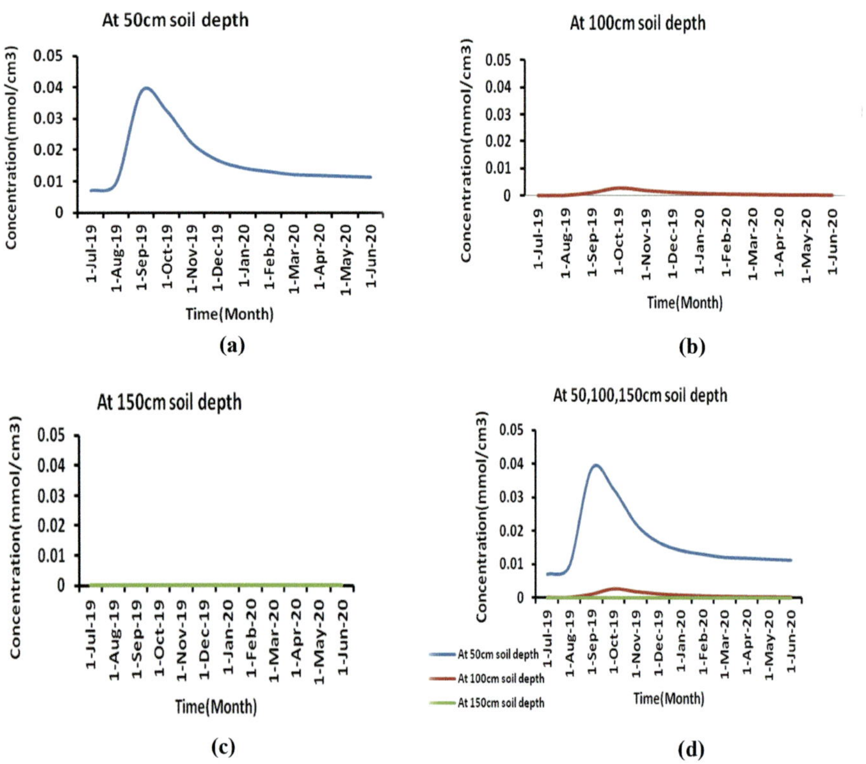

Fig. 12.3 Variation of nitrogen concentration at various depth of soil column for AAP

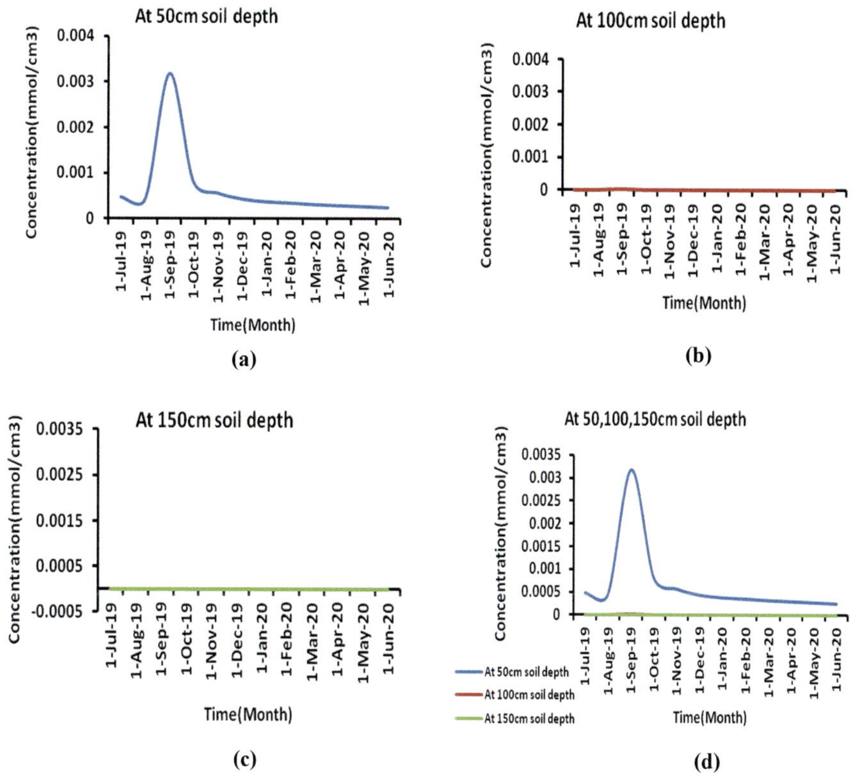

Fig. 12.4 Variation of phosphorous concentration at various depth of soil column for AAP

displacement through the unsaturated zone down to the ground surface which is forced by infiltrated rainfall.

Soil Potassium content peaked at the depth of 50 cm about 0.0005746 mmol cm^{-3}, then concentration decreased from 100 to 150 cm soil depth about 0.000000354 and 0 mmol cm^{-3}, it continues to decrease thereafter as shown in Fig. 12.5. The higher Potassium content on the top soils (0–50 cm) was apparently attributable to fertilization on soil surface. After 50 cm depth of soil, the decrease in the concentration of Potassium content is due to the Potassium displacement through the unsaturated zone down to the ground surface which is forced by infiltrated rainfall. The simulation results showed the variation in NPK concentration at various depths of soil column. Even though quantity of fertilizer applied for phosphorus and potassium were not varied but variations were observed due to it dependency on several factors such as solute flux in the liquid phase which represents different transport processes like advection, diffusion and hydrodynamic dispersion and also the water and chemical movement, chemical reactions, and microbial transformations.

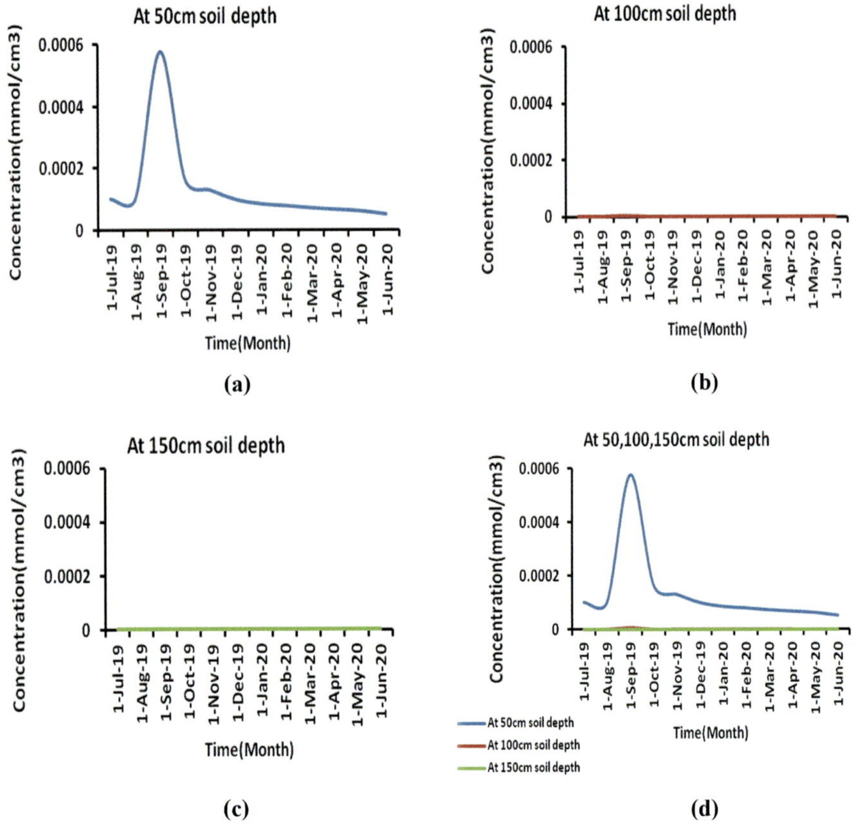

Fig. 12.5 Variation of potassium content at various depth of soil column for AAP

12.3.2 Scenario-2: Impact of 20% Decreased AAP on NPK Transport

Model calibration pertaining to solute content up to 150 cm depth of soil column for the average annual rainfall of 52.6 cm which was 20% less than the above, resulted that soil nitrogen content peaked at the depth of 50 cm about 0.02708 mmol cm^{-3}, then concentration decreased from 100 to 150 cm soil depth about 0.00021 and 0.000000133 mmol cm^{-3}, it continues to decrease thereafter as shown in Fig. 12.6. For the decreased rainfall about 13.3 cm from average rainfall of 65.9 cm the variation in nitrogen concentration decreased by 29.68% at 50 cm soil depth and continues to decrease along the soil depth when compared to the concentration content of average rainfall condition at various depths. The decreased solute concentration is due to less rainfall which in response resulted in less availability of water to infiltrate and displace the solute through the unsaturated zone down to the ground surface.

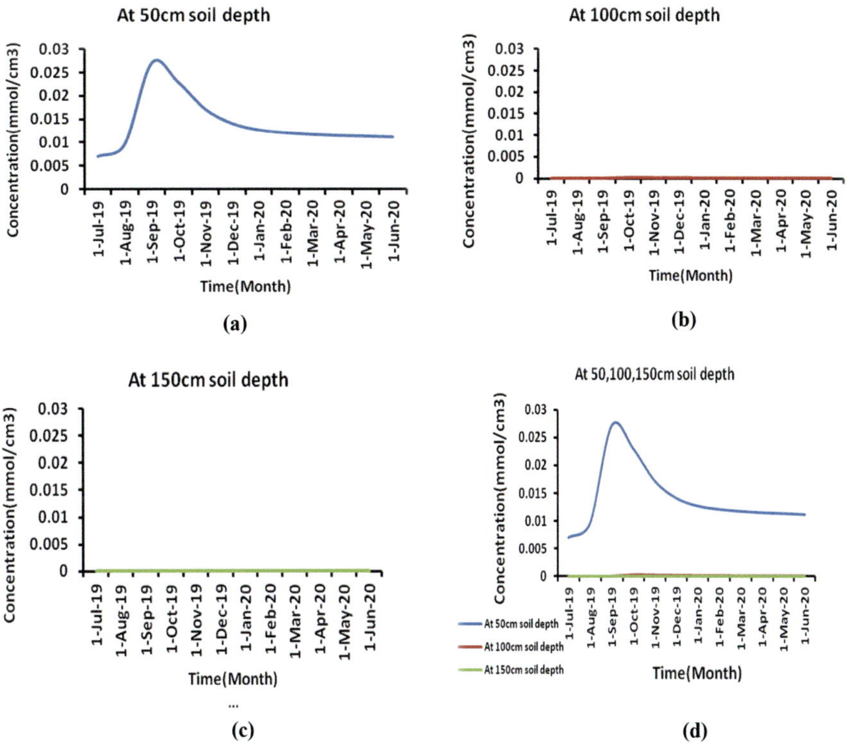

Fig. 12.6 Variation of nitrogen content at various depth of soil column for decreased 20% AAP

Similarly, Soil Phosphorous content peaked at the depth of 50 cm about 0.002238 mmol cm^{-3}, then concentration decreased from 100 to 150 cm soil depth about 0.0000021 and 0 mmol cm^{-3}, it continues to decrease thereafter as shown in Fig. 12.7. For the decreased rainfall about 13.3 cm from average rainfall of 65.9 mm the variation in concentration decreased by 29.60% at 50 cm soil depth and continues to decrease along the soil depth when compared to the concentration content at various depth during average rainfall condition. The decreased solute concentration is due to less rainfall which resulted in less availability of water to infiltrate and displace the solute through the unsaturated zone down to the ground surface.

Soil Potassium content peaked at the depth of 50 cm about 0.000497 mmol cm^{-3}, then concentration decreased from 100 to 150 cm soil depth about 0.000000451 and 0 mmol cm^{-3}, it continues to decrease thereafter as shown in Fig. 12.8. For the decreased rainfall about 13.3 cm from average rainfall of 65.9 cm the variation in concentration decreased by 13.50% at 50 cm soil depth and continues to decrease along the soil depth when compared to the concentration content at various depth during average rainfall condition. At 150 cm depth of soil column the concentration of both potassium and phosphorus have reached to the value of 0 mmol cm^{-3} due to their less mobile in soil when compared to nitrogen which is mobile in nature.

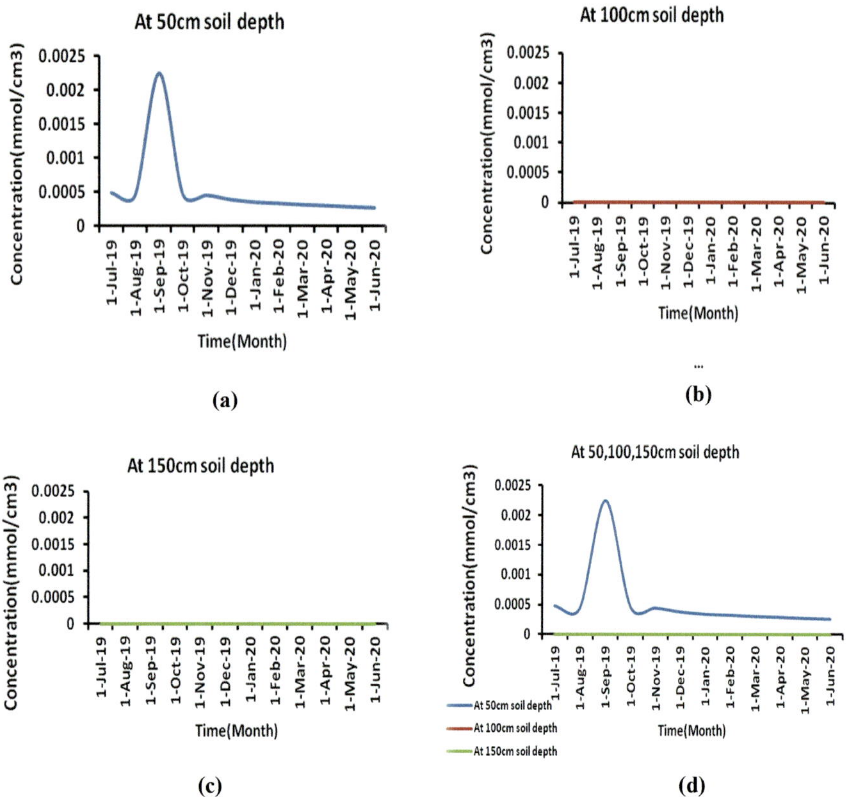

Fig. 12.7 Variation of phosphorous concentration at various depth of soil column for 20% decreased AAP

12.3.3 Scenario-3: Impact of 20% Increased AAP on NPK Transport

Model calibration pertaining to solute content up to 150 cm depth of soil column for the average annual rainfall of 79.2 cm which was 20% more than the above average rainfall of 65.9 cm resulted that Soil Nitrogen content peaked at the depth of 50 cm about 0.04311 mmol cm^{-3}, then concentration decreased from 100 to 150 cm soil depth about 0.008799 and 0.0006969 mmol cm^{-3}, it continues to decrease thereafter as shown in Fig. 12.9. For the increased rainfall about 13.3 cm from average rainfall of 65.9 cm the variation in nitrogen concentration increased by 11.94% at 50 cm soil depth and continues to increase along the soil depth when compared to the concentration content of average rainfall condition at various depth. The increased solute concentration was due to high rainfall which resulted in more availability of water to infiltrate and displace the solute through the unsaturated zone down to the ground surface.

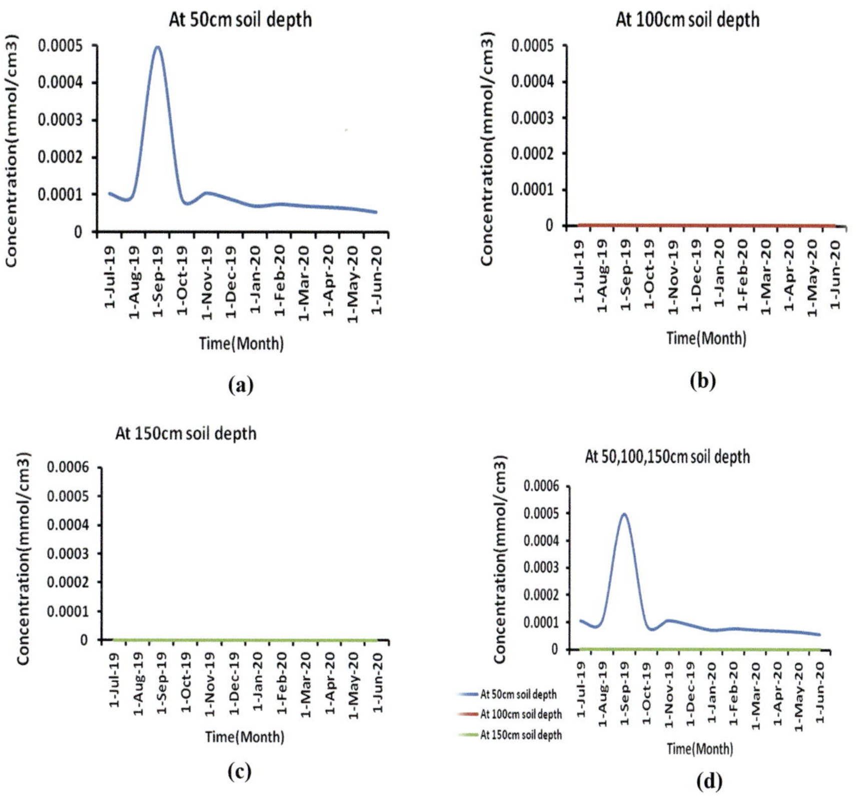

Fig. 12.8 Variation of potassium concentration at various depth of soil column for 20% decreased AAP

Similarly, Soil Phosphorous content peaked at the depth of 50 cm about 0.003625 mmol cm^{-3}, then concentration decreased from 100 to 150 cm soil depth about 0.0001365 and 0.00000029 mmol cm^{-3}, it continues to decrease thereafter as shown in Fig. 12.10. For the increased rainfall about 13.3 cm from average rainfall of 65.9 cm the variation in concentration increased by 14.02% at 50 cm soil depth and continues to increase along the soil depth when compared to the concentration content at various depths during average rainfall condition.

Soil Potassium content peaked at the depth of 50 cm about 0.0007158 mmol cm^{-3}, then concentration decreased from 100 to 150 cm soil depth about 0.00002907 and 0.0000000625 mmol cm^{-3}, it continues to decrease thereafter as shown in Fig. 12.11. For the increased rainfall about 13.3 mm from average rainfall of 65.9 cm the variation in concentration increased by 24.57% at 50 cm soil depth and continues to increase along the soil depth when compared to the concentration content at various depth during average rainfall condition.

It has been observed from Fig. 12.12 which represents the simulations for the compared solute concentration (NPK) at different rainfall dynamics. From this we

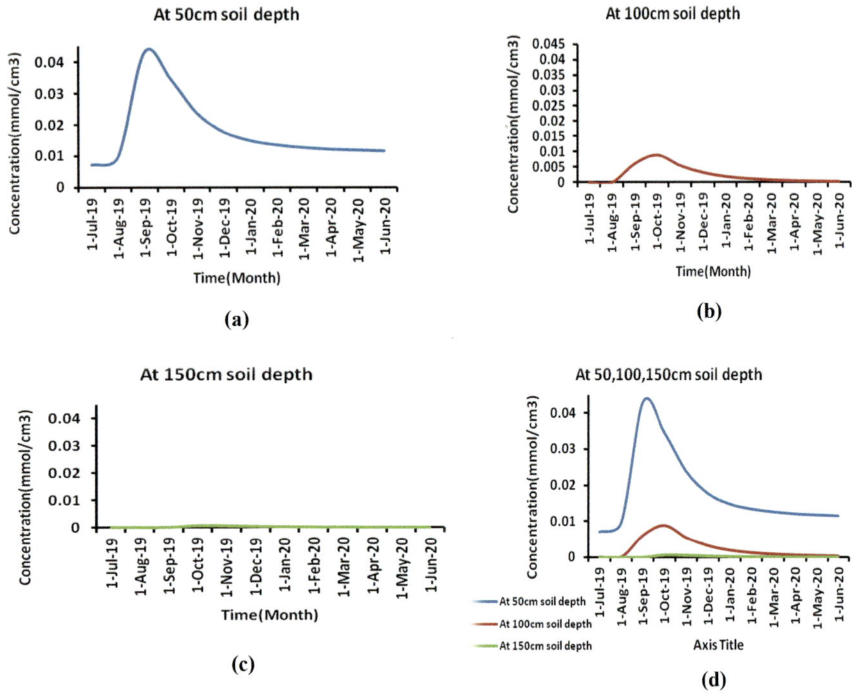

Fig. 12.9 Variation of nitrogen concentration at various depth of soil column for 20% increased AAP

can infer that as the rainfall increased about 20% from avg. ppt the solute concentration also increased at 150 cm depth of soil column and reached to its peak about 0.000697 mmol cm^{-3} of nitrogen as shown in Fig. 12.12a, 0.00000029 mmol cm^{-3} of phosphorus as in Fig. 12.12b, and 0.0000000625 mmol cm^{-3} of potasssium as in Fig. 12.12c. But when rainfall decreased about 20% from avg. precipitation the solute concentration had negligible amount of solute concentration.

From the above simulated results, it has been observed that the concentration of nitrogen was higher in the soil. It might be due to the hydrolysis of urea under proper supply of oxygen undergoes nitrification. The distribution of availability of nitrogen varied both horizontally and vertically. The values of available nitrogen in the bottom layers were more as compared to the Phosphorus and Potassium as mentioned in Table 12.6 at various depth of soil column, due to the leaching of nitrogen from the upper layer. N is a highly mobile element in soil when compared to the phosphorus, which is immobile in soil and potassium is less mobile. The accumulation of N, P and K as per Table 12.6 was only up to 150 cm. Since, groundwater is at more vertical depth, no direct inference can be obtained about groundwater. But, it might be sure that if the concentration of fertilizer will increase with depth, it would hamper the aquifer system also. In the study area, maximum N, P and K leaching at 50 cm was 0.03851, 0.003179, 0.0005746 mmol cm^{-3}.

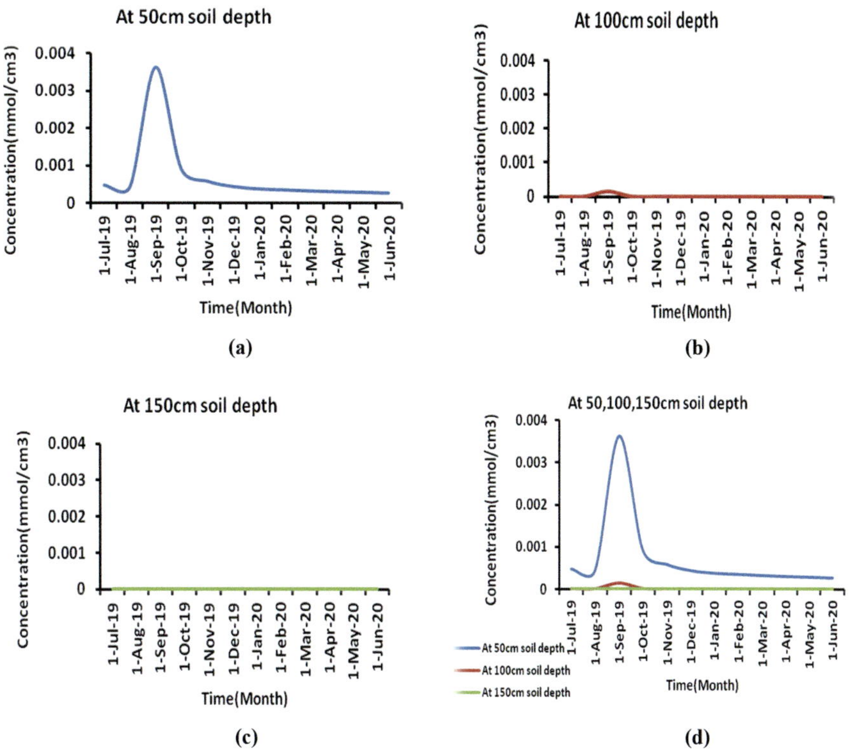

Fig. 12.10 Variation of phosphorous concentration at various depth of soil column for 20% increased AAP

12.4 Conclusions

In the present study, HYDRUS-1D model was used to evaluate the impacts of rainfall dynamics on solute transport. Among all the elements needed for plant growth, nitrogen (N), phosphorous (P), potassium (K) is considered the most important fertilizer element applied to soils. The results of average annual data of 5 years from 2015 to 2020 of NPK fluxes showed that too much of precipitation or water application have induced solute to percolate faster and contaminate the groundwater resources, among these nutrients nitrate leaching could be significant. Simulations showed that drainage fluxes of NPK from the soil column increased after introducing irrigation/fertilization. The NPK concentrations increased at all depths after rainfall/water application. The common feature at all depths was that NPK concentrations increased with each new precipitation/irrigation event. Solute concentration (NPK) at the end of simulation period for average precipitation, 20% decreased precipitation and 20% increased precipitation at 150 cm depth of soil column was that for N was 0.000109, 0.00000000424, 0.000371 mmol cm^{-3}, P was 0, 0, 0.00000014 $mmol^{-3}$ and K was 0,

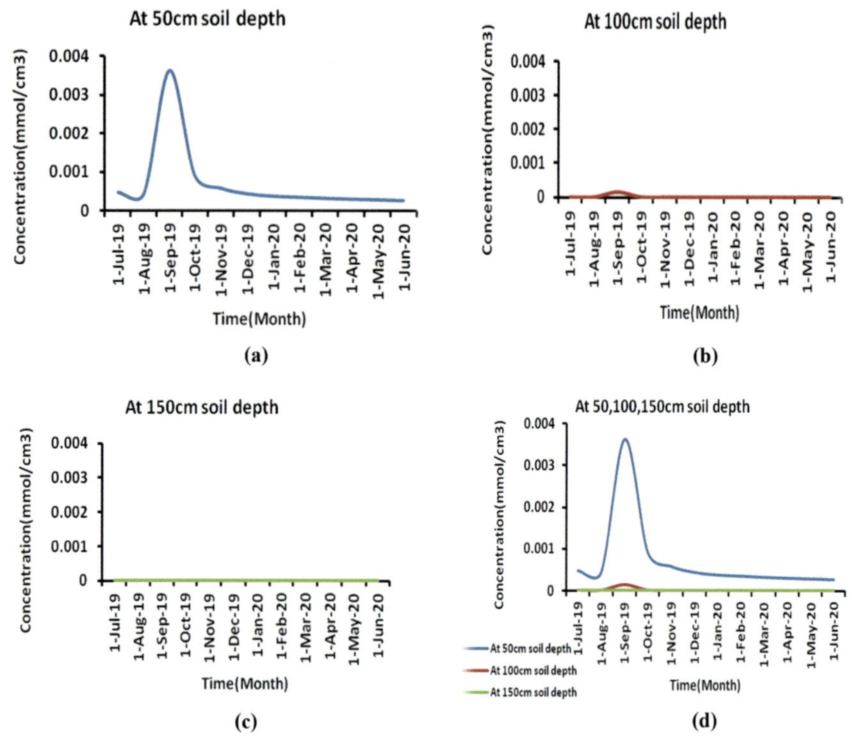

Fig. 12.11 Variation of potassium concentration at various depth of soil column for 20% increased AAP

0, 0.0000000329 mmol^{-3} respectively. From this, we can infer that increased precipitation causes solutes to move faster than usual. Conversely, significantly reduced or decreased precipitation leads to solutes accumulating on the soil surface, thereby contaminating it. Therefore, understanding the optimal amount of fertilizer application is crucial to minimize groundwater quality impairment. Another mitigation strategy would be to lower the NPK input by reducing the amount of manure or fertilizer applied. For future research, researcher can couple Hydrus1 D and 2D to study the effect of climate change on solute transport of heavy metals as well as can model pathogen movement in vadose zone as well as groundwater.

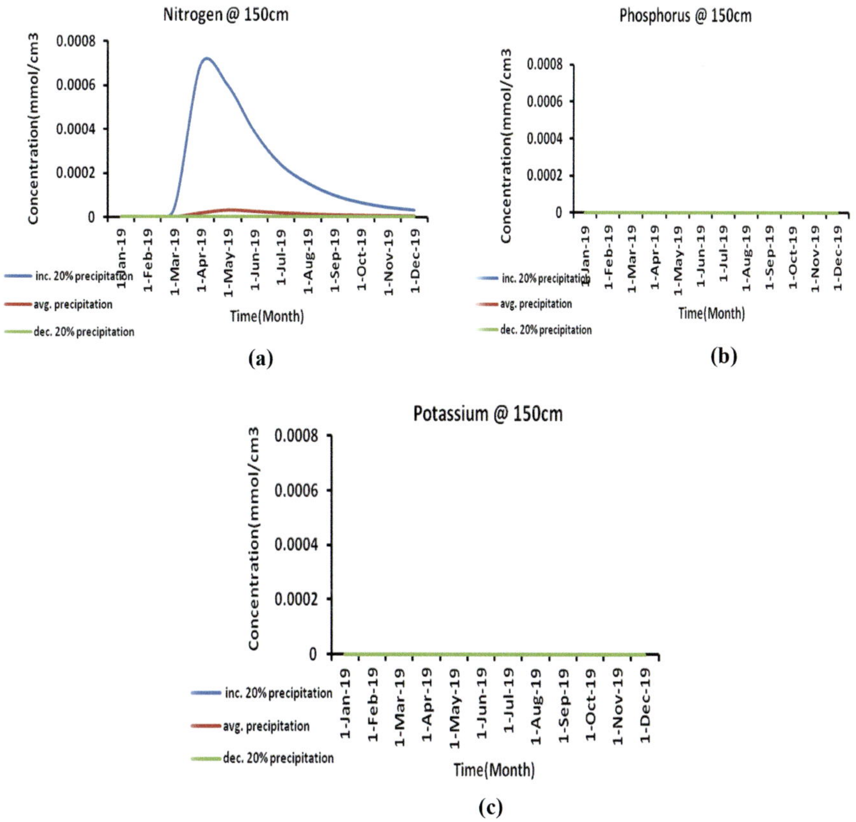

Fig. 12.12 Compared solute concentration (NPK) at different rainfall dynamics

Table 12.6 Variation of nitrogen, phosphorous, and potassium content at various depth of soil column

S. no	Precipitation (mm)	Solute type	At 50 cm depth (mmol cm^{-3})	At 100 cm depth (mmol cm^{-3})	At 150 cm depth (mmol cm^{-3})
1	659 (AAP)	**Nitrogen**	0.03851 (ANC)*1	0.0024778	0.0000321
2	792 (20% increased AAP)		0.04311 (+11.94% of ANC)	0.008799	0.0006969
3	526 (20% decrease AAP)		0.02708 (−29.68% of ANC)	0.00021	0.000000133
4	659 (AAP)	**Phosphorus**	0.003179 (APC)*2	0.0000165	0
5	792 (20% increased AAP)		0.003625 (+14.02% of APC)	0.0001365	0.00000029
6	526 (20% decrease AAP)		0.002238 (−29.60% of APC)	0.0000021	0
7	659 (AAP)	**Potassium**	0.0005746 (AKC)*3	0.00000035	0
8	792 (20% increased AAP)		0.0007158 (+24.57% of AKC)	0.00002907	0.0000000625
9	526 (20% decrease AAP)		0.000497 (−13.50% of AKC)	0.00000045	0

*1—Actual nitrogen concentration; *2—actual phosphorus concentration; *3—actual potassium concentration

References

Ch S, Kumar D, Prasad RK, Mathur S (2013) Optimal design of an in-situ bioremediation system using support vector machine and particle swarm optimization. J Contam Hydrol 151:105–116

Chowdary VM, Rao NH, Sarma PBS (2004) A coupled soil water and nitrogen balance model for flooded rice fields in India. Agric Ecosyst Environ 103(3):425–441

Chen XP, Zhang FS, Cui ZL, Li F, Li JL (2010) Optimizing soil nitrogen supply in the root zone to improve maize management. Soil Sci Soc Am J 74(4):1367–1373

Deb SK, Shukla MK, Šimůnek J, Mexal JG (2013) Evaluation of spatial and temporal root water uptake patterns of a flood-irrigated pecan tree using the HYDRUS (2D/3D) model. J Irrig Drain Eng 139(8):599–611

Deb SK, Sharma P, Shukla MK, Ashigh J, Šimůnek J (2016) Numerical evaluation of nitrate distributions in the onion root zone under conventional furrow fertigation. J Hydrol Eng 21(2):05015026

Feddes RA, Kowalik K, Zaradny H (1978) Model for simulating soil-water content considering evapotranspiration—comments. J Hydrol 37(3–4):393–397

Gärdenäs AI, Hopmans JW, Hanson BR, Šimůnek J (2005) Two-dimensional modeling of nitrate leaching for various fertigation scenarios under micro-irrigation. Agric Water Manage 74(3):219–242
Gupta PK, Kumari B, Gupta SK, Kumar D (2020) Nitrate-leaching and groundwater vulnerability mapping in North Bihar, India. Sustain Water Resour Manage 6:1–12
Kukal SS, Aggarwal GC (2002) Percolation losses of water in relation to puddling intensity and depth in a sandy loam rice (Oryza sativa) field. Agric Water Manage 57(1):49–59
Kumar D, Ch S, Mathur S, Panigrahi BK (2013) Hybrid algorithm performance with varying population size for multi-objective optimisation of in-situ bioremediation of groundwater. Int J Bio Inspired Comput 5(3):164–174
Kumari B, Gupta PK, Kumar D (2019) In-situ observation and nitrate-N load assessment in Madhubani District, Bihar, India. J Geol Soc India 93:113–118
Liu X, Sun S, Ji P, Šimůnek J (2013) Evaluation of historical nitrate sources in groundwater and impact of current irrigation practices on groundwater quality. Hydrol Sci J 58(1):198–212
Mmolawa K, Or D (2003) Experimental and numerical evaluation of analytical volume balance model for soil water dynamics under drip irrigation. Soil Sci Soc Am J 67(6):1657–1671
Mualem Y (1976) A new model for predicting the hydraulic conductivity of unsaturated porous media. Water Resour Res 12(3):513–522
Simunek J, Sejna M, Van Genuchten MT (1999) The HYDRUS-2D software package. Int Ground Water Model Cent 1:742–764
Šimůnek J, Van Genuchten MT, Šejna M (2006) The HYDRUS software package for simulating two-and three-dimensional movement of water, heat, and multiple solutes in variably-saturated media. Tech Manual 1:241–253
Šimůnek J, Van Genuchten MT, Šejna M (2011) The HYDRUS software package for simulating two-and three-dimensional movement of water, heat, and multiple solutes in variably-saturated media. Technical Manual, Version 2.0, PC- Progress, Prague, Czech Republic
Singh KB, Gajri PR, Arora VK (2001) Modelling the effects of soil and water management practices on the water balance and performance of rice. Agric Water Manage 49(2):77–95
Skaggs TH, Trout TJ, Šimůnek J, Shouse PJ (2004) Comparison of HYDRUS-2D simulations of drip irrigation with experimental observations. J Irrig Drain Eng 130(4):304–310
Tafteh A, Sepaskhah AR (2012) Application of HYDRUS-1D model for simulating water and nitrate leaching from continuous and alternate furrow irrigated rapeseed and maize fields. Agric Water Manage 113:19–29
Wang H, Ju X, Wei Y, Li B, Zhao L, Hu K (2010) Simulation of bromide and nitrate leaching under heavy rainfall and high-intensity irrigation rates in North China Plain. Agric Water Manage 97(10):1646–1654

Chapter 13
Health Risk Assessment of Heavy Metal Toxicity in Groundwater—A Comparative Study on the Industrial Corridors of Integrated Vellore District, Tamil Nadu

K. Loganathan and L. Vignesh Rajkumar

Abstract Groundwater has become a vital natural resource for humans and other species to sustain life. Every day, groundwater gets polluted due to population growth and industrialization. Many industrial centers in northern Tamil Nadu draw more groundwater for the efficient operation of their industries. Groundwater quality in the integrated Vellore District of Tamil Nadu is being impacted by population growth and industrial pollution. This study evaluates the heavy metal toxicity and health risk assessment of three industrial corridors, namely Ambur, Ranipet, and Vaniyambadi. The evaluation matrices utilized in this study include multivariate statistical analysis, contamination factor (CF), and geographic information systems (GIS). Groundwater samples from 42 open and bore wells were collected and five heavy metals chromium (Cr), cadmium (Cd), iron (Fe), lead (Pb), and cobalt (Co) were considered in all three industrial corridors. The t-test indicates that the sample averages of the heavy metals Cd (0.003 mg/l) and Cr (0.05 mg/l) are substantially higher than the WHO limit at $p < 0.05$. CF examines Cd and Cr pollution, and the findings reveal the highest CF value at Ranipet with 78% and 22%, respectively. The principal component analysis (PCA) results at Ambur show that Electrical Conductivity (EC), Total Dissolved Solids (TDS), pH, Total Hardness (TH), and Fluoride (F) have a total variance of 29.33% and a strong positive correlation with Cd and Cr. The PCA at Ranipet shows that Cd and Cr positively correlate with Magnesium (Mg^{2+}), Chloride (Cl^-), TH, and SO_4^{2-}, which comprise 36.64% of the total variance. EC, Potassium (K^+), TH, and Sulphate (SO_4^{2-}) positively correlate with Cd and Cr with a total variance of 27.76% at Vaniyambadi whereas in Ambur EC, TDS, pH, TH, and F positively correlate with 29.33% cumulative variance. The Principal Component Analysis (PCA) results at all three industrial corridors manifest that hydrogeochemistry is impacted by anthropogenic and geogenic forces. According to the correlation analysis result,

K. Loganathan · L. Vignesh Rajkumar (✉)
Department of Environmental and Water Resources Engineering, School of Civil Engineering, Vellore Institute of Technology, Vellore, Tamil Nadu, India
e-mail: vigneshrajkumar.l@vit.ac.in

A. Nanda et al. (eds.), *Navigating the Nexus*, Water Science and Technology Library 102, https://doi.org/10.1007/978-3-031-76532-2_13

the uninformed discharge of household waste, tannery effluent, e-waste burning, and agricultural waste in an open environment or water bodies significantly contributes to groundwater contamination by heavy metals. So, the groundwater in the industrial corridors is unsuitable for drinking, irrigation, and residential uses. Health risk assessment shows Cr and Cd significantly impact infants and children for carcinogenic risk (> 75%), while adults and children have non-carcinogenic impacts (> 65%). This study suggests that it is essential to assess the quality of wastewater discharged from industries to prevent groundwater pollution. The study revealed that Ranipet residents have the highest health risk impacts from groundwater contamination through uninformed industrial wastewater discharge. However, heavy metals (Cd and Cr) have a greater impact in Ranipet than Ambur and Vaniyambadi. This study serves as a baseline information for water-resources governing bodies to take the necessary steps to clean up the environment to prevent the spread of heavy metal-associated disorders through contaminated groundwater.

Keywords Contamination factor · Human health risks · Environmental impacts · Hydrogeochemistry · Heavy metals contamination · Multivariate statistical analysis

13.1 Introduction

Water is an important natural resource on the earth because it empowers the life of all species. Humans, animals, and all other living creations require an adequate level of water quantity and quality to sustain life in the world. In the past few decades, surface water contamination has drastically increased, so people are depending on groundwater as their primary drinking source (Vijayakumar et al. 2022). Water is regarded as one of the most valuable and vital natural resources. Still, it has been degraded globally, and several researchers recently reported contamination of drinking water due to various toxic metals (Krishan et al. 2021). Groundwater is a significant water source for irrigation, drinking, and industrial uses in Southern India, approximately 85% of rural and 50% of urban water criteria are still heavily reliant on groundwater supplies, with agriculture being the most significant usage proportion, preceded by industrialization and residential managements (Tyagi and Sarma 2021). Due to the expansion of cities and industries, human exposure to heavy metals has rapidly increased. Natural factors like soil erosion, rock weathering, rainfall, and anthropogenic activities like disposal of industrial and municipal waste are the major sources of heavy metals (Kazemi Moghaddam et al. 2022). Groundwater is becoming contaminated with heavy metals with the utilization of untreated industrial and municipal effluent in agricultural fields that poses a threat of carcinogenic health issues (Iqbal et al. 2022). The agricultural activities, as well as the extensive usage of agrochemicals, are also contributing more to the impacts on the alteration of groundwater hydrochemistry (Xu et al. 2023). The residuals from agricultural and industrial effluents greatly affect the groundwater characteristics through infiltration (Grema et al. 2022). Further, usage of indiscriminate agrochemicals, improper waste disposal in landfills, and removal of untreated or partially treated sewage, industrial, and domestic effluents have made groundwater unfit for drinking and agriculture purposes (Uyo et al. 2021). The next element in groundwater contamination is improper municipal solid waste handling practices (Dritsas and Trigka 2023). The release of leachate, which can contaminate the surrounding environment, including groundwater, is one of the consequences of landfill systems (Egbueri et al. 2023). The abandoned landfills revealed the possibility of environmental pollution and groundwater contamination (Afolabi et al. 2022). Correspondingly, municipal and hospital waste disposal/dumping contributes to the contamination of groundwater aquifers (Hafiz et al. 2022). Furthermore, the mixing of municipal waste, industrial waste, and agricultural waste into the river affects the quality of water resources (Satish Kumar et al. 2016). Because of its penetration through the soil, industrial effluent has increased contamination in groundwater (Thenmozhi et al. 2020). A large amount of untreated and unsorted municipal waste contributes significantly to heavy metal contamination of groundwater (Afolabi et al. 2022). Due to low living standards, residents of rural areas rely on tube wells and nearby springs for water, which may be a leading cause of their health and illness problems (Qasemi et al. 2019). Due to high toxicity and stability, heavy metals are one of the most severe and harmful environmental contaminants (Mohammadi et al. 2018). Groundwater contaminated with heavy metals can cause various health issues, so drinking it without proper treatment is not advisable (Jain et al. 2021). This toxic element is commonly and primarily released into the environment by burning Cd-containing materials, possibly polluting both surface and groundwater (Lucon et al. 2020). Heavy metals Pb and Cd are primarily introduced into groundwater through agricultural and industrial activities and landfilling, mining, and transportation (Balli and Leghouchi 2018). The use of cosmetics is one way that people are exposed to heavy metals. Since ancient times to the modern era, Pb and Cd pollution due to cosmetic products has been found frequently (Ghaderpoori et al. 2020). Cd is quickly mobilized due to its low sorption affinity compared to other heavy metals, potentially causing higher Cd intensities in groundwater (Kubier et al. 2020). Cd is excreted slowly, primarily in the kidneys and liver, making them the most vulnerable organs to chronic poisoning (Biela et al. 2019). One another most dangerous heavy metal is Cr, utilized in various standard industrial processes like dyeing, tanning leather, electrolysis, porcelain, pyrotechnic, electronic parts, and woodworking (Andrade et al. 2022). Cr heavy metal interferes with cellular processes such as growth, proliferation, differentiation, damage-repair, and apoptosis (Filippini et al. 2022). Toxic metal can also cause epigenetic changes, which can alter gene expression (Balali-Mood et al. 2021). Because of its high water solubility, oxidation potential, and ability to enter cell membranes, Cr(VI) is toxic in biological systems (Perraki et al. 2021). Most tanning industrial waste is easily transported into wastewater containing organic and inorganic contaminants. Organic species such as phenol and metal ions such as Cr(VI), Cd(II), are toxic to humans, either carcinogenic or impacting the physical-motor of the body (Othmani et al. 2021). The integrated Vellore district is one of the major industrial sectors in Tamil Nadu, especially for leather, automotive, explosives, surgical tools, and service sectors. In recent decades, industrial wastewater has been discharged into the open environment without proper treatment, which significantly causes groundwater contamination and poses a huge threat to human health (Manoj et al. 2021). In this context, the groundwater quality, human health risks, and heavy metal impacts have not been extensively studied in the industrial corridors of integrated Vellore district. There is no research that compares the substantial impact on groundwater quality in and around the industrial corridors due to anthropogenic activities in northern Tamil Nadu. So there is an urgent need for an assessment of groundwater quality in the integrated Vellore district due to anthropogenic activities. This study serves as a preliminary study to solve problems brought on by poor groundwater quality in the industrial corridors. Implementation of effective management strategies in industrial corridors should help raise the standard of living for residents. In addition, policy-makers and farmers might consider it beneficial to comprehend the consequences of groundwater heavy metal pollution in the study regions while developing corrective measures.

13.2 Study Area Description

The industrial corridors of the northern region of Tamil Nadu are selected as the study area, namely Ambur, Vaniyambadi, and Ranipet, as depicted in Fig. 13.1. The river Palar and groundwater are the major sources of drinking water supplies in all three industrial corridors. Ambur and Vaniyambadi have small-scale enterprises in the northern region of Tamil Nadu, which experiences hot summers due to its tropical wet-and-dry environment and rainy winters. Ambur and Vaniyambadi, located in the Eastern Ghats, are significant areas for the leather industry, focusing on leather tanning and goods production. Ranipet is a major industrial hub in southern India and it houses around 400 leather production, processing, and related firms. The Ranipet industrial corridors' primary businesses are leather tanning and creating leather goods such as shoes, gloves, and outerwear. These three industrial corridors have been contaminating groundwater resources rapidly by improperly disposing of the industry's effluents for the past few decades. This has posed human health risks and is also hazardous to the environment.

13.3 Materials and Methods

13.3.1 Sample Collection

This study was carried out by considering three industrial zones: Ambur, Ranipet, and Vaniyambadi, in the northern part of Tamil Nadu. To evaluate groundwater quality exclusively in detail, 14 groundwater samples were collected from each zone during January and February 2022 from bore and open wells in and around the tanneries

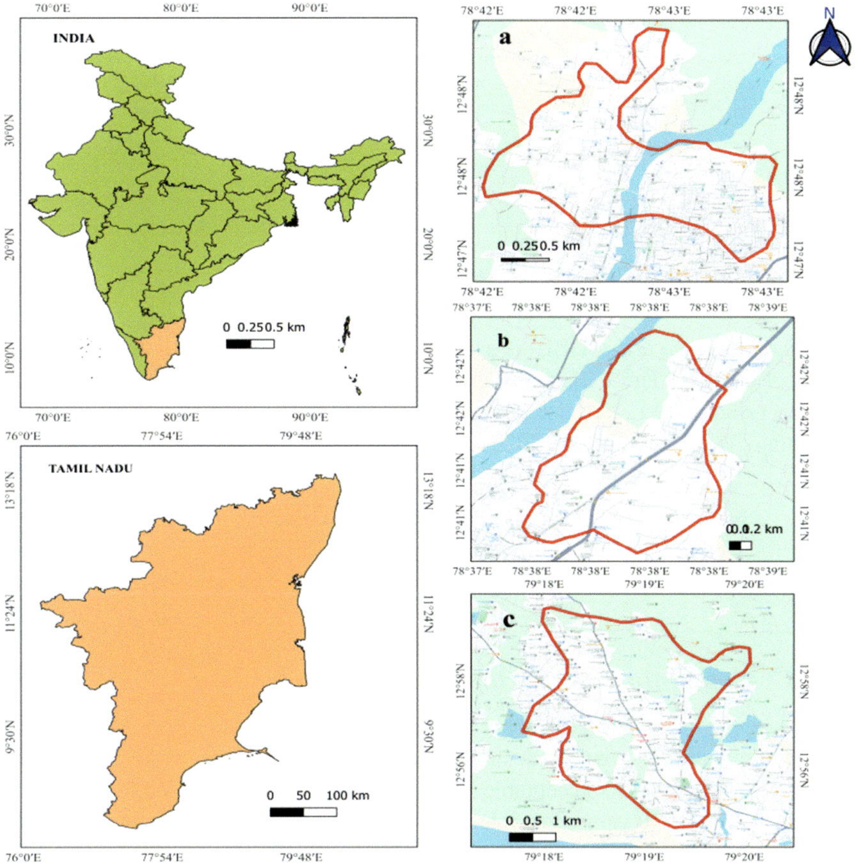

Fig. 13.1 Industrial corridors of northern part of Tamil Nadu **a** Ambur **b** Vaniyambadi **c** Ranipet

and industries. Ambur had 8 open wells and 6 bore wells; Ranipet had 9 open wells and 5 bore wells; and Vaniyambadi had 7 open wells and 7 bore wells. Freshwater samples were collected in 500-ml bottles thoroughly pre-cleansed with distilled water to protect them from contamination before the sample collection. The collected groundwater samples are immediately transported to the laboratory, where they are quantified for physicochemical parameters and heavy metals contamination. All the basic parameters of groundwater properties, pH, Total Dissolved Solids (TDS), Electrical Conductivity (EC), Sodium (Na^+), Potassium (K^+), Calcium (Ca^{2+}), Sulphate (SO_4^{2-}), Chloride (Cl^-), carbonates (CO_3^-), bicarbonates (HCO_3^-), Magnesium (Mg^{2+}), and Total hardness (TH) were determined by standard Laboratory methods (BIS 2012). Groundwater heavy metals such as Cd, Cr, Fe, Pb, and Co were quantified by an Atomic absorption spectrophotometer (AAS).

Table 13.1 CF classification

Sl. No.	Scale	Contamination class
1	< 1	Low contamination
2	1 < CF < 3	Moderate contamination
3	3 < CF < 6	Considerable contamination
4	> 6	High contamination

13.3.2 Charge Balance Error

The charge balance error (CBE) is used to confirm that the groundwater has neutral electrical charges. The CBE was calculated using Eq. 13.1. The CBE is less than 10% for groundwater, so the results were accurate for groundwater samples in the study area (Vijayakumar et al. 2022).

$$\%CBE = \frac{\sum \text{Cations} - \sum \text{Anions}}{\sum \text{Cations} + \sum \text{Anions}} \tag{13.1}$$

13.3.3 Contamination Factor

The Contamination factor (CF) is the proportion between the quantity of heavy metal in a water sample and its acceptable limit concentration as presented in Table 13.1 (Shen et al. 2019). CF values less than 1 (< 1) as low contamination, between 1 and 3 as moderate contamination (1 < CF < 3), Considerable contamination if the CF value falls between 3 and 6 (3 < CF < 6), CF values more than six as high contamination (CF > 6). CF is calculated using the expression provided in Eq. 13.2.

$$CF = \frac{\text{samples concentration mean}}{\text{WHO acceptable limit}} \tag{13.2}$$

13.3.4 Spatial Distribution Map Preparation—GIS

Spatial interpolation is a technique for estimating unknown point values using known point values. This study uses the inverse distance weight (IDW) method to prepare an interpolation map using QGIS v3.28.3 software. IDW interpolation assigns a value to sample points, minimizing the impact of one point on the other as the location of the point is estimated.

13.3.5 Multivariate Statistical Analysis

PCA is a method for determining the dynamics of all variables in an understudied system. PCA aims to reduce the dimensionality of a multivariate dataset by separating information into a small number of critical components that resemble common environmental factors. As a result, by removing data omissions, the vital principal features include all essential data (Raheja et al. 2023). This study identifies the relationship between or among heavy metal using Pearson's correlation. This study considered five heavy metals: Cr, Cd, Fe, Pb, and Co. The positive correlation among the parameters indicates a strong relationship between them and a level of dependence. If not, the relationship is inappropriate or the dependency between the parameters does not exist. The groundwater characteristics and their impacts in all three study regions are assessed using multivariate statistical analysis using the mean, minimum, maximum, and standard deviation (Pundir et al. 2022).

13.3.6 Health Risk Assessment

This study conducted a health risk assessment for infants, children, and adults based on USEPA guidelines. The health risk assessment was carried out only considering the ingestion pathway. Chronic Daily Intake (CDI) using Eq. 13.3, Hazard Quotient (HQ) using Eq. 13.4 and Incremental Lifetime Cancer Risk (ILCR) using Eq. 13.5 were used for the calculation of health risk for the heavy metals (Cr and Cd).

$$CDI_{Ing} = \frac{\mathrm{C_w \times DI \times F \times EP}}{\mathrm{BW \times AT}} \tag{13.3}$$

$$HQ_{Ing} = \frac{CDI_{Ing}}{RfD_{Ing}} \tag{13.4}$$

$$ILCR = (CDI * CSF) \tag{13.5}$$

where, C_w—heavy metal concentration in groundwater (mg/l), DI—Dosage intake (L/day), F—frequency duration (Days/year), EP—Exposure frequency (years), B_W—body weight (kg), AT—average time (Days).

13.4 Result and Discussions

13.4.1 t-Tests Analysis of Heavy Metals

Five toxic heavy metals are considered for the groundwater quality assessment: Cr, Cd, Fe, Pb, and Co. Among the five heavy metals, Cd and Cr are significantly contaminating the groundwater of the Vaniyambadi. The WHO 2011 recommendation limits for Cd and Cr concentrations are 0.003 mg/l and 0.05 mg/l; Fe, Pb, and Co are within the permissible limit of WHO, whereas Cd (max. 0.499 mg/l) and Cr (max. 0.160 mg/l) exceed the permissible limit of WHO. The sample mean, standard deviation, Degree of freedom, and t-static value for Cd and Cr were calculated for the Ambur, Ranipet, and Vaniyambadi industrial corridors, as represented in Tables 13.2 and 13.3. Tables 13.2 and 13.3 show that the Cd and Cr values range from 2.13, 3.5, 4.06, and 3.12, 2.3, and 3.1 (mg/l) at a significance level less than $p < 0.05$. According to the t-test results, the mean values of heavy metal Cd and Cr concentrations are very high. Cd and Cr are carcinogenic elements, and all three tannery industrial corridors significantly impact the groundwater quality with these carcinogenic elements, i.e. Vaniyambadi and Ambur industrial corridors are primarily influenced by Cd contamination.

In contrast, Ranipet and Ambur are affected by Cr contamination. Hexavalent Cr is the most common toxic heavy metal released from tannery wastewater mixed with surface and groundwater resources (Singh 2016).

Since heavy metal exposure is a significant external component, reducing it through drinking water will lower people's chances of developing cancer (Iqbal et al. 2022). Tables 13.4, 13.5, and 13.6 present the sample concentration, mean, standard deviation, and standard error for Ambur, Ranipet, and Vaniyambadi, respectively, for

Table 13.2 t-test for Cd

Group	Sample mean (x)	Standard deviation	Degree of freedom	t-static value	WHO limit of Cd (mg/l)
Ambur	0.064	0.113	13	2.13	0.003
Ranipet	0.02	0.021	13	3.50	0.003
Vaniyambadi	0.16	0.143	13	4.06	0.003

Table 13.3 t-test for Cr

Group	Sample mean (x)	Standard deviation	Degree of freedom	t-static value	WHO limit of Cr (mg/l)
Ambur	0.061	0.0714	13	3.20	0.05
Ranipet	0.091	0.15	13	2.29	0.05
Vaniyambadi	0.053	0.065	13	3.061	0.05

the heavy metals Cr and Cd. According to WHO permissible limits, all three industrial corridor locations have high concentrations of the toxic heavy metals Cd and Cr. Environmental agencies and governing authorities have to identify the type, level, extent, and sources of heavy metal pollution to reduce heavy metal contamination and human health risks (Qasemi et al. 2019). The Cr and Cd contamination order is Ranipet > Ambur > Vaniyambadi and Vaniyambadi > Ambur > Ranipet based on this study. In India, open-environment garbage collectors are highly exposed to heavy metals, as heavy metals are found in large quantities in household waste, office waste, and industrial waste. Low-skilled workers need more knowledge of electronic waste handling and disposal. Further, the heavy metal reaches the groundwater through leaching and contaminates the groundwater. So it is suggested that the residents in the industrial corridors (Ambur, Ranipet, and Vaniyambadi) take more precautionary measures to safeguard their health as well as groundwater quality because these areas are more industrialized and densely populated. Ranipet industrial corridors are those of leather tanneries, electronic industries, ultramarine and pigment industries, chemical industries, metal fabricators, etc. The improper disposal of wastewater from these industries causes Cd leaching into groundwater; the sources of Cd leaching are electronic chips and cathode ray tubes, among other electronic goods. A massive proportion of Cr and Cd ions from broken glass and electronic waste may mix with water and become polluted; this scenario is widespread in landfills as well. As a result, extremely poisonous by-products spread into the groundwater, impacting both local environments and global scales (Sivakumar et al. 2012).

Table 13.4 Distribution of Cd and Cr in Ambur (TN), India

Sample number	Cr (mg/l)	Mean ± S.D.	Std. error	Cd (mg/l)	Mean ± S.D.	Std. error
1	0.002	0.061 ± 0.071	0.019	0	0.064 ± 0.113	0.03
2	0.170			0.141		
3	0			0.002		
4	0			0.025		
5	0.080			0.014		
6	0.004			0.311		
7	0.200			0		
8	0.110			0.001		
9	0.050			0.020		
10	0.090			0.007		
11	0.140			0.043		
12	0			0.001		
13	0.001			0.320		
14	0.010			0.010		

Table 13.5 Distribution of Cd and Cr in Ranipet (TN), India

Sample number	Cr (mg/l)	Mean ± S.D.	Std. error	Cd (mg/l)	Mean ± S.D.	Std. error
1	0.06	0.091 ± 0.15	0.04	0.01	0.02 ± 0.02	0.006
2	0.55			0.06		
3	0			0		
4	0.01			0.03		
5	0.22			0.02		
6	0.06			0.02		
7	0.19			0		
8	0.10			0.02		
9	0.01			0.01		
10	0.06			0.01		
11	0			0.07		
12	0.02			0.01		
13	0			0.01		
14	0			0		

Table 13.6 Distribution of Cd and Cr in Vaniyambadi (TN), India

Sample number	Cr (mg/l)	Mean ± S.D.	Std. error	Cd (mg/l)	Mean ± S.D.	Std. error
1	0.06	0.091 ± 0.15	0.04	0.01	0.02 ± 0.02	0.006
2	0.55			0.06		
3	0			0		
4	0.01			0.03		
5	0.22			0.02		
6	0.06			0.02		
7	0.19			0		
8	0.10			0.02		
9	0.01			0.01		
10	0.06			0.01		
11	0			0.07		
12	0.02			0.01		
13	0			0.01		
14	0			0		

13.4.2 Spatial Analysis of Cd and Cr

The spatial distribution of Cd and Cr for Ambur is shown in Fig. 13.2a, b. According to interpolation results, Cr concentration exceeds the WHO limits (0.05 mg/l) and significantly affects the groundwater quality. According to CF analysis, 42% of

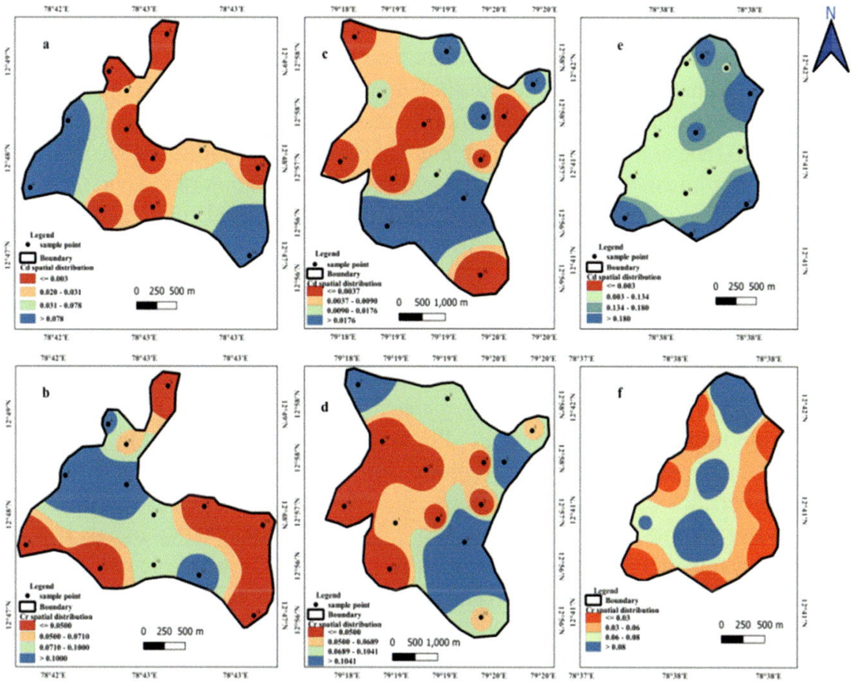

Fig. 13.2 Spatial distribution of heavy metals **a** Cd-Ambur **b** Cr-Ambur **c** Cd-Ranipet **d** Cr-Ranipet **e** Cd-Vaniyambadi **f** Cr-Vaniyambadi

samples fall under the low contamination classification, 28% fall under moderate contamination, and 14% fall under considerable contamination. Based on interpolation results, it is observed that wastewater from the Ambur industrial corridor is severely impacting the groundwater quality.

The WHO permissible limit for Cd is 0.003 mg/l; according to the spatial analysis result of Ambur, the Cd concentration of all the groundwater samples ($n = 14$) is beyond the WHO permissible limit. The mean and standard deviation of the Cd and Cr samples in the Ambur industrial corridor groundwater samples are shown in Table 13.4 as 0.064, 0.113 (mg/l), and 0.061, 0.071 (mg/l), respectively. Based on the CF results, 43% of samples are in the low contamination, 21% are classified as moderate, 15% as considerable, and 21% as high. However, the Ambur industrial corridor polluted by Cr and Cd is a significant environmental threat, and these heavy metals are contaminating surface and groundwater sources. The interpolation map for Cd and Cr for the Ranipet industrial corridor is shown in Fig. 13.2c, d, respectively. Totally 14 samples were collected in the Ranipet industrial corridor; 7 samples are beyond the permissible limit of the WHO. The mean and standard deviation for groundwater samples of the Ranipet industrial corridor for Cd (0.02 and 0.021) and Cr (0.091 and 0.15) are presented in Table 13.5. In the Ranipet industrial corridor, the tannery industries constitute a significant industry discharging their wastewater without proper treatment. Results indicate that tannery industrial wastewater mixed with groundwater resources causes human health risks (Srinivasa Gowd and Govil 2008). The Ranipet industrial corridor's Cd and Cr exceed the WHO permissible limits, with the minimum and maximum concentrations of Cd being 0.009 mg/l and 0.5 mg/l, respectively. Cr minimum and maximum concentrations are 0.01 mg/l and 0.55 mg/l, respectively. The mean and standard deviation of Cd and Cr in the Vaniyambadi industrial corridor are 0.02 mg/l, 0.021 mg/l, 0.091 mg/l, and 0.15 mg/l, respectively, as presented in Table 13.6.

The Vaniyambadi industrial corridor spatial distribution results reveal that Vaniyambadi groundwater is contaminated by Cd and Cr toxic heavy metals. The interpolation map for Cd and Cr for the Vaniyambadi industrial corridor is shown in Fig. 13.2e, f, respectively. However, the study area's groundwater quality was affected by two primary toxic heavy metals, Cd and Cr (which exceeded WHO permissible limits); the remaining Fe, Pb, and Co heavy metals had no impact on groundwater quality as the concentrations were within the permissible limit. The industrial corridor of the Ambur, Ranipet, and Vaniyambadi CF was calculated for groundwater samples. CF is shown in Fig. 13.3 for the toxic heavy metals Cd and Cr of Vaniyambadi, Ambur, and Ranipet, respectively. CF was calculated for the heavy metals Cr, Cd, Fe, Pb, and Co. Figure 13.3 shows only Cr and Cd concentrations because other heavy metal concentrations fall within WHO recommendations. Cd and Cr are major contributing contaminants to the groundwater at the northern part of Tamil Nadu. Among the three corridors, Cd is highly contaminated in the industrial corridor of Vaniyambadi groundwater, with a CF of 80%. The Cd order is based on CF: Vaniyambadi > Ambur > Ranipet. Hence, high Cd concentrations pose a considerable threat to the groundwater quality and environment because they are carcinogens and cause many human health disorders.

13.4.3 Statistical Analysis of Heavy Metals

Pearson's correlation method was used to identify the relationship among the heavy metal concentrations for Cr, Cd, Co, Pb, and Fe. According to the correlation analysis, there was a strong positive correlation (Table 13.7) between the heavy metals Cr, Cd, and Co in the samples from Ambur. The positive correlation among heavy metals is likely due to the tannery effluent discharge into an open area. The correlation matrix for Ranipet groundwater samples of 15 geochemistry and heavy metals is shown in Table 13.8. Ca^{2+}, Mg^{2+}, CO_3^-, and HCO_3^- had a positive correlation, indicating that the hardness of the groundwater has increased due to Na^+, Mg^{2+}, and carbonates high concentrations in the groundwater. In particular, in the Ranipet industrial corridor, all five heavy metals have a solid positive correlation. All heavy metals (Cr, Cd, Fe, Pb, and Co) correlation coefficients are considerably high. The correlation results show that anthropogenic activities contribute significantly to groundwater quality by improperly discharging agricultural waste, tannery effluent, and domestic waste into open environments or water bodies. TDS improves water's capacity to carry positive

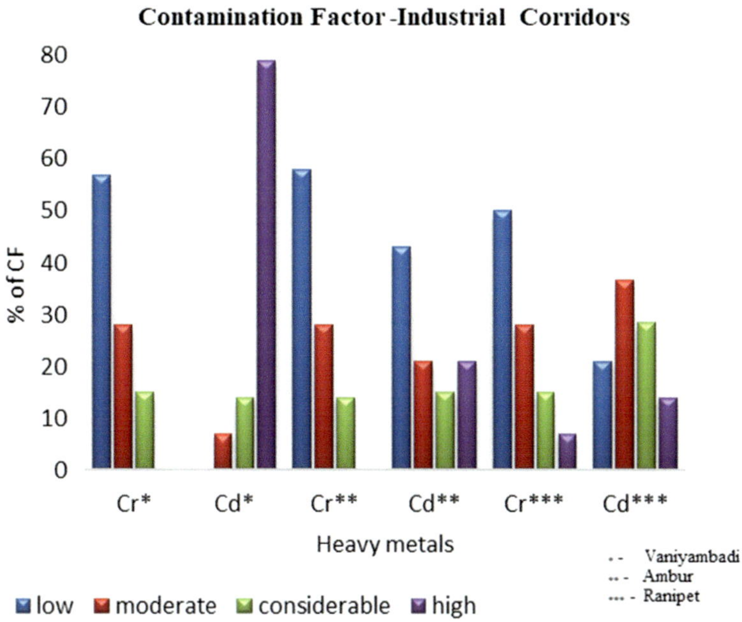

Fig. 13.3 CF representation for industry corridors of Vaniyambadi, Ambur and Ranipet

and negative ions, establishing a relationship between EC and water conductivity. In groundwater samples, a positive correlation exists between EC with other ions, notably Ca^{2+}, Cd, and Co, as shown in Table 13.9. This positive correlation indicates that TDS has influenced the hydrochemistry of the Vaniyambadi, as presented in Table 13.9.

Total hardness also exhibits a strong positive correlation with Cr and Fe. Hence, it is evident that they have a positive relationship because Ca^{2+} and Mg^{2+} carbonates and bicarbonates cause high hardness in the groundwater. The SO_4^{2-} is highly correlated with Cl^-, K^+, and Cd, showing that sewage was observed flowing through open drains, leading to groundwater contamination in the study area. Among the three locations, the impact of Cr is found to be very high at Ambur and Ranipet, whereas Vaniyambadi industrial corridors have a lower impact of Cr. The Cd is another heavy metal that causes a considerable impact in all three study locations. Therefore, the residents of industrial corridors have the potential to cause carcinogenic effects by ingesting heavy metals such as Cr and Cd through contaminated groundwater.

13.4.4 Health Risk Assessment

The health risks are assessed for Cr and Cd considering the ingestion pathway for infants, children, and adults. According to the CDI and ILCR results, three industrial

Table 13.7 Correlation matrix—Ambur

	pH	EC	TDS	Ca	Mg	SO4	TH	F	Cl	K	Cr	Cd	Co	Pb	Fe
pH	1.00														
EC	− 0.40	1.00													
TDS	− 0.38	0.97	1.00												
Ca	0.37	− 0.11	− 0.11	1.00											
Mg	0.61	− 0.26	− 0.28	0.22	1.00										
SO_4	0.51	− 0.32	− 0.34	0.84	0.59	1.00									
TH	0.07	0.21	0.25	− 0.06	− 0.05	− 0.22	1.00								
F	− 0.13	− 0.10	− 0.17	0.09	− 0.27	0.03	0.19	1.00							
Cl	0.59	0.39	− 0.43	0.50	0.89	0.78	− 0.04	− 0.19	1.00						
K	0.06	0.32	0.32	0.26	0.38	0.38	0.42	− 0.30	0.43	1.00					
Cr	− 0.40	0.39	0.32	− 0.39	− 0.31	− 0.43	− 0.03	− 0.16	− 0.33	− 0.08	1.00				
Cd	0.62	− 0.29	− 0.22	− 0.16	0.21	0.05	0.02	− 0.02	0.00	− 0.12	− 0.20	1.00			
Co	− 0.12	− 0.16	− 0.23	0.27	− 0.09	0.27	− 0.21	− 0.05	0.17	0.07	0.51	− 0.41	1.00		
Pb	− 0.01	0.25	0.26	0.04	− 0.02	− 0.16	0.47	0.30	− 0.06	− 0.03	− 0.39	− 0.29	− 0.37	1.00	
Fe	0.38	0.16	0.20	0.31	0.24	0.18	− 0.11	− 0.31	0.24	− 0.02	− 0.15	− 0.17	0.01	0.15	1.00

Table 13.8 Correlation matrix—Ranipet

	pH	EC	TDS	Ca	Mg	SO4	TH	F	Cl	K	Cr	Cd	Co	Pb	Fe
pH	1.00														
EC	0.00	1.00													
TDS	0.03	0.96	1.00												
Ca	0.04	0.91	0.88	1.00											
Mg	0.11	0.95	0.97	0.84	1.00										
SO_4	0.26	0.06	0.05	0.32	0.00	1.00									
TH	0.01	0.89	0.88	0.86	0.89	0.12	1.00								
F	− 0.01	0.92	0.93	0.87	0.92	0.04	0.93	1.00							
Cl	0.03	− 0.28	− 0.26	− 0.32	− 0.30	0.13	− 0.29	− 0.31	1.00						
K	0.23	0.07	0.01	− 0.02	− 0.02	− 0.01	− 0.11	− 0.13	− 0.38	1.00					
Cr	0.20	− 0.01	0.12	− 0.12	0.09	− 0.30	− 0.07	0.00	− 0.08	0.21	1.00				
Cd	0.11	0.20	0.25	0.00	0.32	0.12	0.26	0.27	0.32	− 0.15	0.14	1.00			
Co	− 0.26	− 0.05	0.03	− 0.14	− 0.01	− 0.12	0.17	0.04	0.33	− 0.22	0.23	0.43	1.00		
Pb	0.11	0.24	0.24	0.10	0.34	0.05	0.20	0.12	0.27	− 0.06	0.05	0.71	0.20	1.00	
Fe	0.13	0.12	0.16	0.03	0.23	0.31	0.36	0.24	0.11	0.03	− 0.01	0.79	0.34	0.49	1.00

Table 13.9 Correlation matrix—Vaniyambadi

	pH	EC	TDS	Ca	Mg	SO4	TH	F	Cl	K	Cr	Cd	Co	Pb	Fe
pH	1.00														
EC	− 0.40	1.00													
TDS	− 0.38	0.97	1.00												
Ca	0.37	− 0.11	− 0.22	1.00											
Mg	0.51	− 0.26	− 0.28	0.22	1.00										
SO_4	0.51	0.32	− 0.34	0.74	0.59	1.00									
TH	0.07	0.21	0.25	− 0.06	− 0.05	− 0.22	1.00								
F	− 0.13	0.10	− 0.17	0.09	− 0.27	0.03	0.19	1.00							
Cl	0.59	0.39	− 0.43	0.50	0.77	0.78	− 0.04	− 0.19	1.00						
K	0.06	0.32	0.32	0.26	0.48	0.52	0.36	− 0.40	0.43	1.00					
Cr	− 0.40	0.39	0.32	− 0.39	− 0.31	− 0.43	− 0.03	− 0.16	− 0.33	− 0.08	1.00				
Cd	0.62	− 0.29	− 0.22	− 0.16	0.21	0.05	0.02	− 0.02	0.00	− 0.12	− 0.20	1.00			
Co	− 0.12	− 0.16	− 0.23	0.17	− 0.09	0.27	− 0.21	− 0.05	0.17	0.07	0.51	− 0.41	1.00		
Pb	− 0.01	0.25	0.26	0.04	− 0.02	− 0.16	0.47	0.30	− 0.06	− 0.23	− 0.39	− 0.49	− 0.47	1.00	
Fe	0.38	0.16	0.20	0.31	0.24	0.18	− 0.11	− 0.31	0.24	− 0.02	− 0.25	− 0.17	0.01	0.25	1.00

corridors are exposed to carcinogenic impacts as the results are higher than the USEPA recommendation limit, i.e., less than 1×10^{-6} is considered non-carcinogenic and above 1×10^{-6} is considered carcinogenic (USEPA 2004). The infant, child, and adult age groups are likely to have carcinogenic risks. Children and infants in the Ranipet region have a more significant impact than adults in the Ambur and Vaniyambadi regions, according to the CDI result as presented in Table 13.7.

All three age groups are likely to have carcinogenic risks, with children and infants in the Ranipet industrial corridors having a higher impact based on the CDI and ILCR results as presented in Tables 13.10 and 13.11. According to the HQ results, 78% of Cr heavy metals in the Ambur region are more affected by infant, child, and adult age groups. The Cd has a greater impact on the child and infant age groups (64%). The Ranipet region has impacted 71% of the child, infant, and adult age groups by Cr and 78% of the child and infant age groups. According to HQ results for Vaniyambadi, the child age group had more impact than the infant and adult age groups. In the case of Cd, infants and children have a high hazard compared to adult groups. According to the ILCR results, Cr and Cd cause high cancer impacts ($< 50\%$) for children and adults in the industrial corridors, as presented in Fig. 13.4.

13.4.5 Strength and Major Findings of the Study

- The highlight of this study is the comparison of the impact of heavy metal distribution in the three industrial corridors, its impacts on groundwater quality, and health risk assessment. The current research helps to identify the suitability of groundwater for both domestic and agricultural purposes. The uniqueness of the study is that there is no study carried out to compare the tannery industrial corridors in the northern part of Tamil Nadu and their significant impact on human health risks among the residents.
- The sampling locations considered in this study are in and around the industrial corridors to have more precise and accurate information about the impacts on groundwater quality. The results of investigations help identify the most contaminated industrial corridors in northern Tamil Nadu. Among the three industrial corridors, Vaniyambadi had low contamination, Ambur had moderate contamination, and Ranipet had very high contamination based on the concentrations of heavy metals such as Cr and Cd.
- In this study, three industrial corridors were assessed for groundwater quality and heavy metal impacts. In the Ranipet industrial corridor, there are 256 functioning tanneries available, out of which less than 100 are treating their wastewater through common effluent treatment plant regularly. Most of the tanneries discharge their wastewater into the open environment. So there is a higher possibility of heavy metals being present in groundwater due to the uninformed industrial wastewater discharge.
- Ambur has a considerably smaller number of tanneries, and the common effluent treatment plant does better wastewater treatment processing compared to Ranipet

Table 13.10 Health risk assessment of Cr and Cd (CDI)

	Ambur (CDI-Cr)			Ambur (CDI-Cd)			Ranipet (CDI-Cr)		
	Adults	Children	Infants	Adults	Children	Infants	Adults	Children	Infants
1	5.71E−05	1.50E−04	1.60E−04	0.00E+00	0.00E+00	0.00E+00	1.71E−03	4.50E−03	4.80E−03
2	4.86E−03	1.28E−02	1.36E−02	4.03E−03	1.06E−02	1.13E−02	1.57E−02	4.13E−02	4.40E−02
3	0.00E+00	0.00E+00	0.00E+00	5.71E−05	1.50E−04	1.60E−04	0.00E+0	0.00E+00	0.00E+00
4	0.00E+00	0.00E+00	0.00E+00	7.14E−04	1.88E−03	2.00E−03	2.86E−04	7.50E−04	8.00E−04
5	2.29E−03	6.00E−03	6.40E−03	4.00E−04	1.05E−03	1.12E−03	6.29E−03	1.65E−02	1.76E−02
6	1.14E−04	3.00E−04	3.20E−04	8.89E−03	2.33E−02	2.49E−02	1.71E−03	4.50E−03	4.80E−03
7	5.71E−03	1.50E−02	1.60E−02	0.00E+00	0.00E+00	0.00E+00	5.43E−03	1.43E−02	1.52E−02
8	3.14E−03	8.25E−03	8.80E−03	2.86E−05	7.50E−05	8.00E−05	2.86E−03	7.50E−03	8.00E−03
9	1.43E−03	3.75E−03	4.00E−03	5.71E−04	1.50E−03	1.60E−03	2.86E−04	7.50E−04	8.00E−04
10	2.57E−03	6.75E−03	7.20E−03	2.00E−04	5.25E−04	5.60E−04	1.71E−03	4.50E−03	4.80E−03
11	4.00E−03	1.05E−02	1.12E−02	1.23E−03	3.23E−03	3.44E−03	0.00E+0	0.00E+00	0.00E+00
12	0.00E+00	0.00E+00	0.00E+00	2.86E−05	7.50E−05	8.00E−05	5.71E−04	1.50E−03	1.60E−03
13	2.86E−05	7.50E−05	8.00E−05	9.14E−03	2.40E−02	2.56E−02	0.00E+0	0.00E+00	0.00E+00
14	2.86E−04	7.50E−04	8.00E−04	2.86E−04	7.50E−04	8.00E−04	0.00E+0	0.00E+00	0.00E+00
	Ranipet (CDI-Cd)			Vaniyambadi (CDI-Cr)			Vaniyambadi (CDI-Cd)		
	Adults	Children	Infants	Adults	Children	Infants	Adults	Children	Infants
1	2.86E−04	7.50E−04	8.00E−04	0.00E+00	4.50E−03	0.00E+00	1.71E−03	4.50E−03	4.80E−03
2	1.71E−03	4.50E−03	4.80E−03	3.43E−03	1.58E−02	9.60E−03	6.00E−03	1.58E−02	1.68E−02
3	0.00E+00	0.00E+00	0.00E+00	5.14E−03	9.75E−03	1.44E−02	3.71E−03	9.75E−03	1.04E−02
4	8.57E−04	2.25E−03	2.40E−03	0.00E+00	3.38E−03	0.00E+00	1.29E−03	3.38E−03	3.60E−03
5	5.71E−04	1.50E−03	1.60E−03	0.00E+00	2.00E−02	0.00E+00	7.63E−03	2.00E−02	2.14E−02

(continued)

Table 13.10 (continued)

	Ranipet (CDI-Cd)			Vaniyambadi (CDI-Cr)			Vaniyambadi (CDI-Cd)		
	Adults	Children	Infants	Adults	Children	Infants	Adults	Children	Infants
6	5.71E−04	1.50E−03	1.60E−03	2.29E−03	6.75E−04	6.40E−03	2.57E−04	6.75E−04	7.20E−04
7	0.00E+00	0.00E+00	0.00E+00	3.71E−04	4.50E−03	1.04E−03	1.71E−03	4.50E−03	4.80E−03
8	5.71E−04	1.50E−03	1.60E−03	2.86E−05	1.57E−02	8.00E−05	5.97E−03	1.57E−02	1.67E−02
9	2.86E−04	7.50E−04	8.00E−04	1.71E−04	3.74E−02	4.80E−04	1.43E−02	3.74E−02	3.99E−02
10	2.86E−04	7.50E−04	8.00E−04	3.06E−03	1.79E−02	8.56E−03	6.80E−03	1.79E−02	1.90E−02
11	2.00E−03	5.25E−03	5.60E−03	0.00E+00	2.48E−02	0.00E+00	9.43E−03	2.48E−02	2.64E−02
12	2.86E−04	7.50E−04	8.00E−04	0.00E+00	5.93E−03	0.00E+00	2.26E−03	5.93E−03	6.32E−03
13	2.86E−04	7.50E−04	8.00E−04	4.57E−03	1.88E−03	1.28E−02	7.14E−04	1.88E−03	2.00E−03
14	0.00E+00	0.00E+00	0.00E+00	1.71E−03	1.50E−03	4.80E−03	5.71E−04	1.50E−03	1.60E−03

Table 13.11 Health risk assessment of Cr and Cd (ILCR)

	Ambur ILCR-Cr			Ambur ILCR-Cd			Ranipet ILCR-Cr		
	Adults	Children	Infants	Adults	Children	Infants	Adults	Children	Infants
1	5.71E−05	1.50E−04	1.60E−04	0.00E+00	0.00E+00	0.00E+00	1.71E−03	4.50E−03	4.80E−03
2	4.86E−03	1.28E−02	1.36E−02	4.03E−03	1.06E−02	1.13E−02	1.57E−02	4.13E−02	4.40E−02
3	0.00E+00	0.00E+00	0.00E+00	5.71E−05	1.50E−04	1.60E−04	0.00E+0	0.00E+00	0.00E+00
4	0.00E+00	0.00E+00	0.00E+00	7.14E−04	1.88E−03	2.00E−03	2.86E−04	7.50E−04	8.00E−04
5	2.29E−03	6.00E−03	6.40E−03	4.00E−04	1.05E−03	1.12E−03	6.29E−03	1.65E−02	1.76E−02
6	1.14E−04	3.00E−04	3.20E−04	8.89E−03	2.33E−02	2.49E−02	1.71E−03	4.50E−03	4.80E−03
7	5.71E−03	1.50E−02	1.60E−02	0.00E+00	0.00E+00	0.00E+00	5.43E−03	1.43E−02	1.52E−02
8	3.14E−03	8.25E−03	8.80E−03	2.86E−05	7.50E−05	8.00E−05	2.86E−03	7.50E−03	8.00E−03
9	1.43E−03	3.75E−03	4.00E−03	5.71E−04	1.50E−03	1.60E−03	2.86E−04	7.50E−04	8.00E−04
10	2.57E−03	6.75E−03	7.20E−03	2.00E−04	5.25E−04	5.60E−04	1.71E−03	4.50E−03	4.80E−03
11	4.00E−03	1.05E−02	1.12E−02	1.23E−03	3.23E−03	3.44E−03	0.00E+0	0.00E+00	0.00E+00
12	0.00E+00	0.00E+00	0.00E+00	2.86E−05	7.50E−05	8.00E−05	5.71E−04	1.50E−03	1.60E−03
13	2.86E−05	7.50E−05	8.00E−05	9.14E−03	2.40E−02	2.56E−02	0.00E+0	0.00E+00	0.00E+00
14	2.86E−04	7.50E−04	8.00E−04	2.86E−04	7.50E−04	8.00E−04	0.00E+0	0.00E+00	0.00E+00
	Ranipet ILCR-Cd			Vaniyambadi ILCR-Cr			Vaniyambadi ILCR-Cd		
	Adults	Children	Infants	Adults	Children	Infants	Adults	Children	Infants
1	2.86E−04	7.50E−04	8.00E−04	0.00E+00	4.50E−03	0.00E+00	1.71E−03	4.50E−03	4.80E−03
2	1.71E−03	4.50E−03	4.80E−03	3.43E−03	1.58E−02	9.60E−03	6.00E−03	1.58E−02	1.68E−02
3	0.00E+00	0.00E+00	0.00E+00	5.14E−03	9.75E−03	1.44E−02	3.71E−03	9.75E−03	1.04E−02
4	8.57E−04	2.25E−03	2.40E−03	0.00E+00	3.38E−03	0.00E+00	1.29E−03	3.38E−03	3.60E−03
5	5.71E−04	1.50E−03	1.60E−03	0.00E+00	2.00E−02	0.00E+00	7.63E−03	2.00E−02	2.14E−02

(continued)

Table 13.11 (continued)

	Ranipet ILCR-Cd			Vaniyambadi ILCR-Cr			Vaniyambadi ILCR-Cd		
	Adults	Children	Infants	Adults	Children	Infants	Adults	Children	Infants
6	5.71E−04	1.50E−03	1.60E−03	2.29E−03	6.75E−04	6.40E−03	2.57E−04	6.75E−04	7.20E−04
7	0.00E+00	0.00E+00	0.00E+00	3.71E−04	4.50E−03	1.04E−03	1.71E−03	4.50E−03	4.80E−03
8	5.71E−04	1.50E−03	1.60E−03	2.86E−05	1.57E−02	8.00E−05	5.97E−03	1.57E−02	1.67E−02
9	2.86E−04	7.50E−04	8.00E−04	1.71E−04	3.74E−02	4.80E−04	1.43E−02	3.74E−02	3.99E−02
10	2.86E−04	7.50E−04	8.00E−04	3.06E−03	1.79E−02	8.56E−03	6.80E−03	1.79E−02	1.90E−02
11	2.00E−03	5.25E−03	5.60E−03	0.00E+00	2.48E−02	0.00E+00	9.43E−03	2.48E−02	2.64E−02
12	2.86E−04	7.50E−04	8.00E−04	0.00E+00	5.93E−03	0.00E+00	2.26E−03	5.93E−03	6.32E−03
13	2.86E−04	7.50E−04	8.00E−04	4.57E−03	1.88E−03	1.28E−02	7.14E−04	1.88E−03	2.00E−03
14	0.00E+00	0.00E+00	0.00E+00	1.71E−03	1.50E−03	4.80E−03	5.71E−04	1.50E−03	1.60E−03

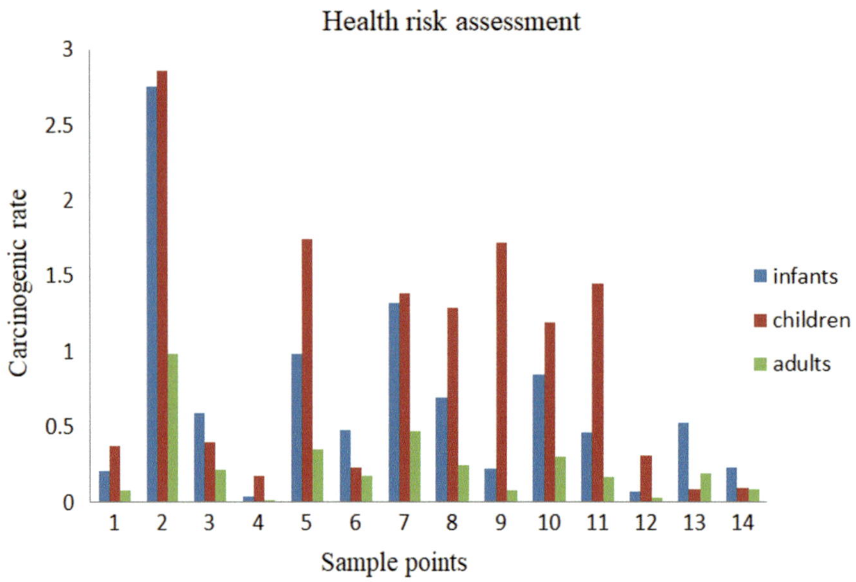

Fig. 13.4 Health risk assessment for age groups

and Vaniyambadi, as identified by the spatial distribution of heavy metals in respective corridors. Out of the three industrial corridors, heavy metals (Cd and Cr) had a greater impact and the possibility of spreading carcinogenic diseases in the Ranipet industrial corridor compared to the Ambur and Vaniyambadi industrial corridors.

- The tannery industrial wastewater discharge into an open environment causes groundwater contamination and poses a risk to the quality of the groundwater, which is not suitable for residential and agricultural purposes. Wastewater needs to be cleaned and tested before being discharged into aquatic systems. This health risk assessment can be used to address serious health hazards, provide appropriate treatment technologies, improve current regulatory standards, and also address the Sustainable Development Goal (SDG6): clean water and sanitation.

13.4.6 Limitations of the Study

The limitation of this study is that it considers only the dry season for the investigation of groundwater quality and more conclusive results can derived when both wet and dry seasons are analyzed. The study considered only ingestion as a pathway for human health risk assessment of Cr and Cd.

- This study assessed the integrated Vellore district industrial corridors with only 42 samples, more conclusive results will be obtained by increasing the number of samples in each industrial corridor.
- This study considered only the groundwater samples in and around the industrial corridors. Significant results can be explored when samples are collected far away from the industrial corridors; this assessment will be very helpful in studying the movement and behavior of contaminant.
- This study considered only the dry season samples; instead, a season wise (pre-monsoon, post-monsoon, and monsoon) assessment will yield precise results for understanding the spatial distribution of contaminants.
- Groundwater quality is very important in the integrated Vellore district because people rely on groundwater for all their needs due to non-perennial river flow and seasonal precipitation. So the seasonal trend analysis will give the heavy metal contamination scenario, which will help policymakers refine their existing policies to prevent further groundwater contamination by anthropogenic activities.

13.5 Conclusion

The industrial corridors of Ambur, Ranipet, and Vaniyambadi were studied to assess groundwater quality and heavy metal contamination considering five heavy metals: Cr, Cd, Fe, Pb, and Co. According to the t-test results, Cr and Cd pose the greatest threat to groundwater quality, implying that their levels are significantly higher than the WHO's permissible limits for Cd (0.003 mg/l) and Cr (0.05 mg/l), causing degradation of the groundwater quality in those study areas. A positive correlation among the heavy metals in the industrial corridor confirms that anthropogenic activities like untreated industrial wastewater, municipal waste, and agricultural waste are reasons for the high concentration of heavy metals in the groundwater, which causes human health risks. According to CF results, Cd and Cr had a significant negative influence on the environment in the study area, and the order of the heavy metals is Cd > Cr > Fe > Pb > Co. According to the CF investigation, the Cd contamination was higher in Vaniyambadi, Ambur, and Ranipet at 78%, 20%, and 15%, respectively. The most substantial Cr contamination rates were found in Ranipet, Vaniyambadi, and Ambur at 22%, 15%, and 15%, respectively. To prevent further heavy metal poisoning of the groundwater, Vaniyambadi and Ranipet need immediate attention. The PCA results show a strong positive correlation between EC, TDS, Ca^{2+}, Mg^{2+}, K^{+}, and Cl^{-} and the heavy metals Cr and Fe. These are likely human-made sources of untreated industrial effluent discharge. According to health risk assessment results, Cr-induced cancer risks among infant, child, and adult age groups are greater than 78%. The Cd-induced health risks are higher for infants and children, at more than 90% compared to adults. The CDI and ILCR results show that 78% of Cr heavy metals affect the infant, child, and adult age groups, while Cd has a more significant impact on child and infant age groups (about 64%). According to HQ results, the Ranipet region has a higher influence (92%) on children and infants, which is more than adults in the Ambur and Vaniyambadi regions. So the residents in industrial corridors must know the quality of the groundwater before utilizing it to protect them from health hazards. In Vaniyambadi, the child age group has more impact than the infant and adult age groups. The Ambur region impacts 78% of child, infant, and adult age groups with Cr, and 64% of infants and children have higher hazards than adults due to Cd. Groundwater contamination by heavy metals (Cr and Cd) impacts human health risks through physiological effects on the liver, intestines, respiratory system, nerves, and many organelles and systems within the body. According to the heavy metal spatial distribution analysis and health risk assessment, it is identified that there is a lack of industrial wastewater handling practices leading to groundwater contamination and human health risks. So the recommendation is that industries must connect their industrial wastewater to a common treatment plant maintained by the water-regulating authorities. Further, unsafe wastewater discharge should be avoided by common treatment plants for wastewater management. In addition, the industrial wastewater physicochemical composition should be verified before going to a common treatment plant to verify the type and concentration of contaminants in the industrial wastewater. The governing authority should check the wastewater discharge record at least once a month to verify the quality and quantity of wastewater. The refinement of existing water regulation policies should be modified based on the current scenario of industrial wastewater handling practices to safeguard groundwater sources and human health.

References

Afolabi OO et al (2022) Potential environmental pollution and human health risk assessment due to leachate contamination of groundwater from anthropogenic impacted site. Environ Challenges 9(October):100627. https://doi.org/10.1016/j.envc.2022.100627

Andrade DC, Đolić MB, Martínez-Huitle CA, dos Santos EV, Silva TFCV, Vilar VJP (2022) Coupling electrokinetic with a cork-based permeable reactive barrier to prevent groundwater pollution: A case study on hexavalent chromium-contaminated soil. Electrochim Acta 429(June). https://doi.org/10.1016/j.electacta.2022.140936

Balali-Mood M, Naseri K, Tahergorabi Z, Khazdair MR, Sadeghi M (2021) Toxic mechanisms of five heavy metals: mercury, lead, chromium, cadmium, and arsenic. Front Pharmacol 12 (April):1–19. https://doi.org/10.3389/fphar.2021.643972

Balli N, Leghouchi E (2018) Assessment of lead and cadmium in groundwater sources used for drinking purposes in jijel (Northeastern algeria). Glob Nest J 20(2):417–423. https://doi.org/10.30955/gnj.002747

Biela R, Kucera T, Konecny J (2019) Use of sorption materials for removing cadmium from water. IOP Conf Ser Earth Environ Sci 221(1). https://doi.org/10.1088/1755-1315/221/1/012133

BIS (2012) Indian standard drinking water specification (second revision). Bur Indian Stand IS 10500(May):1–11. [Online]. Available: http://cgwb.gov.in/Documents/WQ-standards.pdf

Dritsas E, Trigka M (2023) Efficient data-driven machine learning models for water quality prediction. Computation 11(2). https://doi.org/10.3390/computation11020016

Egbueri JC, Agbasi JC, Ayejoto DA, Khan MI, Khan MYA (2023) Extent of anthropogenic influence on groundwater quality and human health-related risks: an integrated assessment based on selected physicochemical characteristics. Geocarto Int 38(1). https://doi.org/10.1080/10106049.2023.2210100

Filippini T, Wise LA, Vinceti M (2022) Cadmium exposure and risk of diabetes and prediabetes: a systematic review and dose-response meta-analysis. Environ Int 158:106920. https://doi.org/10.1016/j.envint.2021.106920

Ghaderpoori M et al (2020) Health risk assessment of heavy metals in cosmetic products sold in Iran: the Monte Carlo simulation. Environ Sci Pollut Res 27(7):7588–7595. https://doi.org/10.1007/s11356-019-07423-w

Grema HM, Hamidu H, Suleiman A, Kankara AI, Umaru AO, Abdulmalik NF (2022) Cadmium geochemistry and groundwater pollution status evaluation using indexing and spatial analysis for Keffe community and Environs Sokoto Basin, North Western Nigeria. Niger J Basic Appl Sci 30(1):05–23. https://doi.org/10.4314/njbas.v30i1.2

Hafiz IZ, Singla S, Hafiz BZ (2022) Physio-chemical and heavy metal assessment of groundwater near dumping sites in Srinagar District, Jammu & Kashmir, India. Water Supply 22(8):7006–7013. https://doi.org/10.2166/ws.2022.236

Iqbal Z et al (2022) Human health risk of heavy metal contamination in groundwater and source apportionment. Int J Environ Sci Technol 19(8):7251–7260. https://doi.org/10.1007/s13762-021-03611-9

Jain CK, Sharma SK, Singh S (2021) Assessment of groundwater quality and determination of hydrochemical evolution of groundwater in Shillong, Meghalaya (India). SN Appl Sci 3(1):1–14. https://doi.org/10.1007/s42452-020-03993-4

Kazemi Moghaddam V et al (2022) Heavy metal contaminated soil, water, and vegetables in northeastern Iran: potential health risk factors. J Environ Heal Sci Eng 20(1):65–77. https://doi.org/10.1007/s40201-021-00756-0

Krishan G et al (2021) Occurrences of potentially toxic trace metals in groundwater of the state of Punjab in northern India. Groundw Sustain Dev 15. https://doi.org/10.1016/j.gsd.2021.100655

Kubier A, Hamer K, Pichler T (2020) Cadmium background levels in groundwater in an area dominated by agriculture. Integr Environ Assess Manage 16(1):103–113. https://doi.org/10.1002/ieam.4198

Lucon TN, Costa AT, Galvão P, Leite MGP (2020) Cadmium behavior in a karst environment hydrological cycle. Environ Sci Pollut Res 27(9):8965–8979. https://doi.org/10.1007/s11356-020-07894-2

Manoj S, RamyaPriya R, Elango L (2021) Long-term exposure to chromium contaminated waters and the associated human health risk in a highly contaminated industrialised region. Environ Sci Pollut Res 28(4):4276–4288. https://doi.org/10.1007/s11356-020-10762-8

Mohammadi AA, Yousefi M, Soltani J, Ahangar AG, Javan S (2018) Using the combined model of gamma test and neuro-fuzzy system for modeling and estimating lead bonds in reservoir sediments. Environ Sci Pollut Res 25(30):30315–30324. https://doi.org/10.1007/s11356-018-3026-7

Othmani A, Magdouli S, Senthil Kumar P, Kapoor A, Chellam PV, Gökkuş Ö (2022) Agricultural waste materials for adsorptive removal of phenols, chromium (VI) and cadmium (II) from wastewater: a review. Environ Res 204 (August 2021). https://doi.org/10.1016/j.envres.2021.111916

Perraki M, Vasileiou E, Bartzas G (2021) Tracing the origin of chromium in groundwater: current and new perspectives. Curr Opin Environ Sci Heal 22:100267. https://doi.org/10.1016/j.coesh.2021.100267

Pundir S, Kandari V, Singh P, Singh R, Siddique MA (2022) Assessment of groundwater quality of Dehradun Valley, Uttarakhand, India, for drinking purposes using multivariate statistical techniques, no 0123456789. Springer Netherlands. https://doi.org/10.1007/s10668-022-02784-8

Qasemi M et al (2019) Cadmium in groundwater consumed in the rural areas of Gonabad and Bajestan, Iran: occurrence and health risk assessment. Biol Trace Elem Res 192(2):106–115. https://doi.org/10.1007/s12011-019-1660-7

Raheja H, Goel A, Pal M (2023) Groundwater quality appraisal using IWQI and PCA for irrigation uses. ISH J Hydraul Eng 00(00):1–10. https://doi.org/10.1080/09715010.2023.2218829

Satish Kumar V, Amarender B, Dhakate R, Sankaran S, Raj Kumar K (2016) Assessment of groundwater quality for drinking and irrigation use in shallow hard rock aquifer of Pudunagaram, Palakkad District Kerala. Appl Water Sci 6(2):149–167. https://doi.org/10.1007/s13201-014-0214-6

Shen F, Mao L, Sun R, Du J, Tan Z, Ding M (2019) Contamination evaluation and source identification of heavy metals in the sediments from the lishui river watershed, southern China. Int J Environ Res Public Health 16(3). https://doi.org/10.3390/ijerph16030336

Singh SK (2019) Modeling of arsenic transport in groundwater using MODFLOW: a case study, October 2016

Sivakumar T, Vijayaraghavan G, Vimal-Kumar A (2012) Global challenges in E-waste management. Int J Adv Eng Technol 2(4):10–15

Srinivasa Gowd S, Govil PK (2008) Distribution of heavy metals in surface water of Ranipet industrial area in Tamil Nadu, India. Environ Monit Assess 136(1–3):197–207. https://doi.org/10.1007/s10661-007-9675-5

Thenmozhi M, Thangamani S, Jayabharathi J (2020) Groundwater quality: estimation of spatial variation using IDW spatial analyst in Cuddalore district, Tamil Nadu. 10(5):49–58

Tyagi S, Sarma K (2021) Seasonal variability, index modeling and spatiotemporal profiling of groundwater usability in semi-urban region of western Uttar Pradesh, India, vol 80, no 22. Springer, Berlin, Heidelberg. https://doi.org/10.1007/s12665-021-10018-9

USEPA (2014) Human health evaluation manual, supplemental guidance: update of standard default exposure factors. U.S. Environ Agency (USEPA) Risk Assess 2004 (July 2004):6. [Online]. Available: https://www.epa.gov/risk/update-standard-default-exposure-factors

Uyo C, Emereibeole E, Mgbeahuruike L, Irene EO (2021) Groundwater contamination from a municipal solid waste dumpsite: case study of Obosi, Southeastern Nigeria

Vijayakumar CR, Balasubramani DP, Azamathulla HM (2022) Assessment of groundwater quality and human health risk associated with chromium exposure in the industrial area of Ranipet, Tamil Nadu, India. J. Water Sanit. Hyg. Dev. 12(1):58–67. https://doi.org/10.2166/washdev.2021.260

Xu J et al (2023) Hydrochemistry, quality, and integrated health risk assessments of groundwater in the Huaibei Plain, China. Environ Sci Pollut Res 30(59):123466–123479. https://doi.org/10.1007/s11356-023-30966-y

Part IV
Advanced Tools and Techniques to Provide Climate Solutions

Chapter 14
Assessing the Impacts of Climate Change on Hydroclimatic Regimes in Beas River Basin

Siddik Barbhuiya, Salil Sharma, Ashish Pathania, and Vivek Gupta

Abstract Effective water resource management under climate change stress necessitates precise impact assessments. Addressing the coarse spatial resolution limitations of General Circulation Models (GCMs), this study presents an advanced regional climate modelling approach, enhancing GCM outputs to a finer resolution of $0.0125^\circ \times 0.0125^\circ$. This enhancement is crucial for understanding the intricate interplay between topography and hydrological processes, particularly in mountainous regions like Himachal Pradesh, India. Utilizing the Variable Infiltration Capacity (VIC) model, known for its variable infiltration rates affecting runoff and evapotranspiration, the study emphasizes meticulous model calibration against high-resolution observational data. This calibration, including iterative parameter adjustments, significantly improves model performance, as demonstrated by our case study's results. The model's robustness, validated through rigorous empirical testing, underscores its potential as a pivotal tool for policy development, aimed at creating resilient water management systems in anticipation of hydrological changes induced by climate change. Furthermore, this research introduces a novel methodological framework for hydrological modelling, ensuring that regional climate change assessments are both precise and actionable for policymakers. Our findings highlight stark differences in hydrological impacts under various Shared Socioeconomic Pathways, emphasizing the need for adaptive strategies in water resource management to accommodate anticipated climatic shifts.

Keywords Climate change · General circulation models (GCMs) · Variable infiltration capacity (VIC) model · Downscaling techniques

S. Barbhuiya (✉) · S. Sharma · A. Pathania · V. Gupta
School of Civil and Environmental Engineering, Indian Institute of Technology Mandi, Suran, Himachal Pradesh, India
e-mail: siddikbarbhuiya@gmail.com

V. Gupta
e-mail: vivekgupta@iitmandi.ac.in

A. Nanda et al. (eds.), *Navigating the Nexus*, Water Science and Technology Library 102, https://doi.org/10.1007/978-3-031-76532-2_14

14.1 Introduction

The reality of climate change, driven by human activities, is incontrovertible. A plethora of scientific evidence underscores the significant and accelerating alteration of global climate patterns, primarily attributed to anthropogenic greenhouse gas emissions. This phenomenon, characterized by an increase in atmospheric concentrations of gases such as carbon dioxide (CO_2), methane (CH_4), and nitrous oxide (N_2O), disrupts the natural greenhouse effect, leading to a warming of the Earth's surface—a process known as radiative forcing (Knutti and Hegerl 2008; Sellers 1969). The implications of these changes are profound, affecting precipitation patterns, evapotranspiration rates, and the frequency and intensity of extreme weather events, which in turn have significant consequences for water resources management, agricultural productivity, and ecosystem health (Middelkoop et al. 2001). The role of the Intergovernmental Panel on Climate Change (IPCC) has been pivotal in synthesizing and disseminating knowledge on the dynamics of climate change. Since its inception, the IPCC has highlighted the overwhelming influence of CO_2 and other greenhouse gases in driving climate change over the past two centuries. This has brought to light the nuanced understanding of climate sensitivity—how the Earth's climate responds to changes in radiative forcing—and its implications for global and regional climate patterns (IPCC reports). Such an understanding is critical as it underscores the complex feedback mechanisms within the climate system, necessitating sophisticated modelling techniques to accurately predict future climatic conditions and their impacts.

General Circulation Models (GCMs) have emerged as indispensable tools in the scientific investigation of climate change, offering the capability to simulate the long-term effects of increased greenhouse gas concentrations on global climate systems. These models encompass the physical processes of the atmosphere, oceans, cryosphere, and land surface, thereby projecting future climatic variables across a range of temporal and spatial scales. However, the utility of GCMs in regional hydrological assessments is hampered by inherent limitations, including coarse spatial resolution and the models' limited capacity to accurately simulate local hydrological processes. This gap between the global projections provided by GCMs and the detailed requirements of hydrological assessments at the catchment or river basin level highlights the necessity for integrated hydrological modelling approaches (Thompson et al. 2016). The discrepancy between the coarse resolution of GCMs and the need for fine-scale hydrological analysis has catalysed the development of downscaling techniques. These methods, which include both dynamic and statistical approaches, aim to refine GCM outputs, making them suitable for hydrological assessments at more localized scales. Dynamic downscaling, achieved through the deployment of regional climate models, generates high-resolution meteorological fields. Meanwhile, statistical downscaling establishes empirical relationships between large-scale atmospheric conditions and local climate or hydrological responses, providing a cost-effective means of generating localized climate scenarios (Dibike and Coulibaly 2007; Hewitson and Crane 1996). Although these downscaling methodologies have significantly enhanced the capacity to assess regional hydrological impacts of climate change, they introduce additional layers of uncertainty. These uncertainties stem from the selection of appropriate models, the accurate representation of physical processes, and the interpretation of projected changes in climate variables, complicating the task of climate impact assessment (Schoof 2013). The IPCC has been at the forefront of developing long-term emission scenarios, which are crucial for understanding the potential futures of greenhouse gas emissions. These scenarios, based on comprehensive assumptions about demographic, technological, and socioeconomic changes, have evolved from the initial IPCC IS92 set to the more refined Representative Concentration Pathways (RCPs) introduced in the Fifth Assessment Report. RCPs, with pathways labelled 2.6, 4.5, 6.0, and 8.5 watts per square meter, extend until the year 2100 and result from detailed collaborations among experts in various fields. Each RCP reflects a different narrative about the future, ranging from optimistic scenarios with strong environmental policies to more challenging futures with rapid population growth and lower environmental focus (van Vuuren et al. 2011). However, RCPs focus solely on the amount of radiative forcing, not the societal conditions that lead to those emissions. To address this, Shared Socioeconomic Pathways (SSPs) were introduced. These scenarios consider how both climate and society might change in the future. This integrated approach allows researchers to assess the potential impacts of climate change alongside possible adaptation and mitigation strategies (Gupta et al. 2020; Nazarenko et al. 2022; O'Neill et al. 2017).

Amid these complex narratives and modelling challenges, in this chapter, we delve into the methods and models used to assess the impacts of climate change on hydroclimatic regimes. Alongside this exploration, we present a detailed case study focused on the calibration of the Variable Infiltration Capacity (VIC) model and its application in studying climate change impacts. This case study not only demonstrates the practical application of these models but also provides insights into the specific effects of climate change on water resources, highlighting the significance of accurate modelling in developing strategies for water resource management and adaptation to climate change.

Climate Change Models and Projections

General Circulation Model (GCM)

General Circulation Models (GCMs) are instrumental in the scientific investigation and forecasting of climate patterns. These models, grounded in the fundamental principles of physics, including thermodynamics, radiative heat transfer, and fluid dynamics, enable the simulation of complex interactions within the Earth's climate system (Kour et al. 2016). The development and refinement of GCMs have been pivotal in advancing our understanding of climate dynamics and the potential impacts of climate change. GCMs utilize mathematical equations that respect the conservation laws of physics—energy, mass, and momentum. These laws govern the exchanges between the Earth's atmosphere, oceans, land surfaces, and cryosphere, facilitating a comprehensive simulation of global climate processes (Kour et al. 2016). Through these simulations, GCMs can replicate large-scale climate patterns, such as temperature distributions and precipitation patterns, and predict the effects of anthropogenic emissions on global temperatures and ocean heat content. The evolution of GCMs has seen a significant expansion in the scope of their simulations. Originally focused on the atmosphere's general circulation (Phillips 1956), these models now encompass a wider array of climate system components. This includes atmospheric chemistry, aerosols, interactions between the land surface (including soil and vegetation), sea ice, and even the carbon cycle and biogeochemistry in more advanced Earth System Models (ESMs). Such expansions have enhanced the models' capacity to provide a more detailed and comprehensive view of climate dynamics.

Technological advancements have allowed for an increase in both the number and the spatial resolution of GCMs, especially with the advent of the Coupled Model Intercomparison Project (CMIP) (Meehl et al. 2000). The progression from CMIP3 to CMIP5 and most recently to CMIP6, reflects not just an improvement in computational capabilities but also a broader ambition to capture climate phenomena with greater fidelity. Despite these advancements, the relationship between model complexity, resolution, and accuracy is not straightforward. While some improvements in climate simulation have been noted—such as a better representation of the El Niño-Southern Oscillation (ENSO) (Bellenger et al. 2014), certain cloud characteristics, and Arctic Sea ice loss—the overall enhancement in model performance is nuanced (Wang and Overland 2012). The increase in model resolution and complexity does not uniformly translate into more accurate climate projections. The most recent phase, CMIP6, builds on the foundation laid by its predecessors with a more ambitious scope and refined objectives (O'Neill et al. 2016). CMIP6 aims to address a broader range of scientific questions, including better understanding of climate sensitivity, feedback mechanisms, and regional climate predictions. It introduces new scenarios (Shared Socioeconomic Pathways, SSPs) that combine socioeconomic factors with greenhouse gas emissions trajectories, providing a more comprehensive framework for climate impact assessments (O'Neill et al. 2016). CMIP6's enhancements in model resolution, complexity, and scenario planning are designed to provide more accurate and detailed climate projections. General Circulation Models (GCMs) play a critical role in climate change research, as they simulate the interactions between various components of the Earth's system, such as the atmosphere, oceans, land surface, and sea ice. These models can also be integrated with hydrological models to examine the impact of climate change on water cycles (Fowler et al. 2007a; Larsen et al. 2014). However, the coarse resolution of GCM outputs often fails to capture detailed regional phenomena, especially those related to hydrological processes, limiting their direct applicability for studying land surface impacts of climate change (Kidson and Thompson 1998; Wilby et al. 1999; Wilby and Wigley 1997). The challenge of bridging the scale gap between general circulation models and regional hydrological impacts is addressed through downscaling. Downscaling generates local or regional climate information from broader-scale data, whether modelled or observed, and can be applied across both spatial and temporal dimensions (Trzaska and Schnarr 2014). Spatial downscaling refines GCM outputs to a finer spatial resolution, while temporal downscaling transforms broader temporal data (e.g., from monthly to daily meteorological variables), making it more suitable for hydrological and climate variability studies (Dibike and Coulibaly 2006; Mendes and Marengo 2010).

There are two primary downscaling approaches: statistical and dynamical. Statistical downscaling establishes empirical relationships between large-scale atmospheric conditions (predictors) and local climate variables (predictands), using methods ranging from straightforward interpolation to complex statistical models, including neural networks and weather generators. On the other hand, dynamical downscaling involves embedding Regional Climate Models (RCMs) within GCMs. This method, which leans on the detailed physical processes simulated by RCMs, tends to be computationally intensive. Despite the reliance on dynamical methods in areas lacking detailed observational data, both downscaling techniques have their advantages and limitations.

Different Downscaling Method

Statistical Downscaling Methods

Statistical downscaling serves as a pivotal technique in bridging the gap between the coarse-resolution outputs of Global Climate Models (GCMs) and the finer-scale climatic information necessary for regional climate assessments. This method leverages empirical relationships between large-scale atmospheric predictors and local-scale climatic responses, known as predictands, to enhance our understanding of climate variability and change at a more localized level. The literature delineates various categorizations and methodologies within statistical downscaling, reflecting its diverse applications and the complexity of modelling climate systems.

The methodologies within statistical downscaling have been systematically classified based on different criteria, including the nature of the technique used and the choice of predictor variables. Rummukainen (1997) and Wilby et al. (2004) provide foundational frameworks for understanding these classifications, subsequently expanded upon by Maraun et al. (2010). These classifications illuminate the range of approaches used to derive fine-scale climatic information from broader-scale data. Based on Techniques Used

1. **Regression Methods**: Regression techniques in statistical downscaling serve to define both linear and nonlinear correlations between larger atmospheric predictors and more localized climatic variables, known as predictands. The selection of a specific regression method, be it multiple regression analyses, canonical correlation analysis, or singular value decomposition, hinges on the chosen mathematical transfer functions, the statistical fitting mechanisms employed, or the array of predictor variables considered. Methods like Multiple Regressions, as discussed by (Murphy 2000) and (Huth 2002, 2004), provide a direct and effective means for delineating the relationship between predictors and predictands, particularly when linear representations suffice. Meanwhile, Canonical Correlation Analysis and Singular Value Decomposition (VonStorch et al. 1993; Wigley et al. 1990), excel in dissecting the complex, multivariate interactions inherent

in climate dynamics, offering a deeper insight into the mechanisms of climate variability. Neural Networks (Hewitson and Crane 1996; Wilby et al. 1998), offer a dynamic approach to modeling by accommodating nonlinear relationships and intricately mapping climate data patterns. It's important to note the limitations of linear models in fully capturing the nonlinear complexity of climate systems and the dependency of neural networks and similar nonlinear methods on extensive, high-quality observational data for their accuracy.

2. **Weather pattern approaches**: Weather pattern approaches excel by organizing local weather data into specific categories. This is done using two main strategies: subjective and objective. Subjective methods draw on the rich experience of meteorologists, who use their detailed knowledge of local weather to offer deep insights into how the climate behaves (Hess and Brezowsky 1977). This technique benefits from the seasoned weather experts' profound understanding of the local climate. On the other hand, objective methods focus on a data-driven, consistent way to classify weather, which is vital for accurately studying how the climate changes over time (Bardossy et al. 1995). These methods depend on analysing data to ensure that studies can be replicated and are reliable. Both strategies have their drawbacks. Subjective methods, despite being insightful, might not be suitable for all areas or over time because they rely heavily on individual expertise, which can vary. Meanwhile, objective methods, though strict in their process, depend greatly on the quality and nature of the data they use. This means they might not always be effective, especially when predicting future climate scenarios, as the historical data they are based on might not accurately reflect future conditions.
3. **Stochastic Weather Generators**: Stochastic Weather Generators use statistical techniques to create synthetic sequences of weather data, drawing upon the statistical characteristics of historical weather records. This category encompasses various methodologies, such as the Markov chain approach, the spell length technique, and mixture models, each contributing to the generation of comprehensive weather data series. Such methods are invaluable for extending our understanding of climate variability and its impacts on crucial sectors like hydrology and agriculture (Semenov and Barrow 1997; Wilks 1992). Particularly in areas where observational data are scarce, these generators offer a practical solution for simulating climate conditions under diverse scenarios, thereby enhancing our ability to predict climate-related changes and challenges. However, the effectiveness of these stochastic models is heavily dependent on the quality and completeness of the input data they use. Inaccuracies or gaps in the data can significantly affect the calibration of these models (Soltani and Hoogenboom 2003), potentially undermining their reliability. Moreover, despite their utility, these generators often struggle to fully represent the extremities of weather events or the breadth of climate variability that might occur under future climate scenarios, pointing to a critical area for further development in climate modelling.

Based on Predictor Variable Selection

1. **Downscaling with Surface Variables**: This technique develops statistical relationships between averaged large-scale surface variables and local-scale surface variables.
2. **PP Method**: Here, statistical relationships are established between large-scale atmospheric variables and local surface variables, utilizing observed quantities from a historical sample (Chen and Chen 2003; Klein et al. 1959).
3. **MOS Method**: This approach develops statistical relationships between large-scale atmospheric variables from GCM outputs and local surface variables (Glahn and Lowry 1972).

Each statistical downscaling method comes with its own set of advantages and limitations. For instance, regression methods and weather pattern approaches offer straightforward applications and physically interpretable linkages to surface climate. Conversely, they might not effectively represent observed variance or extreme events and require substantial historical data for calibration. Stochastic weather generators, while providing detailed synthetic weather data, demand extensive daily data series and are sensitive to data gaps in their calibration datasets.

Dynamic Downscaling Method

The dynamical downscaling method, also known as the 'nested' Regional Climate Model (RCM) approach, integrates RCMs within General Circulation Models (GCMs) to refine climate simulations at a regional scale (Teutschbein and Seibert 2013). This integration can occur through either a one-way or two-way nesting approach (Harris and Durran 2010). In the one-way nesting approach, the RCM utilizes output from the GCM to set its initial and lateral boundary conditions without providing any feedback to the GCM. Conversely, the two-way nesting approach (Anthes and Warner 1978), allows for feedback from the RCM to influence the GCM, enhancing the interaction between regional and global climate simulations. This method has seen application in various studies (Debreu et al. 2012; Warner et al. 1997; Zhang et al. 1986) as has the one-way approach (Colle et al. 2005; Deng and Stull 2005). Teutschbein and Seibert (2010) noted that while RCMs can incorporate hydrological components like surface and subsurface runoff, the simulations often do not align with observed streamflow data. This discrepancy arises because RCM runoff schemes are generally designed to capture water balance trends rather than to calculate discharge accurately. Consequently, hydrological variables simulated by RCMs may not be directly applicable for impact studies at the catchment scale (Graham et al. 2007b). Nevertheless, RCM outputs, such as temperature, precipitation, and snowpack, are frequently used as inputs for hydrological modelling (Graham et al. 2007a, b; Teutschbein and Seibert 2010). However, it has been observed that RCM-simulated meteorological variables often exhibit significant biases, which can lead to inaccurate simulations when directly applied in hydrological models (Graham et al. 2007a; Teutschbein and Seibert 2010; Varis et al. 2004). One strategy to mitigate these issues involves employing multiple RCMs to generate a range of simulations, known as 'ensemble simulations' (Teutschbein and Seibert 2013). Despite the benefits of this multi-model approach, deviations from observed data persist, highlighting the need for 'bias correction' techniques to adjust the simulated data for more accurate hydrological modelling.

Bias Correction Methods

1. Linear Scaling (LS):

This is one of the simplest forms of bias correction (Lenderink et al. 2007). For a variable X, the bias-corrected value X_{LS} is computed as:

$$X_{LS} = X_{\text{mod}} + \left(\overline{X_{\text{obs}}} - \overline{X_{\text{mod}}}\right) \quad (14.1)$$

where $\overline{X_{obs}}$ is the observed long-term mean and $\overline{X_{mod}}$ is the modeled long-term mean.

2. Delta Change (DC):

For temperature, it adds the difference between future and present climate simulations to the observed data. For precipitation, it multiplies observed data by the ratio of future to present simulations (Räty et al. 2014):

$$T_{DC} = T_{\text{obs}} + \left(T_{\text{future}} - T_{\text{present}}\right) \quad (14.2)$$

$$P_{DC} = P_{obs} \times \frac{P_{\text{future}}}{} P_{\text{presert}} \quad (14.3)$$

where T denotes temperature and P denotes precipitation.

3. Quantile Mapping (QM):

This method adjusts the entire distribution of the modelled variable to match the observed distribution. For a given quantile q, the bias-corrected value X_{QM} is:

$$X_{QM} = F_{\text{obs}}^{-1}(F_{\text{mod}}(X_{\text{mod}})) \quad (14.4)$$

where F_{obs}^{-1} is the inverse cumulative distribution function (CDF) of the observed data and F_{mod} is the CDF of the modelled data (Qian and Chang 2021).

4. Quantile Delta Mapping (QDM):

An enhanced version of quantile mapping that maintains changes in quantiles between a historical simulation and a future simulation:

$$X_{QDM} = F_{\text{obs}}^{-1}(F_{\text{hist}}(X_{\text{hist}})) \times \frac{F_{\text{future}}^{-1}(q)}{F_{\text{hist}}^{-1}(q)} \quad (14.5)$$

where F_{future}^{-1} is the inverse CDF of the future modeled data and F_{hist}^{-1} is the inverse CDF of the historical modelled data for the same quantile q (Fauzi et al. 2020).

5. Local Intensity Scaling (LOCI):

This approach corrects both the mean and the variance of precipitation. First, it sets a threshold to separate wet and dry days, then it adjusts the intensity of precipitation:

$$\begin{cases} 0 & \text{if } P_{\text{mod}} < P_{th} \\ P_{\text{mod}} \times s & \text{otherwise} \end{cases} \tag{14.6}$$

where P_{mod} is the modelled precipitation, P_{th} is the threshold, and s is the scaling factor based on the wet-day intensities (Schmidli et al. 2006).

6. Empirical Quantile Mapping (EQM):

Similar to QM, EQM uses empirical CDFs and is often applied to daily data:

$$X_{EQM} = F_{\text{obs}}^{-1}\left(F_{\text{mod}}^{emp}(X_{\text{mod}})\right) \tag{14.7}$$

where F_{mod}^{emp} is the empirical CDF of the modelled data (Holthuijzen et al. 2022).

Intercomparison of Downscaling Methods

Various methodologies have been explored to downscale climate model outputs for hydrological impact assessments. Fowler et al. (2007b), Hewitson and Crane (1996), Yarnal et al. (2001) have reviewed different downscaling techniques, focusing on comparison and limitations. Chen et al. (2011) undertook an in-depth analysis of six distinct downscaling strategies to assess their performance in capturing the hydrological impacts of climate change within the Manicouagan 5 river basin in Quebec, Canada, for the period between 1970 and 1999. This assessment encompassed the Canadian Regional Climate Model (CRCM) with and without bias correction, the Change Factor (CF) method applied at both Canadian GCM (CGCM) and CRCM scales, a weather generator-based approach at both CGCM and CRCM levels, the Statistical Downscaling Model (SDSM) at the CGCM scale, and the Discriminant Analysis coupled with Stepwise Regression (DASR) at the CGCM scale. The findings highlighted that, aside from the CGCM-DASR approach, each method presented merits, with CRCM data that underwent bias correction and the SDSM standing out for their superior accuracy in simulating river discharge. The DASR method was identified as less effective, particularly because it underestimated precipitation, both in mean and variability, leading to underestimations in discharge. Conversely, the SDSM achieved notable accuracy in discharge simulations by employing a variance inflation scheme to align the variance of precipitation and temperature more closely with observed data, alongside performing bias corrections. The Change Factor method, known for its simplicity in adjusting model outputs to mirror observed data by modifying temperatures and precipitation based on future and current climate differences projected by models, was critiqued for its inability to adjust the spatial and temporal dynamics of meteorological data. This limitation, noted in studies by Diaz-Nieto and Wilby (2005), Fowler et al. (2005), Minville et al. (2008), underscores the method's constraint in refining the data granularity necessary for hydrological analysis. Wilby et al. (1999) contrasted various techniques for generating rainfall-runoff scenarios, revealing that downscaled GCM outputs typically exhibit higher temperatures compared to those from downscaled NCEP models. This disparity underscores the coarse resolution of GCMs, which poses challenges in capturing regional-scale processes accurately. The comparison between statistical and dynamical downscaling by Kidson and Thompson (1998) and Murphy (1999) for Europe and New Zealand, respectively, suggested that each method has its strengths, with statistical downscaling being more reliable for temperature predictions in summer and dynamical downscaling for winter precipitation. Recent advancements have seen the application of bias correction methods such as Equidistant Quantile Mapping (EDQM), Nested Bias Correction (NBC), and Monthly Bias Correction (MBC) (Sachindra et al. 2014), highlighting the nuanced approaches needed to refine precipitation forecasts. Nevertheless, the effectiveness of these methods varies, with EDQM showing potential in balancing the over-prediction of precipitation. As the demand for precise hydrological predictions grows, the quest for optimal downscaling techniques continues. The integration of high-resolution meteorological data into hydrological models remains crucial for climate impact assessments. This narrative underscores the evolution of downscaling methodologies, from dynamical to statistical approaches, each with its specific applications, strengths, and limitations. The emphasis on bias correction across various studies (Mudbhatkal and Mahesha 2018; Teutschbein and Seibert 2012) highlights the ongoing efforts to enhance the fidelity of hydrological predictions, reaffirming the importance of methodological diversity and adaptability in addressing the challenges posed by climate change on water resources.

Hydrological Modelling

Hydrological modelling is an essential aspect of environmental management, offering insights into the behaviour of water within ecosystems. These models vary in complexity and scale, and are designed to simulate various hydrological processes such as precipitation, evaporation, infiltration, runoff, and storage. They are indispensable tools for predicting the impacts of natural and anthropogenic changes on water systems, and for planning sustainable water management strategies. Among the prominent hydrological models, the Distributed Hydrology Soil Vegetation Model (DHSVM) is particularly notable. Developed by Wigmosta and Burges (1997), Wigmosta and Lettenmaier (1999), DHSVM offers a detailed representation of hydrological processes at both continental and regional scales, capable of simulating the impacts of climatic variation and land use changes on streamflow and snowpack. The H08 model, crafted by Hanasaki and Yamamoto (2010), Saito and Hanasaki (2012), is employed at the river basin level, providing quantitative estimates of the impacts of reservoir operations on the hydrological cycle and facilitating global water resource assessments. Another river basin model is the Variable Infiltration Capacity (VIC) model by Liang et al. (1994), acclaimed for capturing sub-grid variability in land surface vegetation and soil moisture, as well as for its dynamic storm routing capabilities. The Soil and Water Assessment Tool (SWAT), developed by Rosenthal et al. (1995), is a comprehensive model used for large river basins. SWAT analyses the effects of climate and land management changes on water quality and supply with the aid of continuous time-step simulations. For rainfall-runoff simulation on a smaller scale of 1–30 km, the Hydro-BEAM model by Toshiharu (2005) utilizes digital elevation models and GIS data to provide thorough runoff assessments. Schimming et al. (1995) introduced WASMOD for catchment scale studies, focusing on water balance, soil moisture, river flow, and actual evapotranspiration. On the watershed scale, the Deterministic Watershed Simulation Model (DWSM) by Borah et al. (2004) calculates total rainfall and runoff, along with total outflow volume. Similarly, MIKE BASIN, a product of Ireson et al. (2006), is known for its river basin scale applications and serves as a distributed and deterministic, conceptual modelling system. The TOPKAPI model by Ciarapica and Todini (2002) spans from hill slopes to catchments, generating detailed discharge hydrographs, while the Utah Energy Balance (UEB) model by Tarboton (1994), Tarboton et al. (1995) is a point model that estimates snow and glacier melt and outflow on a grid over a watershed. In addition to these models, the GR4J model (Perrin et al. 2003) stands out as a daily lumped rainfall-runoff model developed for catchment hydrology analysis. Its simplicity and efficiency in modelling the rainfall-runoff process for catchments, often using minimal data, make it a practical choice for hydrologists seeking to understand and predict streamflow under various climatic conditions (Barbhuiya et al. 2023).

Application of Hydrological Model in Climate Change Impact

The nexus between climate change and hydrological regimes has been the focus of extensive research, illustrating the anticipated effects on river basins across the globe. A suite of studies employing diverse hydrological and climate models has sought to predict these impacts over an array of environmental and geographical contexts. In the Jinhua River Basin of China, Xu et al. (2014) leveraged models such as ECHAM5, HadCM3, and PRECIS, analysing data from 1961–1990 with projections extending to 2040. The DHSVM hydrological model predicted a nuanced response of the basin's runoff to climate change, expecting decreases in summer and autumn runoff, coupled with slight increases in the spring and winter periods. Turning to the significant river systems of the Ganges, Brahmaputra, and Meghna, Masood et al. (2014) applied the MRI-AGCM3.2 model for the period between 1979–2003, projecting to the latter parts of the twenty-first century. The H08 model forecasts a 3 °C temperature increase by the century's end, accompanied by heightened mean precipitation and runoff levels across these basins. The Indian River basins were examined by Raje et al. (2014) using models like BCCR-BCM 2.0, CGCM3-T63, and CSIRO Mk3 from 1961 to 1990, with a view toward mid 21st and late 21st-century climatic conditions. The VIC hydrological model anticipated a rise in summer monsoon runoff, evapotranspiration, and soil moisture, particularly in the central regions of India. Sood et al. (2013) focused on the Volta River Basin in West Africa, utilizing ECHAM and COSMO-CLM models for the period from 1983 to 2012, with foresight into 2050 and 2100. The SWAT model showed potential reductions in river flow by up to 40%, attributing this decline to a combination of decreased rainfall and a warming climate. For the White Volta and Pra river basins in Ghana, Kankam-Yeboah et al. (2013) engaged ECHAM and CSIRO models, contrasting conditions from 1983–2000 to 2036–2075. SWAT analyses predicted a decrease in mean annual streamflow, signaling challenges for water resource management in the region. Piniewski et al. (2013)'s work on the Narew River basin in Poland utilized IPSL-CM4 and MIROC3.2 models spanning from 1976 to 2000 with projections up to 2069. The combined use of WaterGAP and SWAT models pointed to subtle but variable changes in runoff, influenced by modest shifts in temperature and precipitation. In Canada, Shrestha et al. (2012) studied the Red and Assiniboine basins within the Lake Winnipeg watershed, drawing on CRCM, RCM3, and HRM3 climate models from 1980–2000, projecting to 2062. SWAT model projections indicated an increase in runoff during late summer and fall, suggesting earlier snowmelt and heightened peak flows. For Central America's Rio Lempa basin, Maurer et al. (2009) utilized an ensemble of 16 GCMs covering the years from 1961 to the end of the twenty-first century. VIC model projections pointed to moderate decreases in reservoir inflows, alongside notable temperature rises and precipitation declines. In Turkey's Seyhan river basin, Fujihara et al. (2008) implemented MRI-CGCM2 and CCSR/NIES/FRCGC-MIROC models for the 1990s with an outlook to the 2070s. The Hydro-BEAM model findings were stark, with a projected 52–61% decrease in annual runoff, reflecting significant shifts in the basin's climate and hydrology. For the Murray-Hotham Catchment in Western Australia, Islam et al. (2014) investigated the region with 11 GCMs from 1961 to 2000, extending predictions to 2100. The LUCICAT model indicated decreases in mean annual rainfall and an upturn in average temperatures, hinting at dryer future conditions. In North Morocco, the Makhazine dam basin's future hydrology was forecasted by Tramblay et al. (2013) using a high-resolution ALADIN-Climate RCM from 1984–2010 to 2062. The GR4J model anticipated a reduction in surface runoff by 20–57%, signalling significant hydrological shifts. Khoi and Hang (2015) study examined the Srepok watershed, a segment of the Mekong River basin located in Vietnam's Central Highlands, utilizing a variety of climate models such as HadCM3, CCCMA CGCM3.1, and several others, with data spanning from 1970 to 2000 to project future climate conditions up to 2100. Their research with the SWAT hydrological model indicated a potential linear decrease in mean annual streamflow by 2–9.8%. Additionally, streamflow variations between SRES emission scenarios were relatively minor, suggesting more significant changes in river discharge could occur with a 2 °C rise in global mean temperature, ranging from a 6% decrease to a 16.1% increase. Shrestha (2014) focused on 25 river basins in Thailand using the B1 and B2 PRECIS RCM climate models, covering a baseline period from 1971 to 2000 and extending projections to the year 2100. Utilizing the HEC-HMS hydrological model, findings suggest a significant rise in precipitation, with variations in water availability expected to range from a 5.5% decrease in the 2010s to a 29% increase by the 2080s. Peleg et al. (2015) analysed the upper Dalaya and Taninim catchments in northwestern Israel using the CMIP5 climate models, with a historical period from 1996 to 2005 and future projections for 2046–2055. Their work predicts a decrease in mean annual rainfall by 15–18% and a substantial reduction in streamflow volumes, alongside a shortened streamflow season, as per the RCP4.5 and RCP8.5 emission scenarios. Hattermann et al. (2014) research spanned diverse geographic regions in Germany, including the northern lowlands and the German Alps, utilizing models like ECHAM5, CCLM, and REMO over a baseline period from 1961 to 2000, extending to the 2100s. The SWIM hydrological model indicated potential increases in flood-related losses in future scenarios.

Application: Case Study in Himachal Pradesh

This section presents a case study focusing on the climate change impact assessment of the Beas River Basin. The study employs the CMIP6 climate model and the Variable Infiltration Capacity (VIC) model. Careful calibration of the VIC model was crucial, and the refined parameters were then used to simulate the hydrological dynamics across the Beas River Basin. The overall aim was to assess the potential impacts of climate variability on the region's water resources.

Study Area and Its Description

The Beas River Basin (Fig. 14.1), selected as the study area for this study, is situated in the northwestern segment of the Indian Himalayan Region. Extending for approximately 470 km, the Beas River courses through the Mandi, Kullu, Hamirpur, and Kangra districts before converging with the Sutlej River. The catchment area, delineated upstream of the Pong Reservoir, encompasses around 12,560 km^2, with the topography marked by a striking elevation gradient that ranges from 335 to 6176 m above sea level. Geographically, the basin is framed by latitudes 31° 28′N to 32° 26′N and longitudes 75° 56′E to 77° 48′E (Raaj et al. 2024). Hydrologically, the Beas River is predominantly sustained by the seasonal snowmelt. During the summer months, the higher reaches of the basin, characterized by their glaciated and snow-capped landscapes, release meltwater, which significantly contributes to the river's streamflow. The basin's climatic conditions are marked by winter temperatures that oscillate between an average minimum of 0.22 °C and an average maximum of 14.1 °C.

Dataset Used in The Study

To ensure the robust application and calibration of the Variable Infiltration Capacity (VIC) model (Liang et al. 1994) for our study, we compiled a large-scale, spatially distributed dataset. This dataset included meteorological data (precipitation, temperature, wind speed), land cover information, topographical data (elevation, slope, aspect), and streamflow measurements. Soil coverage data were acquired from the digitized World Soil Map (http://www.fao.org/soils-portal/soil-survey/soil-maps-and-databases/harmonized-world-soil-database-v12/en/). This provided a granular view of the soil types prevalent within the study area. Topographical characteristics, critical to the delineation of the basin and analysis of its geomorphological features, were extracted from the Shuttle Radar Topography Mission (SRTM) digital elevation data (https://www2.jpl.nasa.gov/srtm/mission.htm), with a high spatial resolution of 30 m. Meteorological inputs, pivotal for driving the VIC model, were sourced from the Indian Monsoon Data Assimilation and Analysis (IMDAA) (Rani et al. 2021). This dataset, covering the period from 2014 to 2020, initially offered data at a resolution of approximately 12 km. Through the application of nearest neighbour interpolation techniques, we refined this data to a higher resolution. To enhance

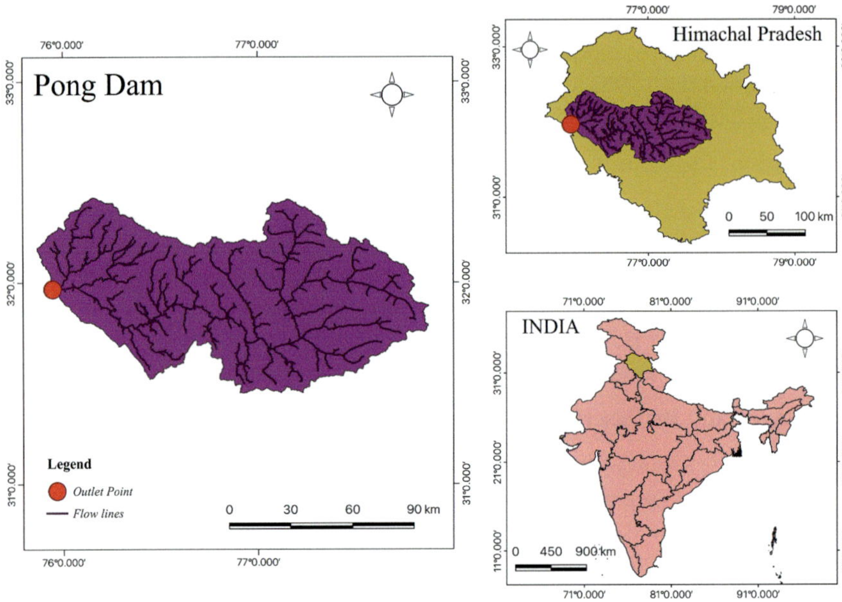

Fig. 14.1 Study area of Beas river basin

the spatial accuracy of our simulations. The dynamics of vegetation, a significant factor in hydrological modelling, were incorporated through the analysis of three MODIS (Moderate Resolution Imaging Spectroradiometer) products: MOD13Q1, MOD11A1, and MOD15A2H, each providing a spatial resolution of 500 m. This inclusion allowed for an in-depth examination of vegetation cover and its temporal fluctuations. Calibration of the VIC model was grounded on discharge data for the Pong Dam, retrieved from http://210.212.64.190/. Utilizing historical and prospective hydrological simulations, we engaged daily scale data from eleven CMIP6 General circulation Models (GCMs)(ACCESS-CM2,ACCESS_ESM1-5,BCC-CSM2-MR,CMXX-CSM2,INM-CM4-8,INM_CM-5–0,MPI_ESM1-2LR,MRI_ ESM2-0,NESM3,NorESM2-LM,NoeESM2-MM,TaiESM1). For the future projections within the Beas River Basin, we selected the SSP245 and SSP585 scenarios, facilitating an analysis of climate change impacts. These GCM outputs were subjected to downscaling and bias correction via the Quintile Delta Mapping method at a resolution of 12 km, with an ensemble mean of all CMIP6 products being utilized for an aggregated analysis. This methodology, cantered on the SSP245 and SSP585 scenarios, aimed to illustrate the anticipated impacts of climate variability on the hydrology of the Beas River Basin. It provided a solid foundation for evaluating the basin's vulnerability and adaptive capacity in response to future climatic challenges, contributing valuable insights into sustainable water resource management within the context of climate change.

Model Development

We employed the Variable Infiltration Capacity (VIC) model (Liang et al. 1994), known for its ability to simulate land-surface water and energy fluxes across diverse scales, at a high spatial resolution of approximately 1 km (1/80°). To optimize the model's performance, we adopted the Dynamic Dimensional Search (DDS) algorithm, recognized for its efficiency and ability to find optimal solutions globally. This algorithm was crucial in refining the model parameters for more precise simulations. Our analysis focused on the Beas River Basin in Himachal Pradesh, encompassing a vast area with over 7691 grids. The study period spanned from 2014 to 2020, offering a comprehensive understanding of the region's hydrological behaviour over a significant timeframe. During calibration, we meticulously compared the model's simulated streamflow against observed records to ensure the simulation's accuracy and the model's reliability. This calibration process involved adjusting VIC's parameters to achieve better agreement between the simulated outputs and the observed natural flow conditions. The effectiveness of this calibration was quantitatively assessed using two metrics: the Nash Sutcliffe Efficiency (NSE) coefficient and the coefficient of determination (R^2). The NSE is a widely acknowledged metric for evaluating hydrological model performance and is calculated as follows:

$$NSE = 1 - \left(\frac{\sum_{t=1}^{T} \left(Q_{obs,t} - Q_{\mathrm{sim},t} \right)^2}{\sum_{t=1}^{T} \left(Q_{obs,t} - \overline{Q}_{obs} \right)^2} \right) \quad (14.8)$$

Here, $Q_{obs,t}$ represents the observed streamflow at time t, $Q_{\mathrm{sim,t}}$ denotes the simulated streamflow at the same time, $\overline{Q}_{obs}$ is the average observed streamflow over the calibration period, and T signifies the total number of observations. An NSE value approaching 1 indicates a model that performs very well, with simulation outputs closely aligning with the observed data (McCuen et al. 2006). In addition to the NSE, the coefficient of determination, R^2, was used to measure the proportion of variance in the observed data that is predictable from the model simulations:

$$R^2 = 1 - \frac{\text{sum squared regression (SSR)}}{\text{total sum of squares (SST)}} \quad (14.9)$$

An R^2 value close to 1 indicates that the model's simulations explain a large proportion of the variance in the observed data (Ozer 1985). Both these statistical metrics, NSE and R^2, were integral in determining the calibration success, enabling us to model the hydrological dynamics of the Beas River Basin with high accuracy. This rigorous calibration underpins our analysis for assessing the potential impacts of climate change and wildfires on the region's water resources, providing a foundation for detailed insight into the complex interactions between hydrological processes and environmental changes.

14.2 Result and Discussion

Calibration of VIC model

The calibration of the VIC model for the Beas River Basin was meticulously carried out on a daily timestep. This involved carefully adjusting nine key parameters of the model to ensure simulated streamflow closely aligned with observed data from the Pong Dam outlet. By fine-tuning these parameters (e.g., infiltration rate, baseflow parameters), we refined the model's ability to represent the physical processes governing water flow within the basin. The parameter Ds, which signifies the fraction of Dsmax where non-linear baseflow commences, was set to 0.89, indicating that a significant portion of maximum baseflow is active before reaching total capacity. The Dsmax, representing the maximum velocity of baseflow, was determined to be 0.78, reflecting a relatively high rate of baseflow response to saturation. The infilt parameter, or the variable infiltration curve parameter (binfilt), was calibrated at 0.35, suggesting a moderate infiltration capacity. Soil moisture layer thicknesses, represented by DEPTH1, DEPTH2, and DEPTH3, were calibrated to values of 0.31, 1.4, and 3.9, respectively, providing a stratified view of soil moisture storage capacity and movement within the soil profile. Ws, the fraction of maximum soil moisture where non-linear baseflow occurs, was set at 0.017, indicating that baseflow processes become non-linear at a very low soil moisture threshold. The temperature parameters, MAX_SNOW_TEMP and MIN_RAIN_TEMP, crucial for the delineation between solid and liquid precipitation, were calibrated to 3.76 °C and − 1.64 °C, respectively. These values define the temperature bounds for snow and rain occurrence, with snowfall not occurring above 3.76 °C and rainfall not occurring below − 1.64 °C. The adopted values for the VIC model parameters reflect the unique hydrological behaviours and climate conditions of the Beas River Basin. By optimizing soil moisture parameters, baseflow constants, and other key parameters, we ensured the model's simulations closely align with the observed hydrological dynamics. The calibrated range of parameters is presented in Table 14.1.

The calibration of the VIC model was conducted using streamflow data from 2014 to 2020. A comparison between the observed and simulated streamflow revealed a strong correlation (Fig. 14.2). The model's performance was further quantified by the Nash–Sutcliffe efficiency coefficient (NSE) of approximately 0.61 and a coefficient of determination (R^2) of around 0.71. These metrics indicate that the model can simulate the hydrological patterns of the study area with acceptable accuracy.

Hydroclimatic Projections

Changes in Climate Variables

Within the framework of the Shared Socioeconomic Pathways (SSPs), specifically under SSP245, which delineates a moderate mitigation scenario aiming for a median level of greenhouse gas (GHG) emissions, climate model projections indicate nuanced changes in precipitation and temperature patterns over the twenty-first century. Under this scenario, it is anticipated that precipitation will experience a

Table 14.1 Acceptable range and adopted values of the calibration parameter in the VIC model

Parameter	Physical meaning	Range	Adopted value
Ds	Fraction of Ds_{max} where non-linear baseflow begins	0–1	0.89
Ds_{Max}	Maximum velocity of baseflow	0–30	0.78
infilt	Variable infiltration curve parameter (binfilt)	0–0.4	0.35
$DEPTH_1$	Thickness of each soil moisture layer	0.01–0.04	0.31
$DEPTH_2$		0.4–1.5	1.4
$DEPTH_3$		1.5–4	3.9
W_s	Fraction of maximum soil moisture where non-linear baseflow occurs	0–1	0.017
MAX_ SNOW_ TEMP	Maximum temperature at which snow can fall	− 0.5 to 5	3.76
MIN_ RAIN_ TEMP	Minimum temperature at which rain can fall	− 3 to − 0.5	− 1.64

progressive increase, with projections estimating an approximate 6% augmentation during the interval from 2025 to 2055, and an enhancement of around 7.7% in the later period from 2070 to 2100. This increment suggests a gradual intensification of the hydrological cycle over the century. In terms of thermal dynamics, the SSP245 scenario forecasts a discernible trend in temperature augmentation. The maximum daily temperatures are projected to ascend by 8.48% within the initial period (2025–2055), with a further increase to 10.5% towards the century's conclusion (2070–2100). This indicates a consistent upward trajectory in daytime temperatures. The projections for minimum daily temperatures under SSP245 reveal a more pronounced escalation, with increases of 12.61 and 12.27% for the respective periods. This trend underscores a significant amplification in nighttime temperatures, suggesting heightened thermal retention within the atmosphere.

SSP585, a pathway assuming high GHG emissions without significant mitigation efforts, presents a starkly different climatological forecast. Within this scenario, precipitation is initially projected to decrease substantially by approximately 17.92% during the period from 2025 to 2055, followed by a partial recovery with a smaller decrement of 6.78% by 2070 to 2100. This trajectory implies increased variability and a potential for exacerbation of extreme hydrological conditions, including droughts. Temperature projections under SSP585 illustrate a pronounced warming trend. Maximum daily temperatures are expected to rise by 16.83% by mid-century and by an alarming 30.71% by the end of the century. Minimum daily temperatures are forecasted to surge even more dramatically, with increases of 31.32% and 35.18% for the two periods, respectively. This indicates an intense amplification of nighttime warmth, significantly exceeding daytime temperature increases. The spatial distribution of these changes is illustrated in Fig. 14.3. These projections illustrate the

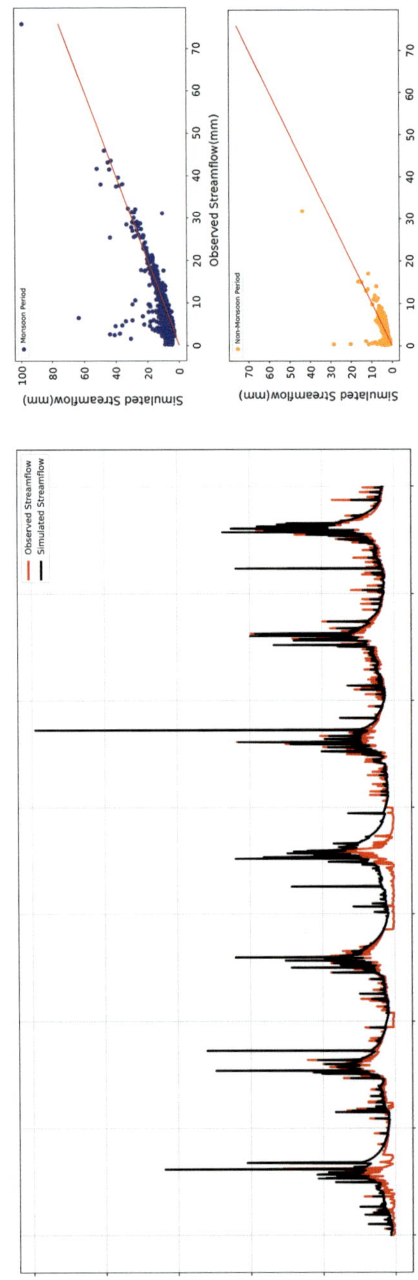

Fig. 14.2 Comparison of observed and simulated streamflow

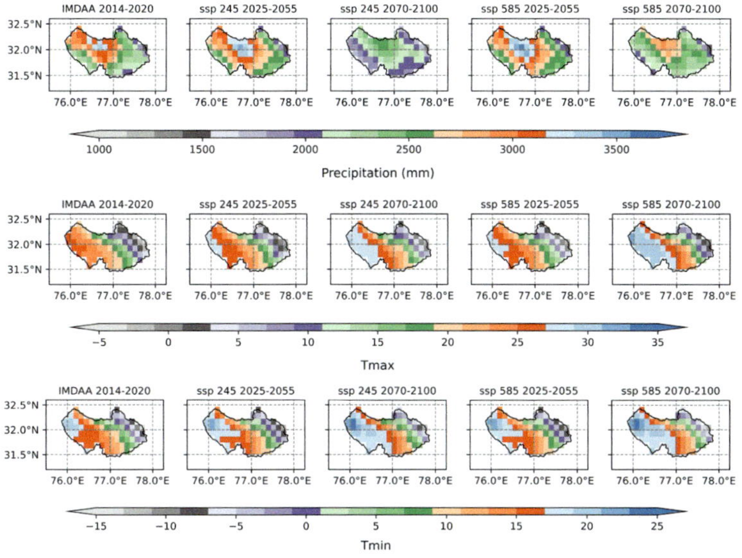

Fig. 14.3 Spatial distribution of projected precipitation and temperature changes in the Beas basin under current and future climate scenarios (SSP 245 and SSP 585)

critical influence of emission trajectories on future climate conditions. The SSP245 scenario demonstrates a steady rise in both precipitation and temperatures, adhering to a path of moderate climate change. On the other hand, SSP585 predicts initial reductions in precipitation leading to heightened drought risk, followed by a recovery and substantial increases in temperatures throughout the century. The pronounced warming, particularly of minimum temperatures under SSP585, portends profound consequences for ecological systems, agriculture, water resource management, and public health. These projections emphasize the imperative for robust mitigation strategies to curb GHG emissions and adaptive measures to improve the impacts of inevitable climatic alterations.

Projection of Hydrological Variables

The impact of climate change on the hydrology of the Beas Basin was explored by assessing changes in water balance components for two future periods (2025–2055 and 2070–2100) under the SSP 245 and SSP 585 scenarios, relative to the current climate conditions. The Variable Infiltration Capacity (VIC) model, fed with Global Circulation Model (GCM) outputs including projected rainfall and daily temperature ranges, was instrumental in simulating these future hydrological conditions. By comparing these simulations to baseline data from the IMDAA dataset (2014–2020)—where 52.83% of precipitation led to runoff, 14.93% to evapotranspiration (ET), and the remainder to various forms of storage—the study offers a nuanced view of potential climate impacts. Under SSP 245, indicative of moderate emissions, the portion of rainfall contributing to runoff and ET for 2025–2055 suggests an

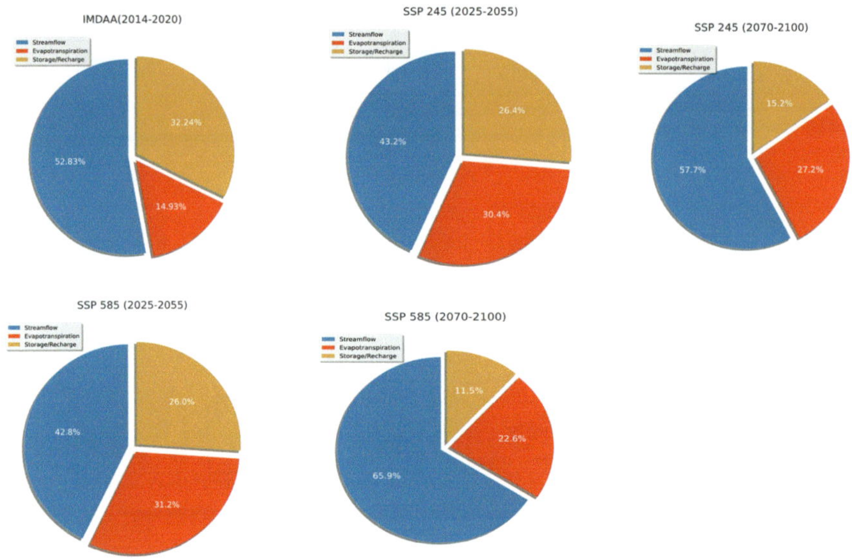

Fig. 14.4 Comparative analysis of water balance components in the Beas basin under current and future climate scenarios (SSP 245 and SSP 585)

increasing trend in ET. This shift becomes more pronounced by 2070–2100, with a notable increase in runoff, highlighting the potential for augmented flood risks and the critical need for strategic water resource management. The SSP 585 scenario, representing a high emissions trajectory, follows a similar pattern in the initial time slice but projects a significant hike in runoff by 2070–2100, alongside changes in ET and storage allocations. This anticipated rise in runoff, potentially fuelled by intensified precipitation or land-use alterations, highlights the urgency for adaptive strategies to combat flood risks and secure water sustainability. To visually encapsulate these findings, Fig. 14.4 presents a series of pie charts illustrating the varied percentage allocations of rainfall into runoff, ET, and storage across the current climate baseline and the future scenarios under SSP 245 and SSP 585.

14.3 Concluding Remarks

The systematic assessment of climate change impacts on the Beas River Basin under SSP245 and SSP585 scenarios reveals critical insights for future water resource management. The meticulously calibrated VIC model provides a reliable framework for projecting changes in hydroclimatic variables. It also highlights the profound implications of different greenhouse gas emission trajectories.

Key findings of the study include:

- Effective calibration of the VIC model is achieved using high-resolution observational data, leading to a methodological advancement in hydrological assessments. The calibration success is indicated by a Nash–Sutcliffe efficiency coefficient (NSE) of approximately 0.61 and a coefficient of determination (R^2) around 0.71, reflecting a good fit between observed and modelled data.
- Under SSP245, the moderate mitigation scenario, projections indicate a notable increase in precipitation and a consistent upward trajectory in temperatures, confirming the hypothesized intensification of the hydrological cycle in line with increased greenhouse gas concentrations.
- SSP585, the high-emission scenario, portends a stark decrease in precipitation followed by a significant uptick, alongside a substantial rise in temperatures. This underpins the theory of exacerbated hydrological extremes under unmitigated emission trajectories.
- Our projections reveal nuanced variations in water balance components, such as runoff and evapotranspiration, across different timeframes and emission scenarios. The VIC model's simulation of future hydrological conditions, including streamflow and storage capacities, informs us of potential changes in the basin's water availability and the risk of extreme hydrological events such as floods and droughts.

Based on these results which provide crucial input for policymakers as they develop integrated water resource management strategies. These strategies must be both robust and flexible enough to accommodate the projected changes in hydroclimate.

References

Anthes RA, Warner TT (1978) Development of hydrodynamic models suitable for air pollution and other mesometerological studies. Mon Weather Rev 106

Barbhuiya S, Raghuvanshi AS, Tiwari HL (2023) Assessment of streamflow in ungauged basin by using physical similarity approach. Arab J Geosci 16(12):672. https://doi.org/10.1007/s12517-023-11786-3

Bardossy A, Duckstein L, Bogardi I (1995) Fuzzy rule-based classification of atmospheric circulation patterns. Int J Climatol. https://doi.org/10.1002/joc.3370151003

Bellenger H, Guilyardi E, Leloup J, Lengaigne M, Vialard J (2014) ENSO representation in climate models: from CMIP3 to CMIP5. Clim Dyn 42(7–8):1999–2018. https://doi.org/10.1007/s00382-013-1783-z

Borah DK, Bera M, Xia R (2004) Storm event flow and sediment simulations in agricultural watersheds using DWSM. Trans ASAE 47. https://doi.org/10.13031/2013.17633

Chen DL, Chen YM (2003) Association between winter temperature in China and upper air circulation over East Asia revealed by canonical correlation analysis. Glob Planet Chang. https://doi.org/10.1016/S0921-8181(02)00199-6

Chen J, Brissette FP, Leconte R (2011) Uncertainty of downscaling method in quantifying the impact of climate change on hydrology. J Hydrol 401:190. https://doi.org/10.1016/j.jhydrol.2011.02.020

Ciarapica L, Todini E (2002) TOPKAPI: a model for the representation of the rainfall-runoff process at different scales. Hydrol Process 16:207. https://doi.org/10.1002/hyp.342

Colle BA, Wolfe JB, Steenburgh WJ, Kingsmill DE, Cox JAW, Shafer JC (2005) High-resolution simulations and microphysical validation of an orographic precipitation event over the Wasatch Mountains during IPEX IOP3. Mon Weather Rev 133:2947. https://doi.org/10.1175/MWR3017.1

Debreu L, Marchesiello P, Penven P, Cambon G (2012) Two-way nesting in split-explicit ocean models: algorithms, implementation and validation. Ocean Model 49–50:1. https://doi.org/10.1016/j.ocemod.2012.03.003

Deng X, Stull R (2005) A mesoscale analysis method for surface potential temperature in mountainous and coastal terrain. Mon Weather Rev 133:389. https://doi.org/10.1175/MWR-2859.1

Diaz-Nieto J, Wilby RL (2005) A comparison of statistical downscaling and climate change factor methods: impacts on low flows in the river Thames, United Kingdom. Clim Chang 69:245. https://doi.org/10.1007/s10584-005-1157-6

Dibike YB, Coulibaly P (2006) Temporal neural networks for downscaling climate variability and extremes. Neural Netw 19:135. https://doi.org/10.1016/j.neunet.2006.01.003

Dibike YB, Coulibaly P (2007) Validation of hydrological models for climate scenario simulation: The case of Saguenay watershed in Quebec. Hydrol Process 21(23):3123–3135. https://doi.org/10.1002/hyp.6534

Fauzi F, Kuswanto H, Atok RM (2020) Bias correction and statistical downscaling of earth system models using quantile delta mapping (QDM) and bias correction constructed analogues with quantile mapping reordering (BCCAQ). J Phys: Conf Ser 1538(1):012050. https://doi.org/10.1088/1742-6596/1538/1/012050

Fowler HJ, Kilsby CG, O'Connel PE, Burton A (2005) A weather-type conditioned multi-site stochastic rainfall model for the generation of scenarios of climatic variability and change. J Hydrol 308:1. https://doi.org/10.1016/j.jhydrol.2004.10.021

Fowler HJ, Blenkinsop S, Tebaldi C (2007a) Linking climate change modelling to impacts studies: recent advances in downscaling techniques for hydrological modelling. Int J Climatol 27(12):1547–1578. https://doi.org/10.1002/joc.1556

Fowler HJ, Blenkinsop S, Tebaldi C (2007b) Linking climate change modelling to impacts studies: recent advances in downscaling techniques for hydrological modelling. Int J Climatol. https://doi.org/10.1002/joc.1556

Fujihara Y, Tanaka K, Watanabe T, Nagano T, Kojiri T (2008) Assessing the impacts of climate change on the water resources of the Seyhan river basin in Turkey: use of dynamically downscaled data for hydrological simulations. J Hydrol 353:33. https://doi.org/10.1016/j.jhydrol.2008.01.024

Glahn HR, Lowry DA (1972) The use of model output statistics (MOS) in objective weather forecasting. J Appl Meteorol 11

Graham LP, Andreasson J, Carlsson B (2007a) Assessing climate change impacts on hydrology from an ensemble of regional climate models, model scales and linking methods—a case study on the Lule river basin. Clim Chang. https://doi.org/10.1007/s10584-006-9215-2

Graham LP, Hagemann S, Jaun S, Beniston M (2007b) On interpreting hydrological change from regional climate models. Clim Chang 81:97. https://doi.org/10.1007/s10584-006-9217-0

Gupta V, Singh V, Jain MK (2020) Assessment of precipitation extremes in India during the 21st century under SSP1–1.9 mitigation scenarios of CMIP6 GCMs. J Hydrol 590:125422. https://doi.org/10.1016/j.jhydrol.2020.125422

Hanasaki N, Yamamoto T (2010) H08 manual user's edition. National Institute for Environmental Studies

Harris LM, Durran DR (2010) An idealized comparison of one-way and two-way grid nesting. Mon Weather Rev 138:2174. https://doi.org/10.1175/2010MWR3080.1

Hattermann FF, Huang S, Burghoff O, Willems W, Osterle H, Buchner M, Kundzewicz Z (2014) Modelling flood damages under climate change conditions—a case study for Germany. Nat Hazards Earth Syst Sci 14:3151. https://doi.org/10.5194/nhess-14-3151-2014
Hess P, Brezowsky H (1977) Katalog der Großwetterlagen Europas (1881–1976). Berichte Des Deutschen Wetterdienstes 15
Hewitson BC, Crane RG (1996) Climate downscaling: techniques and application. Climate Res 7(2):85–95. https://doi.org/10.3354/cr007085
Holthuijzen M, Beckage B, Clemins PJ, Higdon D, Winter JM (2022) Robust bias-correction of precipitation extremes using a novel hybrid empirical quantile-mapping method. Theoret Appl Climatol 149(1–2):863–882. https://doi.org/10.1007/s00704-022-04035-2
Huth R (2002) Statistical downscaling of daily temperature in Central Europe. J Clim 15
Huth R (2004) Sensitivity of local daily temperature change estimates to the selection of downscaling models and predictors. J Clim 17
Ireson A, Makropoulos C, Maksimovic C (2006) Water resources modelling under data scarcity: coupling MIKE BASIN and ASM groundwater model. Water Resour Manag 20:567. https://doi.org/10.1007/s11269-006-3085-2
Islam SA, Bari MA, Anwar AHMF (2014) Hydrologic impact of climate change on Murray Hotham catchment of Western Australia: a projection of rainfall-runoff for future water resources planning. Hydrol Earth Syst Sci 18:3591. https://doi.org/10.5194/hess-18-3591-2014
Kankam-Yeboah K, Obuobie E, Amisigo B, Opoku-Ankomah Y (2013) Impact of climate change on streamflow in selected river basins in Ghana. Hydrol Sci J 58:773. https://doi.org/10.1080/02626667.2013.782101
Khoi DN, Hang PTT (2015) Uncertainty assessment of climate change impacts on hydrology: a case study for the central highlands of Vietnam, pp 31–44. https://doi.org/10.1007/978-3-319-10467-6_2
Kidson JW, Thompson CS (1998) A comparison of statistical and model-based downscaling techniques for estimating local climate variations. J Clim 11
Klein WH, Lewis BM, Enger I (1959) Objective prediction of five-days mean temperatures during winter. J Meteorol 16
Knutti R, Hegerl GC (2008) The equilibrium sensitivity of the earth's temperature to radiation changes. Nat Geosci 1:735. https://doi.org/10.1038/ngeo337
Kour R, Patel N, Krishna AP (2016) Climate and hydrological models to assess the impact of climate change on hydrological regime: a review. Arab J Geosci 9(9):544. https://doi.org/10.1007/s12517-016-2561-0
Larsen MAD, Refsgaard JC, Drews M, Butts MB, Jensen KH, Christensen JH, Christensen OB (2014) Results from a full coupling of the HIRHAM regional climate model and the MIKE SHE hydrological model for a Danish catchment. Hydrol Earth Syst Sci 11:4733. https://doi.org/10.5194/hessd-11-3005-2014
Lenderink G, Buishand A, van Deursen W (2007) Estimates of future discharges of the river Rhine using two scenario methodologies: direct versus delta approach. Hydrol Earth Syst Sci 11(3):1145–1159. https://doi.org/10.5194/hess-11-1145-2007
Liang X, Lettenmaier DP, Wood EF, Burges SJ (1994) A simple hydrologically based model of land surface water and energy fluxes for general circulation models. J Geophys Res 99:14415. https://doi.org/10.1029/94JD00483
Maraun D, Wetterhall F, Ireson AM (2010) Precipitation downscaling under climate change: recent developments to bridge the gap between dynamical models and the end user. Rev Geophys. https://doi.org/10.1029/2009RG000314
Masood M, Yeh PJF, Hanasaki N, Takeuchi K (2014) Model study of the impacts of future climate change on the hydrology of Ganges-Brahmaputra-Meghna (GBM) basin. Hydrol Earth Syst Sci 11:5747. https://doi.org/10.5194/hessd-11-5747-2014
Maurer EP, Adam JC, Wood AW (2009) Climate model based consensus on the hydrologic impacts of climate change to the Rio Lempa basin of Central America. Hydrol Earth Syst Sci 13:183. https://doi.org/10.5194/hess-13-183-2009

McCuen RH, Knight Z, Cutter AG (2006) Evaluation of the Nash-Sutcliffe efficiency index. J Hydrol Eng 11(6):597–602. https://doi.org/10.1061/(ASCE)1084-0699(2006)11:6(597)
Meehl GA, Boer GJ, Covey C, Latif M (2000) The coupled model intercomparison project (CMIP). Bull Am Meteorol Soc 81(2):313–318. https://doi.org/10.2307/26215108
Mendes D, Marengo JA (2010) Temporal downscaling: a comparison between artificial neural network and autocorrelation techniques over the Amazon basin in present and future climate change scenarios. Theor Appl Climatol 100:413. https://doi.org/10.1007/s00704-009-0193-y
Middelkoop H, Daamen K, Gellens D, Grabs W, Kwadijk JCJ, Lang H, Parmet BWAH, Schadler B, Schulla J, Wilke K (2001) Impact of climate change on hydrological regimes and water resources management in the Rhine basin. Clim Chang 49:105. https://doi.org/10.1023/A:1010784727448
Minville M, Brissette F, Leconte R (2008) Uncertainty of the impact of climate change on the hydrology of a Nordic watershed. J Hydrol 358:70. https://doi.org/10.1016/j.jhydrol.2008.05.033
Mudbhatkal A, Mahesha A (2018) Bias Correction methods for hydrologic impact studies over India's Western Ghats Basins. J Hydrol Eng 23:05017030. https://doi.org/10.1061/(ASCE)HE.1943-5584.0001598
Murphy J (1999) An evaluation of statistical and dynamical techniques for downscaling local climate. J Clim 12
Murphy J (2000) Prediction of climate change over Europe using statistical and dynamical downscaling techniques. Int J Climatol 20
Nazarenko LS, Tausnev N, Russell GL, Rind D, Miller RL, Schmidt GA, Bauer SE, Kelley M, Ruedy R, Ackerman AS, Aleinov I, Bauer M, Bleck R, Canuto V, Cesana G, Cheng Y, Clune TL, Cook BI, Cruz CA et al (2022) Future climate change under SSP emission scenarios with GISS-E2.1. J Adv Model Earth Syst 14(7). https://doi.org/10.1029/2021MS002871
O'Neill BC, Tebaldi C, van Vuuren DP, Eyring V, Friedlingstein P, Hurtt G, Knutti R, Kriegler E, Lamarque J-F, Lowe J, Meehl GA, Moss R, Riahi K, Sanderson BM (2016) The scenario model intercomparison project (ScenarioMIP) for CMIP6. Geosci Model Develop 9(9):3461–3482. https://doi.org/10.5194/gmd-9-3461-2016
O'Neill BC, Kriegler E, Ebi KL, Kemp-Benedict E, Riahi K, Rothman DS, van Ruijven BJ, van Vuuren DP, Birkmann J, Kok K, Levy M, Solecki W (2017) The roads ahead: narratives for shared socioeconomic pathways describing world futures in the 21st century. Glob Environ Chang 42:169. https://doi.org/10.1016/j.gloenvcha.2015.01.004
Ozer DJ (1985) Correlation and the coefficient of determination. Psychol Bull 97(2):307–315. https://doi.org/10.1037/0033-2909.97.2.307
Peleg N, Shamir E, Georgakakos KP, Morin E (2015) A framework for assessing hydrological regime sensitivity to climate change in a convective rainfall environment: a case study of two medium-sized eastern Mediterranean catchments, Israel. Hydrol Earth Syst Sci 19:567. https://doi.org/10.5194/hess-19-567-2015
Perrin C, Michel C, Andréassian V (2003) Improvement of a parsimonious model for streamflow simulation. J Hydrol 279(1–4):275–289. https://doi.org/10.1016/S0022-1694(03)00225-7
Phillips NA (1956) The general circulation of the atmosphere: a numerical experiment. Q J R Meteorol Soc 82:123. https://doi.org/10.1002/qj.49708235202
Piniewski M et al. (2013) Effect of modelling scale on the assessment of climate change impact on river runoff, Hydrolog Sci J 58(4):737–754. https://doi.org/10.1080/02626667.2013.778411
Qian W, Chang HH (2021) Projecting health impacts of future temperature: a comparison of quantile-mapping bias-correction methods. Int J Environ Res Public Health 18(4). https://doi.org/10.3390/ijerph18041992
Raaj S, Gupta V, Singh V, Shukla DP (2024) A novel framework for peak flow estimation in the Himalayan river basin by integrating SWAT model with machine learning based approach. Earth Sci Inf 17(1):211–226. https://doi.org/10.1007/s12145-023-01163-9
Raje D, Priya P, Krishnan R (2014) Macroscale hydrological modelling approach for study of large scale hydrologic impacts under climate change in Indian river basins. Hydrol Process 28:1874. https://doi.org/10.1002/hyp.9731

Rani SI, T, A., George JP, Rajagopal EN, Renshaw R, Maycock A, Barker DM, Rajeevan M (2021) IMDAA: high resolution satellite-era reanalysis for the Indian Monsoon Region. J Climate 1–78. https://doi.org/10.1175/JCLI-D-20-0412.1
Räty O, Räisänen J, Ylhäisi J (2014) Evaluation of delta change and bias correction methods for future daily precipitation: intermodel cross-validation using ENSEMBLES simulations. Clim Dyn 42:2287. https://doi.org/10.1007/s00382-014-2130-8
Rosenthal WD, Srinivasan R, Arnold JG (1995) Alternative river management using a linked GIS-hydrology model. Trans ASAE 38. https://doi.org/10.13031/2013.27892
Rummukainen M (1997) SwECL1M SWF.dish regional climate modelling programme. SMHI reports meteorology and climatology methods for statistical downscaling of GCM simulations
Sachindra DA, Huang F, Barton A, Perera BJC (2014) Statistical downscaling of general circulation model outputs to precipitation—part 2: bias-correction and future projections. Int J Climatol. https://doi.org/10.1002/joc.3915
Saito Y, Hanasaki N (2012) H08 manual analyzer's edition. National Institute for Environmental Studies
Schimming CG, Mette R, Reiche EW, Schrautzer J, Wetzel H (1995) Nitrogen fluxes in a typical agroecosystem in Schleswig-Holstein-measurements, budgets, model validation. J Plant Nutr Soil Sci 158
Schmidli J, Frei C, Vidale PL (2006) Downscaling from GCM precipitation: a benchmark for dynamical and statistical downscaling methods. Int J Climatol 26(5):679–689. https://doi.org/10.1002/joc.1287
Schoof JT (2013) Statistical downscaling in climatology. Geography Compass 7(4):249–265. https://doi.org/10.1111/gec3.12036
Sellers WD (1969) A global climate model based on the energy balance of the earth-atmosphere system. J Appl Meteorol 8
Semenov MA, Barrow EM (1997) Use of a stochastic weather generator in the development of climate change scenarios. Clim Chang 35:397. https://doi.org/10.1023/A:1005342632279
Shrestha S (2014) Assessment of water availability under climate change scenarios in Thailand. J Earth Sci Clim Chang 5:9. https://doi.org/10.4172/2157-7617.1000184
Shrestha RR, Dibike YB, Prowse TD (2012) Modelling of climate-induced hydrologic changes in the Lake Winnipeg watershed. J Great Lakes Res 38:83. https://doi.org/10.1016/j.jglr.2011.02.004
Soltani A, Hoogenboom G (2003) Minimum data requirements for parameter estimation of stochastic weather generators. Clim Res. https://doi.org/10.3354/cr025109
Sood A, Muthuwatta L, McCartney M (2013) A SWAT evaluation of the effect of climate change on the hydrology of the Volta river basin. Water Int 38:297. https://doi.org/10.1080/02508060.2013.792404
Tarboton DG (1994) Measurement and modeling of snow energy balance and sublimation from snow. International Snow Science Workshop
Tarboton DG, Chowdhury TG, Jackson TH (1995) A spatially distributed energy balance snowmelt model, in biogeochemistry of seasonally snow-covered catchments. Proceedings of a boulder symposium, IAHS 228
Teutschbein C, Seibert J (2010) Regional climate models for hydrological impact studies at the catchment scale: a review of recent modeling strategies. Geogr Compass. https://doi.org/10.1111/j.1749-8198.2010.00357.x
Teutschbein C, Seibert J (2012) Bias correction of regional climate model simulations for hydrological climate-change impact studies: review and evaluation of different methods. J Hydrol 456:297. https://doi.org/10.1016/j.jhydrol.2012.05.052
Teutschbein C, Seibert J (2013) Is bias correction of regional climate model (RCM) simulations possible for non-stationary conditions? Hydrol Earth Syst Sci 17:5061. https://doi.org/10.5194/hess-17-5061-2013

Thompson JR, Crawley A, Kingston DG (2016) GCM-related uncertainty for river flows and inundation under climate change: the Inner Niger Delta. Hydrol Sci J 61(13):2325–2347. https://doi.org/10.1080/02626667.2015.1117173

Toshiharu K (2005) Hydrological river basin environment assessment model (Hydro-BEAM). CRC Press

Tramblay Y, Ruelland D, Somot S, Bouaicha R, Servat E (2013) High-resolution Med-CORDEX regional climate model simulations for hydrological impact studies: a first evaluation of the ALADIN-climate model in Morocco. Hydrol Earth Syst Sci. https://doi.org/10.5194/hess-17-3721-2013

Trzaska S, Schnarr E (2014) A review of downscaling methods for climate change projections

van Vuuren DP, Edmonds J, Kainuma M, Riahi K, Thomson A, Hibbard K, Hurtt GC, Kram T, Krey V, Lamarque J-F, Masui T, Meinshausen M, Nakicenovic N, Smith SJ, Rose SK (2011) The representative concentration pathways: an overview. Clim Change 109(1–2):5–31. https://doi.org/10.1007/s10584-011-0148-z

Varis O, Kajander T, Lemmela R (2004) Climate and water: from climate models to water resources management and vice versa. Clim Chang 66:321. https://doi.org/10.1023/B:CLIM.0000044622.42657.d4

VonStorch H, Zorita E, Cubasch U (1993) Downscaling of global climate change estimates to regional scales: an application to Iberian rainfall in wintertime. J Clim 6

Wang M, Overland JE (2012) A sea ice free summer Arctic within 30 years: an update from CMIP5 models. Geophys Res Lett 39(18). https://doi.org/10.1029/2012GL052868

Warner TT, Peterson RA, Treadon RE (1997) A tutorial on lateral boundary conditions as a basic and potentially serious limitation to regional numerical weather prediction. Bull Am Meteorol Soc 78

Wigley TML, Jones PD, Briffa KR, Smith G (1990) Obtaining sub-grid-scale information from coarse-resolution general circulation model output. J Geophys Res. https://doi.org/10.1029/JD095iD02p01943

Wigmosta MS, Burges SJ (1997) An adaptive modeling and monitoring approach to describe the hydrologic behavior of small catchments. J Hydrol 202:48. https://doi.org/10.1016/S0022-1694(97)00057-7

Wigmosta MS, Lettenmaier DP (1999) A comparison of simplified methods for routing topographically-driven subsurface flow. Water Resour Res 35:255. https://doi.org/10.1029/1998WR900017

Wilby RL, Wigley TML (1997) Downscaling general circulation model output: a review of methods and limitations. Prog Phys Geogr 21:530. https://doi.org/10.1177/030913339702100403

Wilby RL, Wigley TML, Conway D, Jones PD, Hewitson BC, Main J, Wilks DS (1998) Statistical downscaling of general circulation model output: a comparison of methods. Water Resour Res 34:2995. https://doi.org/10.1029/98WR02577

Wilby RL, Hay LE, Leavesley GH (1999) A comparison of downscaled and raw GCM output: implications for climate change scenarios in the San Juan river basin, Colorado. J Hydrol 35:225. https://doi.org/10.1016/S0022-1694(99)00136-5

Wilby RL, Charles SP, Zorita E, Timbal B, Whetton P, Mearns LO (2004) Guidelines for use of climate scenarios developed from statistical downscaling methods document history

Wilks DS (1992) Adapting stochastic weather generation algorithms for climate change studies. Clim Chang 22:67. https://doi.org/10.1007/BF00143344

Xu Y, Gao X, Zhu Q, Zhang Y, Kang L (2014) Coupling a regional climate model and a distributed hydrological model to assess future water resources in Jinhua river basin, east China. J Hydrol Eng 20:04014054. https://doi.org/10.1061/(ASCE)HE.1943-5584.0001007

Yarnal B, Comrie AC, Frakes B, Brown DP (2001) Developments and prospects in synoptic climatology. Int J Climatol 21:1923. https://doi.org/10.1002/joc.675

Zhang DL, Chang HR, Seaman NL, Warner TT, Fritsch JM (1986) A two-way interactive nesting procedure with variable terrain resolution. Mon Weather Rev 114

Chapter 15
Precision Agriculture Technologies for Climate-Resiliency and Water Resource Management

Gaurav Jha, Udit Debangshi, Sindhu Palla, Farshina Nazrul, Sourajit Dey, Wri Dutta, and Sangeeta Bansal

Abstract This chapter explores how precision farming techniques and technologies are used to manage water resources and tackle the challenges brought by climate change in agriculture. Precision farming relies on automation, sensors, drones, and/or telemetry systems to monitor soil conditions, water availability, and weather patterns. By doing so, farmers can optimize productivity, use resources efficiently, and reduce greenhouse gas emissions. Precision water management helps farmers make learned decisions about irrigation, saving water, and making crops more resilient to extreme weather. Additionally, climate-resilient crops, sensing techniques, and agrivoltaics help manage water efficiently and reduce climate risks in farming. Technologies like remote sensing, wireless sensors, and agrivoltaics provide promising solutions for water resource management, ensuring sustainable farming practices despite water scarcity. These approaches improve water efficiency, boost crop yields, and enhance resilience to climate change, building a sustainable future for agriculture.

Keywords Precision agriculture · Climate-resilient · Crop yield

15.1 Introduction

Imagine earth as a big puzzle where climate, soil or land and water are important segments. Removing any one of these components can throw off the balanced cycle that sustains all living organisms. Climate is continuously changing and as

G. Jha (✉) · U. Debangshi · S. Palla · S. Dey · W. Dutta
Department of Agronomy, Kansas State University, Manhattan, KS, USA
e-mail: gjha@ksu.edu

F. Nazrul
Department of Electrical and Computer Engineering, Montana State University, Bozeman, MT, USA

S. Bansal
Department of Plant Science, California State University, Fresno, CA, USA

A. Nanda et al. (eds.), *Navigating the Nexus*, Water Science and Technology Library 102, https://doi.org/10.1007/978-3-031-76532-2_15

there are more people inhabiting the planet over time, the demand to increase food production by utilizing natural resources is increasing. Simultaneously, the weather is becoming increasingly unpredictable, seasons are shifting, and extreme events are happening more often. At the same time, the water we need for farming is also going through a lot of upheavals leading to—drought, flood or glacial melts. As climate change accelerates, its consequences are reverberating through the foundation of our agricultural systems. Consequently, conventional farming practices are being challenged by abrupt weather patterns, changing growing seasons and crop shifts, and an increasing frequency of extreme events. Concurrently, water resources are being affected by shifts in precipitation, melting glaciers, and changing river patterns. In the context of evolving water resource challenges, it is notable that global water usage has escalated dramatically, nearly six-fold since 1900 (Chellaney 2013). The largest share of this usage is agricultural, consuming 69% of water withdrawals, while industrial uses account for 19%, and municipal needs 12% (Plappally and Lienhard 2012). These trends and the impact of human activities and climate change on water systems highlight the growing need for effective water resource management tools among professionals like hydrologists and engineers, particularly in areas where traditional monitoring methods are insufficient. This chapter sheds light on the complex web of interactions, the cascading effects of climate change, and the delicate balance between precision agriculture, water resources, and optimization strategies.

15.1.1 Climate-Resilient Agricultural Practices

In the context of sustainable water management, resilience is defined as the capacity of agriculture system to maintain or enhance productivity by preventing, mitigating, or managing risks; adapting to changes in water status during stressed conditions, recovering from the impacts of climate changes and extreme weather events. In response to climate change, the concept of climate-smart agriculture has emerged as a comprehensive approach to promote adaptation and food security while effectively addressing climate change issues; often referred to as triple win (Lipper et al. 2014). These extreme weather events involve challenges associated with drought conditions, flooding, heat or cold waves, erratic rainfall patterns, prolonged dry spells, insect or pest population explosions, and other perceived threats as a result of climate change. These factors play a crucial role in determining the threats to global food security and the need to adopt resilience practices for cropping systems. Strengthening resilience involves the implementation of short and long-term climate mitigation and adaptation strategies. Additionally, ensuring transparent and inclusive participation of stakeholders in decision-making processes is crucial for the effectiveness of these strategies. Strengthening the grit of agricultural production systems in the face of climate challenges is important for sustaining global food security.

Climate resilient agriculture offers a multidisciplinary approach to tackle the interdependent issues of food security and climate change, with a focus on three primary goals outlined by Rao (2016). (1) sustainably enhancing agricultural productivity to ensure a fair increase in farm incomes and bolster food security; (2) building an adapting and resilient production system to climate change at multiple levels; (3) minimizing greenhouse gas emissions from agriculture (including livestock, fisheries, etc.), to the extent possible. In the realm of climate-smart agriculture, a diverse range of agricultural technologies and practices comes into play. These practices include conservation agriculture, agroforestry, soil and water conservation, advancements in crop genetics, and integration of precision farming. For instance, concerning water management, resilient **farm-scale** practices may include managed aquifer recharge, in-situ moisture conservation, farm ponds, and implementation of efficient irrigation systems. Moreover, **crop-scale** practices may include the use of drought and flood-tolerant varieties, intercropping systems, etc. Additional climate resilient practices include the adoption of diversified cropping systems for soil conservation, crop-livestock integration, developing disease-resistant crops and livestock breeds, and formulating risk management strategies to adapt to changing climate while enhancing farm profitability.

The adoption of climate smart agriculture yields several positive outcomes. Diversified cropping systems with incorporated cover crops contribute to nitrogen fixation and other nutrients in the soil, potentially eliminating the need for additional synthetic inputs. Crop-livestock integration improves soil health by directly synergistic with biogeochemical nutrient cycling and restoring grasslands, particularly when maintaining an appropriate animal grazing density. Such practices lead to increased resource use efficiency, reduced water, and energy requirements, decreased soil runoff and erosion, enhanced aggregate stability, and carbon sequestration in soils. Moreover, these practices enhance pest and disease resistance, thereby improving yield stability and resilience to extreme weather events.

15.2 Precision Farming Techniques for Resilient Agriculture

Precision farming has revolutionized modern agriculture, enhancing efficiency through automated and control systems, data-processing software, mobile apps, and web-based applications (Balasundram et al. 2023). Precision farming employs technologies such as the use of drone technologies, wireless sensor networks, satellites, sensors, robotics, Internet of Things (IoT) based automation, and mobile apps, which can be used to continuously monitor, evaluate, and manage several abiotic factors to enhance productivity and economics benefit. Satellites and high-resolution drone imagery are utilized to track the quality of crops, their water levels, soil moisture levels, and soil salinity. It also helps to generate yield maps and identify crop stress and soil health. The automation tools contribute to smart irrigation and adapting precision technologies even in remote areas.

Adaptation and mitigation strategies, related to precision tools and management practices, provide solutions to achieving higher yields with fewer inputs. Inappropriate irrigation and nutrient management practices significantly contribute to greenhouse gas emissions. Major anthropogenic greenhouse gases, methane, and nitrous oxide, consist of a global warming potential of 27.9, 273 times that of carbon dioxide in a 100-year timeframe (Arias et al. 2021). To mitigate these gas emissions and improve resource use efficiency, sensible use of water and nutrients through precision farming can be employed. For example, increasing nutrient use efficiency with the exact source, rate, place, and application method using precision input management reduced nitrous oxide emissions, mitigating greenhouse gas emissions.

As the impact of climate change persists, building climate resilience through precision and digital agriculture strategies becomes progressively vital enabling farmers to make informed decisions, conserve water resources, and enhance crop productivity. Resilient to drought and other extreme weather events can be put into force with such technologies. By optimizing irrigation practices, precision farming can help maintain the water levels at various aquifers, reservoirs, rivers, canals, and other water sources with easy access even in drought. Alongside precision farming, the adoption of climate-resilient varieties, rainwater harvesting, groundwater recharge, minimum tillage practices, crop diversification, slow-releasing fertilizers, carbon sequestration, and sustainable land use management plays an important role in moderating greenhouse gas emissions in the current scenario of climate change.

15.3 Technologies for Water Resource Management

Water stands as a fundamental and indispensable resource for agriculture playing an important role in ensuring food security. Conventional tools and techniques of irrigation scheduling and crop water use measurements pose significant challenges due to their time-consuming and labor-intensive nature, rendering them impractical for large-scale agricultural operations. By 2050, over half of the global population (57%) is projected to inhabit regions experiencing at least one month of water scarcity annually (Boretti and Rosa 2019). Consequently, mitigating water scarcity and improving water usage efficiency are critical aspects of contemporary water resource management. Precision agriculture facilitated by technological advancements offers a promising solution to this challenge. Precision irrigation preserves and enhances crop yields while conserving water by at least 15% or more than 50% (Zotarelli et al. 2009). By providing insights into spatiotemporal variability, recent research has explored precision technology as an alternative to conventional field tools for assessing plant stress indicators. Precision technologies include a range of methods including remote sensing, wireless sensor networks, LiDAR irrigation, and agrivoltics. These technologies enhance water use efficiency and are integral to sustainable agricultural goals.

15.3.1 Remote Sensing Tools

Remote sensing is a valuable tool in irrigation water management by providing spatial (space) and temporal (time) information on crop water content, particularly for large fields where ground-level measurements or point-based data are impractical to make decisions for entire fields or farms. It offers various avenues for integration into irrigation management, with the most effective method being the detection of water stress levels within plants to schedule irrigation appropriately (Bello et al. 2014). When plants experience water stress, their temperature rises above absolute zero emitting thermal infrared radiation which can be detectable by specialized sensors. This temperature data, coupled with air temperature and pressure allows the construction of water stress indices providing insights into crop water stress levels. Leaf spectral reflectance is also utilized to estimate crop coefficients for irrigation scheduling, leveraging bands at specific wavelengths affected by stress conditions (Bausch 1995). Several indices, including the normalized difference vegetation index (NDVI), and crop water stress indices (CWSI) play crucial roles in measuring water stress across visible and near-infrared spectral ranges. Additionally, the photochemical reflectance index (PRI) offers a substitute for thermal remote sensing, using affordable image sensors with high spatial resolution capabilities not feasible in the thermal domain (Suárez et al. 2010). With advancements in high-resolution and high-frequency remote sensing technologies, monitoring spatial changes in crop water and evapotranspiration at the field level has become increasingly accessible (Debangshi and Sadhukhan 2023).

15.3.2 Wireless Sensor Network (WSN)

Wireless Sensor Networks (WSNs) are highly efficient in irrigation systems due to their dynamic application, flexibility, self-monitoring capabilities, scalability, accuracy, and inter-sensor node collaboration (Hamami and Nassereddine 2020). WSNs comprise sensor nodes distributed throughout fields, communicating wirelessly via handheld devices. They facilitate long-term irrigation assessment without necessitating sensor removal for field operations, thus minimizing equipment damage and data loss (Debangshi and Sadhukhan 2023). WSNs prioritize intelligent controller valves and pumps during irrigation, aiming to allocate water equitably to meet all demands (Al-ammri and Ridah 2014). Advanced WSNs include a wide-area network comprising underground nodes connected to a single above-ground hub or base station, enhancing efficiency. These systems utilize sensor arrays and web-based decision support tools to gather real-time soil moisture data from farmers. This data is then transmitted to a server, enabling farmers to monitor their fields in real time. Moreover, the system employs irrigation management indicators to evaluate water efficiency with smart programming facilitating rapid irrigation recommendation calculations (Jha et al. 2022).

15.3.3 Light Detection and Ranging (LiDAR)

Light Detection and Ranging, a remote sensing technology that measures distances by sending laser pulses and analyzing the light that is reflected, is similar to radar but uses light pulses instead of radio waves; and operates by measuring the time it takes for light pulses to travel to collision points and back to sensors with precision (Rivera et al. 2023). These systems emit light pulses periodically, generating a map of the surrounding environment based on detected collisions.

Various types of LiDAR find applications in agriculture, with bathymetric LiDAR being notable for its use of green light to penetrate water for detection and mapping purposes. This technology aids in identifying water sources and suitable areas for well digging. Moreover, it enables the creation of maps indicating where irrigation may be unnecessary, such as in flood-prone areas or where soils retain moisture at saturation levels. Recent advancements in 3-dimensional LiDAR imagery coupled with natural color offer precise estimation of water stress and its location. The short-band laser light effectively penetrates crop canopies facilitating the detection of water stress within plants (Ahmad et al. 2021).

15.3.4 Agrivoltaics

Agriculture plays a critical role in ensuring economic stability and ensuring global food security. However, this sector is currently facing issues such as climate change, with restricted water resources being the most crucial limitation today. A potential solution to address this challenge is agrivoltaics, which integrates solar photovoltaic (PV) systems with agricultural practices. Agrivoltaics, also known as solar sharing or agri-photovoltaics, involves using land for both generating solar energy and doing agricultural activities. It can help reduce climate change by increasing agricultural production and reducing water usage, while also meeting the growing needs for food, energy, and renewable technology. The National Renewable Energy Laboratory (NREL) foresees that by 2030 utility-scale solar installations in the USA will cover 2 million acres, offering farmers opportunities to cultivate crops beneath solar modules, thus enhancing efficiency, conserving water, and promoting sustainable food production (Macknick et al. 2013). In addition to water storage and recycling, the relationship between solar panels and crops can be leveraged to store water in the shelter provided by the panels and to return the water used for panel cleaning for agricultural irrigation. Agrivoltaics not only enhances water management but also increases land productivity by as much as 70% through the integration of agricultural production and energy generation on a single site. By employing this dual land-use technique, the overall agricultural output is increased while resource utilization is optimized. Numerous studies demonstrate a positive correlation between agrivoltaics and water use efficiency (Zainol et al. 2021). For instance, a recent project in the arid regions of Arizona revealed that cultivating crops under solar panels reduces soil evaporation thus reducing irrigation needs by 50% alongside a more than 300% increase in crop yield through agrivoltaics implementation leading to more efficient water use in agriculture (Bello et al. 2014) (U.S. Department of Energy). Moreover, agrivoltaics systems contribute to increased crop resilience and productivity, particularly when adverse weather conditions such as drought prevail. Solar panels generate a favorable micro-climate through the provision of shade, which serves to mitigate heat stress and enhance the water efficiency of crops. In addition, the vegetation situated beneath the panels experiences enhanced energy efficiency, resulting in reduced water consumption for irrigation purposes (Time et al. 2024). Sometimes agrivoltics systems reduce crop yield by providing shade on the crop so to encounter this issue there are agrivoltaics systems based on Concentrated Lighting Agrivoltaics Systems (CAS) (Fig. 15.1) that transmit red, blue, and far-red light from the sun to plants for photosynthesis. Power is generated by the concentration and reflection of the residual sunlight onto photovoltaic panels (Liu et al. 2018). Alternately, grooved glass plates can be positioned between conventional PV panels (Even-lighting Agrivoltaics System) to promote optimal plant growth and uniform sunlight dispersion. Both approaches, which employ Agrivoltaics Spectral Differentiation (AVSD) effectively, have exhibited productive agricultural development and streamlined electrical production (Ali Abaker Omer et al. 2022). By implementing these two technologies, the even-lighting Agrivoltaics System effectively decreased accumulative water evaporation from pan and soil surfaces by 33% and 19%, respectively, compared to the Concentrated-lighting Agrivoltaics System's 11% and 14%. Subsequently, the moisture content of the soil is greater in areas unexposed to sunlight, including the water and soil surfaces beneath the systems (Liu et al. 2018).

15.4 Data Analytics and Decision Support Systems

The most efficient use of irrigation water is a multifaceted operation that depends on several variables and numerous data layers (such as plant development stage, weather parameters, soil condition, etc.), most of which are linked to different degrees of uncertainty (Jha et al. 2022). In regions that are vulnerable to drought, when water becomes scarce, the difficulties of this work become increasingly stressful (Singh Rawat et al. 2019). Since crop evapotranspiration (ET) is primarily used to calculate irrigation volume and schedule, an accurate prediction of the spatial ET (plant-scale) can help prevent overwatering or underwatering and so ensure optimal irrigation practices (Ghiat et al. 2021). Precision agriculture has advanced to the point that it is now feasible to anticipate many plant health monitoring indicators, including high-resolution spatial ET such as, Normalized Difference Vegetation Index (NDVI), Crop Coefficients (Kc), Land Surface Temperature (LST), Enhanced Vegetation Index (EVI), Normalized Difference Water Index (NDWI), Chlorophyll Content, Leaf Area Index (LAI) and many more (Abdulmumin et al. 1987). The primary sources of this data are in-situ radiometric and telemetric sensors (like Arable Mark III, Tule, and CropX), optical remote sensing technology (like satellite data, multispectral

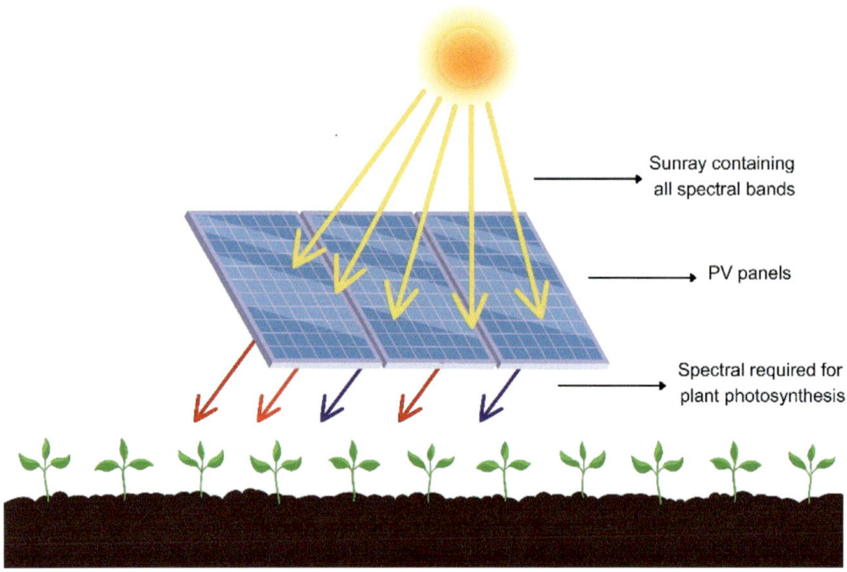

Fig. 15.1 Agrivoltaics systems based on concentrated lighting agrivoltaic systems (CAS)

imagery, etc.), and publicly accessible meteorological data (like Mesonet, the California Irrigation Management Information System (CIMIS), the National Weather Service (NWS), etc.) (Gavilán et al. 2019). The majority of these data are currently accessible to the public via PRISM Climate, Climate Data Online (CDO), and the National Oceanic and Atmospheric Administration (NOAA), among other websites. These resources, easily accessible by farmers, are weather-dependent temporarily and contribute to lower irrigation expenses overall (Ghiat et al. 2021; Guo et al. 2020).

Automated data-driven irrigation strategies can be conceptualized as a structured pipeline incorporating processed based modeling statistical and machine learning techniques, as illustrated in Fig. 15.2. The data collected from optical, telemetric, and radiometric devices including satellites, drones, local weather stations, ground sensors and it serves as the inputs to this pipeline. These inputs undergo sequential processing through distinct stages of the pipeline, outlined as follows.

15.4.1 Feature Selection

The process of selecting a subset of pertinent predictors, that are essential for precisely characterizing the target variable, is known as feature selection (Liu 2011). The input data types are the predictors in this scenario, and ET is the target variable. Numerous

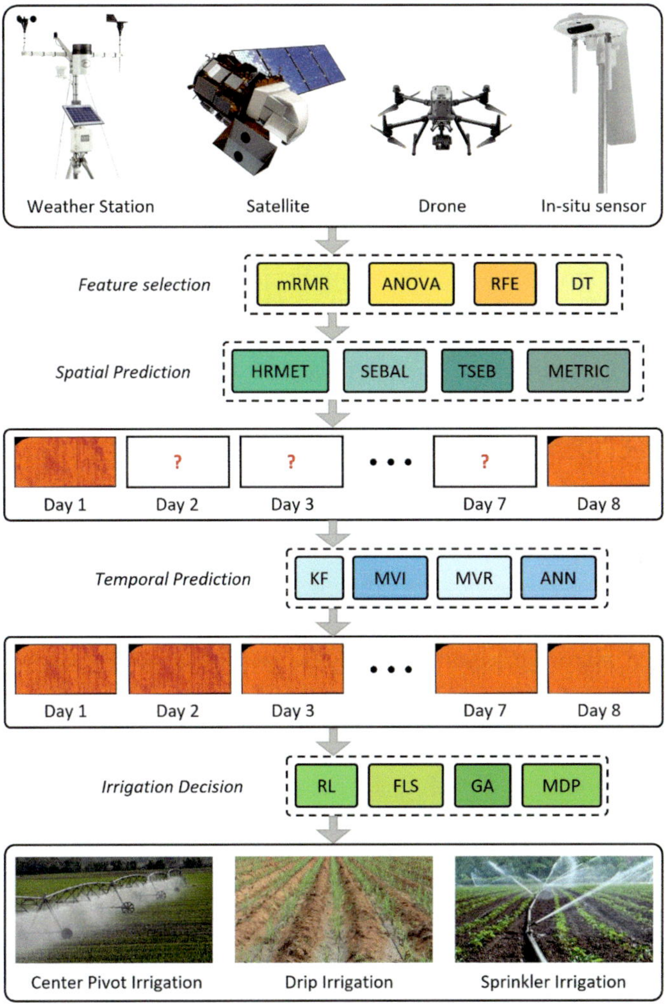

List of abbreviations

ANN: Artificial Neural Network	METRIC: Mapping EvapoTranspiration at high Resolution with Internalized Calibration
ANOVA: Analysis of Variance	mRMR: Minimum Redundancy Maximum Relevance
DT: Decision Tree based Methods	MVI: Multi Variate Interpolation
FLS: Fuzzy Logic System	MVR: Multi Variate Regression
GA: Genetic Algorithm	RFE: Recursive Feature Elimination
HRMET: High Resolution Mapping of EvapoTranspiration	RL: Reinforcement Learning
KF: Kalman Filtering	SEBAL: Surface Energy Balance Algorithm for Land
MDP: Markov Decision Process	TSEB: Two-Source Energy Balance

Fig. 15.2 A pipeline of data-driven automated irrigation system

metrics about crop, weather, and soil factors are accessible, similar to most real-world applications. However, not every one of these metrics has a strong correlation with ET. Incorporating extraneous information into the model might lead to needless complexity in computations and perhaps lower prediction accuracy (Guyon and Elisseeff 2003). Techniques for selecting features address this by retaining only the most important elements (Chandrashekar and Sahin 2014).

Several feature selection algorithms work better in certain situations. The kind of data, model assumptions, and quantity of the dataset all influence the method selection (Chandrashekar and Sahin 2014). For instance, model-specific techniques (such Recursive Feature Elimination (RFE) and Decision Tree (DT) based Methods) exist if the prediction model is predetermined (Guyon and Elisseeff 2003). However, there are several model-free techniques (such as Analysis of Variance (ANOVA), Minimum Redundancy Maximum Relevance (mRMR), etc.) that provide users the flexibility to use whatever model they like (Liu 2011; Witten et al. 2017).

15.4.2 Spatial Prediction

Even within a very small agricultural area, ET demonstrates significant spatial heterogeneity based on crop kinds, soil variability, and geographical topography (Wigmosta et al. 1994). Because of this variability, different parts of the field may require different irrigation levels, resulting in overwatering in some places and underwatering in others (Singh Rawat et al. 2019). Although radiometric sensors located on the ground have a good degree of accuracy in their specific region, they frequently do not record field fluctuation, which makes them ineffective for effective micro irrigation (Wang et al. 2020). Thus, the spatial ET prediction at the plant size is an essential component for irrigation water optimization (Ahmad et al. 2021).

With the most significant set of characteristics chosen (or all at once, depending on the feature selection technique), the remotely sensed input data can be used as an appropriate spatial prediction model (Singh Rawat et al. 2019). Other methods use empirical equations and iterative surface energy balance equation solutions to estimate the field ET (Ershadi et al. 2013). The Surface Energy Balance Algorithm for Land (SEBAL), Two-Source Energy Balance (TSEB), Mapping Evapotranspiration at High Resolution with Internalized Calibration (METRIC), and High-Resolution Mapping of Evapotranspiration (HRMET) are a few of the widely used models in this area (Hoffmann et al. 2016). These models use the field's thermal and multispectral reflectance photos, meteorological data, and accessible surface energy to calculate the sensible heat flux iteratively (Abdulmumin et al. 1987; Ershadi et al. 2013). The difference between the sensible heat flow and the available energy is then utilized to compute the latent heat flux, which is ultimately used to estimate surface ET (Ghiat et al. 2021). As shown in Fig. 15.2, the surface ET images provide a nice visualization and useful metric of plant behaviors.

15.4.3 Temporal Prediction

The data that the spatial ET predictors may produce is only relevant to the day and time of the picture capture (Wang et al. 2023). Daily observations and forecast of plants with ET, help growers to identify plant stress early on and adjust watering schedules appropriately (Ahmad et al. 2021). Frequent and regular gathering of thermal and multispectral spatial data is impractical due to technical limitations (Abdulmumin et al. 1987). For example, Aguilar et al. (2018) assessed the MOD16 algorithm, which estimates evapotranspiration using satellite data. They compared its results with on-ground eddy covariance measurements in five arid or semiarid locations in Northwestern Mexico used for wheat cultivation or hosting natural shrub vegetation. The study found variability in MOD16 performance across these sites, often underestimating evapotranspiration. While MOD16 provided reasonable crop water need estimations for these areas, the lack of extensive ground measurements limits the broader applicability of this satellite-based method.

On the other hand, operating aerial imagery is expensive and requires human effort and hence is not affordable for small-scale farmers (Jha et al. 2022). Consequently, temporal predictions (for each pixel or mini-batches per image) might assist in overcoming the challenge and contribute to a significant reduction in the expenses associated with collecting data (Guo et al. 2020).

Figure 15.2 demonstrates how the days between aerial imagery may be filled up with information via temporal prediction. Assume for the moment that the spatial ET data was available on days 1 and 8, and that the data for days 2 through 7 were absent. In this instance, temporal prediction may be used to estimate these data for the missing data days using the matching weather, soil, and on-the-ground sensor properties. Many intriguing methods exist in this regard, such as deep learning techniques like Support Vector Machine (SVM) regression, Long Short-Term Memory (LSTM), Feedforward Neural Network (FNN), etc.—the latter two models are members of the Artificial Neural Network (ANN) family—or statistical methods like Kalman Filtering (KF), Multi Variate Regression (MVR), Multi Variate Interpolation (MVI), etc. (Witten et al. 2017).

15.4.4 Irrigation Decision

The aforementioned approaches have thus far been relatively straightforward; they merely require the provision of certain input data to an established method. The matching output is either found by the model using empirical and/or statistical calculations, or it is learned via training instances. But in decision-making, the methods might become more inventive and difficult simultaneously (Torres-Sanchez et al. 2020). First, the amount of irrigation may be calculated using linear or polynomial regression approaches by integrating the pertinent data (weather and soil conditions, plant physiology and phenology, availability, and quality of water, etc.) (Torres-Sanchez et al. 2020). These techniques might not be particularly successful or adaptable to real-world circumstances since they oversimplify the irrigation approximation function (Puerto et al. 2013). Second, it is possible to train well-known machine learning models—like Random Forest (RF), Fuzzy Logic System (FLS), Support Vector Machine (SVM), and ANN—to determine the necessary irrigation quantity (Li et al. 2019). When enough training data is provided, the models will ultimately be able to mimic human decision-making. The training data can consist of either historical irrigation practices or expert human judgment (Pluchinotta et al. 2018). Naturally, these abovementioned approaches are mostly directly or indirectly dictated and, hence are limited by anthropogenic solutions (Chen et al. 2021).

The complexity of the process and the vast array of potential alternatives never guarantee a particular solution is the optimal one (Knuth 1974). At the same time, constant observation and quick decision-making are necessary due to the physical systems' dynamism and unpredictability (Torres-Sanchez et al. 2020). Modern artificial intelligence systems, like the Markov Decision Process (MDP), Genetic Algorithm (GA), and Reinforcement Learning (RL), can be used singly or in combination to achieve the highest possible level of water usage efficiency. These systems can even outperform human-provided solutions (Chen et al. 2021; Ding and Du 2022). However, these methods need detailed planning (quite specific to the irrigation system in question) and difficult physical system calibrations, which call for a high degree of human skill (Chen et al. 2021).

After they are set up, smart irrigation controllers (such as those for subsurface drip irrigation and micro-irrigation) may be immediately integrated with the outputs of these automated irrigation decision support systems (IDSS) (Ding and Du 2022). Towards notable contributions to optimizing irrigation and soil moisture management, Gutiérrez et al. (2013) introduced an automated irrigation system that employs a wireless sensor network and a GPRS module for real-time monitoring and control of soil moisture, aiming to conserve water in irrigation practices. For predicting soil moisture Gill et al. (2006) proposed a method using support vector machines relying on air temperature, relative air humidity, and soil temperature data. These studies collectively highlight the potential of integrating advanced technology, such as sensor networks and machine learning algorithms in enhancing agricultural efficiency and resource management. A novel irrigation sensor was developed by Jagüey et al. (2015) using a smartphone camera to estimate soil moisture. This system processes images from RGB to grayscale to determine the wet and dry areas of soil and sends this data to a water motor controller for efficient irrigation management.

15.5 Drones and Satellites; and Their Applications in Agriculture and Water Management

In the last two decades, the agricultural sector has seen a significant rise in the adoption of modern technologies aimed at enhancing various aspects of crop management. These advancements are designed to assist farmers in optimizing their net earnings by adopting recommendations from consultants and researchers, particularly in the sustainable utilization of limited resources such as water and raw materials. The water cycle, supply and demand for water, and the stress placed on the water systems are impacted by anthropogenic disruptions and climate change (Haddeland et al. 2014). With the global water withdrawal and usage in industries, agriculture and other municipal purposes, the amount of pure available water is reduced, making it essential for efficient utilization and management. Monitoring and assessment are essential tools for managing and evaluating the success of investments and agricultural initiatives. When combined with local data gathered through conventional methods that address data volume limits, remote sensing techniques are a potent tool for processing real-time data. Rainfall and evapotranspiration are two of the many aspects where different indices are correlated with water budget and management.

15.5.1 Rainfall for Water Budget and Management

Utilizing remote sensing data, an integrated automated application for the Water Resources and Agriculture Spatial Indicators System (WRASIS) was created to track and evaluate various agricultural development projects, analyzing remotely sensed time series data of (Agriculture Moderate Resolution Imaging Spectra-radiometer MODIS NDVI—Rain-Fall Estimate RFE 2.0) (Zahran et al. 2022). With its unique features of moderate spatial resolution, better spectral resolution, frequent observations, improved atmospheric calibration, and free download for end users, MODIS data packages present an amazing opportunity for land-cover and land-use change research (Xiao et al. 2003). Though most satellite image analysis and GIS software packages now in use lack a defined procedural setup for handling time series image analysis, and they are not tailored to the requirements of agriculture monitoring operations, several comparative analyses of crop monitoring were performed for better understanding (Ji-hua and Bing-fang 2008). The NDVI-MOD13-Q1 classification process comprises discriminating the quantity of vegetation at each pixel level and creating reflection maps of chlorophyll. To distinguish between fields with and without vegetation, the NDVI uses the difference in spectral reflectance in the Red (RED) and Near-Infra Red (NIR) bands, as per the following equation:

$$\mathrm{NDVI} = (\mathrm{NIR} - \mathrm{RED})/(\mathrm{NIR} + \mathrm{RED})$$

Every 16 days, 250 m spatial resolution global MOD13-Q1 data are made available as a unique mesh product in the Sinusoidal projection. To create a composite NDVI, algorithms must process data on a per-pixel level and require several observations over several days in a given area (Zahran et al. 2022).

15.5.2 Evapotranspiration-Based Irrigation Scheduling

For a long time, remote sensing has been acknowledged as the most practical way to deliver geographically dispersed regional evapotranspiration (ET) data on terrestrial surfaces (Moran 1994). The data is contemporary and often collected regularly from air or space platforms. The two main approaches are (i) land surface energy balance (EB) approach which makes use of surface temperature (radiometric) from an infrared (IR) thermal band and remotely detected surface reflectance in the visible (VIS) and near-infrared (NIR) regions of the electromagnetic spectrum and, (ii) Crop coefficient (Kc) based on reflectance (Kcr) and reference ET method, in which vegetation indices obtained from canopy reflectance values are linked to the Kc (Gowda et al. 2008).

(i) To estimate ET as a residual of the land surface energy balance equation, EB models translate satellite-detected radiances into land surface attributes such as leaf area index, vegetation indices, surface emissivity, and surface temperature as:

For local estimation: LE = Rn − G − H (Allen et al. 1998).

(where Rn is the net radiation resulting from the short and long-wave incoming and emitted radiation respectively,)

LE is the latent heat flux from evapotranspiration,

G is the soil heat flux, and

H is the sensible heat flux (W m^{-2} units)).

For regional estimation: Rn $= (1-\alpha)Rs + \varepsilon_a \sigma T_a^4 - \varepsilon_s \sigma T_s^4$ (Jackson et al. 1985) (where a is surface albedo, Rs is incoming short-wave radiation (W m^{-2}), σ is the Stefan–Boltzmann constant (5.67 E^{-08} W m^{-2} K^{-4}), ε is emissivity and T temperature (K) with subscripts "a" and "s" for air and surface, respectively.)

The second method uses measurements of R and NIR reflectance to calculate a vegetation index like the soil-adjusted vegetation index (SAVI) or normalized differentiation vegetation index (NDVI); where data from satellites or local meteorological stations along the timeframe forms the basis of computing ET. Thermal data obtained from satellites to estimate ET using the formula:

$$\mathrm{LE} = \mathrm{R_n} - G - \frac{r_a\ C_p\ (T_s - T_a)}{\mathrm{r}_{ah}}$$

where r_a is air density (kg m^{-3}), C_p is the specific heat capacity of the air (J kg^{-1} K^{-1}), and rah is aerodynamic resistance for heat transfer (s m^{-1}). T_s and T_a are expressed in K.

It is challenging to get accurate estimations of H, mainly when T_s is utilized in place of the (aerodynamic temperature) and when surface emissivity and atmospheric effects are not counted. Hence later, Rosenberg et al. (1983) incorporated the term surface aerodynamic temperature (To) to calculate H. Based on the manner H is estimated, several algorithms and models have been formulated from time to time. These models included the surface energy balance algorithm for land (SEBAL), which uses hot and cold pixels to develop an empirical temperature difference equation, the surface energy balance index (SEBI), which is based on the contrast between wet and dry areas, and the two-source model (TSM), which models the energy balance of soil and vegetation separately and then combines them to estimate total LE (Gowda et al. 2008). Mapping Evapotranspiration with Internalized Calibration (METRICTM); and ET mapping algorithm (ETMA) are some of the recent approaches to calibration (Loheide and Gorelick 2005). SEBI and SEBAL based on the crop water stress index (CWSI) are replaced with planetary boundary layer (PBL) scaling. Two independent, high-quality datasets were collected at the field scale during the soil moisture-atmosphere coupling experiment (SMACEX), which was used to evaluate SEBS, in the Walnut Creek agricultural watershed near Ames, Iowa, USA (Su et al. 2005).

(ii) For the reflectance-based crop coefficient method crop coefficients (Kc) have shown relatedness to vegetation indices like NDVI, and SAVI, where the crop coefficient is calculated mathematically using surface roughness, aerodynamic resistance, and albedo obtained from remote sensing (from LAI), without the requirement of crop development stage. Using multi-temporal, multi-spectral, digital, aerial video images collected during the growth season of a cotton crop, crop coefficients based on reflectance were determined and linked to the SAVI (Neale et al. 1996). In Kenya, crop ET was estimated using the Priestley-Taylor equation to derive crop coefficient (Kcr) based on reflectance from Landsat imagery (Michael and Bastiaanssen 2000).

One of the basic remote sensing platforms is satellites (Bansod et al. 2017). In the early 1970s with the launching of LANDSAT 1, satellite use was started for remote sensing in agriculture (Acker et al. 2014). After passing years of development in satellites for remote sensing, the EOS SAT-1 satellite is the latest one launched at the beginning of 2023. This satellite is unique with the most advanced tools for precision agriculture to date. The Wyoming Game and Fish Department utilizes the EOSDA Crop Monitoring, employing a scouting app for near-real-time field monitoring, streamlining data collection, enhancing efficiency, and enabling prompt response to vegetation issues, thereby reducing costs and improving the overall management of crop damage claims through the analysis of satellite imagery data (Elijah 2022). However, satellites and piloted aircraft record distant images that lack substantial information regarding the spatial and temporal dynamics of crop responses and might not match the resolution occupied by drones or UAVs (Table 15.1). To overcome these drawbacks, the use of drones has evolved in recent years for water management. The sensing systems, spectral and thermal cameras are often integrated or can be mounted on a drone using a gimbal system that is developed specifically for high-resolution mapping of crop water stress on spatial and temporal scales. It is reported that the worldwide agriculture drones market is anticipated to experience substantial growth, increasing from USD 4.5 Billion in 2023 to USD 17.9 billion by 2028, with a compounded annual growth rate (CAGR) of 31.5% over the specified period (Markets 2023).

Drones excel in capturing detailed data from complex environments, overcoming limitations of traditional methods like satellite and aerial imagery. It provides high-resolution spatial and temporal data crucial for managing surface and subsurface water resources, supporting various applications such as water quality monitoring and flood extent mapping. The increased use reflects its value in providing detailed environmental insights and shaping future hydrological studies and water resource management strategies. Advancements in technology have propelled UAVs for water resource management on the spatiotemporal scale (Elmeseiry et al. 2021). The drones offer a quick, accurate, and economically viable means for on-site analysis of water resources, particularly in complex environments. Koparan et al. (2018) developed a hexacopter-based system for on-site water quality measurements using open-source electronic sensors. This system was tested on a 1.1 ha agricultural pond and produced

Table 15.1 Contrasting the pros and cons of satellites and drones

Pros of satellites	Cons of drones
Fully autonomous	Often needs an operator
Unlimited accessibility along its orbit	Limited accessibility in areas which are inaccessible for the operator
Consistent imaging; continuous mapping possible	Inconsistent mapping or loss of continuous mapping due to limited flight time
Image pricing based on size of the area, resolution, and time of capture	Pricing based on duration of operation which is comparatively higher than satellites
Pros of drones	**Cons of satellites**
High-quality and resolution images due to proximity to the ground	Comparatively lower image quality
Weather condition (like cloudy) does not hinder imaging	Bad weather hinders imaging
Ease of timely deployment; can be operated close to land	Reliant on orbital parameters
Works better to record NDVI, MSAVI and GNDVI	Comparatively lower efficiency
Less revisiting time and ease of data extraction	High revisiting time with difficulty in data extraction
Low price with better flexibility of use	High price with lack of flexibility to use

Source Bansod et al. (2017)

accurate spatial distribution maps of parameters like water temperature and dissolved oxygen. However, limitations such as limited flight duration exist.

Ramírez Cuesta et al. (2019) introduced a user-friendly tool integrated with ArcGIS for estimating crop water needs using satellite imagery. This tool, utilizing Landsat 7 and 8, and Sentinel 2A images alongside the dual crop coefficient approach, effectively maps spatial variability in crop water demands with minimal inputs (Serrano et al. 2019) focused on monitoring pasture quality in the Montado ecosystem, crucial for extensive grazing systems. They compared two methods: proximal sensing with optical sensors and remote sensing using Sentinel 2 satellite images. Both methods computed NDVI and PDQI, revealing significant correlations. Values below 0.6 for NDVI and above 0.7 for PDQI indicated a need for supplementary feeding. However, issues like spring cloudiness can hinder remote sensing effectiveness, prompting the recommendation to combine both techniques for optimal pasture management.

Multispectral sensors capture light from the visible to near-infrared spectrum, typically between 400 to 1000 nm. They have a narrower bandwidth compared to RGB sensors but may experience geometric distortions depending on lens characteristics. They can also have spectral and radiometric distortions, requiring additional processing for accurate data interpretation, such as correcting atmospheric interference. Advanced multispectral sensors like the MicaSense RedEdge, Micasenses Altum PT (panchromatic, multispectral—Red, Green, Blue, Red Edge, Near Infrared; and Long Wave Infrared or thermal in the same camera system), and Slant Range 3P often include calibration systems but are costlier and heavier than RGB sensors. They can be integrated into UAVs separately or can be integrated as part of the system.

Hyperspectral sensors represent a significant advancement in remote sensing, offering remarkable spectral resolution across a broad spectral range. While UAV-based hyperspectral imaging provides a cost-effective alternative to satellite methods, challenges include higher costs and compatibility issues with standard UAV software (Adão et al. 2017). Notable hyperspectral sensors include the Resonon Pika-L and Micro-Hyperspec X-series NIR. Gao et al. (2019) explored a novel technique using UAV photogrammetry and image recognition to measure reservoir water levels, yielding reliable results at a Chinese hydropower station after introducing a correction method to counteract UAV drift. Multispectral (multiband) images (MSI) capture discrete and specific wavelengths, often in the visible (VIS) and near-infrared (NIR) regions, through individual images taken at specific wave numbers or wavelengths using filters or LEDs across the electromagnetic spectrum (Vidal and Amigo 2012). MSI consists of individual channel images, each capturing specific wavelengths. Unlike hyperspectral images, MSI discrete wavelength measurements require different treatments (Amigo 2019). Hyperspectral images (HSI) involve the measurement of one continuous spectrum for each pixel, typically expressed in nanometers or wave numbers (Zhang et al. 2022). These images can be acquired through various electromagnetic measurements, including visible, near-infrared, middle infrared, and Raman spectroscopy, showcasing versatile applications beyond traditional imaging modalities (Tran and Fei 2023). Emerging applications of hyperspectral imaging extend to technologies such as confocal laser microscopy scanners, Terahertz spectroscopy, X-ray spectroscopy, 3D ultrasound imaging, and magnetic resonance, showcasing its increasing popularity and diverse usage across various fields (Amigo 2019). In hyperspectral images, spectral resolution refers to the interval or gap between different wavelengths in a specific range, with a higher resolution achieved by acquiring more bands or spectral channels within a smaller wavelength range (Morales et al. 2021).

Thermal cameras capture far-infrared radiation emitted by objects on Earth, correlating directly with their surface temperatures, typically falling within the 3–100 μm range (Reach et al. 1995). The infrared portion of the spectrum can be categorized as near-infrared (NIR, 0.7–1.5 μm), short-wave infrared (SWIR, 1.5–3 μm), medium-wave infrared (MWIR, 3–8 μm), long-wave infrared (LWIR, 8–14 μm), and far-infrared (FIR, 14–100 μm). Thermal sensors operate in the long-wave spectrum, usually between 7.5 and 13.5 μm, allowing them to detect temperatures ranging from − 25 to 135 °C. The landscape, including vegetation, soil, water, and individuals, emits thermal infrared radiation in the 3.0–14 μm range of the electromagnetic spectrum, comprising of medium-wave infrared (MWIR) and long-wave infrared (LWIR), at its normal temperature of ~ 300 K (Sabins Jr and Ellis 2020). They boast a pixel sensitivity as precise as 0.04 °C (Kaplan 2007). Thermal cameras are categorized as uncooled or cooled based on the sort of image detector they use. A cooled infrared thermal camera typically uses mid-wave infrared wavelengths of 3–5 μm. It can offer enhanced resolution and the ability to detect minute temperature variations of the target. Cooled thermal cameras can detect smaller temperature variations for extended periods compared to uncooled cameras. However, its big size, high cost, and significant energy consumption make it unsuitable for use in field environments. Uncooled thermal cameras are lighter and cheaper than cooled ones, making them suitable for deployment on ground-based and UAV-based platforms. Despite causing significant inaccuracies in orthoimage production due to the poor contrast of thermal pictures, this issue can be resolved through calibration, utilizing a thermally controlled flat plate blackbody. Calibration methods were created to precisely calibrate the uncooled thermal camera for increased accuracy. The use of validation panels and temperature-controlled references to calibrate the thermal imagery involved extracting median pixel values from high- and low-temperature references, creating a linear equation to correlate pixel values with known temperatures, and calibrating all other pixels in the mosaic (Han et al. 2020). Thermal imagery has many uses in water management, including irrigation systems, crop water stress (Zhou et al. 2021), canopy water content (Elsherbiny et al. 2021), groundwater (USGS 2019), and plant water relation (Jones and Leinonen 2003). Thermal sensing methods have a non-contact nature, reduced labor requirements, and the ability to provide non-destructive monitoring for assessing crop stress based on leaf canopy temperatures (Grant et al. 2006). Temperature-based indices have demonstrated notable correlations between crop canopy temperature, stomatal conductance, and leaf water potential, with correlations strengthening as stress intensity increases. The most commonly used low-cost

Table 15.2 Optimal spectral values are important for developing spatial and temporal imagery for water management

Sensor	Spectral range	Parameters for water management and ancillary canopy parameters	References
1. RGB	Visible spectrum (450 nm–750 nm) 1. Red (620–750 nm) 2. Green (495–570 nm) 3. Blue (450–495 nm)	Sensitivity for red, green, and blue. Green pixels represent vegetation	Lu et al. (2024)
2. Multispectral	Wavelength ranges from (450–1000 nm) Visible spectrum (450–750 nm) Near infrared (NIR) (700–1000 nm)	Chlorophyll content, stress detection, vegetation cover Vegetation health and biomass	Wang et al. (2022)
3. Hyperspectral	Spectral wavelength from (350–2500 nm) Visible spectrum (450–750 nm) Near infrared (NIR) (700–1000 nm) Short wave infrared (SWIR) (1000–2500 nm)	Plant pigment detection (450–650 nm), nitrogen detection (405 nm), Chlorophyll absorption (687 nm) Biophysical Quantity and Biomass (687 nm–1100 nm) Water content in canopies (970 nm–2100 nm), plant stress (2295 nm), Heavy metal stress (1650 nm)	Chen et al. (2021)

thermal sensors are FLIR, where it combines the accurate temperature measurement with an image up to 1030 °C (Wu et al. 2019) (Table 15.2).

15.5.3 Limitations of Data

- *Spatial and temporal resolution*: For the acquisition of data from satellite images spatial (pixel size) and temporal (repeat frequency) resolution varies from different forms of spectral bands obtained from their respective satellite, varying between Landsat, MODIS, and sentinel images. While Landsat 5 has a 16-day repetition cycle with 30–120 m spatial resolution, MODIS uses the same satellite platform to provide thermal images with a 1000 m resolution, in contrast to the images collected in other bandwidths with a 250 m resolution. Most of the time the pixel size of these maps is larger than a single field, resulting in substantial errors in ET estimation, ET maps derived from remote sensing data obtained by daily-coverage satellite-based sensors like MODIS, GOES (geostationary environmental satellite), and AVHRR (advanced very high-resolution radiometer) are too coarse to be of any use in agricultural regions (Tasumi et al. 2006).

- Accuracy of data: Gaps in data along a timeframe or the efficacy of data based on pixels, the true potential can at times lead to errors in spatial estimation. When TSM hardly requires extreme temperature (T_s) pixels, METRIC, SEBAL, and SEBS are largely dependent on T_s (Gowda et al. 2008)
- Processing time: Instantaneous real-time imagery and data with rapid processing calls the aid of artificial intelligence for increasing efficacy and simplicity of the data processing. More research in this area is needed, even though several studies are now being conducted to examine the transferability of crop coefficients to different locales (Howell et al. 2012).
- Model implementation: Different remote sensing-based ET methods are available for calculating the amount and trends of regional ET, varying in different ranges of complexity. To accurately estimate using observed data, it is necessary to properly utilize the integration of LE pixels while using different heat flux source models (Chávez et al. 2005).
- Governmental policies and limitations: Variations in worldwide data access and their limitations for better resolution images like Worldview data restricting the use of specific algorithms and models globally (Harris and Baumann 2015).

15.6 Emerging Technologies

15.6.1 Nanosensors and Nanobionics

Nanosensors and biosensors strengthen agricultural productivity and safeguard agroecosystems by sensing soil nutrients, microclimate, temperature, moisture, heavy metals, pesticides, plant diseases, and harmful substances. Nanosensors can assist farmers in controlling their farms precisely and identifying plant requirements on time. It transforms conventional farming practices into smart farming, promoting energy efficiency and sustainability. Nanosensors can also detect soil and plant moisture, pesticide residue, and nutrient requirements. Nanosensors outperform traditional sensors due to their high surface-to-volume ratio, quick response time and consistent results. This emerging area of science termed **nanobionics** is an interdisciplinary field that combines nanotechnology with biology to create innovative solutions for various applications. For agriculture water management, nanobionics plays a crucial role in developing advanced technologies that integrate nanomaterials with biological systems to enhance water efficiency, monitor water quality and optimize irrigation practices (Fig. 15.3).

Nanobionics pioneers the use of nanomaterials to optimize water absorption and retention within plants. By delicately modifying plant structures at the nanoscale, nanobionics enhances the efficiency of water transport within crops. Photosynthesis

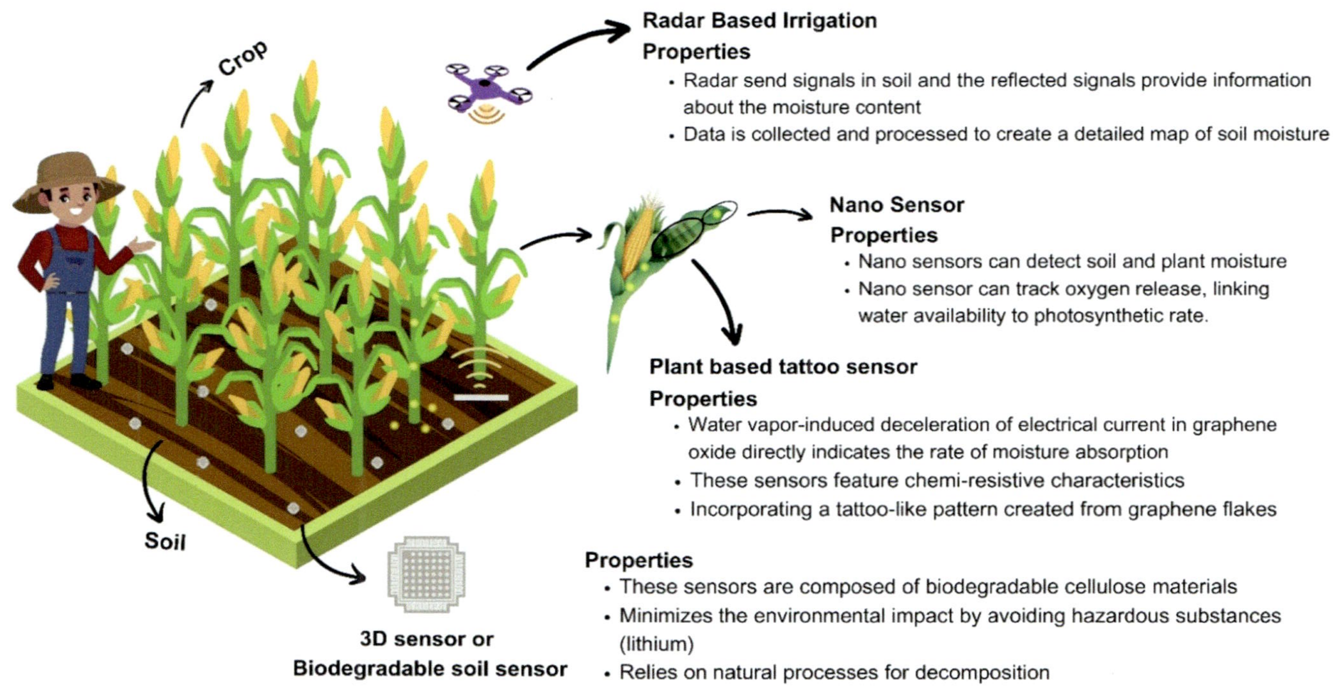

Fig. 15.3 Emerging precision water management technologies for data intensive irrigation

transforms sunlight energy into food with the help of water. The release of oxygen from plants is linked to photosynthesis. In addition, water indirectly influences the photosynthetic rate. When there is not enough water, the stomata close, lowering photosynthesis. Consequently, a correlation can be observed between the availability of water and the emission of oxygen, which can be monitored through the utilization of a nanosensor. Clark-type polarographic electrode sensors have been widely employed in the detection of oxygen current flow.

The gravimetric method of soil water content measurement is reliable, but it is time-consuming and not suitable for continuous monitoring of large fields. On the other hand, modern technologies such as nanosensors can be utilized to continuously monitor soil water. It is affordable and produces rapid results. An optics-based nanosensor developed using an aluminum oxide nano-porous ceramic plate is used for continuous monitoring of soil water (Chen et al. 2019). The sensor is connected to a silicon diaphragm which is sensitive to water stress. When buried in unsaturated soil it experiences a negative pressure which bends the diaphragm. This displacement can be used to measure dry soil saturation (Chen et al. 2019). Leone et al. demonstrated a small, affordable sensor that measures the water content of soil by utilizing an optical fiber with near-infrared-based detection. The sensor is placed in a protective tube buried in a soil tank for soil testing. The probe is generally inserted at different depths of soil which measure the reflected spectrum. The reflected spectrum decreases as the amount of water increases (Leone et al. 2022).

Nanobionics contributes significantly to the development of advanced water filtration systems tailored for agricultural use. Nano-enabled filters act as sentinels, efficiently removing impurities, sediments, and contaminants from irrigation water. The deployment of clean and high-quality water to crops becomes a reality, fostering robust plant health and crop yield. Additionally, nanobionics coatings minimize water evaporation and act as barriers against water loss. The result is a landscape where water retention is optimized, water conservation is prioritized, and efficient irrigation water use becomes a hallmark of sustainable agriculture.

15.6.2 Plant Tattoo Sensors

The ingenuity and fundamental principle underlying plant tattoo sensors reside in their capacity to efficiently and non-invasively monitor water dynamics within crops. Conventional methods employed on the ground to evaluate plant water stress are frequently arduous and time-intensive, and they might fail to furnish comprehensive spatial-scale data which is sometimes essential for large-scale agricultural operations. This sensor allows researchers to gain valuable insights into crop health and water use efficiency by measuring how long it takes for water to move from the roots to different parts of the plant (water dynamics).

In large-scale farm operations, conventional ground-based techniques are insufficient for efficiently and effectively monitoring the water in crops due to their time-consuming and labor-intensive nature. These generate large amounts of data and occasionally produce data points that are not representative of the entire field or are incorrect. Plant tattoo sensors are used for monitoring the dynamics of plant water stress in a non-invasive and efficient way. The plant tattoo sensors possess intricate and chemi-resistive characteristics which are infused with a tattoo-like pattern created from several layers of graphene flakes. Graphene is made up of a monolayer of carbon and oxygen atoms that exhibits excellent electrical conductivity (Hossain et al. 2024). It possesses excellent electrical conductivity and high mechanical strength, ensuring its resistance to breakage. Tattoo sensors designed for plants are attached to the leaf with adhesive tape and connected to small wires (Fig. 15.3). A tiny battery is connected to a device that measures electrical conductivity.

Plants absorb water from the soil and release it through stomata in their leaves. These water vapors are the primary indicator in the plant tattoo sensor. The presence of water vapor reduces the speed of electrical current passing through graphene oxide. The magnitude of the electrical conductivity varies according to the velocity at which water ascends within the plant. This determines the rate at which the plant absorbs moisture from the soil (Oren et al. 2017). With the deceleration of electrical current following the attachment of a sensor to a leaf, researchers can determine the rate at which the plant absorbs water. Researchers from Iowa State University (United States of America) have shown that applying a plant tattoo sensor induces substantial development in corn plants. An accurate result was obtained by measuring the time required for two varieties of corn plants to transport water from their roots to their lower leaves and upper leaves (predictivephenomicsinplants.iastate.edu).

15.6.3 Biodegradable Soil Moisture Sensors

The in-situ monitoring of soil volumetric water content is critical owing to its substantial impact on the hydrological, biochemical, and economic aspects of agriculture (Gopalakrishnan et al. 2022). The most effective instruments for measuring in-situ soil moisture are wireless sensors with telemetry. However, the majority of such sensors are equipped with embedded electronic components and active onboard batteries which increase production and assembly expenses and restrict the number of nodes that can be deployed in the field (Fig. 15.3). Additionally, hazardous substances such as lithium and lead can escape from batteries in sensors which is also not sustainable over time. To address these issues, the integration of radio-transmitting biodegradable sensors in conjunction with antennas and drones is now used in the agricultural field (Oren et al. 2017). All sensor tags are embedded in the soil, and a flying drone collects reflected signals from the sensors while flying. The reflected signals dependent on soil properties, such as volumetric water content (VWC), can be effectively transmitted to signal stations for action. These sensors, composed of biodegradable cellulose materials (paper-based substrates), gradually degrade after a few months to years of operation (Oren et al. 2017). For example, researchers developed a wirelessly powered biodegradable soil sensor. The sensor is made up of a tin conductive line and a nanopaper substrate that has been coated with a natural wax for protection. The sensor is mostly decomposed by soil bacteria and the other components are also environmentally friendly (Kasuga et al. n.d.).

15.6.4 Radar-Based Variable Rate Irrigation (VRI) Systems

Radar technology, known as radio detection and ranging, works by transmitting and receiving electromagnetic waves that reflect from objects. By analyzing the time, frequency, and angle of these reflected waves, radar systems can provide characteristics and changes in objects, including their distance, speed, shape, and movement. These systems are adaptable, functioning within different frequency bands like X-band, C-band, L-band, or P-band, tailored to specific requirements for resolution, range, and penetration.

The integration of radar sensors, such as ground-penetrating radar (GPR), has revolutionized agricultural practices, particularly in precision irrigation (Fig. 15.3). The GPR enables continuous and non-invasive monitoring of soil moisture content, allowing farmers to implement variable rate irrigation based on real-time and in-situ soil moisture data. Farmers can precisely adjust water application rates, minimizing water wastage, reducing operational costs, and optimizing crop yields. Additionally, for remote sensing methods for soil characterization, radar technology has emerged as a valuable tool for mapping soil conditions over large areas. Techniques like satellite imaging, Synthetic Aperture Radar (SAR), radiometry, Time Domain Reflectometry (TDR), and GPR, radar offer unique advantages in measuring soil water content and topsoil thickness as soil moisture changes in real time. The radar technology in agricultural systems like Autonomous Pivot (https://www.autonomouspivot.com/) further improves water use efficiency and management efforts. By integrating a GPR system into the moving pivot, farmers can accurately detect soil moisture status and adjust water application rates accordingly. Radar-equipped drones also play a pivotal role in augmenting water management practices in agriculture. These drones outfitted with radar sensors such as P-band radar, efficiently survey large agricultural areas, providing detailed insights into underground water profiles and soil moisture distribution (Frid and Frid 2024). Leveraging cloud computing and intelligent software, radar-equipped drones process and analyze data in real-time, enables the generation of tailored irrigation plans suited to specific field conditions (Sinchi et al. 2023).

15.7 Challenges and Opportunities

Precision water management represents a promising approach to address the complex challenges posed by climate change and water resource management. However, its successful implementation is accompanied by a set of challenges that need to be navigated effectively. One of the primary hurdles lies in the availability and quality assurance of data validation for informed decision-making. Remote sensing technologies offer extensive datasets, but sometimes ensuring their accuracy, reliability, and consistency can be a significant challenge. Data acquired from drones and satellites may be subject to errors, noise, and inconsistencies, if it does not have validation and calibration processes accounted during photogrammetry or image processing steps.

Moreover, the technical complexity for farmers associated with precision water management presents another obstacle. Integrating diverse datasets, sensor technologies, and analytical tools into existing water management frameworks can be daunting. Challenges such as interoperability, data compatibility, and system integration need to be addressed to ensure seamless operation. Stakeholders may require specialized expertise and resources to navigate and overcome these technical barriers effectively. The cost of adopting precision water management practices poses a significant financial challenge, especially for resource-constrained regions and small-scale irrigated farms impacted by drought. High initial investment costs, subscription fees for data services, and the need for specialized equipment and software contribute to the financial burden.

Furthermore, human capacity and institutional readiness are critical factors influencing the successful implementation of precision water management strategies. Effective utilization of remote sensing data and technology requires a skilled workforce with expertise in water resource management, data analytics, and technology deployment. However, there is often a shortage of trained professionals and institutional capacity to train for adoption of emerging technologies effectively. Investing in human capacity through training programs, knowledge exchange initiatives, and academic collaborations is essential to address this challenge. Technological innovations and advancements in remote sensing technologies, machine learning algorithms, and data analytics hold potential for improving the accuracy and efficiency of water management data, strategies and decisions.

15.8 Summary

Precision farming has offered a scientific approach to modern agriculture, using sensors and drones to monitor soil, water, and weather conditions. By making informed decisions on *per drop per crop* and adopting resilient crop varieties, farmers can enhance crop resilience and conserve water resources. Innovations in sensing systems and aerial imagery have further contributed to sustainable farming practices enriching soil health, and maximizing productivity while minimizing environmental impact. Technologies and telemetry have enabled precise water management, optimizing crop growth and resource use efficiency. To sum up, the ideas covered in this chapter, precision farming has the potential to help with climate change issues and provide food security in the future. Increased adoption of these technologies can help farmers nurture resilient and eco-friendly agricultural systems, safeguarding the welfare of both humanity and the planet.

References

Abdulmumin S, Myrup LO, Hatfield JL (1987) An energy balance approach to determine regional evapotranspiration based on planetary boundary layer similarity theory and regularly recorded data. Water Resour Res 23(11):2050–2058. https://doi.org/10.1029/WR023i011p02050

Acker J, Williams R, Chiu L, Ardanuy P, Miller S, Schueler C, Vachon PW, Manore M (2014) Remote sensing from satellites. In: Reference module in earth systems and environmental sciences. Elsevier, p B9780124095489094409. https://doi.org/10.1016/B978-0-12-409548-9.09440-9

Adão T, Hruška J, Pádua L, Bessa J, Peres E, Morais R, Sousa JJ (2017) Hyperspectral imaging: a review on UAV-based sensors, data processing and applications for agriculture and forestry. Remote Sensing 9(11):1110

Aguilar AL, Flores H, Crespo G, Marín MI, Campos I, Calera A (2018) Performance assessment of MOD16 in evapotranspiration evaluation in Northwestern Mexico. Water 10(7):901. https://doi.org/10.3390/w10070901

Ahmad U, Alvino A, Marino S (2021) A review of crop water stress assessment using remote sensing. Remote Sensing 13(20):4155

Al-ammri AS, Ridah S (2014) Smart irrigation system using wireless sensor network. Int J Eng Res 3(1)

Ali Abaker Omer A, Liu W, Li M, Zheng J, Zhang F, Zhang X, Osman Hamid Mohammed S, Fan L, Liu Z, Chen F, Chen Y, Ingenhoff J (2022) Water evaporation reduction by the agrivoltaic systems development. Solar Energy 247:13–23.https://doi.org/10.1016/j.solener.2022.10.022

Allen RG, Pereira LS, Raes D, Smith M (1998) Crop evapotranspiration-Guidelines for computing crop water requirements-FAO irrigation and drainage paper 56. Fao, Rome 300(9):D05109

Amigo JM (2019) Hyperspectral and multispectral imaging: setting the scene. In: Data handling in science and technology, vol 32. Elsevier, pp 3–16. https://doi.org/10.1016/B978-0-444-63977-6.00001-8

Arias P, Bellouin N, Coppola E, Jones R, Krinner G, Marotzke J, Naik V, Palmer M, Plattner G-K, Rogelj J (2021) Climate change 2021: the physical science basis. Contribution of working group I to the sixth assessment report of the intergovernmental panel on climate change; technical summary. https://elib.dlr.de/137584

Balasundram SK, Shamshiri RR, Sridhara S, Rizan N (2023) The role of digital agriculture in mitigating climate change and ensuring food security: an overview. Sustainability 15(6):5325

Bansod B, Singh R, Thakur R, Singhal G (2017) A comparision between satellite based and drone based remote sensing technology to achieve sustainable development: a review. J Agric Environ Int Develop (JAEID) 111(2). https://doi.org/10.12895/jaeid.20172.690

Bausch WC (1995) Remote sensing of crop coefficients for improving the irrigation scheduling of corn. Agric Water Manag 27(1):55–68

Bello M, Nasidi N, Shanono N (2014) Remote sensing as a tool for irrigation water management

Boretti A, Rosa L (2019) Reassessing the projections of the world water development report. Npj Clean Water 2(1), Article 1. https://doi.org/10.1038/s41545-019-0039-9

Chandrashekar G, Sahin F (2014) A survey on feature selection methods. Comput Electr Eng 40(1):16–28. https://doi.org/10.1016/j.compeleceng.2013.11.024

Chellaney B (2013) Water, peace, and war: confronting the global water crisis. Rowman & Littlefield. https://books.google.com/books?hl=en&lr=&id=U75ehIi6uvUC&oi=fnd&pg=PR5&dq=Chellaney,+B.,+2013.+Water,+peace,+and+war:+Confronting+the+global+water+crisis.+Rowman+%26+Littlefield.&ots=8wWmk33ZDS&sig=KqTYVtUUVev3NKmlXOpNbwMPnYM

Chen M, Cui Y, Wang X, Xie H, Liu F, Luo T, Zheng S, Luo Y (2021) A reinforcement learning approach to irrigation decision-making for rice using weather forecasts. Agric Water Manag 250:106838. https://doi.org/10.1016/j.agwat.2021.106838

Chen Y, Tian Y, Wang X, Dong L (2019) Miniaturized soil sensor for continuous, in-situ monitoring of soil water potential. In: 2019 20th international conference on solid-state sensors, actuators

and microsystems & eurosensors XXXIII (TRANSDUCERS & EUROSENSORS XXXIII), 2025–2028. https://doi.org/10.1109/TRANSDUCERS.2019.8808562
Debangshi U, Sadhukhan A (2023) Crop water stress monitoring through precision technologies: a review. Pharma Innov 12:803–808. https://doi.org/10.22271/tpi.2023.v12.i11Sk.24052
Ding X, Du W (2022) DRLIC: deep reinforcement learning for irrigation control. In: 2022 21st ACM/IEEE international conference on information processing in sensor networks (IPSN), pp 41–53. https://doi.org/10.1109/IPSN54338.2022.00011
Elijah R (2022) EOSDA solution improves operations in nature agency. Retrieved from EOSDA Blog: https://eos.com/blog/wyoming-game-and-fish-department-case-study/
Elmeseiry N, Alshaer N, Ismail T (2021) A detailed survey and future directions of unmanned aerial vehicles (uavs) with potential applications. Aerospace 8(12):363
Elsherbiny O, Zhou L, Feng L, Qiu Z (2021) Integration of visible and thermal imagery with an artificial neural network approach for robust forecasting of canopy water content in rice. Remote Sensing 13(9):1785. https://doi.org/10.3390/rs13091785
Elshikha DE (n.d.) Using drones for management of crops
Ershadi A, McCabe MF, Evans JP, Walker JP (2013) Effects of spatial aggregation on the multi-scale estimation of evapotranspiration. Remote Sens Environ 131:51–62. https://doi.org/10.1016/j.rse.2012.12.007
French AN, Hunsaker DJ, Thorp KR (2015) Remote sensing of evapotranspiration over cotton using the TSEB and METRIC energy balance models. Remote Sens Environ 158:281–294. https://doi.org/10.1016/j.rse.2014.11.003
Frid M, Frid V (2024) Advancements in drone-based (UAV) ground penetrating radar for accurate boundary mapping between disturbed clayey soil and natural rock. https://www.preprints.org/manuscript/202401.1034
Gao A, Wu S, Wang F, Wu X, Xu P, Yu L, Zhu S (2019) A newly developed unmanned aerial vehicle (UAV) imagery-based technology for field measurement of water level. Water 11(1):124
Gavilán V, Lillo-Saavedra M, Holzapfel E, Rivera D, García-Pedrero A (2019) Seasonal crop water balance using harmonized landsat-8 and sentinel-2 time series data. Water 11(11):2236. https://doi.org/10.3390/w11112236
Ghiat I, Mackey HR, Al-Ansari T (2021) A review of evapotranspiration measurement models, techniques and methods for open and closed agricultural field applications. Water 13(18):2523. https://doi.org/10.3390/w13182523
Gill MK, Asefa T, Kemblowski MW, McKee M (2006) Soil moisture prediction using support vector machines. J Am Water Resour Assoc 42(4):1033–1046. https://doi.org/10.1111/j.1752-1688.2006.tb04512.x
Gopalakrishnan S, Waimin J, Zareei A, Sedaghat S, Raghunathan N, Shakouri A, Rahimi R (2022) A biodegradable chipless sensor for wireless subsoil health monitoring. Sci Rep 12(1):8011. https://doi.org/10.1038/s41598-022-12162-z
Gowda PH, Chavez JL, Colaizzi PD, Evett SR, Howell TA, Tolk JA (2008) ET mapping for agricultural water management: present status and challenges. Irrig Sci 26(3):223–237. https://doi.org/10.1007/s00271-007-0088-6
Grant OM, Chaves MM, Jones HG (2006) Optimizing thermal imaging as a technique for detecting stomatal closure induced by drought stress under greenhouse conditions. Physiol Plant 127(3):507–518. https://doi.org/10.1111/j.1399-3054.2006.00686.x
Guo F, Bao L, Wang H, Larson SL, Ballard JH, Knotek-Smith HM, Zhang Q, Su Y, Wang X, Han F (2020) A simple method for the synthesis of biochar nanodots using hydrothermal reactor. MethodsX 7:101022
Gutiérrez J, Villa-Medina JF, Nieto-Garibay A, Porta-Gándara MÁ (2013) Automated irrigation system using a wireless sensor network and GPRS module. IEEE Trans Instrum Meas 63(1):166–176
Guyon I, Elisseeff A (2003) An introduction to variable and feature selection. J Machine Learn Res 3(Mar):1157–1182

Haddeland I, Heinke J, Biemans H, Eisner S, Flörke M, Hanasaki N, Konzmann M, Ludwig F, Masaki Y, Schewe J, Stacke T, Tessler ZD, Wada Y, Wisser D (2014) Global water resources affected by human interventions and climate change. Proc Natl Acad Sci USA 111(9):3251–3256. https://doi.org/10.1073/pnas.1222475110

Hamami L, Nassereddine B (2020) Application of wireless sensor networks in the field of irrigation: a review. Comput Electron Agric 179:105782. https://doi.org/10.1016/j.compag.2020.105782

Han X, Thomasson JA, Swaminathan V, Wang T, Siegfried J, Raman R, Rajan N, Neely H (2020) Field-based calibration of unmanned aerial vehicle thermal infrared imagery with temperature-controlled references. Sensors (Basel, Switzerland) 20(24):7098. https://doi.org/10.3390/s20247098

Harris R, Baumann I (2015) Open data policies and satellite Earth observation. Space Policy 32:44–53. https://doi.org/10.1016/j.spacepol.2015.01.001

Hoffmann H, Nieto H, Jensen R, Guzinski R, Zarco-Tejada P, Friborg T (2016) Estimating evaporation with thermal UAV data and two-source energy balance models. Hydrol Earth Syst Sci 20(2):697–713. https://doi.org/10.5194/hess-20-697-2016

Hossain NI, Noushin T, Tabassum S (2024) Tattoo-like flexible ethylene sensor for plant stress monitoring in real-time. IEEE Sens J. https://doi.org/10.1109/JSEN.2023.3327547

Howell TA, Evett SR, Tolk JA, Copeland KS, Dusek DA, Colaizzi PD (2012). Crop Coefficients Developed at Bushland, Texas for Corn, Wheat, Sorghum, Soybean, Cotton, and Alfalfa. https://doi.org/10.1061/40856(200)291

Jackson RD, Pinter PJ, Reginato RJ (1985) Net radiation calculated from remote multispectral and ground station meteorological data. Agric Meteorol 35(1):153–164. https://doi.org/10.1016/0168-1923(85)90081-4

Jagüey JG, Villa-Medina JF, López-Guzmán A, Porta-Gándara MÁ (2015) Smartphone irrigation sensor. IEEE Sens J 15(9):5122–5127

Jha G, Nicolas F, Schmidt R, Suvočarev K, Diaz D, Kisekka I, Scow K, Nocco MA (2022) Irrigation decision support systems (IDSS) for California's water–nutrient–energy Nexus. Agronomy 12(8), Article 8. https://doi.org/10.3390/agronomy12081962

Ji-hua M, Bing-fang W (2008) Study on the crop condition monitoring methods with remote sensing

Jones HG, Leinonen I (2003) Thermal imaging for the study of plant water relations. J Agric Meteorol 59(3):205–217. https://doi.org/10.2480/agrmet.59.205

Chávez JoséL, Neale CMU, Hipps LE, Prueger JH, Kustas WP (2005) Comparing aircraft-based remotely sensed energy balance fluxes with Eddy covariance tower data using heat flux source area functions. J Hydrometeorol 6(6):923–940. https://doi.org/10.1175/JHM467.1

Kaplan H (2007) Practical applications of infrared thermal sensing and imaging equipment (vol 75). SPIE press. https://books.google.com/books?hl=en&lr=&id=vuJ8dalst-kC&oi=fnd&pg=PR13&dq=Kaplan,+H.,+2007.+Practical+applications+of+infrared+thermal+sensing+and+imaging+equipment+(Vol.+75).+SPIE+press.&ots=A6EKFLsIPs&sig=3F3EDVveKVrzreTVfaD6dXSi7-A

Kasuga T, Mizui A, Koga H, Nogi M (n.d.) Wirelessly powered sensing fertilizer for precision and sustainable agriculture. Adv Sustain Syst n/a(n/a):2300314. https://doi.org/10.1002/adsu.202300314

Knuth DE (1974) Postscript about NP-hard problems. ACM SIGACT News 6(2):15–16. https://doi.org/10.1145/1008304.1008305

Koparan C, Koc AB, Privette CV, Sawyer CB (2018) In situ water quality measurements using an unmanned aerial vehicle (UAV) system. Water 10(3):264

Leone M, Consales M, Passeggio G, Buontempo S, Zaraket H, Youssef A, Persiano GV, Cutolo A, Cusano A (2022) Fiber optic soil water content sensor for precision farming. Opt Laser Technol 149:107816. https://doi.org/10.1016/j.optlastec.2021.107816

Li M, Sui R, Meng Y, Yan H (2019) A real-time fuzzy decision support system for alfalfa irrigation. Comput Electron Agric 163:104870. https://doi.org/10.1016/j.compag.2019.104870

Lipper L, Thornton P, Campbell BM, Baedeker T, Braimoh A, Bwalya M, Caron P, Cattaneo A, Garrity D, Henry K (2014) Climate-smart agriculture for food security. Nat Clim Chang 4(12):1068–1072

Liu H (2011) Feature selection. In: Sammut C, Webb GI (eds) Encyclopedia of machine learning. Springer US, pp 402–406. https://doi.org/10.1007/978-0-387-30164-8_306

Liu W, Liu L, Guan C, Zhang F, Li M, Lv H, Yao P, Ingenhoff J (2018) A novel agricultural photovoltaic system based on solar spectrum separation. Sol Energy 162:84–94. https://doi.org/10.1016/j.solener.2017.12.053

Loheide SP, Gorelick SM (2005) A local-scale, high-resolution evapotranspiration mapping algorithm (ETMA) with hydroecological applications at riparian meadow restoration sites. Remote Sens Environ 98(2–3):182–200. https://doi.org/10.1016/j.rse.2005.07.003

Lu Y, Gong M, Li J, Ma J (2024) Investigating the effects of full-spectrum LED lighting on strawberry traits using correlation analysis and time-series prediction. Plants 13(2):149. https://doi.org/10.3390/plants13020149

Macknick J, Lee C, Mosey G, Melius J (2013) Solar development on contaminated and disturbed lands (NREL/TP--6A20-58485, 1260337; p. NREL/TP--6A20-58485, 1260337). https://doi.org/10.2172/1260337

Markets R (2023) Global agriculture drones market 2023–2028: increased awareness of digital agriculture drives demand for advanced imaging sensors. GlobeNewswire Newsroom. https://www.globenewswire.com/en/news-release/2023/08/17/2727136/28124/en/Global-Agriculture-Drones-Market-2023-2028-Increased-Awareness-of-Digital-Agriculture-Drives-Demand-for-Advanced-Imaging-Sensors.html

Michael MG, Bastiaanssen WGM (2000) A new simple method to determine crop coefficients for water allocation planning from satellites: results from Kenya. Irrig Drain Syst 14(3):237–256. https://doi.org/10.1023/A:1026507916353

Morales G, Sheppard JW, Logan RD, Shaw JA (2021) Hyperspectral dimensionality reduction based on inter-band redundancy analysis and greedy spectral selection. Remote Sensing 13(18):3649. https://doi.org/10.3390/rs13183649

Moran MS (1994) Irrigation management in Arizona using satellites and airplanes. Irrig Sci 15(1):35–44. https://doi.org/10.1007/BF00187793

Neale CU, Ahmed RU, Moran M, Pinter Jr P, Qi J, Clarke T (1996) Estimating seasonal cotton evapotranspiration using canopy reflectance. In: International evapotranspiration irrigation scheduling conference Ed.^ Eds. https://www.ars.usda.gov/research/publications/publication/?seqNo115=73479

Oren S, Ceylan H, Schnable PS, Dong L (2017) High-resolution patterning and transferring of graphene-based nanomaterials onto tape toward roll-to-roll production of tape-based wearable sensors. Adv Mater Technol 2(12):1700223. https://doi.org/10.1002/admt.201700223

Plappally AK, Lienhard V (2012) Energy requirements for water production, treatment, end use, reclamation, and disposal. Renew Sustain Energy Rev 16(7):4818–4848. https://doi.org/10.1016/j.rser.2012.05.022

Pluchinotta I, Pagano A, Giordano R, Tsoukiàs A (2018) A system dynamics model for supporting decision-makers in irrigation water management. J Environ Manage 223:815–824. https://doi.org/10.1016/j.jenvman.2018.06.083

Puerto P, Domingo R, Torres R, Pérez-Pastor A, García-Riquelme M (2013) Remote management of deficit irrigation in almond trees based on maximum daily trunk shrinkage. Water relations and yield. Agric Water Manag 126:33–45. https://doi.org/10.1016/j.agwat.2013.04.013

Ramírez Cuesta J, Miras-Avalos J, Rubio-Asensio JS, Intrigliolo D (2019) ArcGIS toolbox for estimating crop water demands by integrating the dual crop coefficient approach with multi-satellite imagery. Application manual for sentinel 2A/B and Landsat 7/8 satellite imagery

Rao S (2016) National innovations in climate resilient agriculture (NICRA) research highlights (2015–16). https://krishi.icar.gov.in/jspui/handle/123456789/22487

Reach WT, Dwek E, Fixsen DJ, Hewagama T, Mather JC, Shafer RA, Banday AJ, Bennett CL, Cheng ES, Eplee J, Leisawitz D, Lubin PM, Read SM, Rosen LP, Shuman FGD, Smoot GF, Sodroski

TJ, Wright EL (1995) Far-infrared spectral observations of the galaxy by Cobe. Astrophys J 451:188. https://doi.org/10.1086/176210

Rivera G, Porras R, Florencia R, Sánchez-Solís JP (2023) LiDAR applications in precision agriculture for cultivating crops: a review of recent advances. Comput Electron Agric 207:107737. https://doi.org/10.1016/j.compag.2023.107737

Rosenberg NJ, Blad BL, Verma SB (1983) Microclimate: the biological environment. Wiley. https://books.google.com/books?hl=en&lr=&id=c6MI6VvTqTsC&oi=fnd&pg=PR17&dq=Rosenberg+NJ,+Blad+BL,+Verma+SB+(1983)+Microclimate,+the+biological+environment,+2nd+edn.+Wiley,+New+York,+495p&ots=GruSbgeT32&sig=ZsqH1MtkzqX7U_QkEd6wbTjoPB4

Sabins Jr FF, Ellis JM (2020) Remote sensing: principles, interpretation, and applications. Waveland Press. https://books.google.com/books?hl=en&lr=&id=rAnaDwAAQBAJ&oi=fnd&pg=PR1&dq=Sabins+Jr,+Floyd+F.,+and+James+M.+Ellis.+2020.+Remote+Sensing:+Principles,+Interpretation,+and+Applications.+Waveland+Press.+&ots=_GYtbzuA2j&sig=BsNP-UxlsAXt2NZ160Wndho-BTU

Serrano J, Shahidian S, Marques Da Silva J (2019) Evaluation of normalized difference water index as a tool for monitoring pasture seasonal and inter-annual variability in a Mediterranean Agro-Silvo-pastoral system. Water 11(1):62. https://doi.org/10.3390/w11010062

Sinchi KS, Calderon D, Aziz I, Watts A, Soltanaghai E, Alipour M (2023) Under-Canopy biomass sensing using UAS-mounted radar: a numerical feasibility analysis. In: IGARSS 2023–2023 IEEE international geoscience and remote sensing symposium, pp 3292–3295. https://ieeexplore.ieee.org/abstract/document/10282771/

Singh Rawat K, Kumar Singh S, Bala A, Szabó S (2019) Estimation of crop evapotranspiration through spatial distributed crop coefficient in a semi-arid environment. Agric Water Manag 213:922–933. https://doi.org/10.1016/j.agwat.2018.12.002

Su H, McCabe MF, Wood EF, Su Z, Prueger JH (2005) Modeling evapotranspiration during SMACEX: comparing two approaches for local- and regional-scale prediction. J Hydrometeorol 6(6):910–922. https://doi.org/10.1175/JHM466.1

Suárez L, Zarco-Tejada PJ, González-Dugo V, Berni JAJ, Sagardoy R, Morales F, Fereres E (2010) Detecting water stress effects on fruit quality in orchards with time-series PRI airborne imagery. Remote Sens Environ 114(2):286–298

Tasumi M, Allen RG, Trezza R (2006) Calibrating satellite-based vegetation indices to estimate evapotranspiration and crop coefficients. In: Proceedings of the 2006 USCID water management conference, ground water and surface water under stress: competition, interaction, solutions, Denver. Proceedings… Denver, USA: USCID, pp 103–112. https://mountainscholar.org/bitstream/handle/10217/46560/101_2006-USCID-Boise.pdf?sequence=1#page=111

Time A, Gomez-Casanovas N, Mwebaze P, Apollon W, Khanna M, DeLucia EH, Bernacchi CJ (2024) Conservation agrivoltaics for sustainable food-energy production. Plants, People, Planet, ppp3.10481. https://doi.org/10.1002/ppp3.10481

Torres-Sanchez R, Navarro-Hellin H, Guillamon-Frutos A, San-Segundo R, Ruiz-Abellón MC, Domingo-Miguel R (2020) A decision support system for irrigation management: analysis and implementation of different learning techniques. Water 12(2):548. https://doi.org/10.3390/w12020548

Tran MH, Fei B (2023) Compact and ultracompact spectral imagers: technology and applications in biomedical imaging. J Biomed Optics 28(04). https://doi.org/10.1117/1.JBO.28.4.040901

USGS (2019) What is groundwater? Retrieved from USGS-science for a changing world: Retrieved from: https://www.usgs.gov/faqs/what-groundwater

Vidal M, Amigo JM (2012) Pre-processing of hyperspectral images. Essential steps before image analysis. Chemom Intell Lab Syst 117:138–148. https://doi.org/10.1016/j.chemolab.2012.05.009

Wang K, Sun Y, Tang J, He J, Sun H (2020) Aqueous Cr (VI) removal by a novel ball milled Fe0-biochar composite: role of biochar electron transfer capacity under high pyrolysis temperature. Chemosphere 241:125044

Wang Z, Tang J, Zhu L, Feng Y, Yue L, Wang C, Xiao Z, Chen F (2022) Nanomaterial-induced modulation of hormonal pathways enhances plant cell growth. Environ Sci Nano 9(5):1578–1590

Wang L, Wang J, Ding J, Li X (2023) Estimation and spatiotemporal evolution analysis of actual evapotranspiration in Turpan and Hami cities based on multi-source data. Remote Sensing 15(10):2565. https://doi.org/10.3390/rs15102565

Wigmosta MS, Vail LW, Lettenmaier DP (1994) A distributed hydrology-vegetation model for complex terrain. Water Resour Res 30(6):1665–1679. https://doi.org/10.1029/94WR00436

Witten IH, Frank E, Hall MA, Pal CJ (2017) Practical machine learning tools and techniques. Data mining. Fourth Edition, Elsevier Publishers. https://doc1.bibliothek.li/acb/FLMF040119.pdf

Wu Y, Liu H, Li B, Kosonen R (2019) Prediction of thermal sensation using low-cost infrared array sensors monitoring system. IOP conference series: materials science and engineering, vol 609(3), p 032002. https://iopscience.iop.org/article/10.1088/1757-899X/609/3/032002/meta

Xiao X, Braswell B, Zhang Q, Boles S, Frolking S, Moore B (2003) Sensitivity of vegetation indices to atmospheric aerosols: continental-scale observations in Northern Asia. Remote Sens Environ 84(3):385–392. https://doi.org/10.1016/S0034-4257(02)00129-3

Yu K, Fang S, Zhao Y (2021) Heavy metal Hg stress detection in tobacco plant using hyperspectral sensing and data-driven machine learning methods. Spectrochim Acta Part A Mol Biomol Spectrosc 245:118917. https://doi.org/10.1016/j.saa.2020.118917

Zahran S, Saeed RA-H, Elazizy IM (2022) Remote sensing-based water resources and agriculture spatial indicators system. Egypt J Remote Sensing Space Sci 25(2):515–527. https://doi.org/10.1016/j.ejrs.2022.02.002

Zainol A, MA, Mahyuddin MN, Mohd Zainuri MAA (2021) Solar photovoltaic architecture and agronomic management in agrivoltaic system: a review. Sustainability 13(14), Article 14. https://doi.org/10.3390/su13147846

Zhang J, Su R, Fu Q, Ren W, Heide F, Nie Y (2022) A survey on computational spectral reconstruction methods from RGB to hyperspectral imaging. Sci Rep 12(1):11905. https://doi.org/10.1038/s41598-022-16223-1

Zhou Z, Majeed Y, Diverres Naranjo G, Gambacorta EMT (2021) Assessment for crop water stress with infrared thermal imagery in precision agriculture: a review and future prospects for deep learning applications. Comput Electron Agric 182:106019. https://doi.org/10.1016/j.compag.2021.106019

Zotarelli L, Scholberg JM, Dukes MD, Munoz-Carpena R, Icerman J (2009) Tomato yield, biomass accumulation, root distribution and irrigation water use efficiency on a sandy soil, as affected by nitrogen rate and irrigation scheduling. Agric Water Manag 96(1):23–34

Chapter 16
Hydrometeorological Water Budget and It's Relationship with Extreme Over Alaknanda River Basin: An Application of ERA5-Land

Pramod Kumar, Sudheer Yadav, Khushboo Sharma, Prem Kumar, Jaya Rai, and Ravindra Pratap Singh

Abstract Extreme precipitation is one of the major challenges for mountainous regions imposing threats to the sustenance of the inhabitants. The total precipitation (TP), soil–water (Sm), total evaporation (TE), and total runoff (TR) variations in a microclimatic region, particularly one with challenging topography, are key components of the hydrometeorological water budget. This study is an assessment of the water budget of the Alaknanda River basin (using ERA5-land reanalysis datasets), which is a major tributary of the Ganga River feeding to the most populated region of India. The temporal trends of total precipitation for the study period 1981–2020 is found negative over north-western part of Alaknanda River basin, however southern part is positive (similar for total runoff) whereas, the opposite is found for the total evaporation and soil–water. The change in storage (ΔS) is calculated by subtracting the output from the input. Thereafter, the storage has declining trend over entire basin (except northern-most region) with range value − 20 to − 60 mm/year. Furthermore, the precipitation extremes, along with number of consecutive dry days (CDD) is found to be rising through lower section of the basin. The strong (negative/positive) correlation (spatial/temporal) between extreme precipitation and storage is pragmatic.

Supplementary Information The online version contains supplementary material available at https://doi.org/10.1007/978-3-031-76532-2_16.

P. Kumar (✉) · K. Sharma
Department of Geology, Sikkim University, Gangtok, Sikkim, India
e-mail: promnagar1@gmail.com

P. Kumar
IESD, Banaras Hindu University, Varanasi, UP, India 221005

S. Yadav · R. P. Singh
Department of Earth and Planetary Sciences, University of Allahabad, Prayagraj, Uttar Pradesh, India

P. Kumar · J. Rai
Ram Lal Anand College, University of Delhi, New Delhi, India

A. Nanda et al. (eds.), *Navigating the Nexus*, Water Science and Technology Library 102, https://doi.org/10.1007/978-3-031-76532-2_16

Keywords Weather/climate extremes · Hydrometeorological water budget · Storage · Microclimate

16.1 Introduction

Himalayas is the major sources of water for the most of north Indian human and welfare. The topographic uniqueness and land use changes limits our ability to understand the hydrological balance over the Western Himalayas (Dimri 2012). The central Himalayan region's repeated experiences of precipitation and temperature imbalances throughout 1960 and 1990, had a significant effect on the studied glacier's melting during 1976 and 1990 (Aggarwal et al. 2016). Whereas the Garhwal Himalayan region's rise in the mean temperature and decline in mean precipitation imposes a direct impact on the melting process since 1990 (Singh and Pandey 2021; Kumar et al. 2021). According to Li et al. (2017) atmospheric-hydrology dynamic could be clearly simulated by using the WRF-Hydro modelling approach. Moreover, conferring to WRF simulation, winter season's precipitation has a significant bias for higher elevation. For the Bhagirathi Ganga and Vishnu Ganga basins, total water storage (TWS) results from Gravity Recovery and Climate Experiment (GRACE) and National Oceanic and Atmospheric Administration (NOAA) both reveal a negative trend in TWS of 0.54 cm/year and 0.2 cm/year, respectively, due to the region's high evaporation rate and reduced precipitation (Haq et al. 2013). Furthermore, it is demonstrated that GRACE and the water balance approach, which make use of NOAA (Observation and reanalysis) data, hold great promise for providing hydrological data to watershed management in Uttarakhand (Haq et al. 2013).

To comprehend the spatiotemporal variance of snow accumulation and melting in the subbasins of Yamuna, snowpack indicators were analyzed by Rathore et al. (2015) for water resources management utilizing (AWiFS) satellite datasets. Singh et al. (2010) observed that the discharge-temperature relationship declined throughout all years as the time-lag increased. However, snow melt will have a substantial effect on the water budget in Himalayan basins, and during particular periods of the year, which is the most significant contributor to drainage for the western Himalayas River basin systems (Tanmoyee and Abdul 2015). The influence of irrigation on water budget has been assessed by Tuinenburg et al. (2014). The total amount of rainfall in the Ganga basin is considerably reduced due to decreased moisture convergence into the basin amounting to retention of only 35% of the evaporated moisture (Tuinenburg et al. 2014). As the world's temperature rises, greater evaporation from land and waterbodies occurs as well (Mbaye et al. 2019), indicating intertwined nature of the water balance and rising temperatures. The water balance influences evaporation, surface runoff, and groundwater recharge (Muhammad et al. 2019; Sok et al. 2022). As a result, the hydrological cycle is more vulnerable to hydrometeorological interruptions since the conversion of sensible heat and potential energy into latent and kinetic energy will also result in the occurrence of weather extremes (de Silva Tavares et al. 2023). The entire region of Alaknanda and Mandakini became more delicate and exposed to risks following the recent catastrophic events, which altered the landscape of the areas (Mehta et al. 2016). The Alaknanda basin has been identified as being at a greater risk of flash floods determined by morphometric parameters such as its elongated shape, high relative relief, and steep mean slope of its sub-watersheds (Patel et al. 2022). The various flash flooding incidents triggered by extreme precipitation on Alaknanda and Mandakini basins have been discussed in some of the studies (Joshi and Kumar 2006; Das 2013; Rana et al. 2013; Sundriyal et al. 2015; Mehta et al. 2016; Pant et al. 2020; Khanduri 2021; Patel et al. 2022; Verma et al. 2022).

The influence of extremes on regional hydrological cycle has a very peculiar role in controlling the annual hydrometeorological water budget (Trenberth and Asrar 2014). According to Kumar et al. (2021), the total amount of precipitation over Uttarakhand declined from 1980 to 2020, while the amount of extreme precipitation and extremely rainy days rose. There are very little studies reported on the relationship between extreme precipitation and the Himalayan hydrometeorological budget. Policymakers must understand the consequences of climate change on the drainage basin in order to mitigate its effects and establish coping techniques (Delgado et al. 2009; Sok et al. 2022). Most of the studies have primarily focused on the hydrological perspective. However, the present study aims to bridge the gap by providing a deeper understanding of the relationship between climate extremes and the hydrometeorological water budget in the western Himalaya region. The study's wider goal is to focus on reanalysis data that can be utilized to understand hydrometeorological behavioral patterns under extreme climates and vis versa.

Therefore, the objectives of the current study are: (1) To examine the hydrometeorological water budget over Alaknanda basin, (2) to acknowledge the mechanisms operating in this microclimate zone, and (3) and to investigate the influence of hydrometeorological water budget on extreme climatic events occurring in the Alaknanda basin.

16.2 Materials and Methodology

The present work is a computation and assessment of hydrometeorological water budget using ERA-5 reanalysis data set for 1980 to 2020 over the Alaknanda River basin depicted in Fig. 16.1. The details of the datasets are given in Table 16.1.

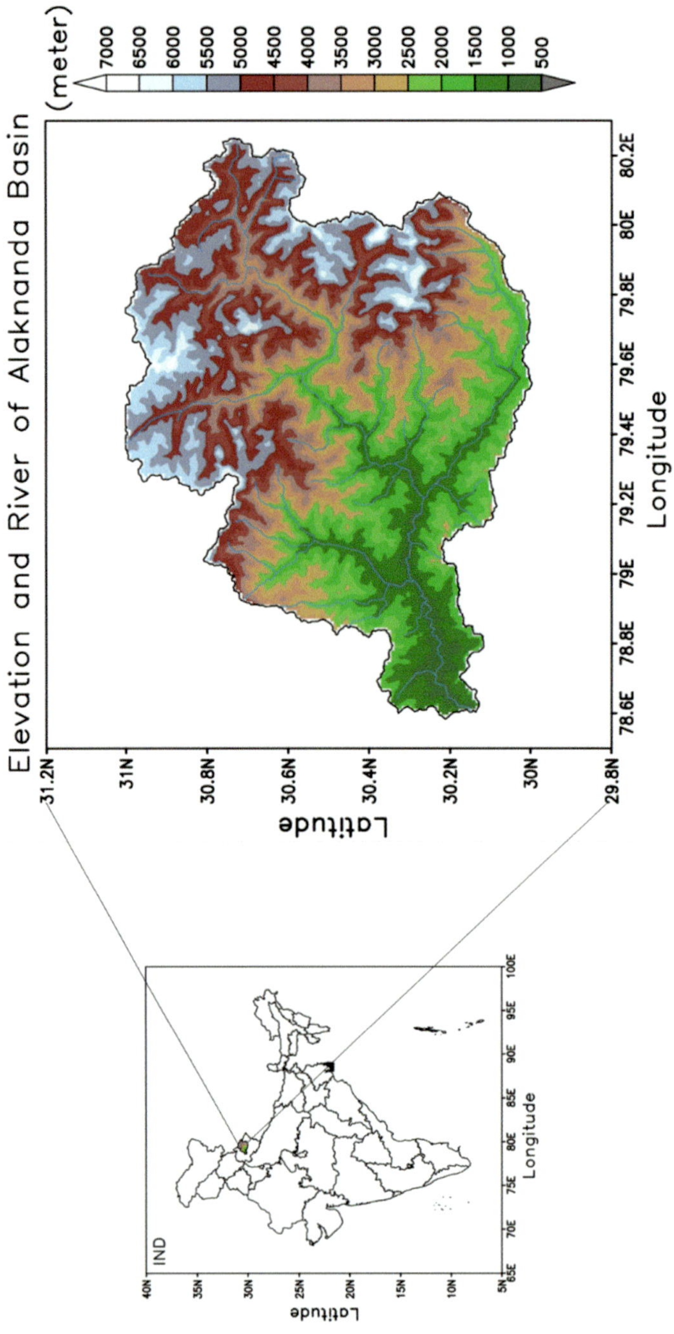

Fig. 16.1 Alaknanda basin topography and major tributaries

Table 16.1 A brief of data sets used for the study

Variables	Data sets	Years (span)	Resolution		Source	Reference
			Temporal	Horizontal		
Total precipitation	ERA5-land (ECMWF)	1981 to 2020	Daily	0.1° * 0.1°	ECMWF https://cds.climate.copernicus.eu/cdsapp#!/dataset/reanalysis-era5-land?tab=form	Hans et al. (2020)
Total evaporation						
Total runoff						
Volumetric soil water for layer 1 to 3 (up-to 100 cm)						
GTOPO30 (topography)	USGS	Current (2010)	2010	30 arc-sec	USGS open data portal	USGS

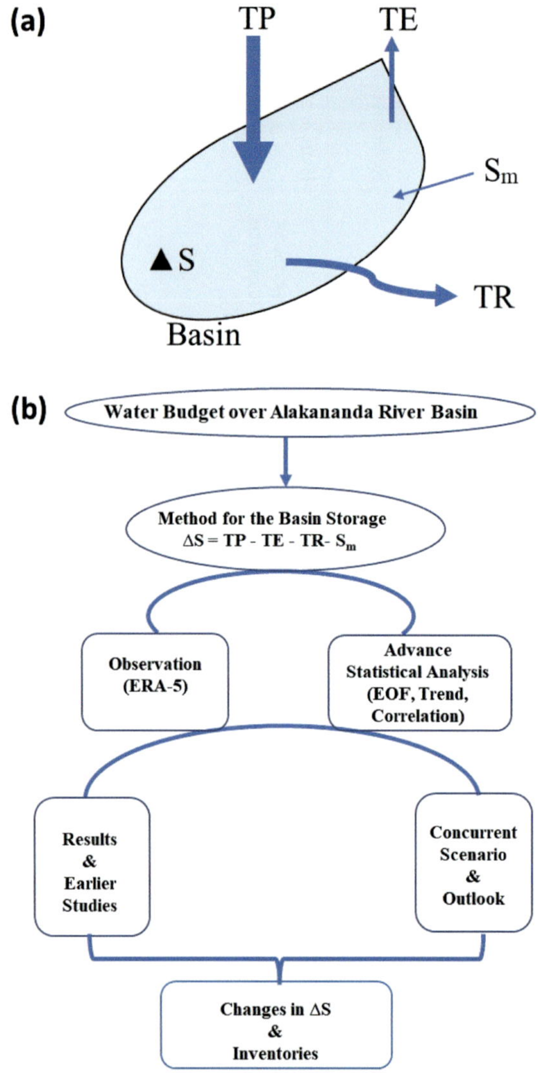

Schematic 16.1 The flow chart shows the methodology outline of the present work. Where TP is total precipitation, TE total evaporation, TR Total runoff, S_m soil–water content, and ΔS storage

Therefore, water storage in the basin can be calculated by subtracting the input and output variables from the basin, which are represented by TP (total precipitation), TE (total evaporation), TR (total runoff), S_m (soil–water content), and ΔS (storage) (Tables 16.2 and 16.3). Moreover, TR contains subsurface runoff and surface runoff, and for minimizing the infiltration and percolation error we used soil–water content (Dimri et al. 2019; Deshmukh and Singh 2016).

$$\text{Storage } \Delta S = (TP - TE - TR - S_m) \tag{16.1}$$

16.2.1 Climatology

The daily value of total precipitation, evaporation, runoff, and soil water content are calculated by summing the hourly data for each day. These daily values are then further summed to calculate the monthly cumulative values. Finally, the annual total is calculated using the given formula (Eq. 16.2). Where, i = 1 to n and X is hourly value of the variables used for the study.

$$\text{cumulative} = \sum_{i}^{n} X \tag{16.2}$$

The climatology (long-term average) has been computed using mean formula as Eq. (16.3) (Thornthwaite 1948).

$$Average = \frac{\sum_{j}^{m} X}{m} \tag{16.3}$$

where, j = 1 to m, variables such as total precipitation, evaporation, runoff, and soil water content.

Table 16.2 Monthly averaged for 1981–2020 (40 years) for entire area of Alaknanda basin, Uttarakhand, ERA5 reanalysis data sets (units: m/month)

Months	Variables				
	Soil water (Sm)	Total evaporation (TE)	Total precipitation (TP)	Total runoff (TR)	Storage (ΔS) = (TE + TR + Sm) – (TP)
January	0.250	– 0.240	1.193	0.180	– 0.521
February	0.237	– 0.318	1.424	0.216	– 0.652
March	0.261	– 0.558	1.388	0.372	– 0.195
April	0.249	– 0.762	1.020	0.557	0.549
May	0.256	– 0.966	1.008	1.257	1.471
June	0.257	– 1.091	1.985	1.869	1.233
July	0.303	– 1.109	5.118	3.557	– 0.148
August	0.310	– 1.102	4.649	3.810	0.573
September	0.282	– 1.118	2.128	1.815	1.087
October	0.255	– 0.868	0.571	0.578	1.130
November	0.229	– 0.465	0.347	0.287	0.634
December	0.235	– 0.278	0.617	0.202	0.101

Table 16.3 Annual averaged for 1981–2020 (40 years) over entire area of Alaknanda basin, Uttarakhand, ERA5 reanalysis data sets (units: m/year)

Years	Soil water (Sm)	Total evaporation (TE)	Total precipitation (TP)	Total runoff (TR)	Storage (ΔS) = (TE + TR + Sm) – (TP)
1981	0.2632	– 0.763	1.745	1.134	0.415
1982	0.2670	– 0.703	1.838	1.309	0.441
1983	0.2685	– 0.710	2.015	1.501	0.464
1984	0.2553	– 0.753	1.759	1.201	0.450
1985	0.2589	– 0.723	1.933	1.249	0.298
1986	0.2691	– 0.731	1.800	1.341	0.541
1987	0.2550	– 0.778	1.278	0.824	0.579
1988	0.2570	– 0.737	2.282	1.616	0.329
1989	0.2576	– 0.719	1.759	1.162	0.379
1990	0.2651	– 0.751	2.021	1.406	0.401
1991	0.2580	– 0.740	1.571	1.103	0.529
1992	0.2624	– 0.717	1.703	1.197	0.473
1993	0.2632	– 0.758	1.826	1.304	0.499
1994	0.2561	– 0.714	2.054	1.491	0.408

(continued)

16.2.2 Extremes

The term consecutive dry days (CDDs) refers to the longest stretch of days with rainfall below a specific threshold, which varies depending on the sources (Frich et al. 2002; Duan et al. 2017). Numerous researchers have used highest number of CDDs with precipitation less than 1 mm, as a reliable indicator of peak season droughts (Duan et al. 2017). Thus, to enable regional comparisons over Alaknanda, considering the significant climatic differences, a lower threshold of 1 mm/day (Frich et al. 2002) has been chosen, which is given as Eq. (16.4a).

$$CDD = \text{Maximum number of consecutive dry days } (R_{day} < 1 \text{ mm}) \tag{16.4a}$$

Similarly, very wet periods are considered as 90th percentile of precipitation values over all timesteps (1980–2020) and are calculated as Eq. (16.4b). The reference period with respect to the computation is conducted is from 1960 to 1990.

$$pctl_{90}(1, x) = 90th \text{ percentile}\left[j\left(t', x\right), t_1 < t' \leq t_n\right] \tag{16.4b}$$

Table 16.3 (continued)

Years	Soil water (Sm)	Total evaporation (TE)	Total precipitation (TP)	Total runoff (TR)	Storage (ΔS) = (TE + TR + Sm) − (TP)
1995	0.2594	– 0.717	1.976	1.408	0.409
1996	0.2635	– 0.700	1.936	1.410	0.438
1997	0.2669	– 0.711	1.861	1.126	0.242
1998	0.2729	– 0.758	1.845	1.398	0.585
1999	0.2614	– 0.770	1.697	1.195	0.530
2000	0.2654	– 0.758	1.942	1.366	0.448
2001	0.2532	– 0.779	1.417	0.868	0.483
2002	0.2552	– 0.748	1.642	1.089	0.450
2003	0.2574	– 0.751	1.770	1.175	0.413
2004	0.2597	– 0.794	1.408	0.799	0.444
2005	0.2653	– 0.716	1.729	1.246	0.498
2006	0.2587	– 0.761	1.567	0.951	0.404
2007	0.2641	– 0.769	1.830	1.294	0.496
2008	0.2668	– 0.734	1.929	1.345	0.417
2009	0.2554	– 0.742	1.428	0.833	0.402
2010	0.2575	– 0.722	2.127	1.571	0.424
2011	0.2612	– 0.721	1.824	1.326	0.485
2012	0.2544	– 0.710	1.667	1.117	0.415
2013	0.2641	– 0.716	2.055	1.506	0.431
2014	0.2628	– 0.732	1.683	1.120	0.432
2015	0.2574	– 0.747	1.661	1.125	0.469
2016	0.2438	– 0.761	1.495	0.987	0.497
2017	0.2605	– 0.775	1.861	1.199	0.374
2018	0.2610	– 0.732	2.020	1.504	0.477
2019	0.2692	– 0.724	1.880	1.245	0.358
2020	0.2610	– 0.736	1.660	1.070	0.407

16.2.3 Regression, Correlation and EOF

We have used different parameters like precipitation, evaporation, temperature, runoff, soil moisture, extreme precipitation and storage to understand the linear regression trend and correlation coefficient for each parameter with time. This was computed using the following formula: for i = 1, …, n.

$$y = a + bx_i \tag{16.5}$$

The regression line intercept (on the y-axis) is represented by the regression parameter a, while the slope of the regression line is represented by the regression parameter b. The R-square (R^2) or coefficient of dependence refers to the Pearson correlation coefficient R, which is given significance by squaring it. This measure is based on the percentage of variability in y that can be explained by the variability in x via their linear relationship, or vice versa, ranging from 0 to 1. The sum of squares (SS) is denoted by SS, and the formula for R^2 is $R^2 = SS_{regresion}/SS_{total}$. It is crucial to note that R^2 is computed solely using the Pearson correlation coefficient in this linear regression test (Kenney and Keeping 1962; Altman and Krzywinski 2015; Lane et al. 2017).

Where, $SS_{regression}$ is the sum of squares of the regression (explained) errors, and SS_{total} is the total sum of squares of the model (both explained and unexplained errors). R^2 provides a measure of how well the regression line fits the observed data. It ranges from 0 to 1, with higher values indicating a better fit. The statistical descriptions of datasets are represented in Table 16.4.

To ascertain the correlation coefficient (Spearman-Pearson) between the storage and hydrometeorological parameters, two real-valued random variables, x and y, are required. The Coefficient of Pearson Correlation (CPC) is defined by Benesty et al. (2009) and Pearson (1909).

$$CPC = \frac{n\left(\sum xy\right) - \left(\sum x\right)\left(\sum y\right)}{\sqrt{\left[n\sum x^2 - \left(\sum x\right)^2\right]\left[n\sum y^2 - \left(\sum y\right)^2\right]}} \tag{16.6}$$

where 'n' is the population size of the variables used for the study (Kenney and Keeping 1962; Lane et al. 2017).

Table 16.4 Statistical descriptions of data sets for 1981–2020

Descriptives	Soil water (m)	Evapo (m)	Total precip (m)	Runoff (m)	Storage (m)
Years	40	40	40	40	40
Mean	0.261	0.740	1.79	1.23	0.441
95% Cl mean upper	0.259	0.733	1.72	1.16	0.420
95% Cl mean lower	0.263	0.747	1.85	1.29	0.463
Median	0.261	0.737	1.81	1.22	0.440
Mode	0.244	0.701	1.28	0.799	0.243
Standard deviation	0.00549	0.0239	0.213	0.206	0.0687
Variance	3.02e−5	5.72e−4	0.0453	0.0425	0.00472
Minimum	0.244	0.701	1.28	0.799	0.243
Maximum	0.273	0.794	2.28	1.62	0.585
Skewness	− 0.388	0.305	− 0.236	− 0.283	− 0.383
Kurtosis	1.16	− 0.830	0.133	− 0.323	1.16
Shapiro–Wilk p	0.436	0.232	0.913	0.452	0.370

The eigenvalues of the scatter matrix (covariance matrix) of the data define the Empirical Orthogonal Functions (EOFs). To simplify the process, the samples are considered as anomalies time series, for more details the present study followed to Rudolph (1988) (Von Storch and Zwiers 2002). The Empirical Orthogonal Function (EOF) approach is employed to analyze geopotential height, zonal wind, and meridional wind variability. This approach identifies the spatial pattern of variability as well as its time fluctuation. EOFs are spatial patterns, while Principal Components (PCs) are temporal series (Kutzbach 1967; Wilks 2011; Jolliffe 2002; Midhuna et al. 2020). The main pattern accounts for the majority of the variance, while the remaining patterns contribute to a significant portion of the remaining variance, and so on. The calculation and figure of Standardized Precipitation Evapotranspiration Index (SPEI) and Standardized Precipitation Index (SPI) are also utilized for measuring wet and dry conditions (details are given in the supplementary methodology S1.0) (Vicente-Serrano et al. 2010).

16.3 Results

Figure 16.2 illustrates the spatial distribution of the long-term average annual cumulative total precipitation, total runoff, total evaporation, total soil–water content, and storage, which are significant factors in the climatology of the region, across the Alaknanda basin. Precipitation and runoff appear to be analogous however, evaporation, soil–water, and storage seems to be alike. The cumulative annual value of the total precipitation is 3–30 m to, total runoff 3–25 m, total evaporation 1–15 m, soil–water 3–12 m to, and storage 1–9 m.

The cumulative monthly spatial distribution of the precipitation indicates the highest values in the month of July–August (JA) and the lowest values in October–November (ON) (Fig. 16.3). The months of January-March (JFM) exhibit winter precipitation as the dominant factor controlling the hydrology of the region (Fig. 16.3). However, the months of June–September (JJAS) demonstrate strong influence of monsoon-dominated precipitation over the region. (Fig. 16.3). The highest evaporation rates have been observed during the month of May, surpassing those in June, April, July, and September, and the lowest evaporation levels are observed in the northernmost part of the basin throughout all months of the seasons (Fig. 16.4). The monthly runoff demonstrates a similar pattern to that of the total precipitation. Specifically, during the months of May–June (MJ), higher runoff is observed in the northeastern part of the basin. On the other hand, during July–September (JAS), the western and southern part of the basin exhibits the highest runoff. Whereas, the months of October-February (ONDJF) experience comparably low runoff, while the central part of the region experiences a peak in March and April (MS) (Fig. 16.5).

The long-term climatology of soil–water content reveals a yearly pattern for all months. The soil–water content reaches its lowest levels during November and December (ND), while it reaches its highest levels during July and August (JA) (Fig. 16.6). Similar observations are exhibited by precipitation during the winter

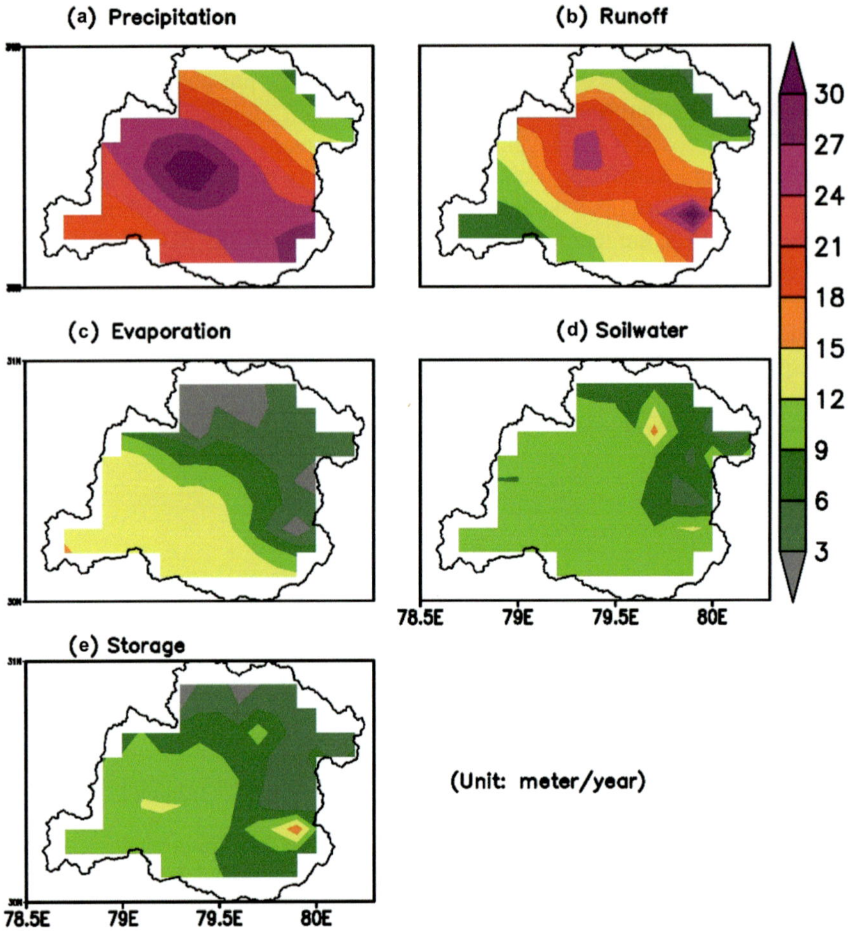

Fig. 16.2 The climatology for 1981–2020 cumulative (m/year) annual total precipitation, runoff, evaporation, soil water content

and monsoon seasons (Figs. 16.6 and 16.3). Throughout the entire study period, the north-eastern part of the basin consistently exhibits the lowest soil–water values (Fig. 16.6). Which is probably high altitude (i.e., above 5000 m elevation), and glaciated area of the study region (Fig. 16.1). Figure 16.7 illustrates the monthly variations of the variables, averaged over the study basin. The month of May demonstrates the highest total water storage levels, followed by June, October, September, November, April, and August. On the other hand, the lowest total water storage levels are observed in February, followed by January, March, and July (Fig. 16.7). The month of July experiences the highest total precipitation, while the maximum runoff occurs in the month of August. To determine the trend across the basin, linear regression has been conducted for all the variables. Figure 16.8 shows the spatial

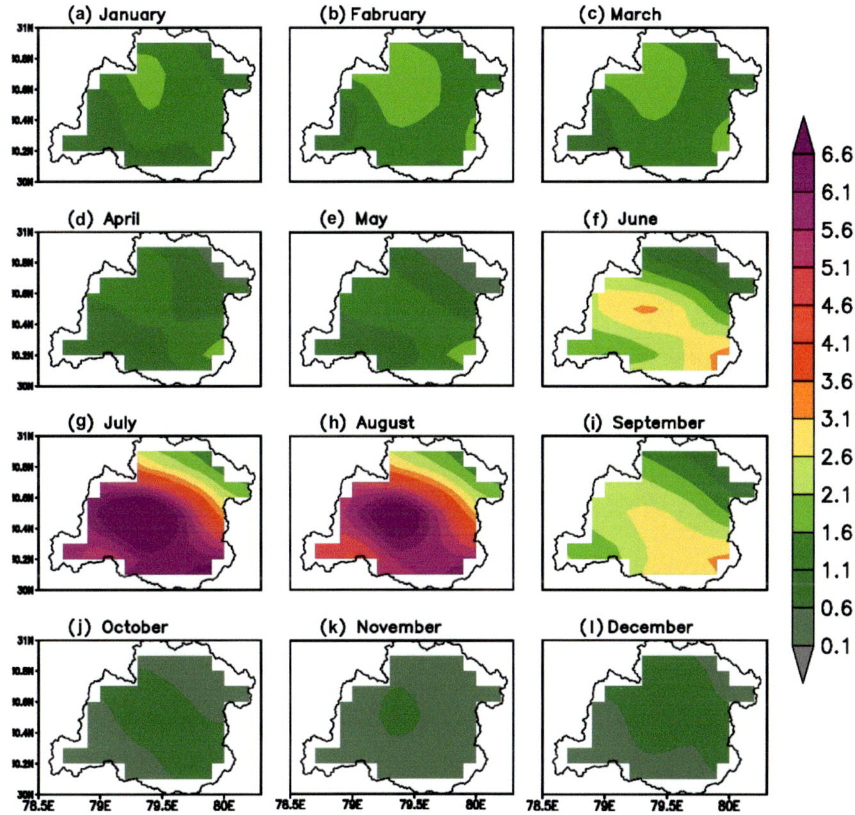

Fig. 16.3 The spatial distribution of total precipitation climatology monthly cumulative for 1981–2020 (units: m/month)

variability in trend from 1980 to 2020. Figure 16.8a shows that the total precipitation has a substantially negative trend value over the northern region of the basin, while a positive trend is observed over the southern part. The runoff is likewise following a similar pattern, with a positive trend in the south and a negative trend in the northwest portion of the basin. Conversely, evaporation shows a contrasting behavior to both precipitation and runoff (Fig. 16.8b, c). In the eastern part of the basin, soil–water content displays a positive trend, while the majority of the basin indicates a negative trend for soil water content (Fig. 16.8d). Additionally, a small portion of the eastern and northern regions of the total water storage exhibit a positive trend, while the remaining part of the basin exhibits a negative trend (Fig. 16.8e). Figure 16.9 illustrates the temporal trend across the entire basin, showing that all variables, except for evaporation, show negative trend lines. The yearly fluctuation of precipitation and runoff is relatively large (Fig. 16.9a, b), whereas that of soil water and evaporation is very small (Fig. 16.9c, d). Additionally, the total water storage is significantly decreasing (Fig. 16.9e). Figure 16.10 displays the spatial distribution of consecutive

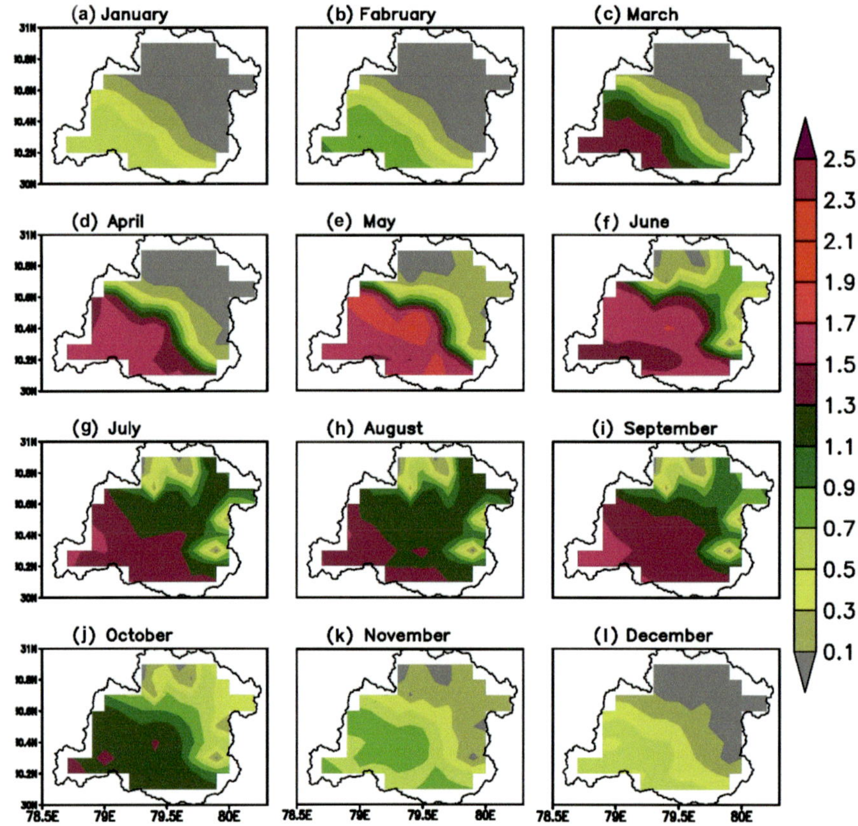

Fig. 16.4 Same as Fig. 16.3, but for total evaporation

dry days (CDD), extreme total precipitation, and trends. The range of CDD over the basin falls between 230 to 275 days. In the western part of the basin, CDD is lower, while the eastern part has the highest CDD (Fig. 16.10a). Extreme precipitation is higher over the western part of the basin and lower over the northern part, with values ranging from 2 to 8% across the entire basin (Fig. 16.10b). However, very wet days ranging from 90 to 99% over entire basin (Fig. 16.10c). A positive trend in CDD has been observed over the central to western part of the basin, while the northern part has a negative trend (Fig. 16.10d). Furthermore, the extreme precipitation exhibits a positive trend pattern over the southern part but a negative trend over the north-western part (Fig. 16.10e) and very wet days declining over entire basin except southernmost part of the basin (Fig. 16.10f). That suggests extreme precipitation is rising over the southern part of the region for a limited time period. Accompanied with higher extreme precipitation, the region also exhibits an increasing trend of dry days. During the study period, the basin is experiencing a higher frequency of drought events compared to floods, as indicated in figure S1 and S2. Prior to the year

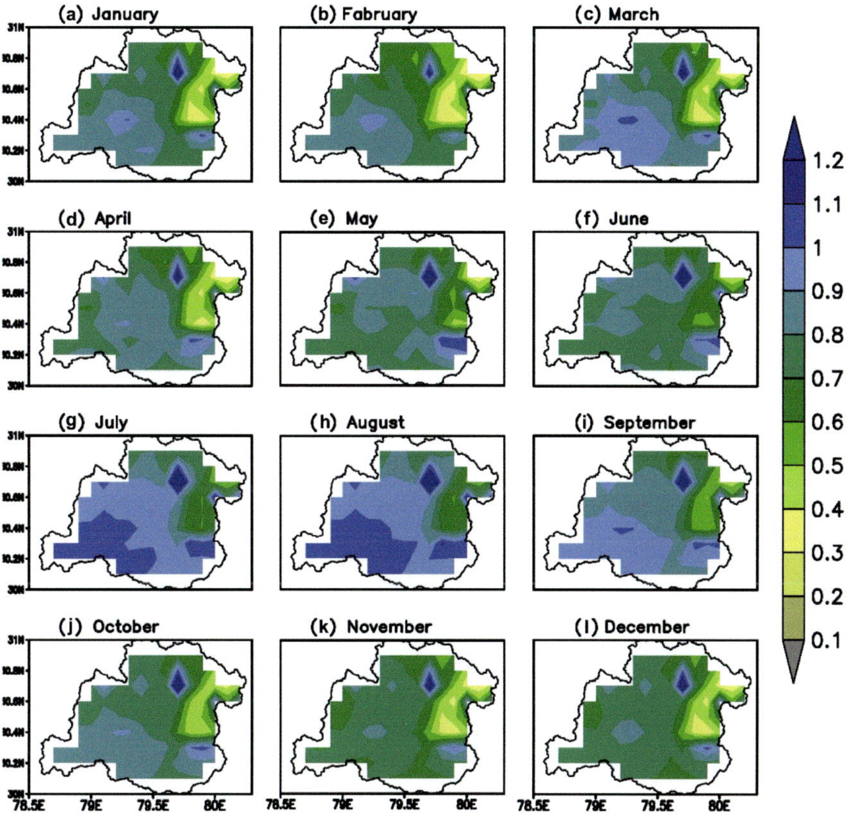

Fig. 16.5 Same as Fig. 16.3, but for total runoff

2000, the basin encountered frequent but relatively weak floods. However, after the year 2000, the basin is experiencing more frequent droughts with stronger intensity. Furthermore, wet years are experiencing more intense rainfall, while dry years are becoming drier and experiencing more pronounced dry conditions (Figs. S1C and S1F).

Empirical orthogonal function (EOF) is applied for all the variables (total precipitation, total evaporation, total runoff, soil–water content) during 1981–2020. Figure 16.11 displays the spatial patterns of the first three Empirical Orthogonal Functions (EOFs), namely EOF1, EOF2 and EOF3, as well as their corresponding Principal Components (PC1, PC2 and PC3) time series. Figure 16.11a–c illustrate that the patterns of EOF1, EOF2, and EOF3, respectively, exhibiting negative values for total evaporation EOF1, positive values for EOF2, and both negative and positive values across the entire domain for EOF3 with pattern explained variance of (EOF1: 21.8%, EOF2: 8.7%, and EOF3: 4.4%). Figures 16.11d–f show the EOF1, EOF2, and EOF3 respectively for total runoff. The EOF1 has positive value for entire domain, EOF2 is positive over eastern part, and negative over western part of the

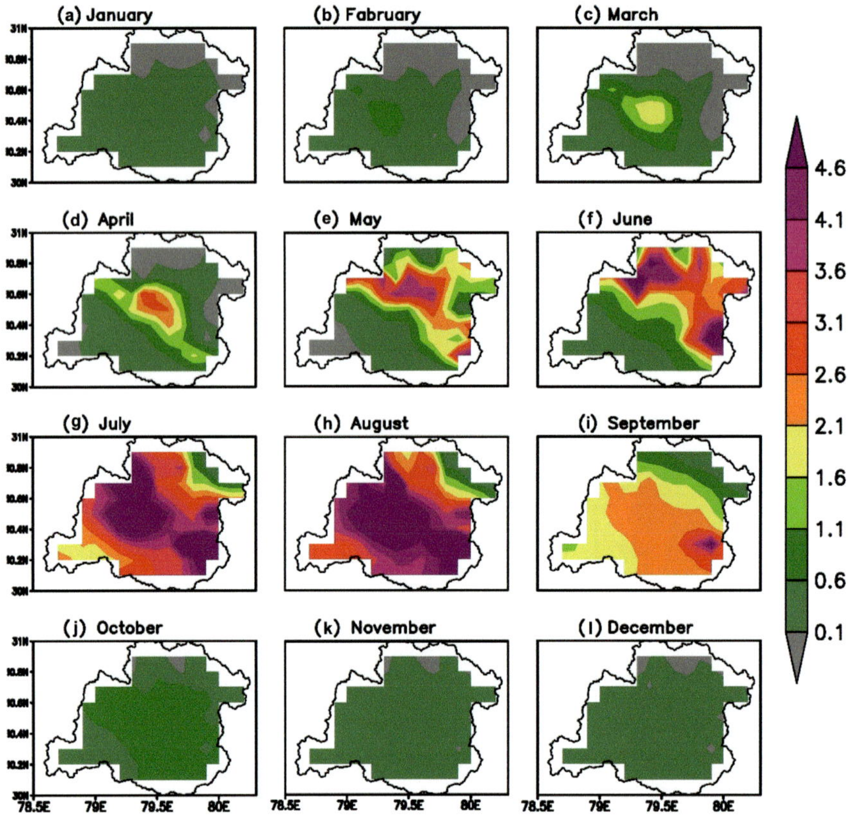

Fig. 16.6 Same as Fig. 16.3, but for soil–water content

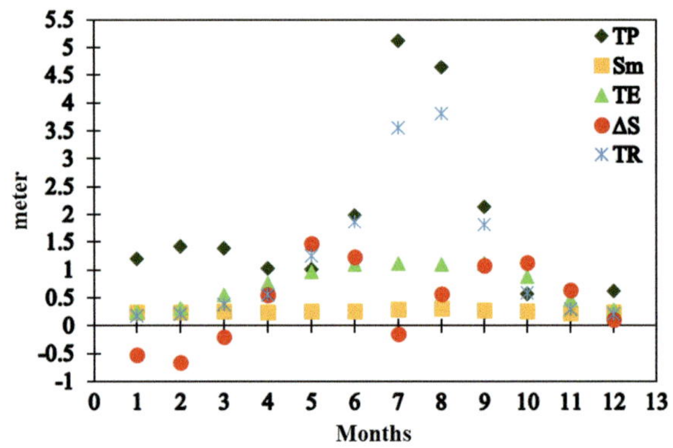

Fig. 16.7 Monthly climatology of area averaged total precipitation (TP), total runoff (TR), total evaporation (TE), soil water (Sm), storage (ΔS) (units: m/month) over Alaknanda basin for 1981–2020

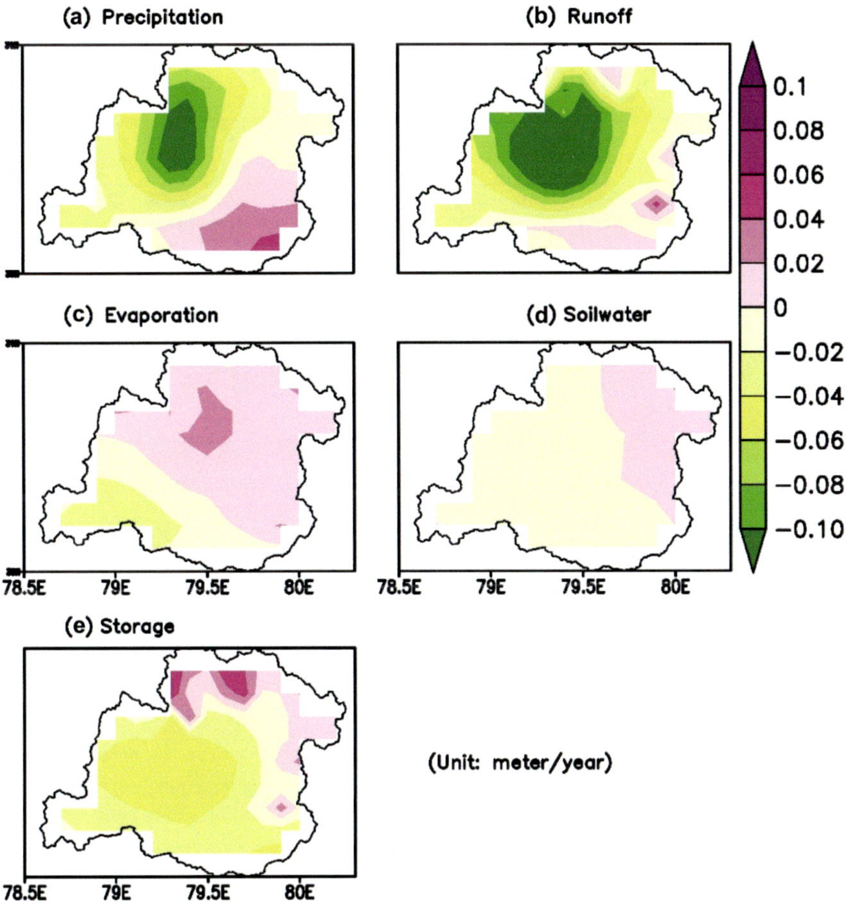

Fig. 16.8 The spatial distribution of linear trend of total precipitation, total runoff, total evaporation, and total soil–water content and storage for the 1981–2020 over Alaknanda basin

basin (Fig. 16.11d, e) with (EOF1: 78.3%, EOF2: 9.7% and EOF3: 7.1%) explain variance. However, EOF3 demonstrates negative values over northern part and positive over southern part (Fig. 16.11f). As shown in Fig. 16.11g–i the EOF1, EOF2, and EOF3 pattern exhibit negative values across the entire domain. This pattern explains the variances of (EOF1: 14.1%, EOF2: 7.1% and EOF3: 1.9%). Total precipitation has nearly alike pattern of explain variance positive values for across entire basin for EOF1 with 72.1% of explained variance, and negative in western part to positive over eastern part of the EOF2 with explained variance of 7.2%. Additionally, EOF3 has negative over northern part and positive values pattern on southern part of the basin with explained variance of 5.4% (Fig. 16.11j–l). PC1, PC2 and PC3 having loading vector − 1.0 to 1.0, are seen for evaporation (Fig. 16.12a–c). Total runoff loading vector varies from − 0.6 to 0.6 for PC1, PC2 and PC3 (Fig. 16.12d–f). Soil–water

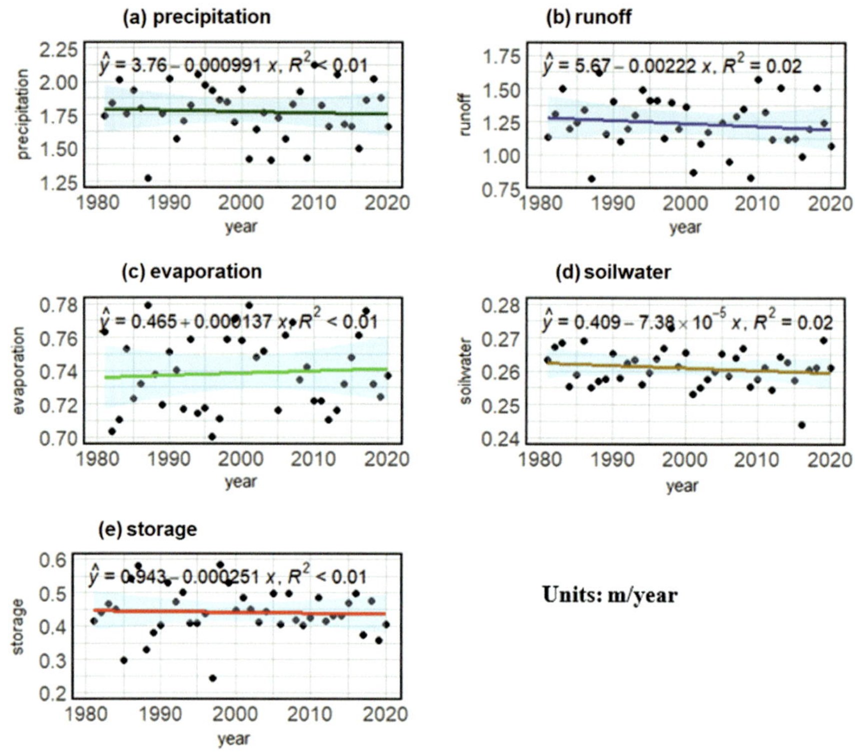

Fig. 16.9 Trend line to area averaged over Alaknanda basin for total precipitation, runoff, evaporation, soil water and storage (units are same as Fig. 16.8). The shading shows confidence level 0.99

content shows loading vector ranging from 0.3 to − 0.2 for PC1, PC2, and PC3 (Fig. 16.12g–i). However, total precipitation loading vectors vary from − 0.6 to 0.6 for PC1, PC2 and PC3 (Fig. 16.12j–l).

The spatiotemporal distribution of correlation among the variables (extreme precipitation, CDD, storage (ΔS), TP, TR, TE, Sm) are depicted in Fig. 16.13a–g. The negative correlation between extreme precipitation and storage has been found over entire basin region (except for the eastern most part of the basin) (Fig. 16.13a). A negative correlation has been observed in extreme precipitation and CDD as well (except for the eastern most part) (Fig. 16.13b). However, CDD and storage have mixed pattern of the positive and negative relation (Fig. 16.13c). The temporal variation of correlation shows positive value for extreme precipitation and storage (Fig. 16.13d), however extreme precipitation versus CDD has negative values observed (Fig. 16.13e). The correlation between storage and CDD is found negative for most of the years over the basin (Fig. 16.13f). In Fig. 16.13g, the significant correlations among the variables used in the study are represented. The results

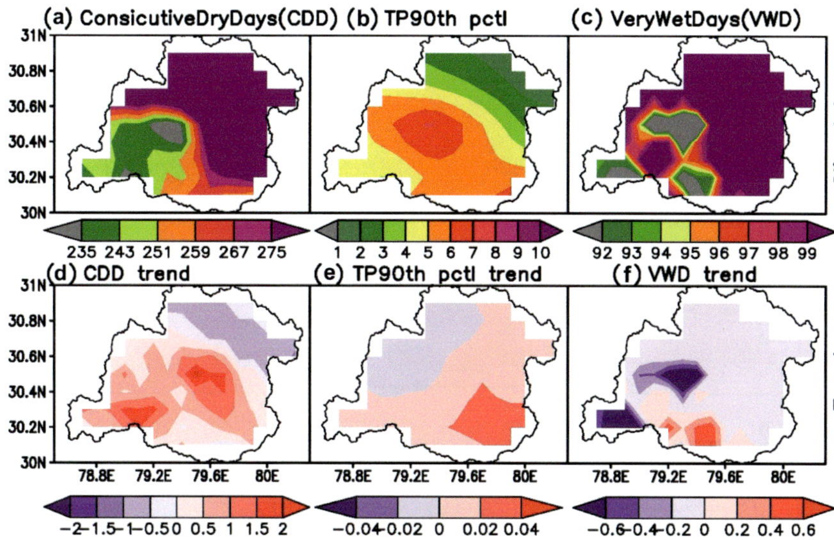

Fig. 16.10 The spatial distribution of climate extremes and its linear trend over the Alaknanda basin for 1981–2020

indicate that extreme precipitation exhibits a significant and strong positive correlation with TP and TR. Additionally, extreme precipitation shows a significant and negative correlation with CDD and TE. Furthermore, there is a negative correlation between extreme precipitation and storage (ΔS). Further, Fig. 16.13g shows that ΔS has positive correlation with TR, Sm, and TE, while CDD, and TP shows negative correlation. CDD has a significant negative correlation with TP, TR, and Sm, and a positive correlation with TE. TP has a strong and significant positive correlation with TR, and Sm, but a negative correlation with TE. Moreover, TR has a good and significant positive correlation with Sm, but a strong negative correlation with TE. Additionally, a negative correlation is found between Sm and TE (Fig. 16.13g).

16.4 Discussion

The overall distribution of precipitation in the basin supports the observed runoff pattern, indicating that the entire basin receives an adequate amount of annual precipitation. Likewise, the basin's storage model can be explained by average annual evaporation and soil–water patterns, leading to higher storage in the region with higher precipitation. The monthly variation highlights the monsoon season as a significant contributor to the highest precipitation, runoff, evaporation and soil–water. The highest evaporation has been observed in the months of April, May, and June. While the highest precipitation is observed in the months of June, July, August, and September, aligning with the summer monsoon. Similarly, the runoff follows a clear

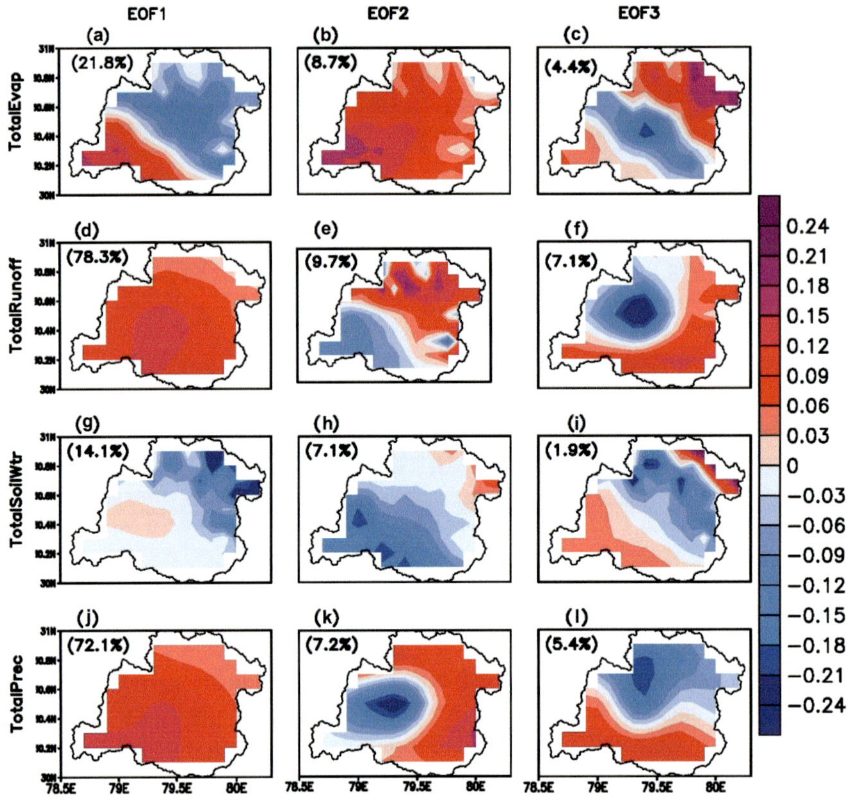

Fig. 16.11 EOF analysis for the annual total evaporation (EOF1: 21.8%, EOF2: 8.7% and EOF3: 4.4%), total runoff (EOF1: 78.3%, EOF2: 9.7% and EOF3: 7.1%), total soil–water content (EOF1: 14.1%, EOF2: 7.1% and EOF3: 1.9%), and total precipitation (EOF1: 72.1%, EOF2: 7.2% and EOF3: 5.4%) for 1981–2020. The spatial pattern of the first mode is EOF1, the spatial pattern of the second mode is EOF2 and the spatial pattern of the third mode is EOF3

sequence from April to September, coinciding with the monsoon season. In the upper region of the basin, the runoff pattern in May and June (summer season) indicates that the melting of the snow cover in that area is a significant factor contributing to the overall runoff. The monthly regional average is clearly suggesting that the highest precipitation and runoff has been observed during July and August (i.e., core monsoon period of the basin). The storage is found higher during the months of May, June, August, September, October, and November. While, the months of January, February, March, and July show negative storage over the basin for the period 1981–2020. The precipitation trend shows a decreasing pattern over northern part along with the runoff, while southernmost region shows positive trend values. However, evaporation, soil–water, and storage have nearly the same pattern. In addition to this the storage shows positive trend values over the northernmost part of

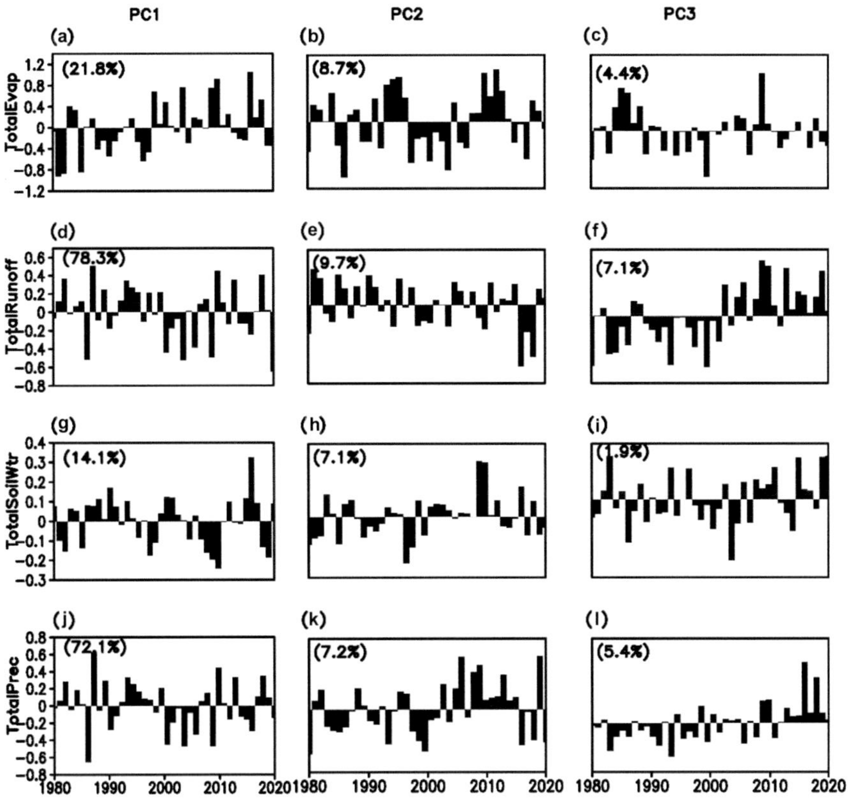

Fig. 16.12 PC analysis for the annual total evaporation, total runoff, total soil–water content, and total precipitation. The first, second and third modes explain (21.8%, 8.7% and 4.4%), (78.3%, 9.7% and 7.1%), (14.1%, 7.1% and 1.9%), and (72.1%, 7.2% and 5.4%) of the total variance for total evaporation, total runoff, total soil–water content and total precipitation respectively

the basin, and rest of the basin evidencing decreasing trend. Thereafter, the evaporation shows a rising trend over most part of the basin. The temporal variation of trend lines for precipitation, runoff, soil–water and storage indicates a declining trend over the study period for the entire basin with 99% confidence. Although the rate is relatively low, but it is significant for all the variables. Evaporation has shown the rising trend, probably due to the influence of warming (i.e., higher the warming, more is the evaporation) This has an impact on the basin's overall budget. The basin is facing more drought than flood during the study period, as indicated by different scale of SPEI and SPI showing significantly decreasing trend over the region. This pattern is also related to the rising CDD and extreme precipitation over the basin, as explained by EOF1, EOF2, EOF3, and PC1, PC2, PC3 of the precipitation. The negative component of total evaporation EOF1 and PC1 indicates the region as dry region. The extreme precipitation has a significant and a strong positive correlation

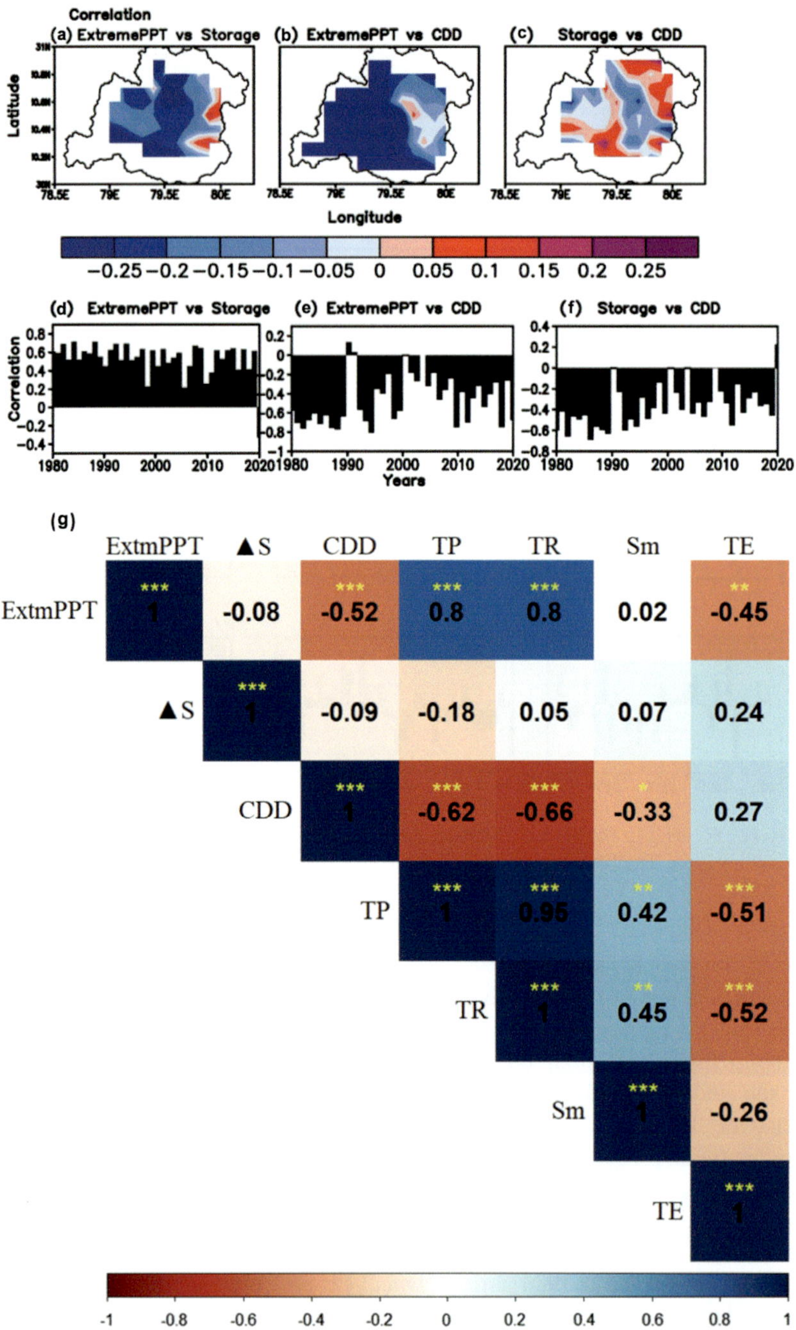

Fig. 16.13 Correlation plot for extreme precipitation (ExtmPPT), storage (ΔS) and CDD: consecutive dry days, **a**–**c** for spatial, **d**–**f** for temporal, and **g** correlation matrix for all the variables (total precipitation (TP), total runoff (TR), total evaporation (TE), soil–water (Sm), ΔS, ExtmPPT, and CDD) used in the study. The significant level for the correlation matrix is defined as (99%: "***", 95%: "**", and 90%: "*")

with TP and TR, likely due to heavy rainfall increasing runoff. Conversely, the negative relationship between extreme precipitation and storage in the lower region of the basin suggests that heavy precipitation leads to flooding and significant runoff. In the high-altitude region a positive relation of storage and CDD indicates that the solid precipitation melts during the dry and warm period. The temporal variation of storage and CDD suggests that storage follows dry days reciprocally and affects each year of the study period. The temporal variation of extreme precipitation and storage indicates a positive relation for each year. The decreasing tendency of very wet days and the growing trend of extreme precipitation suggest that the majority of the precipitation is falling in a shorter period of time.

16.5 Conclusions

The study concludes that the declining storage over the basin is significantly controlled by extremes events. The increasing consecutive dry days and more frequent dry years are contributing to the decline in total precipitation and runoff over the basin. Additionally, the significant decrease in soil–water content, influenced by higher evaporation rates, is contributing to more frequent seasonal drought years. The rise in extreme precipitation in the lower elevation of the basin is one of the key influencing factors in the incidence of intense floods in the basin. The temporal variation of storage and CDD relationship reveals that storage precedes dry days in an inverse way and impacts the entirety of the study period. The chronological fluctuation of excessive precipitation and storage implies a favorable relationship for each year as they accumulate. A strong negative correlation has also been perceived for CDD and extreme precipitation. Empirical Orthogonal Function (EOF) of TE, TP, TR, and Sm suggested that all of the hydroclimatic variables have great variability in hydrometeorological water budget. This research demonstrates that the extremes have a substantial influence on the basin's dropping storage. As the number of consecutive dry days in the basin increases, there is a subsequent decrease in the frequency of dry years, which is attributed to the decline in total precipitation and runoff.

Acknowledgements Authors acknowledge the Department of Earth and Planetary Sciences, University of Allahabad for providing the MS dissertation position to Mr. Sudheer Yadav, which is further to extended and considered to be a research chapter. We acknowledge the ERA5-land data portal for free available of data sets. Authors also acknowledge tools used for the analysis as CDO, R, QGIS, and GrADS.

Funding Source There is no funding provided for this study. However, this work is a part of Master dissertation.

Competing Interest Authors declare that there is no competing interest.

Authors Contribution Dr. Pramod Kumar: conceptualization, figures, writing, editing, and proof-reading.

Mr. Sudheer Yadav: figures, writing, editing, and proof-reading.

Dr. Khushboo Sharma: editing and proof-reading.

Dr. Prem Kumar: editing and proof-reading.
Miss. Jaya Rai: editing and proof-reading.
Dr. Ravindra Pratap Singh: prof-reading, editing, and mentoring.

Conflict of Interest Authors declare that there are no conflicts of interest for the present study.

References

Aggarwal SP, Thakur PK, Garg V, Nikam BR, Chouksey A, Dhote P, Bhattacharya T (2016) Water resources status and availability assessment in current and future climate change scenarios for Beas river basin of Northwestern Himalaya. Int Arch Photogramm Remote Sens Spat Inf Sci 41:1389. https://doi.org/10.5194/isprs-archives-XLI-B8-1389-2016

Altman N, Krzywinski M (2015) Points of significance: simple linear regression. Nat Methods 12(11). https://doi.org/10.1038/nmeth.3627

Benesty J, Chen J, Huang Y, Cohen I (2009) Pearson correlation coefficient. In: Noise reduction in speech processing. Springer, Berlin, pp 1–4. https://doi.org/10.1007/978-3-642-00296-0_5

da Silva Tavares P, Acosta R, Nobre P, Resende NC, Chou SC, de Arruda Lyra A (2023) Water balance components and climate extremes over Brazil under 1.5 °C and 2.0 °C of global warming scenarios. Reg Environ Change 23(1):40. https://doi.org/10.1007/s10113-023-02042-1

Das PK (2013) 'The Himalayan Tsunami'-cloudburst, flash flood & death toll: a geographical postmortem. IOSR J Environ Sci Toxicol Food Technol (IOSR-JESTFT) 7(2):33–45

Delgado JM, Apel H, Merz B (2009) Flood trends and variability in the Mekong River. Hydrol Earth Syst Sci Discuss 14:407–418. https://doi.org/10.5194/hess-14-407-2010

Deshmukh A, Singh R (2016) Physio-climatic controls on vulnerability of watersheds to climate and land use change across the US. Water Resour Res 52(11):8775–8793. https://doi.org/10.1002/2016WR019189

Dimri AP (2012) Atmospheric water budget over the western Himalayas in a regional climate model. J Earth Syst Sci 121:963–973. https://doi.org/10.1007/s12040-012-0204-8

Dimri AP, Kumar D, Chopra S, Choudhary A (2019) Indus River Basin: future climate and water budget. Int J Climatol 39(1):395–406. https://doi.org/10.1002/joc.5816

Duan Y, Ma Z, Yang Q (2017) Characteristics of consecutive dry days variations in China. Theoret Appl Climatol 130:701–709. https://doi.org/10.1007/s00704-016-1984-6

Frich PALV, Alexander LV, Della-Marta P, Gleason B, Haylock M, Tank AK, Peterson T (2002) Observed coherent changes in climatic extremes during the second half of the twentieth century. Climate Res 19(3):193–212. https://doi.org/10.3354/cr019193

Haq MA, Jain K, Shoab M, Menon KPR (2013) Estimation of terrestrial water storage change in the Bhagirathi Ganga and Vishnu Ganga basins using satellite gravimetry. In: 2013 IEEE international geoscience and remote sensing symposium-IGARSS. IEEE, pp 1817–1820. https://doi.org/10.1109/IGARSS.2013.6723153

Hans H et al (2020) The ERA5 global reanalysis. Q J Royal Meteorol Soc 146.730:1999–2049

Jolliffe IT (2002) Principal component analysis for special types of data. Springer, New York, pp 338–372

Joshi V, Kumar K (2006) Extreme rainfall events and associated natural hazards in Alaknanda valley, Indian Himalayan region. J Mt Sci 3:228–236. https://doi.org/10.1007/s11629-006-0228-0

Kenney JF, Keeping ES (1962) Linear regression and correlation. Math Stat 1:252–285

Khanduri S (2021) Flash flood struck Dhauliganga valley on February 7, 2021: a case study of Chamoli district of Uttarakhand Himalaya in India. Acad Platform J Nat Hazards Disaster Manag 2(1):1–15. https://doi.org/10.52114/apjhad.903387

Kumar D, Singh AK, Singh DS (2021) Spatio-temporal fluctuations over Chorabari glacier, Garhwal Himalaya, India between 1976 and 2017. Quatern Int 575:178–189. https://doi.org/10.1016/j.quaint.2020.04.037

Kutzbach JE (1967) Empirical eigenvectors of sea-level pressure, surface temperature and precipitation complexes over North America. J Appl Meteorol 6(5):791–802

Lane D, Scott D, Hebl M, Guerra R, Osherson D, Zimmer H (2017) Introduction to statistics. Independent

Li L, Gochis DJ, Sobolowski S, Mesquita MD (2017) Evaluating the present annual water budget of a Himalayan headwater river basin using a high-resolution atmosphere-hydrology model. J Geophys Res Atmos 122(9):4786–4807. https://doi.org/10.1002/2016JD026279

Mbaye ML, Sylla MB, Tall M (2019) Impacts of 1.5 and 2.0 C global warming on water balance components over Senegal in West Africa. Atmosphere 10(11):712. https://doi.org/10.3390/atmos10110712

Mehta M, Shukla T, Bhambri R, Gupta AK, Dobhal DP (2016) Terrain changes, caused by the 15–17 June 2013 heavy rainfall in the Garhwal Himalaya, India: a case study of Alaknanda and Mandakini basins, Geomorphology.https://doi.org/10.1016/j.geomorph.2016.11.001

Midhuna TM, Kumar P, Dimri AP (2020) A new Western disturbance Index for the Indian winter monsoon. J Earth Syst Sci 129:59. https://doi.org/10.1007/s12040-019-1324-1

Muhammad MKI, Nashwan MS, Shahid S, Ismail TB, Song YH, Chung ES (2019) Evaluation of empirical reference evapotranspiration models using compromise programming: a case study of Peninsular Malaysia. Sustainability 11(16):4267. https://doi.org/10.3390/su11164267

Pant N, Dubey RK, Bhatt A, Rai SP, Semwal P, Mishra S (2020) Soil erosion and flood hazard zonation using morphometric and morphotectonic parameters in Upper Alaknanda river basin. Nat Hazards 103:3263–3301

Patel P, Thakur PK, Aggarwal SP, Garg V, Dhote PR, Nikam BR, Swain S, Al-Ansari N (2022) Revisiting 2013 Uttarakhand flash floods through hydrological evaluation of precipitation data sources and morphometric prioritization. Geomat Nat Haz Risk 13(1):646–666. https://doi.org/10.1080/19475705.2022.2038696

Pearson K (1909) Determination of the coefficient of correlation. Science 30(757):23–25. https://doi.org/10.1126/science.30.757.23

Rana N, Singh S, Sundriyal YP, Juyal N (2013) Recent and past floods in the Alaknanda valley: causes and consequences. Curr Sci 105(9):1209–1212

Rathore BP, Singh SK, Bahuguna IM, Brahmbhatt RM, Rajawat AS, Thapliyal A, Panwar A, Ajai (2015) Spatio-temporal variability of snow cover in Alaknanda, Bhagirathi and Yamuna sub-basins, Uttarakhand Himalaya. Curr Sci 1375–1380. https://www.jstor.org/stable/24905502

Rudolph W (1988) Peisendorfer: principal component analysis

Singh G, Pandey A (2021) Evaluation of classification algorithms for land use land cover mapping in the snow-fed Alaknanda River Basin of the Northwest Himalayan Region. Appl Geomatics 13:863–875. https://doi.org/10.1007/s12518-021-00401-3

Singh P, Kumar A, Kumar N, Kishore N (2010) Hydro-meteorological correlations and relationships for estimating streamflow for Gangotri Glacier basin in Western Himalayas. Int J Water Resour Environ Eng 2(3):60–69. https://doi.org/10.5897/IJWREE.9000023

Sok T, Ich I, Tes D, Chan R, Try S, Song L, Ket P, Khem S, Oeurng C (2022) Change in hydrological regimes and extremes from the impact of climate change in the largest tributary of the Tonle Sap Lake Basin. Water 14(9):1426. https://doi.org/10.3390/w14091426

Sundriyal YP, Shukla AD, Rana N, Jayangondaperumal R, Srivastava P, Chamyal LS, Sati SP, Juyal N (2015) Terrain response to the extreme rainfall event of June 2013: evidence from the Alaknanda and Mandakini River Valleys, Garhwal Himalaya, India. Episodes J Int Geosci 38(3):179–188. https://doi.org/10.18814/epiiugs/2015/v38i3/004

Thornthwaite CW (1948) An approach toward a rational classification of climate. Geogr Rev 38(1):55–94. https://doi.org/10.2307/210739

Trenberth KE, Asrar GR (2014) Challenges and opportunities in water cycle research: WCRP contributions. Earth's Hydrol Cycle 515–532. https://doi.org/10.1007/978-94-017-8789-5_3

Tuinenburg OA, Hutjes RWA, Stacke T, Wiltshire A, Lucas-Picher P (2014) Effects of irrigation in India on the atmospheric water budget. J Hydrometeorol 15(3):1028–1050. https://doi.org/10.1175/JHM-D-13-078.1

Verma S, Sharma A, Yadava PK, Gupta P, Singh J, Payra S (2022) Rapid flash flood calamity in Chamoli, Uttarakhand region during Feb 2021: an analysis based on satellite data. Nat Hazards 112(2):1379–1393. https://doi.org/10.1007/s11069-022-05232-y

Vicente-Serrano SM, Beguería S, López-Moreno JI (2010) A multiscalar drought index sensitive to global warming: the standardized precipitation evapotranspiration index. J Clim 23(7):1696–1718. https://doi.org/10.1175/2009JCLI2909.1

Von Storch H, Zwiers FW (2002) Statistical analysis in climate research. Cambridge University Press

Wilks DS (2011) Statistical methods in the atmospheric sciences, vol 100. Academic press

Foote Notes

Tanmoyee B, Abdul RPH (2015) Climate change impact on snowmelt runoff modelling for Alaknanda river basin. Methodology 5(11). https://www.researchgate.net/profile/Tanmoyee-Bhattacharya/publication/279277999_Climate_Change_Impact_on_Snowmelt_Runoff_Modelling_for_Alaknanda_River_Basin/links/559159a508aed6ec4bf82d5a/Climate-Change-Impact-on-Snowmelt-Runoff-Modelling-for-Alaknanda-River-Basin.pdf

Chapter 17
Bayesian Model Averaging for Multi-model Ensemble Streamflows of the Godavari Basin

Rajas Surlikar, Akshay Pachore, and Renji Remesan

Abstract Streamflow prediction is a crucial aspect of water resource management and environmental control. Projected changes in climate variables from Global Climate Models (GCMs) coupled with Global Hydrological Models (GHMs) can be used to study water regimes during baseline and future hydrological conditions in river basins. Multi-model ensemble averaging methods have been widely used to combine the predictions of multiple GCMs-GHMs to produce more accurate and reliable average estimates. This study employs the Bayesian Model Averaging (BMA) technique to simulate values of monthly discharge from five GHMs coupled with projections from five GCMs from the Inter-Sectoral Impact Model Intercomparison Project (ISIMIP—fast track version) during a retrospective period. The BMA technique effectively selects the best-performing GCM-GHM model combinations, and additionally provides a solid framework for uncertainty analysis of ensemble-averaged predictions concerning observed monthly discharge data obtained from the Polavaram region close to the outlet of the Godavari basin from 1971 to 2000. The results show that BMA significantly outperforms traditional averaging approaches like simple model averaging and weighted averaging in predicting streamflow values. The BMA model effectively captures the seasonal fluctuations in the flow data, displaying minor differences in peak flow values while occasionally overestimating the low points in streamflow. This study during the baseline period suggests that BMA is a promising tool for improving the accuracy and reliability of streamflow projections provided by GCM-driven GHMs.

Keywords Godavari · BMA · GHMs · GCMs · ISI-MIP · Multi-model ensemble

R. Surlikar
Department of Physics, Indian Institute of Technology Kharagpur, Kharagpur, India

A. Pachore · R. Remesan (✉)
School of Water Resources, Indian Institute of Technology Kharagpur, Kharagpur, India
e-mail: renji.remesan@swr.iitkgp.ac.in

A. Nanda et al. (eds.), *Navigating the Nexus*, Water Science and Technology Library 102, https://doi.org/10.1007/978-3-031-76532-2_17

17.1 Introduction

Accurately estimating and understanding streamflow patterns in river basins is essential for several purposes, ranging from water availability assessment, preservation of aquatic ecosystems, and mitigation of disastrous effects of extreme climate events like floods and droughts (Mosavi et al. 2018; Moradkhani and Sorooshian 2008; Xu et al. 2019; Yaseen et al. 2016). As a result, global models have recently become a popular choice among hydrologists for simulating streamflow patterns over vast regions of interest and evaluate the impact of various climatic conditions on the streamflow discharge (Xiang et al. 2021; Somisetty et al. 2022). The future changes in the river water regimes are also a matter of greater concern, and numerous studies have delved into this issue, employing Global Climate Models (GCMs) in combination with global hydrological models (GHMs). These models investigate potential trends in the global climate, taking into account the intricate exchanges of water among the atmosphere, ocean, cryosphere, and land surface, both in the past and projected into the future (Wanders and Van Lanen 2015; Gosling et al. 2011). Effectively utilizing these models requires identifying the optimal GCM-GHM combinations to yield accurate streamflow predictions for a specific region. In this regard, the Inter-Sectoral Impact Model Intercomparison Project (ISI-MIP; https://www.isimip.org/) provides a unified framework that spans multiple state-of-the-art GHMs, that take Coupled Model Intercomparison Project Phase 5 (CMIP5) projections (Zhang and Chen 2021) as inputs to simulate diverse streamflow patterns. In particular, the ISI-MIP fast-track version (ISI-MIP fast-track; https://www.isimip.org/protocol/fast-track-simulation-protocol/) allows for timely evaluations, particularly crucial in the context of rapidly evolving climate scenarios for making real-time decisions from robust streamflow estimates. However, it is important to note that none of these individual models capture the fine details of all the underlying phenomena completely without incurring errors (Chen et al. 2013; Liu et al. 2016). Consequently, relying on a single model often introduces statistical bias into the forecast, hence motivating the need for multi-model ensemble (MME) techniques to reduce the inherent uncertainty and make robust predictions (Yan et al. 2020; Somisetty et al. 2022).

Some studies have highlighted the need for a multi-model approach with combinations of GHMs and GCMs for the assessment of climate and human-induced impacts on water availability at river basin scale (Somisetty et al. 2022). Future Streamflow assessment from the combinations of GCMs-GHMs is a highly complex task as the ensemble mean runoff changes obtained may be accompanied by various kinds of uncertainty ranges (Weiland et al. 2012). A variety of sophisticated multi-model ensemble average models have been proposed for this purpose especially to understand the uncertainties associated with the projections (Yan et al. 2020) which consider different weighted combinations of GCMs. Over the past few decades, researchers have devised a variety of model averaging techniques including deterministic approaches like simple averaging, weighted averaging, Granger-Ramanathan averaging (Granger and Ramanathan 1984), and probabilistic frameworks based on Bayesian principles like Bayesian Model Averaging (BMA) (Hoeting et al. 1999;

Raftery et al. 1997), and Bayesian Model Combination (BMC) (Monteith et al. 2011). Out of all model averaging approaches, BMA is one of the most promising multi-model ensemble techniques in comparison to other methods due to its ability to intricately handle model uncertainties associated with predictions, even from limited data sets (Arsenault et al. 2015). BMA has found applications in diverse fields, ranging from healthcare, meteorology, and finance (Bahrami et al. 2023; Kürüm et al. 2017; Raftery et al. 2005; Ma et al. 2018; Sun et al. 2018; Hasan et al. 2018). In recent years, BMA has also been adopted in various water resources and hydrologic studies, spanning groundwater modeling to flood frequency analysis (Rojas et al. 2008; Zeng et al. 2016; Yan and Moradkhani 2016; Duan et al. 2007; Liang et al. 2013). The BMA estimate is a weighted average of all probability distributions in an ensemble, with each weight being non-negative and summing up to 1 and some studies have used it for generating representative best ensemble GCM projections and to quantify the uncertainties of a combination of multiple GCM-GHM models (Basher et al. 2020). The BMA framework produces posterior predictive probability distribution of ensemble-averaged values, hence providing a solid basis for uncertainty analysis associated with ensemble projection of the streamflow in large river basins (Huang, 2014; Parrish et al. 2012; Rojas et al. 2008; Vrugt and Robinson 2007).

Considering the above aspects, this study aims an evaluation and application of global BMA approach utilizing the ISI-MIP fast-track version to obtain streamflow discharge patterns at the Godavari basin scale to fulfil the following objectives: (i) To employ BMA to combine the predictions of multiple GCMs-GHMs combinations to produce best representative ensemble assessments of monthly streamflow estimates for Godavari basin (ii) To check capability of BMA to develop MME estimates of streamflow and its performance in comparison to traditional approaches (simple arithmetic average).

17.2 Study Area and Dataset

The Godavari basin, which makes up around 9.50% of all of India's land area, is home to the largest peninsular river in the nation and the second-largest river basin overall, following the Ganges basin. The Godavari basin spans majorly across the states of Maharashtra, Telangana, Chhattisgarh, and Madhya Pradesh, with smaller fragments extending into Odisha, Madhya Pradesh, Karnataka, and the Union territory of Puducherry, and encompasses a total length of 1465 km. The basin is located in the Deccan plateau between 73°24′ and 83°4′ East longitudes and 16°19′ to 22°34′ North latitudes (Fig. 17.1). As per the Godavari basin report by Central Water Commission of India, the majority of the rainfall (about 85%) occurs during the Southwest monsoon period, i.e., from July to September, with annual rainfall varying from 881 to 1395 mm and average rainfall being equal to 1110 mm. Contrastingly, the months of January and February are characterized by near-zero rainfall with precipitation of less than 15 mm, and the following three months, up to the end of May, witness rainfall ranging from 20 to 50 mm in most parts of the basin.

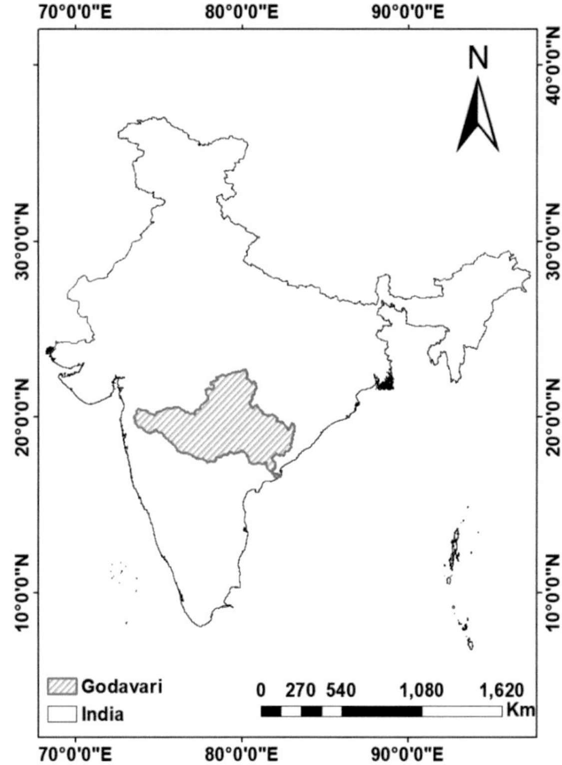

Fig. 17.1 Location map of the Godavari basin within India (CWC, 2014)

17.2.1 Hydro-climatic Datasets

The Inter-Sectoral Impact Model Intercomparison Project (ISI-MIP; https://www.isimip.org/) is a community-driven project that provides a standardized framework for projecting climate change impacts across various sectors and spatial scales in global hydrologic simulations. For our study, we considered five state-of-the-art Global Hydrological Models (GHMs): H08, MPI-HM, PCR-GLOBWB, VIC, and WaterGAP2. Here, we analyze monthly discharge simulations of ISI-MIP fast-track runs from the above five GHMs, which were combined with CMIP5 projections taken from five GCMs: GFDL-ESM2M, HadGEM2-ES, IPSL-CM5A-LR, MIROC-ESM-CHEM, and NorESM1-M. The outputs of the GCMs are bias-corrected and down-scaled using a trend preserving approach, before combining with the GHMs within the ISI-MIP protocol. The selection of GHMs was based on data availability and their ability to accurately simulate hydrological processes, particularly streamflow. The observed monthly discharge data (m^3/s) for a location close to the Godavari basin outlet (Polavaram) was obtained from the India-WRIS website (https://indiawris.gov.in/wris/) for the period 1971 to 2000. The details of the GCM-GHM combination with model number is presented in Table 17.1.

Table 17.1 Details of the GCM-GHM combinations used in the present study

Model number	GHM	GCM	GCM-GHM combination
1	H08	GFDL-ESM2M	H08 GFDL-ESM2M
2		HadGEM2-ES	H08 HadGEM2-ES
3		IPSL-CM5A-LR	H08 IPSL-CM5A-LR
4		MIROC-ESM-CHEM	H08 MIROC-ESM-CHEM
5		NorESM1-M	H08 NorESM1-M
6	MPI-HM	GFDL-ESM2M	MPI-HM GFDL ESM2M
7		HadGEM2-ES	MPI-HM HadGEM2-ES
8		IPSL-CM5A-LR	MPI-HM IPSL-CM5A-LR
9		MIROC-ESM-CHEM	MPI-HM MIROC-ESM-CHEM
10		NorESM1-M	MPI-HM NorESM1-M
11	PCR-GLOBWB	GFDL-ESM2M	PCR-GLOBWB GFDL ESM2M
12		HadGEM2-ES	PCR-GLOBWB HadGEM2-ES
13		IPSL-CM5A-LR	PCR-GLOBWB IPSL-CM5A-LR
14		MIROC-ESM-CHEM	PCR-GLOBWB MIROC-ESM-CHEM
15		NorESM1-M	PCR-GLOBWB NorESM1-M
16	VIC	GFDL-ESM2M	VIC GFDL ESM2M
17		HadGEM2-ES	VIC HadGEM2-ES
18		IPSL-CM5A-LR	VIC IPSL-CM5A-LR
19		MIROC-ESM-CHEM	VIC MIROC-ESM-CHEM
20		NorESM1-M	VIC NorESM1-M
21	WaterGAP	GFDL-ESM2M	WaterGAP GFDL ESM2M
22		HadGEM2-ES	WaterGAP HadGEM2-ES
23		IPSL-CM5A-LR	WaterGAP IPSL-CM5A-LR
24		MIROC-ESM-CHEM	WaterGAP MIROC-ESM-CHEM
25		NorESM1-M	WaterGAP NorESM1-M

17.3 Methodology

17.3.1 Bayesian Model Averaging

The Bayesian framework emphasizes updating beliefs or current knowledge by adding new data in each iteration, in contrast to frequentist approaches, which concentrate on the long-run behavior of hypothetical repeated trials. BMA essentially employs concepts from the total probability theorem and Bayes' theorem to combine forecasts from multiple competing models and return an ensemble-averaged forecast that incorporates information from each one of them. The Bayesian ensemble-averaged predictive density is given by Eq. (17.1).

$$P\left(Y^*|M, \theta, Y\right) = \sum_{i=1}^{k} P\left(Y^*|M_i, Y\right)*P(M_i|Y, \theta) \tag{17.1}$$

where, $P(Y^*|M, \theta, Y)$ represents the probability of observing a future outcome Y^* given the ensemble of models M and the observed data Y; $P(Y^*|M_i, Y)$ represents the likelihood function for the future outcome Y^* given any model M_i; $P(M_i|Y, \theta)$ represents the posterior model probability distribution, which quantifies the relative likelihood of each model given the data. This loosely corresponds to the "weight" of each model.

Posterior Model Probability of given model M_i is further calculated as given in Eq. (17.2).

$$P(M_i|Y) = \int P(Y|M_i, \theta)*P(M_i|\theta) \cdot d\theta \tag{17.2}$$

where, $P(Y|M_i, \theta)$ denotes the likelihood function for the observed data Y given the parameters θ and model M_i; $P(M_i|\theta)$ represents the prior model probability distribution, which reflects our initial beliefs about the relative likelihood of each model before observing any data, based on the parameters θ. Additionally, the individual models are assumed to be independent, meaning that their errors are not correlated, hence leading to Eq. (17.3).

$$\sum_{i=1}^{k} P(M_i|Y) = 1 \tag{17.3}$$

Hence, the term $P(M_i|Y)$ effectively serves as the weight of model M_i during the estimation of the final streamflow value. Therefore, the final formula is expressed as given in Eq. (17.4).

$$P\left(Y^*|M, \theta, Y\right) = \sum_{i=1}^{k} P\left(Y^*|M_i, Y\right) * W_i \tag{17.4}$$

Additionally, to highlight the advantages of BMA over other averaging approaches as well as demonstrate its capabilities to inherently establish the best GCM-GHM combinations, current study has utilized the Taylor Diagram that explicitly identifies the optimal model combinations. Subsequently, the mean value from the predictions is calculated, only for the models selected using Taylor Diagram, as well as for the all models. In this study, the Taylors' diagram statistics from are utilized from the study done by Somisetty et al. (2022).

The flowchart in Fig. 17.2 shows the procedure adopted to systematically carry out BMA and get the final streamflow estimates. To begin with, the monthly data is partitioned into distinct samples in order to preserve the temporal consistency of the initial data and avoid the mixing of streamflow values from various months during the final estimation process. Assuming a linear relationship between each of the 25 GCM-GHM models and the observed data for each month, a Bayesian Linear Regressor is trained (Baldwin and Larson 2017) with the prior probability distributions of each parameter being as depicted in Fig. 17.2. A No-U-Turn (NUTS) Sampler, an extension of Hamiltonian Monte-Carlo (HMC), is employed to generate a trace of samples representing the marginal posterior distribution of the parameters using PyMC (https://www.pymc.io/projects/docs/en/v3/index.html) probabilistic programming language.

This, in turn, is utilized to obtain predictive posterior distribution from a linear model for each month. Following the predictive posterior calculations, Leave-One-Out Cross-Validation (LOO-CV) resampling technique (Vehtari et al. 2017) is utilized that determines the model weights associated with each linear model by removing one observation at a time, fitting the model to the remaining data, and predicting the removed observation. Finally, using BMA, the ensemble-averaged streamflow values of 25 Bayesian linear models are obtained independently for each month, resulting in final monthly predictions over the entire 30-year period.

17.3.2 Evaluation Metrics for Model Performance

Root Mean Square Error (RMSE), Pearson correlation coefficient (R), and Percentage bias (PBIAS) metrics are employed to evaluate the performances of each approach (see Eqs. 17.5–17.9). In general, RMSE and PBIAS values close to zero indicate a "perfect" model, i.e., the model captures every minute aspect of the observed data. RMSE value ranges from $[0, +\infty)$, and PBIAS values range from $(-\infty, ++\infty)$.

$$\text{RMSE} = \sqrt{\frac{\sum_{i=1}^{N}\left(Y_{\text{observed}} - Y_{\text{predicted}}\right)^2}{N}} \tag{17.5}$$

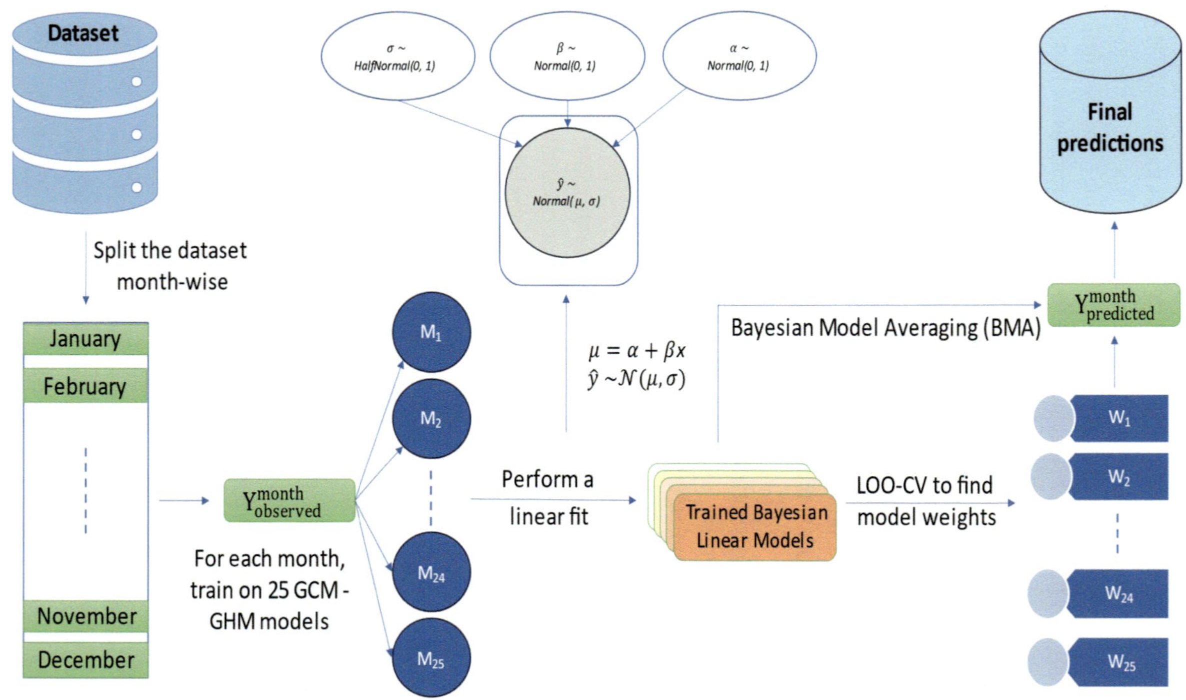

Fig. 17.2 Proposed methodology for streamflow estimation using BMA

$$\sigma_{observed} = \frac{\sum_{i=1}^{N}(Y_{observed})^2}{N} - \left(\frac{\sum_{i=1}^{N} Y_{observed}}{N}\right)^2 \quad (17.6)$$

$$\sigma_{predicted} = \frac{\sum_{i=1}^{N}\left(Y_{predicted}\right)^2}{N} - \left(\frac{\sum_{i=1}^{N} Y_{predicted}}{N}\right)^2 \quad (17.7)$$

$$PBIAS = \frac{\sum_{i=1}^{N}\left(Y^{i}_{observed} - Y^{i}_{predicted}\right)}{\sum_{i=1}^{N}\left(Y^{i}_{observed}\right)} \quad (17.8)$$

$$\text{Correlation (R)} = \frac{N\left(\left(\sum_{i=1}^{N} Y^{i}_{observed}\right)\left(\sum_{i=1}^{N} Y^{i}_{predicted}\right)\right) - \left(\sum_{i=1}^{N} Y^{i}_{observed}\right)\left(\sum_{i=1}^{N} Y^{i}_{predicted}\right)}{\sigma_{observed} \cdot \sigma_{predicted}} \quad (17.9)$$

PBIAS values ranging from $[-10, +10]$ are "Very Good", but if they lie in $(-\infty, -25] \cup [+25, +\infty)$ intervals, they are "Unsatisfactory". Correlation values on the other hand range from $[-1, +1]$, and values close to $\pm$ 1 imply a strong positive (or negative) correlation between 2 variables (Moriasi et al. 2015).

17.4 Results

17.4.1 Data Exploration and Statistics

An examination of observed and simulated streamflow data from GCM-GHM models is as shown in Fig. 17.3, which reveals a common pattern. The distribution of streamflow is largely skewed towards higher values, indicating a significant presence of outliers, i.e., a pronounced positive skewness. This streamflow behavior is expected, originating from the rainfall seasonality in the region. Rainfall events can be highly irregular in space and time with uneven distribution, leading to sudden spikes in streamflow. However, periods of low rainfall can lead to extended periods of low streamflow. Interestingly, data from GCM-GHM models appears to be more skewed than observed data. This can be attributed to the fact that GCM-GHM models may not fully capture the intricacies of the Earth's climate system, leading to some inaccuracies in streamflow predictions. Hence, it is essential to combine predictions from several models to get final averaged predictions.

Figure 17.4 shows that the simulated values from the different GCM-GHMs vary significantly, with some models overestimating streamflow and others underestimating. The observed data lies somewhere in between the simulated values from different models and none of the models are able to capture it with consistency and respectable accuracy. Hence, performing model averaging could potentially return values that are close to the observations.

Figure 17.4 also shows that the GCM-GHMs are generally able to capture the overall trend of streamflow variation throughout the year. However, the peaks that are observed during the monsoon season are displaced with respect to the observations.

The correlation plot (Fig. 17.5) for each model's predicted and observed values suggests that the simulated time series exhibits a fairly strong association with the observed data, although it does not closely resemble the exact behaviour of the system being modelled. Such correlation between simulated and observed data could have arisen from the fact that these are Global models and, hence may not adequately capture the intricacies of the basin-scale climate systems. Apart from the model uncertainty the quality and completeness of the input data used to simulate the streamflow values could be insufficient to fully represent the system's dynamics. However,

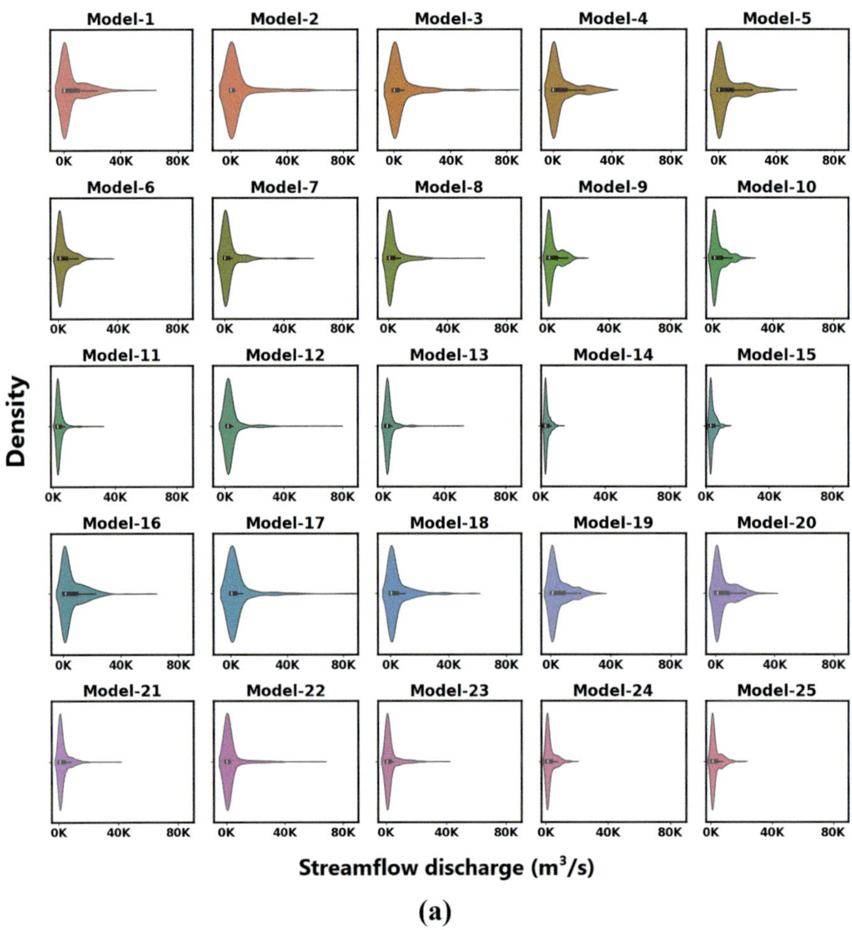

Fig. 17.3 Violin plots showing the distribution of streamflow discharge: **a** GCM-GHM model simulations and **b** observed discharge (here, in *x*-axis, K: $\times 10^3$)

Fig. 17.3 (continued)

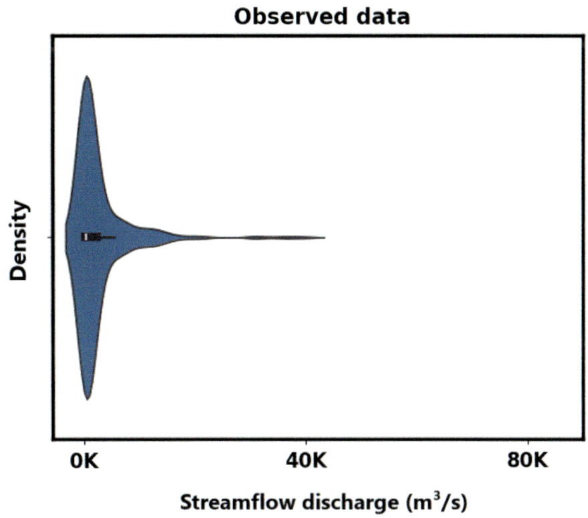

(b)

the system itself may exhibit inherent variability and randomness, making it difficult to predict its behaviour with absolute precision. Hence, averaging multiple models can potentially improve predictive performance by reducing the overall uncertainty.

17.4.2 Periodic Estimation of Streamflow

After obtaining the BMA results from the adapted methodology, Taylor diagram is used to identify the 3 best GCM-GHM combinations (WaterGAP IPSL-CM5A-LR, WaterGAP MIROC-ESM-CHEM, and WaterGAP NorESM1-M) from the set of 25 model combinations considered in the present study (see Fig. 17.6). BMA predicted values are showing the contrasting behaviour with the simple average of best-3 models selected from the Taylors' diagram, as well as average across all models.

Figure 17.6a shows that the observed streamflow data exhibits significant variability over time, with high and low flow occurring on a monthly and annual basis. The BMA forecast generally tracks the observed data well, but it does underestimate some of the high flow values and overestimate some of the low flow values. The mean of the best three GCM-GHMs also tracks the observed data well, but it is slightly more conservative than the BMA forecast. The mean of all GCM-GHMs performs the worst, underestimating the observed streamflow data for most of the time period.

Figure 17.6b reveals that the BMA predictions closely mimicks the trend of observed data as compared to other 2 approaches, but follows a modest and smooth curve. As a result, it falls short on predicting the extreme values and sometimes misplaces its peaks, especially during monsoon period. Meanwhile, the mean of 3 best GCM-GHMs also follows the observed discharge curve to a certain extent, but fails significantly to accurately sync its peaks with observed discharge's peaks. The mean of all models performs the worst of all, often completely out of phase with observed values, as seen in the month of May.

Overall, from Table 17.2, it is obaserved that the BMA forecast provides the most accurate representation of the observed streamflow data. This can be attributed to the fact that in the Bayesian approach, we simulate a large amount of posterior predictive values from each trained linear model. Thus, allowing each model to predict a range of streamflow values for given input. Additionally, utilizing weights to filter out the best predictions from each model further enhances the accuracy of our predictions. However, it is important to note that all of the models exhibit some degree of error (Fig. 17.7).

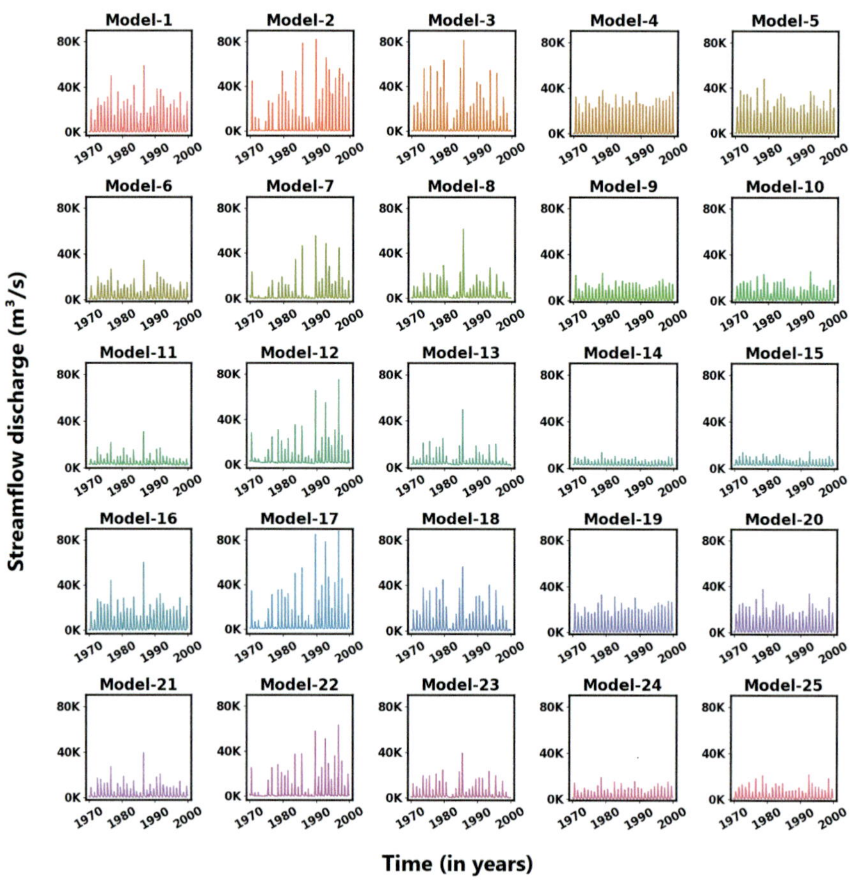

(a)

Fig. 17.4 Time series plots showing the distribution of streamflow discharge across the years: **a** GCM-GHM model simulations and **b** observed discharge

Fig. 17.4 (continued)

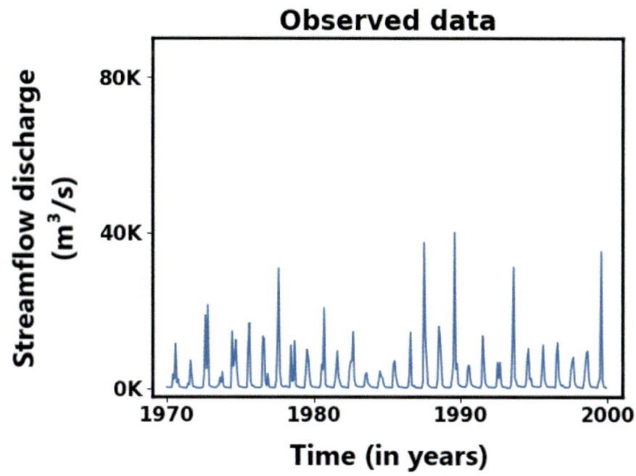

(b)

Fig. 17.5 Lollipop plot to highlight the Pearson correlation coefficient (ρ) between observed discharge and simulated discharge obtained from GCM-GHM models

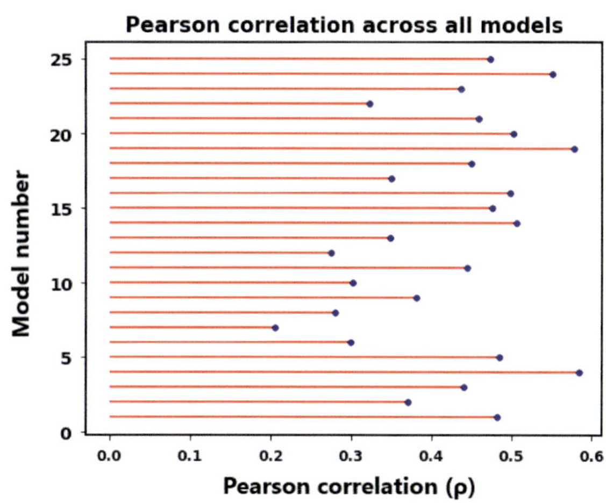

Hence, it is concluded that only 4 of the GCM-GHM models majorly contribute to the final predictions. These 4 models are: PCR-GLOBWB GFDL ESM2M, PCR-GLOBWB HadGEM2-ES, PCR-GLOBWB MIROC-ESM-CHEM, and VIC IPSL-CM5A-LR. Among these 4 models, PCR-GLOBWB HadGEM2-ES has the highest contribution towards the prediction.

17.5 Discussion and Conclusions

The analysis revealed that both observed and simulated streamflow data exhibited a pronounced positive skewness, indicative of frequent outliers and extreme events. This phenomenon possibly stems from the seasonal variability of rainfall and runoff, with the outliers corresponding to the intense rainfall events during monsoon. Moreover, as compared to observed values, the simulated streamflow values from GCM-GHM models displayed even greater skewness, suggesting potential inadequacies within the models. Consequently, due to compromise in data quality and quantity, achieving very high accuracy even after using ensemble-averaged predictions proved to be difficult.

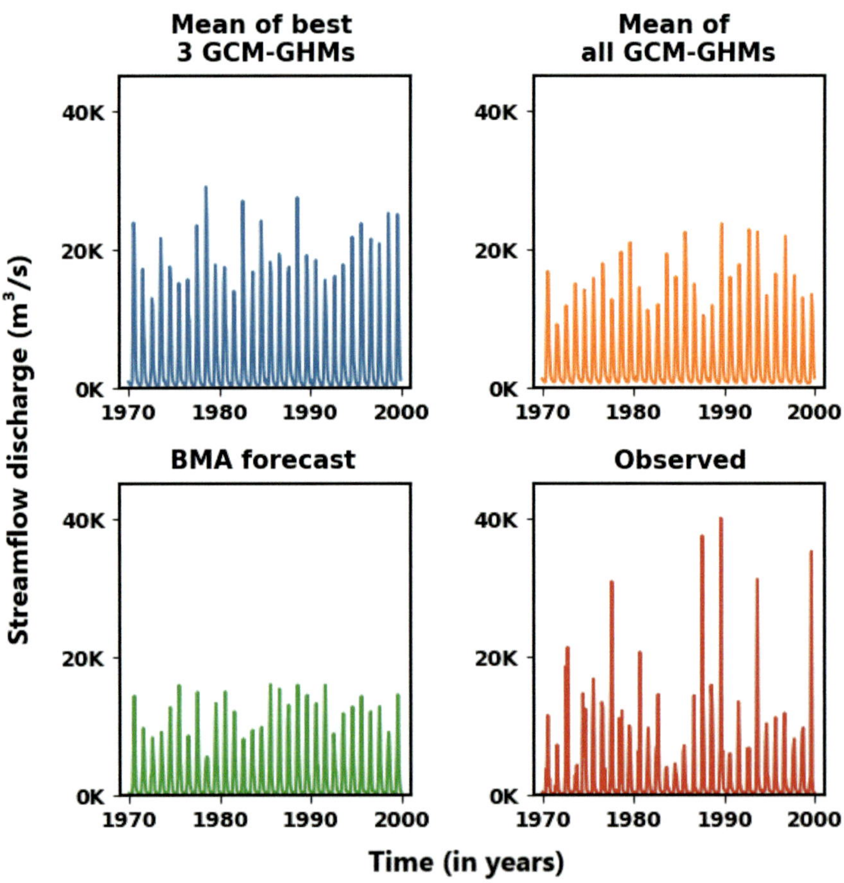

Fig. 17.6 Plots showing the distribution of streamflow discharge: **a** time series plots to contrast the performance of each technique and comparison with observed discharge and **b** time series plots of streamflow discharge for each month across all years

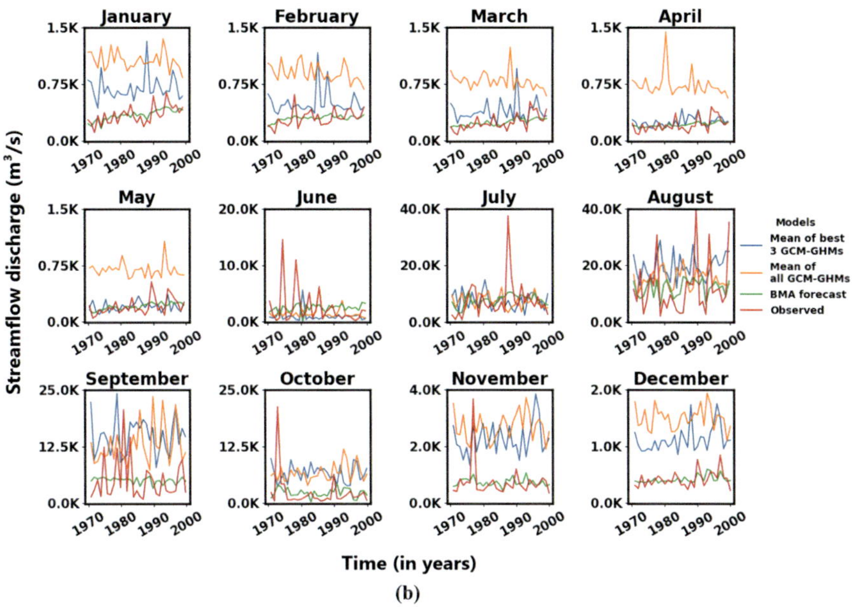

Fig. 17.6 (continued)

Table 17.2 Evaluation metrics for streamflow predictions using different ensemble methods

Method	RMSE	PBIAS	Pearson correlation (R)
BMA	3664.8080	– 0.1051	0.7273
Mean of all models	5347.6300	– 63.1405	0.5376
Mean of best-3 models	5844.8490	– 72.3090	0.5813

The time-series analysis revealed significant variations between simulations from individual GCM-GHM combinations, with none perfectly mirroring the observed patterns. Nevertheless, BMA's ability to leverage the strengths of individual models while also minimizing their weaknesses truly sets it apart. This can be strongly attributed to BMA's ability to generate a large number of posterior predictive values from each trained Bayesian linear regressor. Hence, ultimately granting its forecast the most accurate representation of observed streamflow data, albeit exhibiting some degree of error. Interestingly, the time-series graph in Fig. 17.6a, b reveals a common limitation across all approaches i.e., the inability to effectively predict extremely high values during monsoon months, hinting at an inherent constraint within the dataset itself.

In current work, the normal distribution is assumed to model the prior distributions of the target variable, and the parameters of normal distribution of target variable, namely μ and σ. In future studies, it is aimed to vary the hyperparameters

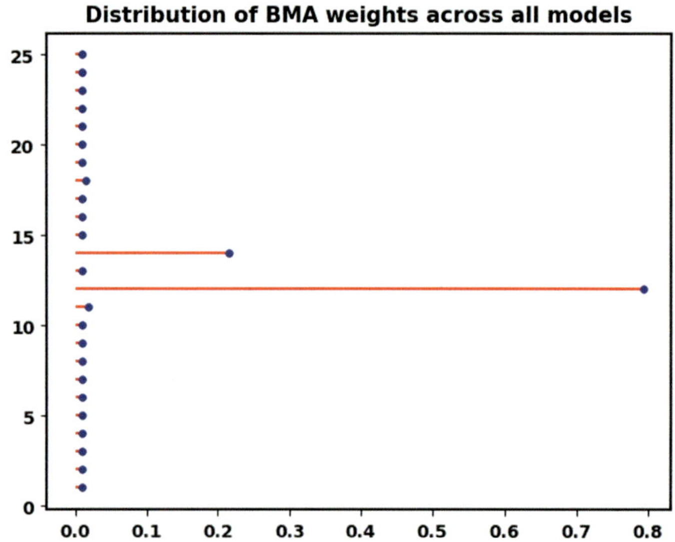

Fig. 17.7 Lollipop plot to visualize the importance of each model's contribution to the BMA predictions using weights obtained through LOO-CV

involved in the prior distributions and experiment with several probability distribution functions. The obtained results can be compared and contrasted in search of the best set of hyperparameters and probability distributions for more accurate streamflow modelling. Gamma distributions are better suited for skewed data, hence they could enhance the model's ability to capture extreme streamflow events and possibly outperform normal distribution (Samadi et al. 2020). Due to BMA's probabilistic nature, many researchers have also carried out studies that quantify the uncertainty in streamflow prediction (Dong et al. 2013; Yan et al. 2020). Several variations have been suggested to the original BMA approach to enhance its prediction capabilities. This includes improvements in the Markov Chains Monte Carlo (MCMC) algorithm (Huang and Merwade 2023; Mu et al. 2019; Speich et al. 2021; Panahi et al. 2021), utilizing Copula Bayesian Model Averaging (Panahi et al. 2021), as well as entropy-based Bayesian Model Averaging (Darbandsari and Coulibaly 2020) for faster convergence.

The present work outlined a detailed assessment for the application of BMA using simulated streamflow values from GCM-GHM combinations. Result from the current analysis also contrasted the performance of BMA with traditional averaging approaches, and found that it significantly outperforms the latter. Thus, utilization of Global models coupled with Bayesian averaging approaches to estimate future streamflow for effective water resource management and environmental control seems a viable approach.

The particular conclusions from the current analysis are as follows:

1. Streamflow from individual GCM-GHMs can capture the overall trend but show considerable variation from the observed streamflow with under and over-estimations.
2. BMA ensembles show weak resemblance to peak streamflow values during the monsoon season, both in terms of timing and magnitude.
3. In the BMA assessment, it is observed that only 4 GCM-GHM model combinations are contributing most towards the ensembled predictions.
4. BMA ensemble obtained from the best-selected GCM-GHMs performs well (Pearson correlation: 0.73) as compared to the simple average (Pearson correlation: 0.58).

However, it is equally vital to acknowledge and address the existing limitations. Underestimation during monsoon seasons and discrepancies in streamflow peaks pose significant challenges that demand further investigation and refinement. By exploring avenues like alternative prior distributions for skewed data, improved data quality and completeness can pave the way for even more robust and reliable streamflow predictions. Incorporating ensemble forecasting techniques for streamflow prediction of future years and contrasting their performance with data-driven models is also an interesting avenue to explore.

References

Arsenault R, Gatien P, Renaud B, Brissette F, Martel JL (2015) A comparative analysis of 9 multi-model averaging approaches in hydrological continuous streamflow simulation. J Hydrol 529:754–767

Bahrami S, Hajian-Tilaki K, Bayani M, Chehrazi M, Mohamadi-Pirouz Z, Amoozadeh A (2023) Bayesian model averaging for predicting factors associated with length of COVID-19 hospitalization. BMC Med Res Methodol 23(1):163

Baldwin SA, Larson MJ (2017) An introduction to using Bayesian linear regression with clinical data. Behav Res Ther 98:58–75

Basher A, Islam AS, Stiller-Reeve MA, Chu PS (2020) Changes in future rainfall extremes over Northeast Bangladesh: A Bayesian model averaging approach. Int J Climatol 40(6):3232–3249

CWC (2014) NRSC. Godavari Basin Report Version 2.0; Ministry of Water Resources, Government of India: New Delhi, India

Chen X, Yang T, Wang X, Xu CY, Yu Z (2013) Uncertainty intercomparison of different hydrological models in simulating extreme flows. Water Resour Manage 27:1393–1409

Darbandsari P, Coulibaly P (2020) Introducing entropy-based Bayesian model averaging for streamflow forecast. J Hydrol 591:125577

Dong L, Xiong L, Yu KX (2013) Uncertainty analysis of multiple hydrologic models using the Bayesian model averaging method. J Appl Math

Duan Q, Ajami NK, Gao X, Sorooshian S (2007) Multi-model ensemble hydrologic prediction using Bayesian model averaging. Adv Water Resour 30(5):1371–1386

Gosling SN, Taylor RG, Arnell NW, Todd MC (2011) A comparative analysis of projected impacts of climate change on river runoff from global and catchment-scale hydrological models. Hydrol Earth Syst Sci 15(1):279–294

Granger CW, Ramanathan R (1984) Improved methods of combining forecasts. J Forecast 3(2):197–204

Hasan I, Horvath R, Mares J (2018) What type of finance matters for growth? Bayesian model averaging evidence. World Bank Econ Rev 32(2):383–409

Hoeting JA, Madigan D, Raftery AE, Volinsky CT (1999) Bayesian model averaging: a tutorial (with comments by M. Clyde, David Draper and EI George, and a rejoinder by the authors). Stat Sci 14(4):382–417

Huang Y (2014) Comparison of general circulation model outputs and ensemble assessment of climate change using a Bayesian approach. Global Planet Change 122:362–370

Huang T, Merwade V (2023) Improving Bayesian model averaging for ensemble flood modeling using multiple Markov Chains Monte Carlo sampling. Water Resour Res 59(10):e2023WR034947

Kürüm E, Warren JL, Schuck-Paim C, Lustig R, Lewnard JA, Fuentes R et al (2017) Bayesian model averaging with change points to assess the impact of vaccination and public health interventions. Epidemiology (Cambridge, Mass.) 28(6):889

Liang Z, Wang D, Guo Y, Zhang Y, Dai R (2013) Application of Bayesian model averaging approach to multimodel ensemble hydrologic forecasting. J Hydrol Eng 18(11):1426–1436

Liu Z, Guo S, Zhang H, Liu D, Yang G (2016) Comparative study of three updating procedures for real-time flood forecasting. Water Resour Manage 30:2111–2126

Ma Y, Hong Y, Chen Y, Yang Y, Tang G, Yao Y et al (2018) Performance of optimally merged multisatellite precipitation products using the dynamic Bayesian model averaging scheme over the Tibetan Plateau. J Geophys Res Atmos 123(2):814–834

Monteith K, Carroll JL, Seppi K, Martinez T (2011) Turning Bayesian model averaging into Bayesian model combination. In: The 2011 international joint conference on neural networks. IEEE, pp 2657–2663

Moradkhani H, Sorooshian S (2008) General review of rainfall-runoff modeling: model calibration, data assimilation, and uncertainty analysis. Springer, Berlin, pp 1–24

Moriasi DN, Gitau MW, Pai N, Daggupati P (2015) Hydrologic and water quality models: performance measures and evaluation criteria. Trans ASABE 58(6):1763–1785

Mosavi A, Ozturk P, Chau KW (2018) Flood prediction using machine learning models: literature review. Water 10(11):1536

Mu Y, See I, Edwards JR (2019) Bayesian model averaging: improved variable selection for matched case-control studies. Epidemiol Biostat Publ Health 16(2)

Panahi F, Ehteram M, Ahmed AN, Huang YF, Mosavi A, El-Shafie A (2021) Streamflow prediction with large climate indices using several hybrid multilayer perceptrons and copula Bayesian model averaging. Ecol Ind 133:108285

Parrish MA, Moradkhani H, DeChant CM (2012) Toward reduction of model uncertainty: integration of Bayesian model averaging and data assimilation. Water Resour Res 48(3)

Raftery AE, Madigan D, Hoeting JA (1997) Bayesian model averaging for linear regression models. J Am Stat Assoc 92(437):179–191

Raftery AE, Gneiting T, Balabdaoui F, Polakowski M (2005) Using Bayesian model averaging to calibrate forecast ensembles. Mon Weather Rev 133(5):1155–1174

Rojas R, Feyen L, Dassargues A (2008) Conceptual model uncertainty in groundwater modeling: combining generalized likelihood uncertainty estimation and Bayesian model averaging. Water Resour Res 44(12)

Samadi S, Pourreza-Bilondi M, Wilson CAME, Hitchcock DB (2020) Bayesian model averaging with fixed and flexible priors: theory, concepts, and calibration experiments for rainfall-runoff modeling. J Adv Model Earth Syst 12(7):e2019MS001924

Somisetty A, Pachore A, Remesan R, Kumar R (2022) Multi-model assessment of streamflow simulations under climate and anthropogenic changes exemplified in two Indian River basins. Water 14(2):194

Speich M, Dormann CF, Hartig F (2021) Sequential Monte-Carlo algorithms for Bayesian model calibration–a review and method comparison. Ecol Model 455:109608

Sun R, Yuan H, Yang Y (2018) Using multiple satellite-gauge merged precipitation products ensemble for hydrologic uncertainty analysis over the Huaihe River basin. J Hydrol 566:406–420
Vehtari A, Gelman A, Gabry J (2017) Practical Bayesian model evaluation using leave-one-out cross-validation and WAIC. Stat Comput 27:1413–1432
Vrugt JA, Robinson BA (2007) Treatment of uncertainty using ensemble methods: comparison of sequential data assimilation and Bayesian model averaging. Water Resour Res 43(1)
Wanders N, Van Lanen HAJ (2015) Future discharge drought across climate regions around the world modelled with a synthetic hydrological modelling approach forced by three general circulation models. Nat Hazard 15(3):487–504
Weiland FS, Van Beek LPH, Weerts AH, Bierkens MFP (2012) Extracting information from an ensemble of GCMs to reliably assess future global runoff change. J Hydrol 412:66–75
Xiang Y, Chen J, Li L, Peng T, Yin Z (2021) Evaluation of eight global precipitation datasets in hydrological modeling. Remote Sens 13(14):2831
Xu J, Anctil F, Boucher MA (2019) Hydrological post-processing of streamflow forecasts issued from multimodel ensemble prediction systems. J Hydrol 578:124002
Yan H, Moradkhani H (2016) Toward more robust extreme flood prediction by Bayesian hierarchical and multimodeling. Nat Hazards 81:203–225
Yan Z, Zhou Z, Liu J, Han Z, Gao G, Jiang X (2020) Ensemble projection of runoff in a large-scale basin: modeling with a global BMA approach. Water Resour Res 56(7):e2019WR026134
Yaseen ZM, Jaafar O, Deo RC, Kisi O, Adamowski J, Quilty J, El-Shafie A (2016) Stream-flow forecasting using extreme learning machines: a case study in a semi-arid region in Iraq. J Hydrol 542:603–614
Zeng X, Wu J, Wang D, Zhu X, Long Y (2016) Assessing Bayesian model averaging uncertainty of groundwater modeling based on information entropy method. J Hydrol 538:689–704
Zhang S, Chen J (2021) Uncertainty in projection of climate extremes: a comparison of CMIP5 and CMIP6. J Meteorol Res 35(4):646–662

Part V
Socio-hydrological Impacts and Policy Implications in a Changing Climate

Chapter 18
Application of AI/ML in Water Resource Management to Resolve Transboundary Water Conflict

Sayantan Sarkar and Prakash Kumar Jha

Abstract Transboundary water conflicts are multifaceted issues characterized by competition over finite resources, environmental considerations, historical complexities, and global repercussions. The imperative of effective management of shared water resources cannot be overstated. Such management not only addresses water scarcity but also promotes cooperation, sustains development, and provides conflict resolution mechanisms, contributing to regional stability, international relations, and overall well-being. AI and ML technologies offer transformative capabilities in this context. They enable predictive modeling, efficient resource allocation, data-driven decision support, and diplomacy among nations. AI and ML play a critical role in achieving sustainable and equitable water resource management by providing early warnings, optimizing resource allocation, and mitigating potential sources of conflict. Furthermore, they foster transparency, data-informed decision-making, and impartiality, fostering trust among stakeholders and enhancing collaboration. This chapter delves into the utilization of Artificial Intelligence (AI) and Machine Learning (ML) techniques in the domain of water resource management, particularly concerning transboundary conflict resolution. It explores the algorithmic and data-driven approaches that underpin the application of AI and ML, resulting in more sustainable and equitable practices for managing water resources. Through illustrative case studies, it showcases the potential of these technologies to predict water scarcity, optimize resource allocation, and foster collaboration among neighboring nations. It underscores the urgency of harnessing AI and ML in resolving the intricate challenges of transboundary water management. By leveraging these technologies, the global community can progress toward a more water-secure and peaceful world where shared water resources become instruments of cooperation and peace, rather than sources of contention and conflict.

Keywords AI/ML · Transboundary water

S. Sarkar (✉) · P. K. Jha
Blackland Research and Extension Center, Texas A&M AgriLife Research, Temple, TX 76502, USA
e-mail: sarkarsayantan@outlook.com

A. Nanda et al. (eds.), *Navigating the Nexus*, Water Science and Technology Library 102, https://doi.org/10.1007/978-3-031-76532-2_18

18.1 Introduction

Water, as a vital component of Earth's ecosystem, disregards geopolitical boundaries, traversing rivers, lakes, and aquifers seamlessly. This hydrological fluidity, both in its physical movement and broader implications, presents nations with the intricate and often contentious realm of transboundary water conflicts (Varady et al. 2023). The significance of such conflicts is paramount. They emerge when multiple nations share access to the same water resources, with implications that extend well beyond the immediate water bodies involved. Transboundary water conflicts encapsulate a complex interplay of ecological, political, economic, and humanitarian factors, rendering them multifaceted challenges of profound global significance (Guo et al. 2016).

A primary driver of transboundary water conflicts is water scarcity. As the world's population grows and demands for water escalate, the competition for limited freshwater resources intensifies (Stuckenberg and Contento 2018). When neighboring nations rely on the same water source for agriculture, industry, and domestic use, equitable allocation becomes a paramount concern. These conflicts can undermine food and water security, impacting the livelihoods of millions (Zeitoun and Mirumachi 2008). Environmental concerns are another dimension of transboundary water conflicts. Shared water bodies sustain ecosystems that span borders, harboring unique biodiversity. Pollution, habitat destruction, and unsustainable water use in one nation can adversely affect the ecosystems and fisheries downstream, sparking disputes and exacerbating ecological degradation (Baten and Titumir 2016). Often, the roots of transboundary water conflicts reach deep into the annals of history and are entwined with geopolitical complexities. Tensions related to sovereignty, national interests, and historical grievances can further complicate these disputes, creating a challenging diplomatic terrain (Ul-Durar et al. 2023).

Beyond national borders, transboundary water conflicts ripple through the international community. They can disrupt regional stability, potentially leading to broader geopolitical tensions. In some cases, water disputes even pose security risks, accentuating their global significance (Al-Muqdadi 2022). At their heart, these conflicts are profoundly humanitarian. Impeded access to clean water can endanger human lives, disrupt communities, and undermine the socio-economic development of entire regions (Dadson et al. 2017). Therefore, recognizing the gravity of transboundary water conflicts is the first step toward addressing them effectively. Cooperative management, sustainable practices, diplomacy, and the use of advanced technologies such as artificial intelligence and machine learning are increasingly critical tools in navigating these challenges (Gemmell et al. 2021). By understanding the significance of transboundary water conflicts, we can work towards a future where shared water resources become a source of cooperation and peace, rather than contention and conflict.

This book chapter will delve into the transformative role of Artificial Intelligence (AI) and Machine Learning (ML) in resolving these conflicts. Through a series of case studies, we will explore how AI and ML can predict water scarcity, optimize resource allocation, and facilitate cooperation among neighboring states. By harnessing the capabilities of these advanced technologies, we aim to present a path forward, where shared water resources become a source of cooperation and peace, rather than contention and conflict.

18.2 Transboundary Water Conflict: Understanding the Landscape

Transboundary water conflicts are multifaceted disputes that arise when two or more nations share access to the same water resources. These conflicts, characterized by their complex interplay of ecological, political, economic, and humanitarian factors, have garnered significant attention in the academic literature. The term "transboundary water conflict" has been defined with precision in scholarly works (Gleick 1998). It is succinctly defined it as disputes between two or more states or entities over the allocation, utilization, or management of shared freshwater resources that cross international boundaries. This definition encapsulates the core elements of these conflicts, emphasizing shared resources, international borders, and disputes arising from competition or differences in resource management.

Transboundary water conflicts exhibit several defining characteristics such as:

- Competition over Scarce Resources: At their core, these conflicts are driven by the scarcity of freshwater resources. As the global population continues to grow and water demands increase, the competition for limited freshwater resources intensifies. Neighboring nations relying on the same water source for agriculture, industry, and domestic use find equitable allocation a paramount concern. These conflicts can undermine food and water security, impacting the livelihoods of millions (Rai et al. 2017; Tayia 2019).
- Environmental Sustainability Concerns: Environmental aspects are an integral dimension of transboundary water conflicts. Shared water bodies sustain ecosystems that traverse borders, harboring unique biodiversity. Pollution, habitat destruction, and unsustainable water use in one nation can adversely affect ecosystems and fisheries downstream, sparking disputes and exacerbating ecological degradation (Watson 2015; Rai et al. 2017).
- Historical and Political Complexity: Often, the roots of transboundary water conflicts are deeply embedded in history and intertwined with geopolitical complexities. Tensions related to sovereignty, national interests, and historical grievances can further complicate these disputes, creating a challenging diplomatic terrain (Susskind 2018; Tayia 2019).
- Global Implications: Beyond their immediate impact, transboundary water conflicts can have far-reaching consequences on the international stage. They can disrupt regional stability and potentially lead to broader geopolitical tensions.

In some cases, water disputes even pose security risks, accentuating their global significance (de Bruin et al. 2018).
- Humanitarian Dimensions: These conflicts are profoundly humanitarian at their heart. Impeded access to clean water can endanger human lives, disrupt communities, and undermine the socio-economic development of entire regions (Dadson et al. 2017).

Transboundary water conflicts and disputes often stem from the competition for limited freshwater resources and can carry far-reaching social, economic, political, and environmental consequences. Grasping the significance of effective water resource management is imperative for upholding regional stability and fostering cooperation among nations (Guo et al. 2016). One of the primary drivers of transboundary water conflicts is water scarcity, which becomes exacerbated as the global population grows, leading to increased water demands and heightened competition for limited freshwater resources. Neighboring nations that share access to the same water source for agriculture, industry, and domestic use often find themselves in fierce competition. In the absence of effective management, ensuring equitable water allocation becomes a challenging task, setting the stage for disputes that can jeopardize food and water security (Samaranayake et al. 2016).

Effective water resource management serves as a fundamental strategy for conflict prevention. By establishing clear guidelines and mechanisms for the allocation of resources, nations can proactively address potential sources of tension. A well-managed system significantly reduces the likelihood of conflicts emerging due to water scarcity and unequal access (Katz and Nagabhatla 2023). Robust water management frameworks play a pivotal role in promoting cooperation among riparian nations. Collaborative endeavors aimed at managing shared resources lead to the development of joint infrastructure, such as dams and irrigation systems, benefitting all involved parties. Effective management also encourages dialogue and information sharing, nurturing trust and diplomatic relations among neighboring states (Wu and Whittington 2006). Furthermore, effective water management contributes to sustainable development by ensuring the long-term availability of freshwater resources. Sustainable practices, including efficient water use, pollution control, and ecosystem preservation, are essential components of effective management. These practices safeguard the environment and support the socio-economic well-being of communities relying on shared water bodies (Cosgrove and Loucks 2015). Even in the presence of conflicts, effective management can serve as a tool for resolution. Well-defined dispute resolution mechanisms, included within management frameworks, provide a structured means for addressing disagreements. Mediation, negotiation, and arbitration become more feasible when nations are already engaged in cooperative water management efforts (Tayia 2019).

Therefore, the importance of effective management extends beyond national borders. Transboundary water conflicts can disrupt regional stability, potentially leading to broader geopolitical tensions. International organizations, such as the United Nations, recognize the significance of water management in global stability and peace (United Nations 2009). In conclusion, effective water resource management is a linchpin for conflict prevention and resolution, fostering cooperation, ensuring sustainable development, and contributing to global peace.

18.3 Examples of Historical Transboundary Water Disputes

Historical transboundary water disputes serve as crucial case studies in understanding the complexities, challenges, and potential resolutions of conflicts over shared water resources. These disputes have left indelible marks on international relations, environmental sustainability, and regional stability (Table 18.1). Examining historical conflicts provides valuable insights into the dynamics of transboundary water management.

1. Nile River Basin: The Nile River, one of the world's longest rivers, has been a focal point of disputes between the 11 riparian nations that share its waters. Historically, Egypt and Sudan, located downstream, have held significant influence over the Nile's resources. This led to tensions with upstream countries, particularly Ethiopia, which sought a more equitable share of the river's waters. The construction of the Grand Ethiopian Renaissance Dam (GERD) in the Blue Nile Basin sparked a major conflict, highlighting the challenges of balancing historical rights with the growing demands of upstream nations (Paisley and Henshaw 2013).

Table 18.1 Historical transboundary water disputes among nations and their consequences

Disputes	Details of dispute	Consequences
Nile river basin	Tensions among 11 riparian nations over Nile's resources	Major conflict due to Grand Ethiopian renaissance dam
Indus river waters treaty	India-Pakistan treaty dividing Indus Basin waters, easing tensions	Cooperative resolution but ongoing regional instability
Colorado river dispute	US-Mexico dispute, overallocation, environmental harm	Environmental damage due to overuse persists
Mekong river basin	Disputes among 6 Southeast Asian nations over Chinese dam	Ecosystem concerns and challenges in transboundary management
Jordan river dispute	Israel, Jordan, Palestine vie for limited resources	Ongoing political complexities affect water management
Amu Darya and Syr Darya Rivers	Disputes among 5 Central Asian nations over water allocation, ecological damage	Environmental degradation and water allocation conflicts continue

2. Indus River Waters Treaty: In South Asia, the Indus River Waters Treaty of 1960 stands as a remarkable example of cooperative conflict resolution. The treaty was brokered between India and Pakistan, two nuclear-armed nations that have historically had tense relations. The treaty divided the waters of the Indus Basin, allocating the three eastern rivers to India and the three western rivers to Pakistan. This agreement has largely withstood the test of time, fostering cooperation between the two nations despite broader geopolitical tensions (Qureshi 2017).
3. Colorado River Dispute: In North America, the Colorado River has been a source of conflict between the United States and Mexico. The river, which crosses the international boundary, experienced overallocation and significant environmental degradation due to overuse. The 1944 Water Treaty between the two countries sought to address these issues. It established water delivery requirements and allocations while committing both nations to address salinity and environmental concerns (Rivera-Torres and Gerlak 2021).
4. Mekong River Basin: Southeast Asia's Mekong River has been a site of contention among the six riparian countries—China, Myanmar, Laos, Thailand, Cambodia, and Vietnam. China's upstream dam construction on the Mekong has raised concerns downstream regarding water flow, sediment deposition, and the health of the river's ecosystem. These disputes underscore the challenges of managing a transboundary river with multiple stakeholders and varying development priorities (Lauridsen 2004).
5. Jordan River Dispute: The Jordan River basin has long been a source of tension in the Middle East. Israel, Jordan, and Palestine have vied for control over the limited water resources in the region. The Oslo II Accord of 1995 attempted to address these issues, establishing a Joint Water Committee to manage shared water resources. However, political complexities continue to impact effective water management in the area (Mukhar 2006).
6. Amu Darya and Syr Darya Rivers: Central Asia's Amu Darya and Syr Darya rivers have been at the center of disputes among the five former Soviet republics in the region. The collapse of the Soviet Union led to conflicts over water allocation, exacerbated by outdated infrastructure and inefficient water management practices. The Aral Sea, fed by these rivers, has faced catastrophic environmental degradation due to excessive withdrawals (Dinar 2005).

In summary, transboundary water conflicts are marked by resource competition, environmental issues, historical and political complexities, global implications, and humanitarian impacts, underscoring the need for comprehensive analysis and resolution strategies. Effective management of shared water resources is crucial for preventing and mitigating these conflicts, contributing to regional stability, international relations, and national well-being. Historical disputes offer valuable insights, ranging from cooperation to enduring tensions and environmental degradation, essential for shaping future conflict prevention strategies. Ultimately, conflict resolution remains of paramount importance, and modern technologies such as AI and ML hold significant promise in this endeavor.

18.4 Role of AI and ML in Water Resource Management

Artificial Intelligence (AI) and Machine Learning (ML) have emerged as transformative technologies with immense potential to revolutionize various sectors, including water resource management (Aggarwal et al. 2022). AI refers to the development of computer systems capable of performing tasks that typically require human intelligence. These tasks include problem-solving, pattern recognition, and decision-making. In the context of water resource management, AI can be employed to develop intelligent systems that collect, analyze, and interpret vast amounts of hydrological and environmental data. AI algorithms, such as neural networks and deep learning, are particularly adept at recognizing complex patterns and trends in water-related data (Adhikari et al. 2023). ML, a subset of AI, focuses on the development of algorithms that enable computer systems to learn from data and improve their performance over time without explicit programming. ML algorithms can process large datasets to develop predictive models, identify anomalies, and optimize resource allocation. In water management, ML is invaluable for predicting water availability, assessing the impact of climate change, and optimizing water distribution systems (Sarkar et al. 2021).

The relevance of AI and ML in water resource management is evident in their ability to analyze data more efficiently and accurately than traditional methods (Sarkar and Jha 2020). They can process real-time data from various sources, such as remote sensing, weather stations, and IoT sensors, to provide insights into water availability, quality, and usage patterns. These technologies also offer the potential to predict water scarcity, optimize allocation, and facilitate cooperation among neighboring states in transboundary water management (Sarkar et al. 2022). The potential impact of AI and ML on resolving water conflicts is profound. These technologies offer several advantages:

- Predictive Modeling: AI and ML enable the development of predictive models that can forecast water availability and scarcity with remarkable accuracy. By analyzing historical data and environmental factors, these models provide early warnings, allowing stakeholders to take preemptive measures to prevent conflicts (Chakraborty et al. 2020).
- Optimized Resource Allocation: Efficient allocation of water resources is crucial in conflict prevention. ML algorithms can consider multiple variables, including supply, demand, environmental factors, and policy constraints, to optimize resource allocation. This ensures equitable distribution and reduces the potential for disputes (Van Anh 2023).
- Data-Driven Decision Support: AI/ML-based decision support systems provide policymakers with real-time insights and scenarios for informed decision-making. This data-driven approach fosters cooperation by enabling stakeholders to explore and evaluate different management options collaboratively (Feizizadeh et al. 2023).
- Environmental Monitoring: AI and ML can continuously monitor the health of ecosystems and detect deviations from ecological norms. This capability helps

prevent ecological degradation and conflicts arising from environmental issues (Mehmood et al. 2020a, b).

- Conflict Resolution Tools: AI-powered negotiation platforms can facilitate dialogue and consensus-building among stakeholders. These platforms provide a structured environment for dispute resolution, offering potential solutions and highlighting areas of compromise (Yan et al. 2019).

Comparing AI/ML with traditional approaches to water management reveals their distinct advantages.

- Efficiency: AI and ML processes data faster and more accurately than traditional manual methods, greatly reducing response times during water crises. Traditional data collection and analysis often involve extensive manual labor, which can be time-consuming and error-prone. AI and ML algorithms, on the other hand, excel in automating data processing, allowing for real-time monitoring and quicker decision-making. This efficiency is critical during water emergencies, where rapid responses can mean the difference between effective mitigation and extensive damage (Nishant et al. 2020).
- Data Volume: Traditional methods can struggle to cope with the vast volumes of data generated in modern water management. With the proliferation of sensors, satellites, and other data sources, the sheer volume of information can overwhelm conventional approaches. AI and ML thrive in this data-rich environment, efficiently handling large datasets. They can sift through massive volumes of information, extracting meaningful insights, patterns, and trends that might be missed by manual analysis. This capacity for comprehensive analysis is invaluable in understanding complex water systems and making informed decisions (Gohil et al. 2021).
- Adaptability: One of the defining strengths of AI and ML systems is their adaptability. These technologies can evolve with changing conditions and learn from new data, constantly improving their performance and accuracy over time. In contrast, traditional approaches often rely on static models and manual calculations that may become outdated as conditions change or new data becomes available. AI and ML's ability to adapt ensures that water management strategies remain effective and up-to-date, even in dynamic environments (Liang et al. 2020).
- Predictive Capabilities: AI and ML have remarkable predictive capabilities that extend far beyond the capabilities of traditional methods. These technologies can forecast water-related issues, such as droughts, years in advance. By analyzing historical data, climate patterns, and various environmental factors, AI and ML models can provide early warnings, allowing for proactive planning and resource allocation. This predictive power is invaluable for managing water resources sustainably and mitigating the impacts of water scarcity (Nova 2023; Patel et al. 2023).
- Cooperation: AI-powered decision support systems foster cooperation among stakeholders by providing impartial, data-driven recommendations. Unlike traditional decision-making processes that can be influenced by biases or subjective

opinions, AI and ML algorithms base their suggestions on empirical data and objective analysis. This impartiality builds trust among stakeholders and promotes collaborative water resource management. It facilitates negotiations by providing a common, scientifically grounded understanding of the issues at hand, thus enhancing the potential for equitable and mutually beneficial agreements (Liang et al. 2020; Mehmood et al. 2020a, b).

Therefore, the capabilities of AI and ML make them indispensable tools for sustainable and equitable water resource management. Their efficiency in data processing, adaptability to changing conditions, predictive prowess, and capacity to foster cooperation position them as transformative technologies that can address the complex challenges of water management in an increasingly interconnected and data-driven world. As the importance of effective water management grows, AI and ML offer the potential to not only prevent and resolve conflicts but also to optimize the use of our most vital resources for the benefit of all.

18.5 Algorithmic Foundations

AI and ML have introduced a range of powerful algorithms tailored for various aspects of water resource management (Alexandra et al. 2023). These algorithms form the backbone of data-driven decision-making processes (Table 18.2).

Table 18.2 AI/ML algorithmic tools in water resource management

AI/ML tool	Advantages	Usability
Neural networks	Exceptional accuracy, recognizing complex patterns, valuable for predicting river flow	Predicting river flow, analyzing historical data
Decision trees	Versatile, transparent, assesses the impact of land-use changes on water quality	Assessing the impact of land-use changes
Random forests	Improves accuracy and mitigates overfitting, applicable in groundwater modeling and climate change impact assessment	Groundwater modeling, climate change impact assessment
Support vector machines	Robust for classification and regression tasks, suitable for predicting water quality with high-dimensional data	Predicting water quality with high-dimensional data
K-means clustering	Segmentation and pattern recognition, identifying regions with similar hydrological characteristics	Identifying regions with similar hydrological characteristics

- Neural Networks: Neural networks are computational models inspired by the structure of the human brain (Han et al. 2018). These algorithms consist of interconnected nodes (neurons) organized into layers. Neural networks excel in recognizing intricate patterns and relationships within vast datasets, making them invaluable for tasks such as predicting river flow, where historical patterns are essential for accurate forecasting (Mumbi et al. 2021; Zagajewski et al. 2021).
- Decision Trees: Decision trees are versatile algorithms employed in classification and prediction tasks. They construct a tree-like structure of decision rules based on the dataset's features (Kotsiantis 2013). In water resource management, decision trees can assess the impact of land-use changes on water quality or predict flooding likelihood by considering various environmental factors. They offer transparency and interpretability, aiding decision support systems (Gunasekara et al. 2014).
- Random Forests: Random forests are an ensemble learning method that combines multiple decision trees to improve accuracy and mitigate overfitting (Nguyen et al. 2021). These algorithms are applied to tasks such as groundwater modeling and assessing climate change impacts on water resources. Random forests excel in handling complex, multidimensional data, enhancing their predictive power (Yan et al. 2019; Zagajewski et al. 2021).
- Support Vector Machines (SVM): SVM is a robust algorithm for classification and regression tasks (Joshi 2022). SVM constructs hyperplanes in multidimensional space to separate data points into distinct classes. In water management, SVM can predict water quality based on various parameters, particularly when dealing with high-dimensional data (Zagajewski et al. 2021).
- K-Means Clustering: K-means clustering is a technique for data segmentation and pattern recognition (Burkardt 2009). It groups similar data points into clusters based on their features. In water resource management, K-means clustering can identify regions with similar hydrological characteristics, facilitating targeted resource allocation and management (Gari et al. 2020).

AI/ML algorithms analyze complex data patterns by systematically selecting relevant features, preprocessing data, recognizing intricate patterns, building predictive models, evaluating model performance, continuously learning from new data, and offering interpretability. These algorithms leverage mathematical and statistical techniques to uncover meaningful insights from vast and multifaceted datasets in the field of water resource management. The processes involve:

- Feature Selection: AI/ML algorithms start by automatically identifying relevant features or variables from the dataset. These features are chosen based on their potential significance in the context of the problem being solved. For instance, in water resource management, relevant features could include precipitation levels, temperature, soil type, land use, and historical water flow data (Ahmad and Aziz 2019; Ghobadi and Kang 2023).
- Data Preprocessing: Before analysis, the data often undergoes preprocessing steps. This can include handling missing data, normalizing data scales, and encoding

categorical variables. Data preprocessing ensures that the dataset is in a suitable format for analysis (Kotsiantis et al. 2006; Ahmad and Aziz 2019).

- Pattern Recognition: AI/ML algorithms excel at recognizing complex patterns and relationships within the selected features. They employ various techniques such as statistical analysis, clustering, and dimensionality reduction to uncover these patterns (Kavhu et al. 2022). For example, they can identify that increased rainfall over several months tends to result in higher river flow during specific seasons.
- Model Building: Once patterns and relationships are identified, AI/ML algorithms build predictive models. These models capture the underlying mathematical relationships between the selected features and the target variable (the variable being predicted). Common model types include regression models, decision trees, and neural networks (Bak and Bae 2023). For instance, a machine learning model can be trained to predict river flow based on historical weather data and river flow records.
- Model Evaluation: To ensure accuracy and reliability, AI/ML models are rigorously evaluated using techniques such as cross-validation and metrics like mean squared error or accuracy (Islam et al. 2021). This evaluation helps determine how well the model performs in predicting water-related trends.
- Continuous Learning: Many ML algorithms have the capacity for continuous learning, also known as online learning or incremental learning. This means they can adapt their models as new data becomes available. For example, if a river's flow patterns change due to climate variations, the model can continuously adjust to these changes to provide up-to-date predictions (Maltoni and Lomonaco 2019).
- Ensemble Techniques: In complex scenarios, ensemble techniques such as random forests or gradient boosting can be employed. These methods combine multiple models to improve prediction accuracy by reducing bias and variance (Islam et al. 2021).
- Interpretability: Some AI/ML models offer interpretability, allowing stakeholders to understand how predictions are made. This transparency is valuable in decision support systems, where human decision-makers need to trust and understand the model's reasoning (Gilpin et al. 2018)

Therefore, it can be understood that in the realm of water resource management, algorithmic approaches have emerged as powerful tools, offering a host of advantages that are transforming the way we understand and manage our vital water resources. These advanced techniques in AI and ML provide remarkable precision, efficiency, adaptability, scalability, automation, early warning capabilities, and resource optimization (Kumar et al. 2023). Together, they have become indispensable in fostering proactive and data-driven water management, promoting the sustainable and equitable utilization of this invaluable resource. The accuracy achieved through algorithmic approaches stands out as a primary benefit. AI and ML algorithms exhibit the capacity to discern subtle patterns, uncover correlations, and reveal nonlinear relationships within vast and intricate datasets, often surpassing the capabilities of human analysis. In the context of water resource management, where even slight deviations can have far-reaching consequences, this precision is paramount for making reliable forecasts (Lowe et al. 2022).

Efficiency is another hallmark advantage of algorithmic approaches. They excel in expediting the analysis of copious data, including real-time information from an array of sources such as sensors, satellites, and monitoring stations. This rapid data processing empowers timely responses to changing conditions within water systems, ultimately enhancing the effectiveness of decision-making during critical situations like water crises or resource allocation (Mount et al. 2016). Water resource management is inherently dynamic, susceptible to influences such as changing weather patterns, climate shifts, and evolving human demands (Mishra et al. 2021). Algorithmic models showcase an inherent adaptability, continuously learning and updating their predictions as new data becomes available (Maltoni and Lomonaco 2019). This adaptability ensures that these models remain accurate in the face of evolving conditions, contributing to proactive and responsive water management. Furthermore, algorithmic approaches are highly scalable and adept at handling vast and multidimensional datasets. This scalability is pivotal in the management of water resources across extensive regions or intricate watersheds, where diverse data sources and numerous variables require consideration (Oyebode et al. 2019). AI and ML algorithms effortlessly manage the inherent complexity of such expansive systems.

Automation is a game-changer in water resource management. Once trained and deployed, algorithmic models automate the analysis of data patterns. This automation diminishes the reliance on manual data processing and analysis, freeing human resources for more strategic and decision-focused tasks (Krishnan et al. 2022). This study found that water resource managers can confidently entrust algorithms for continuous monitoring and analysis, ensuring the system remains vigilant in identifying trends and anomalies. The early warning capabilities of AI and ML algorithms cannot be overstated, as they have the capacity to recognize the early signs of critical water-related events, such as droughts or floods (Muñoz et al. 2021). By analyzing historical data and real-time information, these algorithms provide advance notice, enabling stakeholders to take preemptive measures. These early warnings are essential for minimizing damage and optimizing resource allocation during water crises. Lastly, algorithmic approaches, through optimization techniques, excel in identifying the most efficient allocation of water resources (Abdulbaki et al. 2017). By considering multiple variables, including supply, demand, environmental factors, and policy constraints, these algorithms contribute to achieving equitable distribution, thereby reducing the potential for conflicts over resource allocation.

18.6 Data-Driven Approaches in AI/ML for Water Resource Management

Data plays a fundamental role in the realm of AI and ML applications within water resource management. It serves as the lifeblood of these technologies, providing the foundational material upon which algorithms operate. Data is crucial for various aspects of training in AI and ML models. Data comprising historical and real-time observations stands as a cornerstone (Roh et al. 2019). During the training process, these models learn intricate patterns, relationships, and correlations within the data. The quality and representativeness of the training data are instrumental in shaping the model's accuracy and performance. Furthermore, the prediction and forecasting processes in water resource management heavily rely on historical and real-time data. For instance, predicting river flow necessitates a foundation of historical streamflow data (Ghimire et al. 2021). This data underpins the algorithms' capacity to make well-informed predictions about future water-related events, providing a basis for strategic decision-making. Data-driven decision support systems provide policymakers with real-time insights and scenario assessments, empowering them to make informed and strategic decisions (Charles et al. 2022). These systems depend on continuously updated data streams to offer a comprehensive view of the current state of water resources, enabling timely and responsive decision-making. Moreover, early warning systems are pivotal in water resource management for mitigating water-related crises such as droughts and floods. These systems, grounded in the analysis of historical data and real-time monitoring, rely on data to identify precursors to impending crises and provide sufficient lead time for stakeholders to take preventive measures (Merz et al. 2020).

Despite the paramount importance of data, several challenges exist within water resource management. Data collection can be a complex endeavor, particularly when data sources are dispersed across different agencies, entities, and formats. Ensuring data accessibility and reliability is pivotal to generate meaningful insights (Wilkinson et al. 2016). Data quality is a significant concern, as issues like missing values, errors, and inconsistencies within datasets can jeopardize the accuracy of AI and ML models. Quality assurance and data cleaning processes are vital to address these challenges (Chu et al. 2016). Data integration, combining data from various sources, such as weather stations, remote sensors, and satellite imagery, can present technical difficulties (Zipkin et al. 2021). This process often necessitates standardized formats and protocols to facilitate seamless data exchange. The sheer volume of data generated in modern water management can overwhelm conventional approaches. Managing large datasets poses challenges related to storage, processing, and analysis (Fan et al. 2014).

AI and ML offer several advantages in addressing the challenges associated with processing and analyzing large datasets. AI and ML algorithms efficiently process vast volumes of data, reducing the time required for analysis (Sarker 2021). This swiftness is essential for responding to rapidly changing conditions in water systems. Moreover, ML algorithms excel at recognizing complex patterns and relationships within large datasets. They can identify subtle correlations and trends that might be overlooked in manual analysis (Adi et al. 2020). AI and ML can assist in imputing missing values in datasets, thereby enhancing data quality (Shadbahr et al. 2023). These imputations are made based on patterns and relationships present in the existing data, increasing the completeness of datasets. These technologies can also fuse data from diverse sources, allowing a comprehensive view of water systems. For instance, they can integrate data from sensors, satellite imagery, and weather stations to provide a holistic understanding of a watershed. Furthermore, AI and ML applications provide real-time monitoring of water systems, continually updating models and predictions as new data becomes available (Elmahdi et al. 2006). This real-time capability is crucial for early warning systems and dynamic decision support.

Therefore, we see that, data-driven approaches are central to AI and ML applications in water resource management. While challenges exist in data collection, quality, and integration, AI and ML can assist in overcoming these challenges by efficiently processing and analyzing large datasets. These technologies leverage data to make accurate predictions, provide decision support, and enable early warning systems, ultimately contributing to sustainable and effective water resource management.

18.7 Predictive Modeling Techniques Using AI/ML—Case Study

Predicting water scarcity is a critical facet of water resource management, and AI/ML technologies have emerged as invaluable tools in this endeavor. Predictive modeling using AI/ML involves the development of algorithms that analyze historical and real-time data to forecast future scenarios. These techniques leverage various algorithms, such as neural networks and deep learning, which excel at recognizing intricate patterns and trends in water-related data (Makropoulos and Savić 2019). Through this process, AI/ML algorithms examine historical water usage patterns, environmental factors, climate data, and other relevant variables. They construct predictive models that offer insights into future water availability and potential scarcity. The capacity to understand the dynamics of water supply and demand is fundamental for effective resource allocation and conflict prevention.

Several case studies underscore the effectiveness of AI/ML in predicting water scarcity. For instance, in California, where water scarcity is a recurring concern, AI-based models have been employed to forecast drought conditions (Lynam and Piechota 2021). These models integrate vast datasets, including historical precipitation records, reservoir levels, and climate forecasts, to make accurate predictions about impending droughts. Similarly, in India, the use of AI and ML for predicting water scarcity has yielded promising results. The models incorporate data from various sources, such as satellite imagery, meteorological data, and groundwater levels, to offer early warnings about water stress in specific regions (Mishra et al.

2023). These advanced warning systems provide authorities with the necessary lead time to implement water conservation measures and reduce the potential for conflicts.

Accurate predictions of water scarcity are pivotal for conflict prevention and resolution. These predictions provide decision-makers with the insight needed to take preemptive actions, reducing the likelihood of disputes and the negative impacts of water-related conflicts. In transboundary water management, the early identification of potential scarcity scenarios can foster cooperation among neighboring states. AI/ML-based predictive models enable countries to engage in proactive negotiations and resource-sharing agreements, mitigating the risk of conflicts over shared water resources. For instance, the Mekong River Basin, shared by several Southeast Asian countries, has witnessed successful conflict prevention through the use of AI and ML-based models that forecast seasonal water levels and ensure equitable water distribution (Nguyen et al. 2015). Furthermore, the significance of accurate predictions in local water management cannot be overstated. Timely information on potential water shortages allows communities and local governments to implement conservation measures and equitable distribution strategies. This proactive approach not only prevents conflicts but also ensures the sustainable use of water resources.

Therefore, predictive modeling techniques using AI/ML are crucial for anticipating water scarcity. Case studies from California and India demonstrate the efficacy of these models in making accurate predictions. The significance of these predictions in conflict prevention is profound, as they provide the lead time needed for decision-makers to take proactive measures and ensure the sustainable and equitable management of water resources.

18.8 Optimizing Resource Allocation Using AI/ML—Case Study

Efficient allocation of water resources is a cornerstone of sustainable water management, and AI/ML technologies have proven to be instrumental in this regard. AI/ML plays a pivotal role in optimizing the allocation of water resources by enabling dynamic, data-driven decision-making processes. These technologies utilize algorithms that consider an array of variables, including water supply, demand, environmental factors, and policy constraints, to make optimal allocation decisions (Lowe et al. 2022). AI/ML models, such as reinforcement learning and genetic algorithms, can swiftly adapt to changing conditions and learn from new data, ensuring that allocation strategies remain up-to-date and responsive to the evolving needs of water systems. This adaptability is essential for addressing the dynamic nature of water management.

Several case examples demonstrate the efficiency of AI/ML-based resource allocation practices. For instance, in the Murray-Darling Basin in Australia, an AI-driven decision support system has been employed to allocate water rights among farmers.

The system continuously integrates data from weather forecasts, river flow measurements, and individual water entitlements to make real-time allocation decisions. This approach has enhanced the efficiency of water distribution and ensured that water is allocated in a manner that aligns with environmental sustainability (Yang et al. 2017). In the context of transboundary water management, the Indus Water Treaty between India and Pakistan has utilized AI and ML algorithms to optimize water allocation from the Indus River. These algorithms consider real-time flow data, historical utilization patterns, and climatic conditions to make equitable and efficient allocation decisions (Hammad et al. 2021).

Equitable resource allocation, made possible by AI/ML, is a powerful tool in conflict prevention. When water resources are distributed fairly among stakeholders, the potential for disputes is significantly reduced. AI/ML models can ensure that all parties receive their fair share of water, mitigating grievances and conflicts. In transboundary water management, equitable allocation fosters cooperation among neighboring states. AI/ML algorithms offer impartial, data-driven recommendations that can facilitate negotiations and the development of equitable resource-sharing agreements. This approach has been crucial in preventing conflicts in the Danube River Basin, shared by multiple European countries. AI/ML has enabled efficient allocation of water resources, reducing tensions and fostering collaboration (Krtolica et al. 2022). At the local level, equitable resource allocation benefits communities and regions by ensuring that water is distributed fairly among users. This not only prevents conflicts but also contributes to the overall well-being of the population. In regions with scarce water resources, such as arid areas of the United States, AI/ML-based allocation strategies have led to a reduction in water-related conflicts and improved water security (Ficklin et al. 2013; Rai et al. 2022).

Therefore, AI/ML is a powerful tool for optimizing the allocation of water resources. Case examples from Australia and the Indus Water Treaty demonstrate the effectiveness of AI/ML in resource allocation. Equitable allocation, made possible by AI/ML, plays a pivotal role in preventing conflicts by reducing grievances and fostering cooperation among stakeholders, ultimately contributing to the sustainable and peaceful management of water resources.

18.9 Facilitating Cooperation Among Nations Using AI/ML—Case Study

In the realm of transboundary water management, cooperation between neighboring states is often imperative. AI/ML technologies have emerged as pivotal tools for fostering collaboration and diplomacy. AI/ML plays a crucial role in facilitating cooperation between neighboring states by providing data-driven decision support systems. These systems offer impartial, evidence-based recommendations, reducing biases and subjectivity in decision-making processes (Hu et al. 2023). Moreover, they empower stakeholders to engage in collaborative discussions, explore different management options, and evaluate scenarios collectively. AI/ML algorithms enable the continuous monitoring of water resources across borders, providing real-time insights and scenarios for informed decision-making. By offering a common, scientifically grounded understanding of water-related issues, these technologies bridge gaps in communication and build trust among nations (Aldoseri et al. 2023).

Decision support systems powered by AI/ML serve as vital tools for transboundary water management. These systems integrate data from diverse sources, including remote sensing, weather stations, and IoT sensors, to provide stakeholders with real-time insights into water availability, quality, and usage patterns (Krishnan et al. 2022). The Indus Water Treaty between India and Pakistan provides a notable example of how AI and ML-based decision support systems have enhanced cooperation. This treaty employs AI algorithms that continuously analyze real-time flow data from the Indus River, making equitable allocation decisions (Hammad et al. 2021). These algorithms offer transparent, data-driven insights that have contributed to improved collaboration and conflict prevention.

Several case studies highlight successful cooperation between neighboring states in transboundary water management, facilitated by AI and ML technologies. In the Nile River Basin, shared by multiple African countries, an AI-driven decision support system has been instrumental in fostering collaboration. This system integrates data from various sources, such as satellite imagery and hydrological sensors, to provide real-time information on water availability and quality. It empowers nations to collectively make informed decisions about the use of shared water resources (Kansara and Lakshmi 2022). Additionally, the Danube River Basin, which crosses several European countries, serves as a compelling example of successful cooperation through technology. AI and ML algorithms have been used to optimize water allocation and monitor water quality. These technologies have contributed to a reduction in disputes and tensions among nations, fostering collaboration and sustainable water management (Krtolica et al. 2022).

Therefore, AI/ML technologies play a pivotal role in facilitating cooperation among neighboring states in transboundary water management. Decision support systems offer data-driven, transparent insights, and case studies from the Nile and Danube River Basins illustrate successful cooperation through the use of technology. These advancements contribute to peaceful and sustainable shared water resource management, ultimately benefitting all involved nations.

18.10 Case Studies: Real-World Applications

Real-world applications of AI/ML in transboundary water conflict resolution provide valuable insights into the potential and challenges of these technologies.

18.10.1 Mekong River Basin: Collaborative Management and Data-Driven Decision Support

The Mekong River Basin, shared by several Southeast Asian nations, has grappled with water resource conflicts due to increasing demands for hydropower, agriculture, and domestic water use. To address these challenges, the Mekong River Commission (MRC) implemented AI/ML-driven data analytics. The MRC integrated various data sources, including remote sensing and weather stations, to monitor river flow, water quality, and climate changes in real-time (Polpanich et al. 2022). AI/ML algorithms were deployed to predict water availability, assess potential droughts, and optimize water allocation.

Outcome: The AI/ML system provided stakeholders with a comprehensive understanding of the basin's hydrological conditions, enhancing transparency and cooperation. Predictive models enabled proactive responses to potential water shortages, reducing the risk of conflicts (Gao et al. 2022). The Mekong nations collaboratively developed adaptive strategies based on data-driven insights, contributing to more equitable resource allocation and sustainable water management.

Challenges and Lessons Learned: The major challenge was integrating data from multiple sources and ensuring data quality and consistency. The case emphasizes the importance of collaboration and data sharing among nations, demonstrating the potential for AI/ML to foster cooperation in complex, transboundary water systems (Farinosi et al. 2018).

18.10.2 Danube River Basin: Enhancing Water Quality and Monitoring

The Danube River flows through ten European countries, leading to diverse challenges in water quality monitoring and environmental protection. AI/ML technologies have been applied to address these challenges, using data from water quality sensors and satellite imagery (Mitrović et al. 2019). These technologies continuously analyze water quality, detect pollution sources, and optimize wastewater treatment processes.

Outcome: AI/ML-based water quality monitoring has significantly improved the management of the Danube River. Pollution sources have been identified and mitigated promptly, reducing ecological damage (Georgescu et al. 2023). Data-driven insights fostered cooperation among nations and contributed to the sustainable use of the river's resources.

Challenges and Lessons Learned: Challenges included the need for a robust sensor network and data sharing agreements among the countries sharing the river (Zanoni et al. 2022). The case underscores the importance of collaborative data collection, real-time monitoring, and international cooperation in addressing transboundary water quality issues.

18.10.3 Indus Water Treaty: Data-Driven Allocation and Diplomacy

The Indus Water Treaty between India and Pakistan, one of the most water-stressed regions globally, has relied on AI/ML to optimize water allocation from the Indus River. AI algorithms continuously analyze real-time flow data, historical utilization patterns, and climatic conditions to make equitable allocation decisions (Farinosi et al. 2018).

Outcome: The AI/ML system has contributed to efficient and equitable water allocation between India and Pakistan, reducing the potential for conflicts. Data-driven insights have fostered cooperation, and real-time monitoring ensures timely responses to changing conditions (Farinosi et al. 2018).

Challenges and Lessons Learned: Maintaining the accuracy and reliability of the real-time data was a significant challenge. The case emphasizes the need for trust-building measures and cooperation in transboundary water management (Tahira 2022).

These case studies underscore the potential of AI/ML in transboundary water conflict resolution. They highlight the outcomes, challenges, and lessons learned, emphasizing the importance of data integration, cooperation, and data-driven decision-making in ensuring sustainable and equitable water resource management.

18.11 The Transformative Potential of AI/ML

AI and ML technologies are poised to bring about a profound transformation in the field of water diplomacy, offering innovative solutions to address the complexities of transboundary water conflicts and contribute to global water security and peace. AI/ML technologies have the capacity to revolutionize water diplomacy by offering data-driven, impartial, and predictive solutions. These technologies can process vast amounts of hydrological and environmental data to provide real-time insights, forecast water scarcity, optimize resource allocation, and offer conflict resolution tools (Vaseashta 2022). The transformative impact lies in the ability of AI/ML to address the multifaceted challenges of transboundary water conflicts. They promote informed decision-making, enhance cooperation among nations, and offer early warning systems, thereby reducing the potential for water-related disputes.

AI and ML technologies contribute significantly to water security and peace by promoting sustainable and equitable water management. Their predictive modeling capabilities allow stakeholders to anticipate and prevent water scarcity, reducing the risk of conflicts. Efficient resource allocation ensures that water is distributed fairly, mitigating grievances and enhancing cooperation among nations (Wolf 1999). Furthermore, these technologies foster transparency and data-driven decision-making, reducing biases and subjectivity in negotiations. By providing a common, scientifically grounded understanding of water-related issues, they build trust among stakeholders, ultimately contributing to the peaceful management of shared water resources.

To fully harness the transformative potential of AI/ML, it is essential to encourage their responsible and widespread adoption. This involves investing in the development of AI/ML applications for water resource management, supporting data sharing agreements among nations, and promoting capacity building (Gunasekaran and Boopathi 2023). Additionally, it is crucial to ensure that AI/ML technologies are used ethically and responsibly. This includes addressing concerns related to data privacy, security, and equitable access to technology (McGovern et al. 2022). International cooperation and guidelines can play a pivotal role in ensuring responsible AI/ML adoption. Therefore, AI and ML technologies hold the transformative potential to revolutionize water diplomacy by providing data-driven, predictive solutions. Their impact on water security and peace is profound, as they enable informed decision-making, enhance cooperation, and prevent conflicts. Encouraging their responsible and widespread adoption is essential for realizing their full potential in addressing the complex challenges of transboundary water management.

18.12 Conclusion

In this chapter, we have delved into the critical intersection of AI and ML with water resource management, particularly in the context of transboundary water conflicts. The insights garnered from this exploration underscore several key takeaways:

First, transboundary water conflicts are complex, multifaceted challenges characterized by competition over limited resources, environmental concerns, historical and political intricacies, global implications, and profound humanitarian consequences. These defining characteristics necessitate in-depth analysis and effective resolution strategies.

Second, effective management of shared water resources is essential for preventing and mitigating transboundary water conflicts. By addressing water scarcity, promoting cooperation, supporting sustainable development, and providing mechanisms for conflict resolution, well-managed systems contribute to regional stability, international relations, and the overall welfare of nations.

Third, historical transboundary water disputes serve as valuable case studies, illustrating the multifaceted nature of conflicts over shared water resources. These examples range from cooperative agreements to enduring tensions and environmental degradation, emphasizing the importance of understanding past conflicts in developing effective strategies for the future.

The urgency of leveraging AI and ML in resolving transboundary water conflicts cannot be overstated. These technologies offer transformative capabilities, including predictive modeling, efficient resource allocation, data-driven decision support, and fostering cooperation among nations. They are indispensable tools for achieving sustainable and equitable water resource management. By providing early warnings, optimizing resource allocation, and reducing potential sources of conflict, AI/ML contributes to regional stability and international peace. Moreover, these technologies foster transparency, data-driven decision-making, and impartiality, building trust among stakeholders and enhancing collaboration. In sum, this chapter's contribution lies in highlighting the transformative potential of AI/ML in addressing the complex challenges of transboundary water management. By leveraging these technologies, we can work towards a more water-secure and peaceful world, where shared water resources become a source of cooperation and peace, rather than contention and conflict.

References

Abdulbaki D, Al-Hindi M, Yassine A, Abou Najm M (2017) An optimization model for the allocation of water resources. J Clean Prod 164:994–1006

Adhikari S, Joshi A, Chandra AK, Bharati A, Sarkar S, Dinkar V et al. (2023) SMART plant breeding from pre-genomic to post-genomic era for developing climate-resilient cereals. In: Smart plant breeding for field crops in post-genomics era. Springer, pp 41–97

Adi E, Anwar A, Baig Z, Zeadally S (2020) Machine learning and data analytics for the IoT. Neural Comput Appl 32:16205–16233

Aggarwal K, Mijwil MM, Al-Mistarehi A-H, Alomari S, Gök M, Alaabdin AMZ, Abdulrhman SH (2022) Has the future started? The current growth of artificial intelligence, machine learning, and deep learning. Iraqi J Comput Sci Math 3(1):115–123

Ahmad T, Aziz MN (2019) Data preprocessing and feature selection for machine learning intrusion detection systems. ICIC Express Lett 13(2):93–101

Aldoseri A, Al-Khalifa KN, Hamouda AM (2023) Re-thinking data strategy and integration for artificial intelligence: concepts, opportunities, and challenges. Appl Sci 13(12):7082

Alexandra C, Daniell KA, Guillaume J, Saraswat C, Feldman HR (2023) Cyber-physical systems in water management and governance. Curr Opinion Environ Sustain 62:101290

Al-Muqdadi SW (2022) The spiral of escalating water conflict: the theory of hydro-politics. Water 14(21):3466

Bak G, Bae Y (2023) Deep learning algorithm development for river flow prediction: PNP algorithm. Soft Comput 1–29

Baten MA, Titumir RAM (2016) Environmental challenges of trans-boundary water resources management: the case of Bangladesh. Sustain Water Resour Manag 2:13–27

Burkardt J (2009) K-means clustering. Virginia Tech, Advanced Research Computing, Interdisciplinary Center for Applied Mathematics

Chakraborty M, Sarkar S, Mukherjee A, Shamsudduha M, Ahmed KM, Bhattacharya A, Mitra A (2020) Modeling regional-scale groundwater arsenic hazard in the transboundary Ganges River Delta, India and Bangladesh: infusing physically-based model with machine learning. Sci Total Environ 748:141107

Charles V, Rana NP, Carter L (2022) Artificial intelligence for data-driven decision-making and governance in public affairs. Elsevier, p 101742

Chu X, Ilyas IF, Krishnan S, Wang J (2016) Data cleaning: overview and emerging challenges. In: Paper presented at the proceedings of the 2016 international conference on management of data

Cosgrove WJ, Loucks DP (2015) Water management: current and future challenges and research directions. Water Resour Res 51(6):4823–4839

Dadson S, Hall JW, Garrick D, Sadoff C, Grey D, Whittington D (2017) Water security, risk, and economic growth: insights from a dynamical systems model. Water Resour Res 53(8):6425–6438

de Bruin S, Knoop J, Visser H, Ligtvoet W (2018) Linking water security threats to conflict. Netherlands Environmental Assessment Agency, The Hague

Dinar S (2005) Treaty principles and patterns: selected international water agreements as lessons for the resolution of the Syr Darya and Amu Darya water dispute. Paper presented at the transboundary water resources: strategies for regional security and ecological stability: proceedings of the NATO advanced research workshop on transboundary water resources: strategies for regional security and ecological stability Novosibirsk, Russia, 25–27 Aug 2003

Elmahdi A, Kheireldin K, Hamdy A (2006) GIS and multi-criteria evaluation: robust tools for integrated water resources management. Water Int 31(4):440–447

Fan J, Han F, Liu H (2014) Challenges of big data analysis. Natl Sci Rev 1(2):293–314

Farinosi F, Giupponi C, Reynaud A, Ceccherini G, Carmona-Moreno C, De Roo A et al (2018) An innovative approach to the assessment of hydro-political risk: a spatially explicit, data driven indicator of hydro-political issues. Glob Environ Chang 52:286–313

Feizizadeh B, Omarzadeh D, Kazemi Garajeh M, Lakes T, Blaschke T (2023) Machine learning data-driven approaches for land use/cover mapping and trend analysis using Google Earth Engine. J Environ Plann Manage 66(3):665–697

Ficklin DL, Stewart IT, Maurer EP (2013) Climate change impacts on streamflow and subbasin-scale hydrology in the Upper Colorado River Basin. PLoS ONE 8(8):e71297

Gao J, Castelletti A, Burlado P, Wang H, Zhao J (2022) Soft-cooperation via data sharing eases transboundary conflicts in the Lancang-Mekong River Basin. J Hydrol 606:127464

Gari Y, Block P, Assefa G, Mekonnen M, Tilahun SA (2020) Quantifying the United Nations' watercourse convention indicators to inform equitable transboundary river sharing: application to the Nile River Basin. Water 12(9):2499

Gemmell A, Sterckx A, Vargas CR, Abrahams E, Flügel T, Hugman R, Mosesane B (2021) Big data analytics and transboundary water collaboration. Academic Press

Georgescu P-L, Moldovanu S, Iticescu C, Calmuc M, Calmuc V, Topa C, Moraru L (2023) Assessing and forecasting water quality in the Danube River by using neural network approaches. Sci Total Environ 879:162998

Ghimire S, Yaseen ZM, Farooque AA, Deo RC, Zhang J, Tao X (2021) Streamflow prediction using an integrated methodology based on convolutional neural network and long short-term memory networks. Sci Rep 11(1):17497

Ghobadi F, Kang D (2023) Application of machine learning in water resources management: a systematic literature review. Water 15(4):620

Gilpin LH, Bau D, Yuan BZ, Bajwa A, Specter M, Kagal L (2018) Explaining explanations: an overview of interpretability of machine learning. In: Paper presented at the 2018 IEEE 5th international conference on data science and advanced analytics (DSAA)

Gleick PH (1998) Water in crisis: paths to sustainable water use. Ecol Appl 8(3):571–579

Gohil J, Patel J, Chopra J, Chhaya K, Taravia J, Shah M (2021) Advent of big data technology in environment and water management sector. Environ Sci Pollut Res 28(45):64084–64102

Gunasekara N, Kazama S, Yamazaki D, Oki T (2014) Water conflict risk due to water resource availability and unequal distribution. Water Resour Manage 28:169–184

Gunasekaran K, Boopathi S (2023) Artificial intelligence in water treatments and water resource assessments. In: Artificial intelligence applications in water treatment and water resource management. IGI Global, pp 71–98

Guo L, Zhou H, Xia Z, Huang F (2016) Evolution, opportunity and challenges of transboundary water and energy problems in Central Asia. Springerplus 5:1–11

Hammad M, Shoaib M, Salahudin H, Baig MAI, Khan MM, Ullah MK (2021) Rainfall forecasting in upper Indus basin using various artificial intelligence techniques. Stoch Env Res Risk Assess 35:2213–2235

Han S-H, Kim KW, Kim S, Youn YC (2018) Artificial neural network: understanding the basic concepts without mathematics. Dement Neurocognitive Disord 17(3):83–89

Hu K, Lou B, Baloch B (2023) Reducing human bias in decision making and improving customer satisfaction through artificial intelligence: empirical evidence from a knowledge service company. Available at SSRN

Islam ARMT, Talukdar S, Mahato S, Kundu S, Eibek KU, Pham QB et al (2021) Flood susceptibility modelling using advanced ensemble machine learning models. Geosci Front 12(3):101075
Joshi AV (2022) Support vector machines. In: Machine learning and artificial intelligence. Springer, pp 89–99
Kansara P, Lakshmi V (2022) Water levels in the major reservoirs of the Nile River Basin—a comparison of Sentinel with satellite altimetry data. Remote Sens 14(18):4667
Katz D, Nagabhatla N (2023) Damming disputes: conflict prevention in water management, vol 15. MDPI, p 2476
Kavhu B, Mashimbye ZE, Luvuno L (2022) Characterising social-ecological drivers of landuse/cover change in a complex transboundary basin using singular or ensemble machine learning. Remote Sens Appl Soc Environ 27:100773
Kotsiantis SB (2013) Decision trees: a recent overview. Artif Intell Rev 39:261–283
Kotsiantis SB, Kanellopoulos D, Pintelas PE (2006) Data preprocessing for supervised leaning. Int J Comput Sci 1(2):111–117
Krishnan SR, Nallakaruppan M, Chengoden R, Koppu S, Iyapparaja M, Sadhasivam J, Sethuraman S (2022) Smart water resource management using artificial intelligence—a review. Sustainability 14(20):13384
Krtolica I, Savić D, Bajić B, Radulović S (2022) Machine learning for water quality assessment based on macrophyte presence. Sustainability 15(1):522
Kumar V, Kedam N, Sharma KV, Mehta DJ, Caloiero T (2023) Advanced machine learning techniques to improve hydrological prediction: a comparative analysis of streamflow prediction models. Water 15(14):2572
Lauridsen PE (2004) Transboundary water management in the Mekong: river of controversy or river of promise. From water "wars" to water "riots", pp 47–79
Liang Z, Liu X, Xiong J, Xiao J (2020) Water allocation and integrative management of precision irrigation: a systematic review. Water 12(11):3135
Lowe M, Qin R, Mao X (2022) A review on machine learning, artificial intelligence, and smart technology in water treatment and monitoring. Water 14(9):1384
Lynam L, Piechota T (2021) California drought outlooks based on climate change models' effects on water availability. Water 13(22):3211
Makropoulos C, Savić D (2019) Urban hydroinformatics: past, present and future. Water 11(10):1959
Maltoni D, Lomonaco V (2019) Continuous learning in single-incremental-task scenarios. Neural Netw 116:56–73
McGovern A, Ebert-Uphoff I, Gagne DJ, Bostrom A (2022) Why we need to focus on developing ethical, responsible, and trustworthy artificial intelligence approaches for environmental science. Environ Data Sci 1:e6
Mehmood H, Liao D, Mahadeo K (2020) A review of artificial intelligence applications to achieve water-related sustainable development goals. In: Paper presented at the 2020 IEEE/ITU international conference on artificial intelligence for good (AI4G)
Mehmood H, Mukkavilli SK, Weber I, Koshio A, Meechaiya C, Piman T et al (2020) Strategic foresight to applications of artificial intelligence to achieve water-related sustainable development goals. In: United Nations University Institute for water, environment and health, Hamilton, Canada. UNU-INWEH Report Series
Merz B, Kuhlicke C, Kunz M, Pittore M, Babeyko A, Bresch DN et al (2020) Impact forecasting to support emergency management of natural hazards. Rev Geophys 58(4):e2020RG000704
Mishra BK, Kumar P, Saraswat C, Chakraborty S, Gautam A (2021) Water security in a changing environment: concept, challenges and solutions. Water 13(4):490
Mishra B, Chauhan K, Prasad R (2023) Machine learning opportunities in water scarcity problem
Mitrović T, Antanasijević D, Lazović S, Perić-Grujić A, Ristić M (2019) Virtual water quality monitoring at inactive monitoring sites using Monte Carlo optimized artificial neural networks: a case study of Danube River (Serbia). Sci Total Environ 654:1000–1009

Mount NJ, Maier HR, Toth E, Elshorbagy A, Solomatine D, Chang F-J, Abrahart R (2016) Data-driven modelling approaches for socio-hydrology: opportunities and challenges within the Panta Rhei Science Plan. Hydrol Sci J 61(7):1192–1208
Mukhar RM (2006) The Jordan river basin and the mountain aquifer: the transboundary freshwater disputes between Israel, Jordan, Syria, Lebanon and the Palestinians. Ann Surv Int'l Comp l 12:59
Mumbi AW, Li F, Bavumiragira JP, Fangninou FF (2021) Forecasting water consumption on transboundary water resources for water resource management using the feed-forward neural network: a case study of the Nile River in Egypt and Kenya. Mar Freshw Res 73(3):292–306
Muñoz P, Orellana-Alvear J, Bendix J, Feyen J, Célleri R (2021) Flood early warning systems using machine learning techniques: the case of the Tomebamba catchment at the Southern Andes of Ecuador. Hydrology 8(4):183
Nguyen T-T, Huu QN, Li MJ (2015) Forecasting time series water levels on Mekong river using machine learning models. In: Paper presented at the 2015 seventh international conference on knowledge and systems engineering (KSE)
Nguyen J-M, Jézéquel P, Gillois P, Silva L, Ben Azzouz F, Lambert-Lacroix S et al (2021) Random forest of perfect trees: concept, performance, applications and perspectives. Bioinformatics 37(15):2165–2174
Nishant R, Kennedy M, Corbett J (2020) Artificial intelligence for sustainability: challenges, opportunities, and a research agenda. Int J Inf Manage 53:102104
Nova K (2023) AI-enabled water management systems: an analysis of system components and interdependencies for water conservation. Eigenpub Rev Sci Technol 7(1):105–124
Oyebode O, Babatunde DE, Monyei CG, Babatunde OM (2019) Water demand modelling using evolutionary computation techniques: integrating water equity and justice for realization of the sustainable development goals. Heliyon 5(11)
Paisley RK, Henshaw TW (2013) Transboundary governance of the Nile River Basin: past, present and future. Environ Develop 7:59–71
Patel A, Kethavath A, Kushwaha NL, Naorem A, Jagadale M, Sheetal KR, Renjith PS (2023) Review of artificial intelligence and internet of things technologies in land and water management research during 1991–2021: a bibliometric analysis. Eng Appl Artif Intell 123:106335
Polpanich O-U, Bhatpuria D, Santos Santos TF, Krittasudthacheewa C (2022) Leveraging multi-source data and digital technology to support the monitoring of localized water changes in the Mekong Region. Sustainability 14(3):1739
Programme WWA, UN-Water (2009) Water in a changing world
Qureshi WA (2017) The Indus Basin: water cooperation, international law and the Indus Waters treaty. Mich St Int'l l Rev 26:43
Rai A, Sarkar S, Jha PK (2022) Deficit irrigation: an optimization strategy for a sustainable agriculture. In: Soil-water, agriculture, and climate change: exploring linkages. Springer, pp 163–181
Rai SP, Wolf AT, Sharma N, Tiwari H (2017) Hydropolitics in transboundary water conflict and cooperation. River system analysis and management, pp 353–368
Rivera-Torres M, Gerlak AK (2021) Evolving together: transboundary water governance in the Colorado River Basin. Int Environ Agreements Politics Law Econ 21(4):553–574
Roh Y, Heo G, Whang SE (2019) A survey on data collection for machine learning: a big data-ai integration perspective. IEEE Trans Knowl Data Eng 33(4):1328–1347
Samaranayake N, Limaye S, Wuthnow J (2016) Water resource competition in the Brahmaputra river basin. CNA analysis and solutions: Arlington, VI, USA
Sarker IH (2021) Machine learning: algorithms, real-world applications and research directions. SN Comput Sci 2(3):160
Sarkar S, Jha PK (2020) Is precision agriculture worth it? Yes, may be. J Biotechnol Crop Sci 9(14):4–9
Sarkar S, Ramsey AF, Cazenave A-B, Balota M (2021) Peanut leaf wilting estimation from RGB color indices and logistic models. Front Plant Sci 12:658621

Sarkar S, Rai A, Jha PK (2022) Remote sensing and high-throughput techniques to phenotype crops for drought tolerance. In: Soil-water, agriculture, and climate change: exploring linkages. Springer, pp 107–129

Shadbahr T, Roberts M, Stanczuk J, Gilbey J, Teare P, Dittmer S et al (2023) The impact of imputation quality on machine learning classifiers for datasets with missing values. Commun Med 3(1):139

Stuckenberg D, Contento A (2018) Water scarcity: the most understated global security risk. Harvard Nat Secur J

Susskind L (2018) Complexity of transboundary water conflicts: enabling conditions for negotiating contingent resolutions. Anthem Press

Tahira S (2022) Scale issues in design and implementation of the water apportionment Accord in the Indus Basin. Int J Water Manag Diplomacy 1(6):57–77

Tayia A (2019) Transboundary water conflict resolution mechanisms: substitutes or complements. Water 11(7):1337

Ul-Durar S, Shah M, De Sisto M, Arshed N (2023) Metabolic rift theory and the complexities of water conflict between India and Pakistan: a pathway to effective environmental management. J Environ Manage 347:119164

Van Anh T (2023) Artificial intelligence technique in hydrological forecasts supporting for water resources management of a large river basin in Vietnam. Open J Mod Hydrol 13(4):246–258

Varady RG, Albrecht TR, Modak S, Wilder MO, Gerlak AK (2023) Transboundary water governance scholarship: a critical review. Environments 10(2):27

Vaseashta A (2022) Future of water: challenges and potential solution pathways using a Nexus of exponential technologies and transdisciplinarity. In: Handbook of research on water sciences and society. IGI Global, pp 37–63

Watson JE (2015) Beyond cooperation: environmental justice in transboundary water management

Wilkinson MD, Dumontier M, Aalbersberg IJ, Appleton G, Axton M, Baak A et al (2016) The FAIR guiding principles for scientific data management and stewardship. Sci Data 3(1):1–9

Wolf AT (1999) Criteria for equitable allocations: the heart of international water conflict. Paper presented at the natural resources forum

Wu X, Whittington D (2006) Incentive compatibility and conflict resolution in international river basins: a case study of the Nile Basin. Water Resour Res 42(2)

Yan J, Jia S, Lv A, Zhu W (2019) Water resources assessment of China's transboundary river basins using a machine learning approach. Water Resour Res 55(1):632–655

Yang A, Dutta D, Vaze J, Kim S, Podger G (2017) An integrated modelling framework for building a daily river system model for the Murray-Darling Basin, Australia. Int J River Basin Manag 15(3):373–384

Zagajewski B, Kluczek M, Raczko E, Njegovec A, Dabija A, Kycko M (2021) Comparison of random forest, support vector machines, and neural networks for post-disaster forest species mapping of the krkonoše/karkonosze transboundary biosphere reserve. Remote Sens 13(13):2581

Zanoni MG, Majone B, Bellin A (2022) A catchment-scale model of river water quality by machine learning. Sci Total Environ 838:156377

Zeitoun M, Mirumachi N (2008) Transboundary water interaction I: reconsidering conflict and cooperation. Int Environ Agreements Politics Law Econ 8:297–316

Zipkin EF, Zylstra ER, Wright AD, Saunders SP, Finley AO, Dietze MC et al (2021) Addressing data integration challenges to link ecological processes across scales. Front Ecol Environ 19(1):30–38

Chapter 19
Suitability and its Physicochemical Characterization for Deciphering Surface Water Quality Using Entropy (E) and Fuzzy (F)-AHP Optimization Model in Mahanadi River Basin (MRB), Odisha (India)

Abhijeet Das

Abstract Water is one of the most essential necessities and is employed for a variety of purposes. Surface water needs have increased as a result of dramatic climate change, ongoing population growth, and changing human lifestyles. In addition, the pollution of surface water has become a global environmental issue. Monitoring of surface water is essential to know the current status of water quality and maintain it at certain desirable level. Therefore, this study has highlighted an evaluation of surface water quality (WQ) for drinking purposes in the Mahanadi River Basin, Odisha, by implementing the integrated use of entropy-based WQI (E-WQI), with reliability based MCDMs (Multiple-criteria decision making) such as Fuzzy (F)-Analytic Hierarchy Process (AHP) have been employed for the reliability and practicability for river water quality monitoring and assessment. To perform this, water samples were collected from 16 different locations across the stretch and 20 parameters were analysed for a time frame of 2020–2023 respectively. From the physicochemical results, it is observed that the study area predominantly comprises of slightly alkaline in nature. Parameter TKN and coliform exhibits higher values and crosses the acceptable limits. As per entropy framework, E-WQI revealed that 12.5% (sites X-8 and 16) and 6.25% (site X-9) of samples belong to poor or very poor water quality. It indicated that, due to the influence of rock-water interactions, evaporation, and reverse ion exchange, the chemical composition of surface water varied. Even though it is a crucial component in determining how underutilized water stations are rated, WQI involves contradictory issues. Based on that, MCDM models, such as F-AHP were implemented to resolve conflicts, pertaining to the WQI index. The degree of fuzzy weights were in the order of TC > TKN > EC > TDS > Cl^-.The value at all sampling sites varied between 53.7 and 296.65. Out of 16 water samples, 50% represents

A. Das (✉)
Research Scholar, Department of Civil Engineering, C.V. Raman Global University (CGU), Bhubaneswar, Odisha, India
e-mail: das.abhijeetlaltu1999@gmail.com; abhijeetlaltu1994@gmail.com

A. Nanda et al. (eds.), *Navigating the Nexus*, Water Science and Technology Library 102, https://doi.org/10.1007/978-3-031-76532-2_19

good water, 25% indicate medium water, 25% indicate poor/very poor and finally, 6.25% indicate water unsuitable for domestic purposes. Following this, the proposed approach depicted X-(9) was the most polluted in comparison with other locations, followed by X-(2), (8) and (16) respectively. This was evident from the highest E-WQI values at this location. It also accompanied with high values of TC, TKN, EC, SAR, TDS, TH, Cl^-, SO_4^{2-} and Fe^{2+}, which were likewise the highest of all the areas and greater than their desired concentration. The major sources of contamination included discharge from households, uncovered septic tanks, leachate from waste dump sites, and excess utilization of fertilizers in the agricultural sector. The results of the two models mentioned above allow for the identification of the anthropogenic pressure that corresponds to the anthropogenic impact on WQ, the prioritization and monitoring of restoration measures, and the optimization of monitoring programs that are already in place to account for significant anthropogenic pressures. The study is vital for decision-makers as it promotes sustainable surface water use and ensures water security in the studied area.

Keywords Mahanadi river · Entropy · MCDM · Fuzzy · TKN · Anthropogenic · Optimization

19.1 Introduction

In India, surface water is a naturally occurring, dynamic resource that is essential for home and drinking needs. As a result, it serves as a vital resource for all living things to survive. Surface water is widely distributed, which makes it a potentially important component of social and economic development as well as human adaptation to climatic changes (Das 2022b). Unfortunately, the security of the supply of drinkable water has been jeopardized by overexploitation and surface water pollution (Savari et al. 2022). Over the past few centuries, urbanization and industrialization have also had an impact on human life; nevertheless, over this period, these processes have also greatly harmed the environment, particularly the water system (Abdo et al. 2022; Anusha et al. 2022a). As a result, there is a growing global need for water, and efficiently managing available water supplies has become increasingly difficult, especially in areas with erratic and unpredictable precipitation patterns (Devkota et al. 2022). In the latter scenario, the public is very concerned about the contamination of water by both organic and inorganic substances. However, as a result of urbanization and population growth, there has been a global shortage of freshwater resources in recent years, exposing water supplies to declining quality and quantity (da Silva Monteiro et al. 2022). These days, many regions of the world overuse surface water, a naturally occurring critical resource, to meet the rising demand for drinking, agriculture, urbanization, and industry (Daba and Qureshi 2021). Additionally, managers and planners of metropolitan areas are becoming quite concerned about this surface water. The precipitation, weather, basin physiography, and anthropogenic factors namely, urbanization and industrialization, are all affect the surface water quality in a given area (Zewdie and Yeshanew, 2023). In tropical and sub-tropical regions, the need for water is a major problem because of poor irrigation techniques, overuse, urbanization, and climate change (Basharat et al. 2022). Because of this, it seriously endangers public health, and wastewater from homes, businesses, and farms is the root cause of it (Hassani et al. 2021). Yet, the number of contaminants in river water is largely determined by natural processes such as precipitation, surface runoff, groundwater flow, and water interception and abstraction. These processes also have a substantial impact on river discharge. Thus, evaluating surface water potential zones helps communities and policy makers pinpoint areas with high potential for sustained extraction and utilization and regions where surface water resources are limited or in danger of depletion.

On the other side, anthropogenic factors include domestic and commercial wastewater. In addition, home waste water, municipal solid waste, and industrial wastewater all contribute to water pollution. This makes it easier for the water to become familiar with the microbiological contamination caused by both natural and man-made processes, such as animal dung leaching, septic tank leaks, wastewater used for irrigation, raw sewage discharge, and stormwater runoff (Mukhopadhyay et al. 2020; Anteneh et al. 2022; Asmare et al. 2023)). Thus, agricultural practices, particularly fertilization, wastewater, landfill leaks, and atmospheric nitrogen solubility, are the primary causes of contamination. These elements came together to produce an environmental and public health issue. Indeed, there is global worry over the overabundance of synthetic fertilizers, aquaculture wastes, sewage disposal, and animal wastes (Tadesse et al. 2022). In addition, inadequate wastewater treatment creates tainted drinking water and puts the general public at danger. Therefore, the key strategic policy for the majority of countries in the globe should be to assess possibly accessible surface water resources by conducting efficient technologies in order to meet the demands of rapidly rising urbanization and population (Das 2022c). Surface water flexibility and occurrence are determined by a number of elements, including soil properties, drainage patterns, porosity, weathering, fractures, and climate. On the other hand, this facilitates the identification of the pollutant's source, an essential step in reducing residential water pollution (Das 2023b).

Recently, water management techniques have been put into place, and they mostly depend on surface water and steady runoff (Zhang et al. 2021). Therefore, continuous and frequent monitoring programs are necessary to have accurate information about the intrinsic features of water quality and to comprehend the geographical and temporal variations in the hydro-chemical and biological properties of water (Masoud et al. 2022). Hence, the water quality index (WQI) is regarded as a mathematical instrument that greatly reduces those data sets and yields a single classification value that characterizes the level of pollution or the water quality condition of water bodies (Kaur et al. 2020; Roy et al. 2022b). This encourages the use of well accepted, easy-to-use methods by managers to analyse seasonal and spatial fluctuations as well as the distribution of pollution sources in order to evaluate water quality (Callow et al. 2020). As a result, it serves as a substitute method, utilizing this analysis to complete an evaluation of the environmental state and water quality of the research region. Additionally, in order to assess surface water quality and create improvement plans, WQIs are crucial for comprehending the factors influencing the spatiotemporal variability in water quality. Additionally, it has been used to evaluate the quality of surface water in previous decades, and current studies have shown that these methods may offer source estimation for any contaminant in surface water as well as other ecological topics (Feng et al. 2020). Afterwards, while conducting the procedure of the reservoir's water quality assessment, a thorough value was required. It also describes a method for classifying water quality according to a grade system and a standard approach to determining the overall impact of all water quality parameters (Tarkegn and Jury, 2020). By combining various physical, chemical, and microbiological parameters to produce a numerical value that permits specific pollution levels and to present the current state of the water in rivers or other bodies of water, these WQ indices serve as a useful tool for the quick interpretation of water quality (Das 2023d, e). But calculating WQI is a time-consuming procedure that takes into account a lot of national and international requirements when it comes to water consumption guidelines (Mohamed et al. 2022). This complicates the process even if the calculations are straightforward. Additionally, even a slight alteration in the dataset's weighting will have an impact on how the water quality is interpreted overall, therefore it's critical to develop a new method that makes it simple to choose a set of appropriate weights for the variables that affect water quality (Xia et al. 2022).

Therefore, Entropy (E)-based weights have become a valuable method that leverage information entropy to give water quality criteria weights (Fentahun et al. 2023). In conclusion, this model assesses the potential sources of pollution that have been extensively employed in allocating air and riverine pollution sources. It also proposes the viability and dependability of using entropy theory to the allocation of water pollution sources (Li et al. 2021). The entropy approach was used in this experiment to calculate the parameters since it is a subjective method that assigns weight to chemical parameters based on their value to overall quality and human health concerns (Agarwal et al. 2023). It is difficult to grade water quality indicators since it is dependent on the concentration of the indicator and the rate of relative toxicity (Khan et al. 2022). Although difficult, the method yields a useful result in the form of a comprehensive index value or number that may be used to relate the effects of the major indicators included in the constituent parameter matrix (Xiao et al. 2022). In the end, the weights are combined using the quality rating scale, produces a numerical score that is determined over time and is called as E-WQI. Even though this index shows the quality of the water, the statistics do not help the general people understand the situation of the water quality. It is essential to use the right technologies in order to monitor, access, and preserve potential surface water zones. The most creative and effective technique for determining viable surface water prospects is to use geospatial methods (da Silva Monteiro et al. 2022). Consequently, the data set's integration with a geographic information system (GIS) can aid in the development of strategies for locating and investigating the origins of groundwater contamination as well as suggesting corrective actions (Ifediegwu 2021). As a result, these methodologies have the potential to be applied at both local and regional scales. They do this by employing satellite and ground data analysis, as well as integrated ground and remote sensing data inside a GIS platform, to yield more accurate and comprehensive insights into surface water distribution and prospects (Sarkar and Majumder 2020). Likewise, this method creates different thematic spatial maps to investigate variations in surface water quality employing Inverse Distance Weighted (IDW) interpolation techniques. Within any geographic context, using tools to interpolate spatial parameters like the IDW, including field inventory has made scientific representation more flexible. Thus, when collecting, organizing, storing, and retrieving a range of spatial and non-spatial data, as well as when performing spatial analysis and combining these data to create insightful modelling and outputs, it functions as a useful tool (Jin et al. 2021). So, the IDW is sometimes referred to as one of the spatial interpolation methods that accounts for the creation of the area's thematic layer by estimating values at other places using sample sites with known values (Panneerselvam et al. 2023). It is exhibited that this approach predicts the unknown points and produces a continuous surface of the point data by influencing the predicted value of a point more from neighbouring points than from farther away (Gao et al. 2022). Further, at the later sage, it uses a linearly weighted combination of a series of sample points to calculate the cell values. Ultimately, the study area's index was derived by interpolating the WQI measured at the surface water collection sites (Abdulkerim et al. 2022). However, to remove disparities in error in the entropy method, in previous decades, numerous writers have utilized the Multi-Criteria Decision-Making (MCDM) analysis to a variety of goals in the management of water resources, evaluation of the water's quality in different areas, to tackle issues connected with the environment, sustainability and energy, technology and information management, as well as safety and risk management (Ranganathan et al. 2022). Therefore, when figuring out the specific relationships between the quality metrics and the measurement sites, the assessment of water quality, is needed to be modified with MCDM models (Rashidi et al. 2022). Furthermore, fuzzy sets have been paired with the AHP in a lot of documented research, since the subjective judgments made by the decision-making tools throughout the comparison process, might be imprecise for a variety of reasons. Thus, fuzzy weight is, thus, a scientific approach that also considers the uncertainty of water quality measurements (Choi et al. 2023). In certain cases, Fuzzy logic, a division of artificial intelligence (AI), emphasizes approximation thinking over exact correctness. It offers a mathematical framework for dealing with ambiguity and uncertainty, accepting that things might be partially true or false and enabling the mathematical description of degrees of truth (Hagos et al. 2021). Meanwhile, using Fuzzy (F)-Analytic Hierarchy Process (AHP) method, are generally regarded as appropriate and suitable for demarcating surface water quality zones. Its ability to swiftly and efficiently provide less repetitious, comprehensive, viable spatiotemporal and spectral data over a large region makes it a valuable tool for supplying crucial information about the various variables affecting the occurrence of surface water (Ahmadi et al. 2020). This procedure makes it possible to divide an issue into a hierarchy and guarantees that the evaluation process takes into account both quantifiable and qualitative elements of this innovative approach (Gupta 2022). In recent times, several studies as per Vellaikannu et al. (2021) and Anteneh et al. (2022), have indicated that this method can supply the essential framework for openness and audibility, improving structure and precision of decision-making in the management of water resources (Das, 2023g, h, 2024b).

Meanwhile, it enables one to choose the weightings for specific metrics that are most appropriate and representative of their impact on overall water quality in the context of WQI parameter weightings. It integrates concepts from fuzzy set theory and hierarchical structure analysis with systematic weighing and alternative strategy selection in order to calculate (Das, 2023f). When weight values are represented by fuzzy numbers, the Fuzzy-AHP technique produces results that are realistic and accurate (Singh et al. 2021). Recent case studies have demonstrated encouraging results by combining remote sensing and geographic technologies to map surface water using the fuzzy model. The fuzzy model has demonstrated remarkable predictive accuracy when utilized in conjunction with remote sensing and geospatial approaches. It is also efficient and effective. Therefore, the young researchers have used it because interval judgments are more accurate than fixed value judgments (Laks et al. 2023). Additionally, it lessens ambiguity in the relative weight assigned (Shokri-Kuehni et al. 2020a). The fuzzy of MCDM reduces ambiguities brought about by breaking down the intricate characteristics of surface water controlling factors. The decision-making elements were hierarchically structured, and each element's normalized weight was established to produce a comprehensive and certain image of the work's goals. This further lowers assumptions during the decision-making process and aids in verifying the validity of the decision maker's assessments (Mebrahtu et al. 2021). Finally, the models namely entropy and fuzzy theory shows as an independent, numerical approach of classifying groundwater that allows for the grouping of samples and the establishment of correlations between chemical characteristics and groundwater samples (Naves et al. 2021). Hence, the suggested work investigates intricate processes to create an index and the primary components observable at a larger scale, which impacts surface water quality.

As Mahanadi River, Odisha serves as a source of water for industry, agriculture, aquaculture, energy production, and the habitat of aquatic life; however, the increased urbanization of the area, intense farming practices, and the discharge of domestic and industrial effluent have all had an impact on the water quality. In other words, the use of clean water and proper sanitation was the first line of defence during the recent COVID-19 pandemic, and surface water, as a significant supply of freshwater, was essential to humanity's fight for survival (Das 2022a, 2023a). In brief, prior investigations in the gathering area allocated the sources of pollution within a certain set of pollutants or evaluated the state of a set of pollutants. Nevertheless, little research has been done on the sources of pollution in relation to the combination of heavy metals, salinity, and nutrients. With recent advancements in data science, Remote Sensing and GIS, researchers increasingly use GIS-based MCDM techniques to map potential zones. As a result, it produces more reliable and precise results by reducing overfitting and underfitting, two main concerns with decision-making, and effectively handling high-dimensional data. Thus, it is gaining popularity.

To ensure surface water resources are sustainable, using a combined approach to identify the issue is crucial when choosing a combined surface water-management strategy. In light of this, the study compares various water quality indexing techniques and assesses the annual and regional variations in physicochemical characteristics that impact the water's suitability for consumption. For this purpose, the current work is carried out by analysing datasets that contain 20 water quality parameters collected throughout a three-year monitor period (2020–2023) from 16 sites and then evaluation is carried out by implementing an entropy-WQI, and GIS technologies. The study is to close the knowledge gap about this region's surface water potential, which has not yet been fully explored. Thus, this research can accomplish (1) the examination of the spatiotemporal trends and evaluation of the basin's water quality and (2) the contributions from the identified sources of contamination were allocated to each of the water quality indicators using F-AHP, thus seems to be a successful technique that can reduce model uncertainty and improve the weighting procedure's correctness. In a nutshell this study is a trailblazer in the use of integrated methodologies to evaluate the potable water quality in the specified river segment. Additionally, the causes of contamination are located, leading to the proposal of water treatment methods that protect public health. This approaches offers useful insights for surface water professionals, decision-makers, and government agencies aiming to plan water resources sustainably.

19.2 Description of the Study Area

The study area comprised the Mahanadi River Basin, which is vulnerable and one of India's most densely populated riverbank areas (Das 2023d). Of course, the area contributes significantly to the state and national economies and boasts some of the most productive croplands (Singh and Patra 2024) and was located covering from 83°15'E to 87°00'E and 20°15'N to 22°30'N (Fig. 19.1). The catchment area is roughly 141,600 Km^2 and it expels about 65,580 km^2 area of Odisha city. Also, this area of the river further, extends throughout the major areas, comprising Orissa about 65,600 Km^2 and 74,970 Km^2 belongs to Chhattisgarh, whereas smaller parts of about 650 Km^2 belongs to Jharkhand, Maharashtra as 250 Km^2, and finally, Madhya Pradesh comes out under an area of 130 Km^2. It is regarded as one of the tainted rivers since it immediately takes both household and industrial garbage through multiple drains as it flows into Odisha from the southeast region (Pattanayak et al. 2024). The primary reasons for building this basin in the late 1930s were water supply and flood control. After that, the Hirakud Reservoir serves as a component of Odisha's surface water system for the Mahanadi basin and is utilized for agriculture, hydropower production, and flood control. Moreover, the river's installed reservoir capacity is approximated as 347.5 MW, with a 235,477 ha, that complying under culturable command area. Additionally, it offers refuge to the majority of the Chhattisgarh and Odisha state districts; with a limited of it, supplies to the Jharkhand states comprising of around 0.45% and the rest 0.07% is transferred to Maharashtra state. Thus, plays a significant part in maintaining the national economy of Odisha and the people's access to clean water (Bassi 2023). The basin's annual mean temperature fluctuate in a temperature, recording as 24°–27 °C. The precipitation estimated as 1200–1400 mm, that fell in annually in 2022–2023, as stated by Central Water Commission (CWC). In a brief, the average annual rainfall is in a range of 130 cm, while about 90% of it falling between June and September during the SW monsoon, with occasional cyclonic storms bringing heavy rains (Das 2023c). The climate is categorized as tropical, with a warm temperate temperature in the north and a humid subtropical climate in the south. Likewise, the highest temperature in April and May, the warmest months, when it can get as high as, recorded to be 45 °C (Rajput et al. 2022). When the daily average air temperature falls below 0 °C in December, January, and February, the freshwater in the area may freeze. Research indicates that during the summer monsoon season, around 70% of the yearly rainfall takes place. In the same way, winters are chilly in January, thus, regarded as the coldest month, with frequent lows below freezing. In general, it supplies water to about 38,606,665 people who depend on it for their survival and reasonable needs. The primary crops that are grown in this river are oilseeds, sugarcane, and rice. The interchange of these crops through the nearby, well-developed towns and cities in the state of Cuttack is facilitated by a respectable transportation network. The local farmers who cultivate crops for the sale in and around cities or towns as their primary source of income benefit from this exchange (Samal et al. 2023). Furthermore, the reservoir is mostly used for agriculture, with a combination of commercial and subsistence farming systems, rice production being the main crop. The Arc Map 10.2 program received the image of the land use data for 2022. It demonstrates that there are enormous areas of farmland that holds around 53.43%, villages as 12.16%, and finally, 14.91% teamed up with woodland areas. In other words, industrial areas covers about 0.71% and towns were recorded up to 0.70%, that were scattered in the west. Because of this, the two primary land uses in the basin are agriculture and forestry, which are backed by extensive irrigation infrastructure created by major and medium-sized projects. Farmlands, cities, villages, and industrial areas encircle the reservoir, which causes significant pollution and a large-scale production of unfavourable water bodies dotted with fungus and waterweeds in their eutrophic states (Pati et al. 2023). Thus, identifying the component and source of water pollution is crucial for controlling water contamination, particularly when it comes to nitrate and other nutrient pollution (Mohan and Krishnakumar, 2022).

19.3 Data Sources and Analyses

A first survey has been completed in the Mahanadi River Basin (MRB), and depending on the hydrologic characteristics of rivers and earlier findings, 16 sampling locations in the river basin were chosen, and they were subsequently observed every month for 3 years (2020–2023). The relevant data were obtained from the State Pollution Control Board, Odisha. Out of 45 parameters, 20 were selected based on their sampling continuity at all the selected monitoring sites, to be used in the present analysis (Das 2023d). In order to create a network of sampling stations with enough historical data to build a reliable statistical database of the parameters

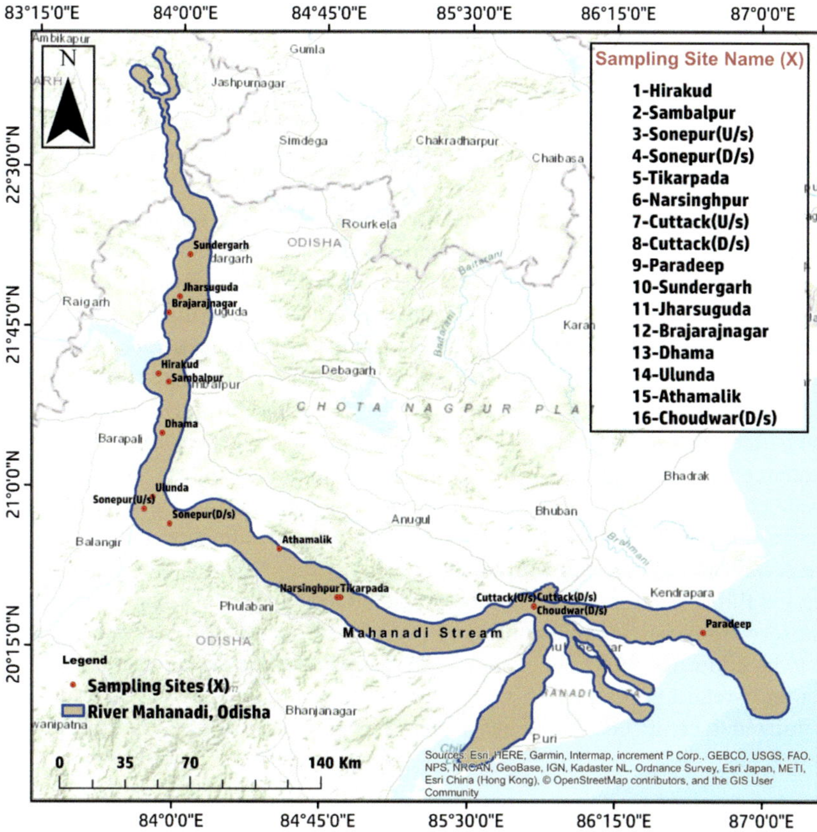

Fig. 19.1 Layout base map of Mahanadi river, Odisha as a study area

under study, taking into account an appropriate spatial distribution on the river, the monitoring stations—which were downloaded from the same source—were superimposed over an ArcGIS map of rivers. The stations' locations were indicated with an adaptable Global Positioning System (GPS). These water sites along the river are selected according to the socioeconomic significance and potential for anthropogenic activity (Pati et al. 2023). However, twenty water quality measures were used for this investigation, including pH, BOD (Biochemical Oxygen Demand), DO (Dissolved Oxygen), TC (coliform), TDS (Total Dissolved Solids), TH (Total Hardness), Alkalinity, Chloride (Cl^-), Sulphate (SO_4^{2-}), NH_3-N (Ammoniacal Nitrogen), Iron (Fe^{2+}), Fluoride (F^-), Boron (B^+), TSS (Total Suspended Solids), Electrical Conductivity (EC), COD (Chemical Oxygen Demand), Free Ammonia (Free NH_3), TKN (Total Kjeldahl Nitrogen), SAR (Sodium Adsorption Ratio), and TH (Total Hardness). The sampling bottles were transported to the laboratory in an airtight container, correctly labelled and stored in the refrigerator for pending further inspection, utilizing the conventional procedures, as stated by American Public Health

Association (APHA 2017). The indicators like EC, pH, DO and TDS were measured at in-situ by potable multi meter, while the Cl^-, $SO_4{}^{2-}$, F^- and $NO_3{}^-$ were measured using Spectrophotometric technique. Alkalinity was examined utilizing titration of H_2SO_4, while COD was ascertained by using digestion followed by titration and Microbial analysis is employed to measure the TC. A total of 2 metals namely iron (Fe^{2+}) and boron (B^+) were analysed by ion chromatograph (IC). BOD is obtained using standard methods. Rest parameters were analysed using standard procedures. As a result, the data pre-treatment technique used determines how well procedures apply to these environmental datasets, and data appropriateness must be guaranteed. Data pre-treatment is necessary for homogenization since environmental datasets differ in their type or units. Moreover, observations that included no data were eliminated. It is important to highlight that all analyses were performed twice and that blank samples or distilled water were analysed in parallel for quality control. This demonstrates the continued ignorance and underutilization of data pre-treatment procedures (da Silva Monteiro et al. 2022). The approach used in this study was preceded by data pre-treatment procedures, such as the removal of non-informative variables, the handling of missing data values, and the identification and handling of outliers (APHA 2017). Afterwards, water quality parameters were compared twice, using Ion Balance Error (IBE), for their precision in analysing chemical data, which is put forward in case of milliequivalent per litre (meq/L). The observed IBE falls below the allowable threshold of $\pm$ 5%. Also, several thematic layers were plotted using the ArcGIS program 10.5's spatial analysis module, while the IDW method was utilized to create GIS-based geographic variation plots of the WQIs categories.

19.4 Research Methodology

Because surface water occurrences are irregular, it is crucial to characterize surface water influencing elements and prospective zone demarcation utilizing GIS, entropy, and F-AHP in order to determine their optimal places. This process is also cost-effective, for the most feasible surface water potential (SWP). With the use of weighted overlay analysis and twenty influencing elements for surface water occurrence and distribution, this study aimed to define the SWP zone. The overall study flowchart that was used to accomplish the suggested goals was specified in Fig. 19.2. Here, a weighted index overlay analysis was used to build the surface water modelling map, which was then verified by adding nineteen locations to the ArcGIS platform. Thus, the construction of the water quality index allowed for the trustworthy evaluation of the dataset comprising several variables examined at various sample sites (Şener et al. 2018). The general methods used in this work were demonstrated in detail as follows:

Therefore, in order to investigate the similitude of variation in water quality parameters between the sampling sites, Entropy (E) weighing method was incorporated in this innovative study. In addition, it is a rating system that shows the combined impact of all the chosen water quality parameters on the overall water

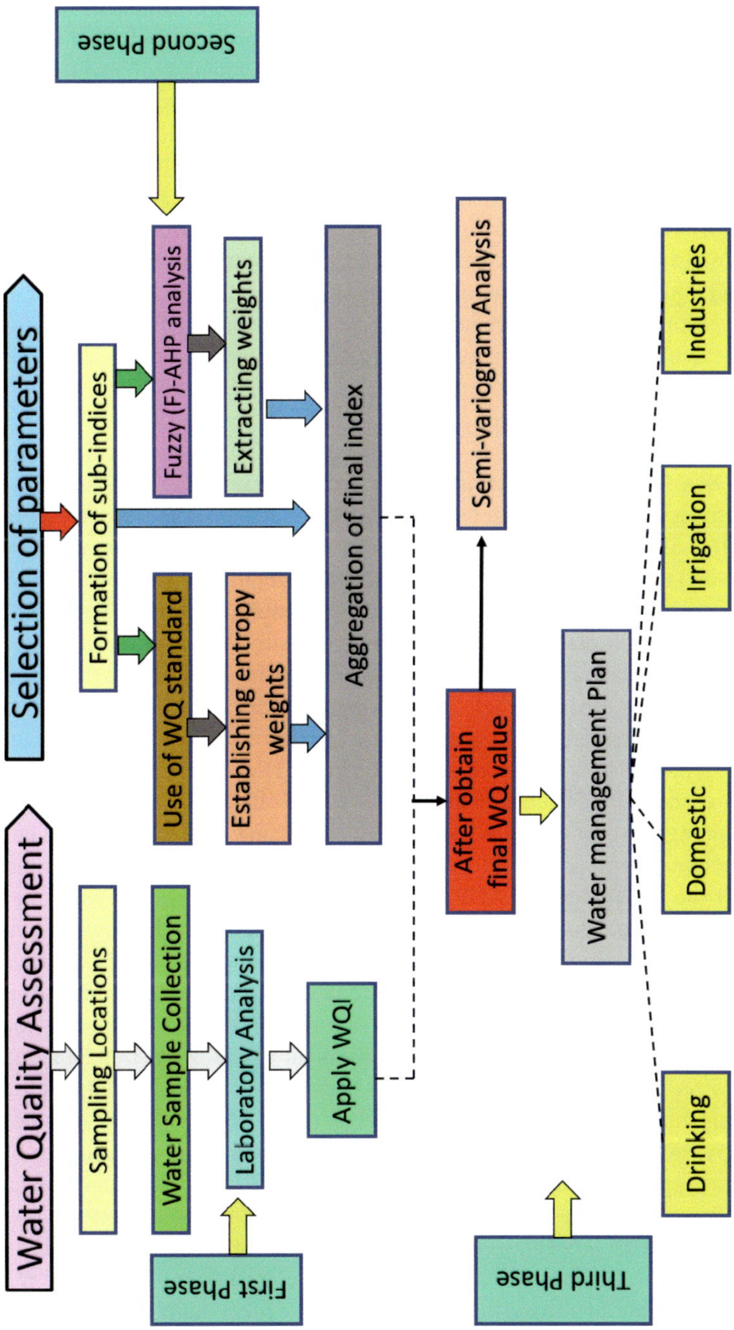

Fig. 19.2 Methodological research framework

quality (Mahammad et al. 2022). The primary goal of this review was to develop a straightforward WQI calculation process based on MCDM techniques for determining the quality of surface and subsurface water. This approach also involves evaluating the temporal and spatial variations of the water quality of a water source with less effort and better accuracy (Verma et al. 2022). Such assessments is simply accomplished with the help of a single numerical score or index. The suggested index's quality rating system, put out by Gautam et al. (2022), illustrates a score of E-WQI < 50, corresponds to excellent; 50–100 as good; 100–150 as medium, a value fluctuate in a range of 150–200, depicted as poor and finally, a score > 200 is put under extremely poor category. In the end, the suggested computational stages include defining and ranking the decision criteria, determining each criterion's comparative importance, estimating each criterion's coefficient, determining the revised weight, and finally calculating the relative weight is addressed by Mohan et al. (2022) and it is explained in the given Fig. 19.3.

Discrepancies frequently occur in decision-making processes when balancing quantitative and qualitative factors. Techniques like MCDM play a critical role in addressing this. These methods entail giving several parameters weights and using expert judgment within a particular thematic framework to rank a subset of the parameters (Trach et al. 2022). It is important to note, nevertheless, that there are certain advantages and disadvantages to this method of developing WQI for water quality evaluation, and that it can be readily streamlined by employing Fuzzy (F)-AHP, that utilized to assess each parameter's weight independently. Still, fuzzy logic is used in many AI fields, such as data analysis, control systems, pattern recognition, and decision-making. It is especially helpful when there is little or conflicting information available, as traditional reasoning might not be able to offer clear answers (Manzar et al. 2022). Since, AHP is frequently employed in MCDM and the F-AHP approach, is thereby expanded by Chauhan et al. (2023), which offers a subjective evaluation of the weighting of several factors, effectively addressing discrepancies that could surface in pairwise comparisons. Meanwhile, this is an advanced analytical technique developed from the traditional AHP model. It compares criteria pairwise among experts in the field to determine their weight, with subjective evaluations playing a major role in determining the weight ratio (Semy and Singh 2021). Therefore, for this approach to be effective, the conceptual framework of assessment methods and the common language adopted to determine and tackle complex water challenges are required. In 1996, this technique was developed by utilizing triangular fuzzy numbers (TFNs) and pairwise comparison has been carried out by utilizing multicriteria analysis models, as instructed by Zeng et al. (2022). Moreover, the estimation of an entropy value will always be difficult due to issues like rigidity, uncertainty, and eclipse. As a result, it may also be expected that fuzzy-AHP approaches can be helpful for lowering uncertainty in steps, such as ranking and parameter selection process, as they further subject the procedures for generating an E-WQI to subjectivity and ambiguity (Uddin et al. 2023). In this method, pairwise comparison matrices are made for every level of the hierarchy tree, and a triangular fuzzy number is assigned to each parameter based on its significance once the decision-making hierarchy has been formed. The adapted steps to compute this model is described in Fig. 19.4. Moreover,

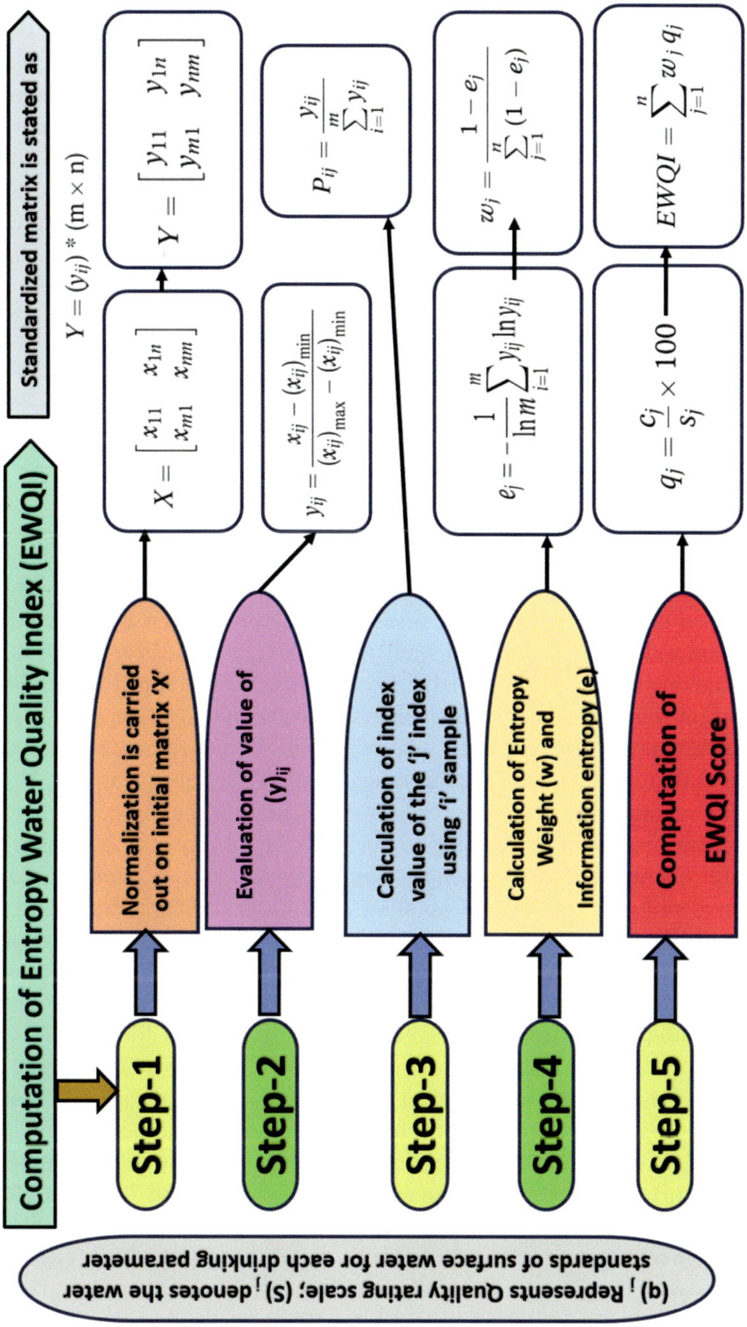

Fig. 19.3 Steps to compute entropy model

this model is categorized into five classes (Şener et al. 2018): < 50 (excellent zone), 51–100 (medium class), 101–150 (medium zone), 151–200 (poor zone), 201–250 (very poor zone) and > 251 (denotes the water to be 'Unsuitable' to use). The steps developed or suggested by Zeng et al. (2022), are represented in a schematic flow chart, which is seen is Figure. The challenge of this technique is based on entropy analysis, which is used to evaluate the weight of the criterion. The F-AHP technique will be used to select the best alternative and establish a ranking preference order of the choices. Later on, the flexibility of this technique in monitoring water quality was confirmed by a combined application of fuzzy technique and entropy weight along with using rough set theory. (Zhao et al. 2022). Hence, after the obtained data is normalized, this model may produce trustworthy ratings for survey locations.

19.5 Results and Discussions

19.5.1 Water Quality Assessment

Water quality is an instrument used to quantify the characteristics of water in relation to human needs. As a result, a variety of physical, chemical, and biological indicators are mostly used to assess it. Therefore, assessing the level of river water pollution is an important and developing field of study that calls for the gathering, analysing, and interpretation of data (Das 2023d). In this case, the aforementioned approaches' effectiveness outweighs that of traditional indexing methodologies (Yan et al. 2022). The investigation reveals that the water quality indexing approach may be impacted by the use of different weights and the theoretical ideal value of any parameter. Thus, it provides a summary of surface water chemistry. As a result, source apportionment with many models is crucial since doing so can boost the validity of a conclusion by comparing the outcomes of various models. Descriptive statistics are often utilized in this study to identify the trends in the variables and their interrelationships based on the parameters that were chosen. In other words, the guidelines of the World Health Organization (WHO 2017) is recommended in this current work, for comparison and also, it is used to assess how each chemical parameter affects agricultural products and human health when surface water is utilized for irrigation and drinking (Wudad et al. 2021). It is observed that pH aids in determining if surface water is acidic or alkaline. Given that the public uses the river for direct consumption like drinking and bathing, so, it is crucial to measure its health. In the present work, the value ranged from 7.741 to 7.913 mg/l, signifying the slightly alkaline nature. It suggests from the dataset that the pH value has a maximum of 7.91 and a minimum of 7.74 with a low standard deviation of 0.05. The pH values in the research region, according to the WHO, do not go over the range of pH values that are advised for drinking water i.e., 6.5 to 8.5. Based on the pH value range, the surface water classification for all locations in the study area maintains a stable pH in the slight alkaline region. All aquatic life rely on DO to metabolize food and produce energy to sustain life and growth,

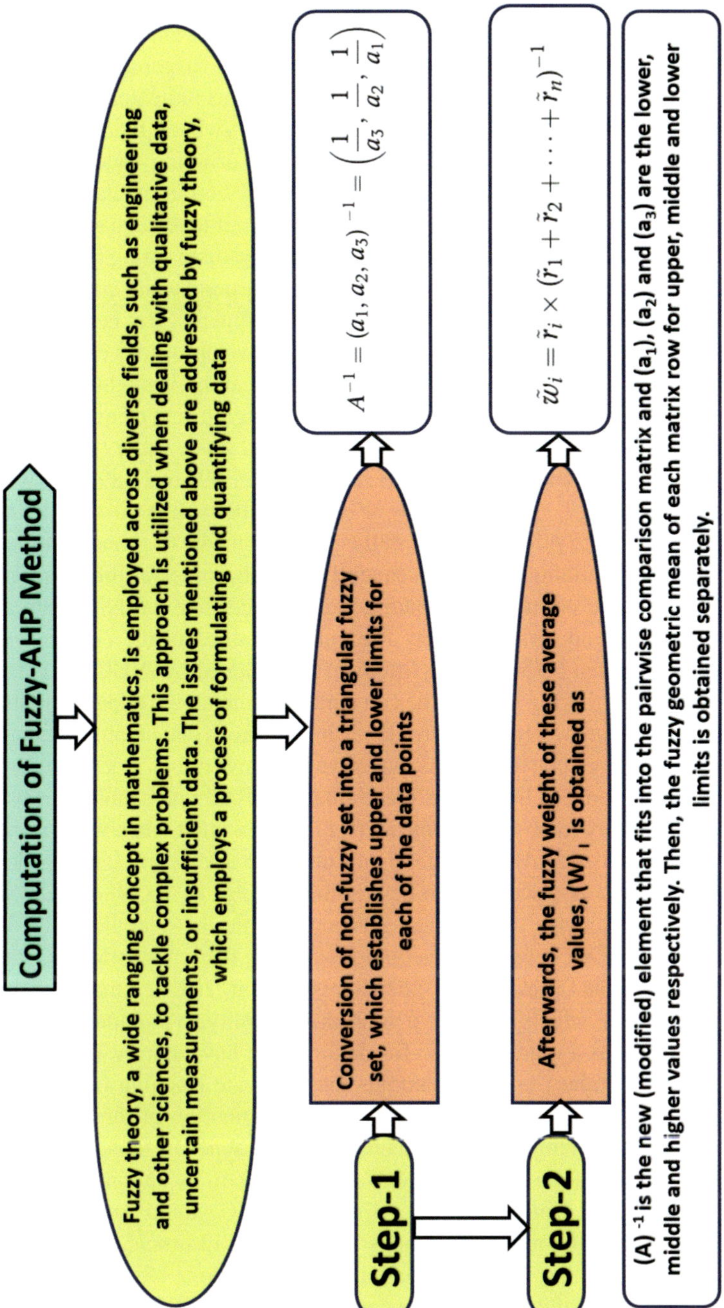

Fig. 19.4 Framework to calculate fuzzy-AHP Model

making it the best index of water quality characteristics for assessing environmental quality (Tolche 2021). A multitude of variables, including temperature, the volume and velocity of the river"s water flow, organic wastes, climate and season, the kind and quantity of organisms present in the water body, height, suspended or dissolved solids, and the amount of nutrients in the water, influence this indicator. In this work, its value spanned from 7.257 to 7.812 mg/l. Almost everywhere, the DO is higher than the allowable limit of 6.0 mg/l, in case of potable drinking water. The main cause may be the seasonal and geographical fluctuations seen in the dataset, which could be explained by rising river water temperatures, heightened aquatic organism biological activity, and rising DO concentrations from organism eating (Das 2023d). The parameter BOD shows how to evaluate surface and ground water contamination caused by the dumping of household and commercial effluents. But because of the high BOD, the DO levels are decreasing, which could have hazardous effects on the biodiversity of the river. Also, elevated BOD demand can be brought on by high nitrate levels, which promote rapid plant growth, and high levels of organic pollutants, which is typically brought on by improperly treated waste water (Tamiru et al. 2021). In the current study, its measurements ranged from 1.095 to 2.389 mg/l, which is under the belt of 5 mg/l, as instructed by WHO regulations. It is possible that the diluting impact of surface runoff from rainfall is the reason why the concentration of the score indicates a declining trend. Subsequently, COD becomes visible as a significant marker of organic pollution originating from sources like partly or untreated urban household and industrial effluent. The value obtained in the current work, exhibits a range between 6.75 and 21.87 mg/l. As mentioned by WHO (2017), the highest threshold for drinking water, to be consumed by humans, is taken as 30 mg/l. When present in drinking water at its stated level, it may have a detectable flavour, and in certain users, very high concentrations may have a laxative effect. According to the current study, the presence of hydroxide, bicarbonate, and carbonate ions in the water is what causes alkalinity (Semy and Romeo, 2021). It shows that when water is more alkaline, it may neutralize acids more effectively and vice versa. Every sampling location's reported value was discovered to be within the range of 70.398–100.88 mg/l, that is being rooted inside the range of 200 mg/l. From the overall results, it is observed that water at X-(9) was relatively more alkaline than other stations, which may be due to additional presence of salts (Shrestra et al. 2023). Another salinity parameter namely EC, which indicates a substance's ability to conduct electricity. As a result, it serves as a crucial metric for determining how salinity-level surface water is. Higher EC values in water samples may indicate leaching or dissolving of aquifer material, mixing of saline sources, or a combination of these activities. Generally, this relies on temperature, concentration, and types of ions present. The restricting value of conductivity is recorded at 250 μS/cm, as authored by WHO guidelines. However, for drinking purposes, the value of EC should not exceed 2250 μ-S/cm. In this survey, the reported findings ranged from 138 to 7779.51 μ-S/cm. The results highlighted that water samples in all the locations were found within the recommended limits of 2250 μS/cm according to WHO (2017) and can be used for drinking water without any further treatment except at site i.e., X-(9). Thus, the salinity factor, which results from mineral breakdown, clarifies it. Also, the addition of anions like Cl^-, PO_4^{3-} and NO_3^-, in sewage discharge or agricultural runoff will also increase conductivity. On the other hand, salinity, which is determined by several unique parameters—defines the amount of dissolved salt in a body of water. On the other hand, it has a direct impact on soil erosion, flood risk, vegetation diversity, water quality, yields in agriculture, and irrigation (Islam et al. 2017). The evaluation of Na^+ concentration is crucial for irrigation because Na^+ interacts with the soil to decrease its structure and permeability. In fact, a high concentration of Na^+ ions in irrigation water causes Na^+ to be absorbed by clay particles, which in turn causes calcium and magnesium ions to be displaced. As a result, the SAR values can be used to calculate the tendency of irrigation water to contribute to the exchange of cations in the soil (Das 2024a). In theory, the ion exchange can become saturated in Na^+ and degrade the soil structure because of the dispersion of clay particles if the irrigated water gets enriched in Na^+ and depleted in Ca^{2+} (Semy and Singh 2021). From the reported results, by applying this index to the samples, that indicates that 93.75% of the surface water belongs to excellent category with SAR value ranging from 0.4 to 16.5. However, one location i.e., X-(9) falling under doubtful category. In most cases, EC and TDS have a direct relationship: the greater the ion concentration, the higher the TDS value and the higher the conductivity. In the end, inorganic salts comprise the majority of TDS, comprising Na^+, Mg^{2+}, Ca^{2+}, Mg^{2+}, K^+, SO_4^{2-}, Cl^-, and HCO_3^-, and in water, organic molecules dissolve in tiny amounts. As stated by WHO guidelines, water is considered unsuitable for consumption by humans if the TDS content is above 200 mg/l. However, if the value recorded to be lower than 100 mg/l, is recommended safe for local people (Das 2023a). In the current work, the TDS values were observed between 82.30 and 13,230. This reports that all samples are within the optimal limit of 100 mg/l except X-(9) and (16). However, it implies that increased evaporation from chemical weathering and human activity raises the TDS value, and these two samples have a tendency to shift from rock dominance to the evaporation zones (Xia and Zeng, 2022). Additionally, the TSS parameter is utilized to determine the optical clarity of water and serves as the most obvious indicator of water quality. Therefore, biological, and inorganic particles, tiny organisms, fine suspended materials, and the muddy or turbid appearance of water are the causes of cloudiness or suspended solids in water. Generally speaking, runoff, algae blooms, disturbed bottom sediments, and soil erosion can all contribute to suspended particles in water. Decomposing material's organic particles may also raise the TSS concentration. Small organic particles can break free from their parent plants and animals' decaying materials and reach the water column as suspended solids (Sachdeva et al. 2020). In this study, the concentration varies from 28.62 to 74.88 mg/l. It therefore, noticed from the results that, the reported readings of TSS in the river water, were in the desirable limits of 100 mg/l. Most of the study area's water quality is excellent throughout. In addition, Ca^{2+} is one of the ions that temporarily hardens surface water, and consuming too much of it through drinking water can be harmful to people's health (Das 2023a). Also, Mg^{2+} functions to activate enzymes properly in cells, but at larger concentrations, it has a laxative effect. Therefore, hardness in water is brought on by a number of different soluble polyvalent metallic ions, mainly

Ca^{2+} and Mg^{2+}. It is noteworthy that, increased levels of TH in water affect physiological functions in humans and can irritate the gastrointestinal tract, particularly in those with renal disease (Rajasekhar et al. 2019). It is noticed that the observed values vary in the range of 51.18–2194.90. As stated by WHO (2017), the limiting threshold is considered as 300 mg/l. On this basis of its findings, the area is found to be within the TH limit, suitable for consumption except X-(9). So, X-(9) was occasionally noted to be higher, which is caused by weathering and carbonate minerals. When its value above permitted limits, it causes boiler and pot scaling and can lead to renal failure in people. Additionally, the parameter namely, TKN is a measurement of the water's organic nitrogen and ammonia-N content. Elevated amounts of nitrogen in freshwater can lead to a rise in the quantity of algae and chlorophyll, although overall, it is not as concerning as other signs. The sources are representations of non-point factors that cause effects, including fertilizers used on fields, atmospheric nitrogen deposition, suburban lawns, and nutrient-rich soil erosion. The levels in the current studied region, ranged between 3.279 and 11.791 mg/l. In spite, its readings should not cross the threshold mark of 5 mg/l, as per WHO criteria. It is exhibited from the findings that, at most places, values are higher. This results from the mingling of human and animal waste, runoff through agricultural land, and manmade contamination. Further, NH_3-N and Free-NH_3 concentration is the most important surface water contaminant when taking agricultural and drinking uses into account. The main causes of the study area's problems were residents' open-land garbage disposal, sewage disposal, and chemical fertilizers, that contributes NH_3 contamination in surface water. Additionally, as nitrate is linked to the breakdown of organic matter from sewage, household waste, animal dung, and nitrogen fertilizers, elevated levels of these markers are primarily indicative of anthropogenic activity (Rawat et al. 2019; Das 2023a). Here, wastes from industry, agriculture, and animals are the main sources. The reported findings of NH_3-N and Free-NH_3 ranged from 0.511 to 1.928 and 0.021–0.059 mg/l respectively. Hence, the recommended limit for drinking water should be 2 mg/l. However, all observations were within the prescribed limits for all sampling locations. During the study period, the value of TC fluctuated from 1219 to 42,530 MPN/100 ml. Among all the variables, it can be observed that TC has the biggest standard deviation. The maximum value of TC is 42530, which surpasses the acceptable limit guidelines. Accordingly, the research shows that the presence of coliform bacteria guarantees faecal contamination in the water body, indicating a higher risk of illness and an unfit water source for drinking. It is depicted that higher values of > 5000 is reflected at X-(8), (9) and (16), that crosses the standard guidelines. But urbanized regions and high-density Environmental Resource Permits (ERP) with more home sewage and industrial effluent have a significant impact on these locations. Therefore, it is advised that greater coliform colonies at three sites be taken into consideration as indicators of a high level of bacterial contamination. As a result, a significant portion of the urban population experienced typhoid, diarrhoea, and dysentery. Therefore, carrying forward, residents' dumping of waste in open areas, sewage disposal, and chemical fertilizers were the main causes of B^+ contamination of surface water. All things considered, this indication finds its way into water systems from a variety of sources, such as sewage, manure, and various artificial fertilizers applied to agricultural fields (Pandey et al. 2020; Laltu 2023). Greater value leads to decreased oxygen in river water, which then affects the consistency of the material and nutrient cycling processes. However, the existing values i.e., 0.029–0.554 in the study area were below the WHO recommendation of 2 mg/l and the concentration in the river was at a level not posing a health concern. Heavy metals namely, Fe^{2+} is an essential trace element required for good health and it was derived from natural and industrial waste water. While it helps the blood carry oxygen, when it is present in high enough concentrations, it can also cause hemochromatosis, exhaustion, shedding pounds, joint pain, stomach issues, vomiting, and damage to DNA (Das 2023a; Barroso et al. 2024). It also states that rainwater that contacted the soil raised the river's Fe^{2+} content. Additionally, highly concentrated iron water can cause turbidity and turn a reddish-brown colour. In the study period, the iron ranges between 0.59 and 2.609 mg/l, which is within of the permissible limit of 1 mg/l. Higher Cl^- has an impact on leachate, saline intrusion, fertilizers, sewage and industrial effluents, natural sources, river ecology, and ecosystem services including fisheries. It is emphasized that the unrestricted movement of Cl^- in the surface water environment was not produced by the breakdown of halite minerals or paleo salinity, which is not being absorbed by the soil, but rather by the scarcity of natural sources identified in the study region. The permissible limit of Cl^- is 250 mg/l as per WHO standards. In the present study, its concentration ranged from 9.648 to 4904.88 and its readings remained below the maximum allowable level at all sampling locations except at X-(9). The area's geological characteristics, runoff from agriculture, wastewater from industry, and domestic sources are the causes of the higher concentration. Later case, Sulphate (SO_4^{2-}) may get into water streams as a result of surface runoff from contaminated areas and sewage discharge. The amount of sulphate in water rises as a result of household and industrial waste. Furthermore, as air temperature rises, more evaporation occurs, which contributes to an increase in this parameter value. It naturally arises in surface water from the weathering of sedimentary and igneous rock. According to WHO guidelines, when its concentration rises above 200 mg/l, its levels may also become unstable and have a laxative effect on human health. Regarding this, the value was thought to be in a variety of 4.97–376.07 mg/l in the present study. However, it is noticed in the current work that the lowest SO_4^{2-} was observed at X-12 (4.97) and the highest at X-9 (376.07 mg/l). It is found that increase in concentration at X-(9). Given the study area's high level of farm-driven activity, it might be connected to runoff from agriculture. The chief sources of nitrate (NO_3^-) in surface water is nitrate seeping into the permeate water as a result of extensive agricultural and human activity. Although surface washing, phytoplankton absorption, and bacterial nitrate denitrification can cause its levels to rise quickly, changes in ground water nitrate levels usually happen gradually. But in most cases, it comes from overusing fertilizers on farmland and animal or human waste pollution since NH_3 oxidizes and releases other compounds that are similar. So, NH_3 penetrates the aquatic environment through indirect processes including nitrogen fixation, air deposition, and runoff from agricultural areas, as well as direct processes like animal excretion of nitrogenous waste and municipal effluent discharge. Furthermore, because of several agricultural and related activities, such as the excessive use of inorganic nitrogenous fertilizers and manures as well as the dumping of wastewater by particular businesses, its concentration level can be noticed in both surface water and ground water. The current region's results showed that its measurements were within the range of 1.289–2.689 mg/l. In all sites, levels are within the criteria limit of 45 mg/l. The weathering of primary rocks and the liquidation of fluoride (F^-)-containing minerals, which are generally associated with low calcium levels and high bicarbonate ions, are the main causes of high fluoride rates in India. It has been shown that persistent water rock percolations of rocks containing fluorite in dry, high-temperature environments also accelerated the evaporation cycle, resulting in a significant concentration of fluoride in surface water. Additionally, this indication is mainly contributed to drinking water through several geological processes. At low concentrations, it prevents and minimizes risks that damage teeth, which has a significant impact on teeth. On account of high F^- levels, it has a connection to dental fluorosis in the area. As a result, it was discovered that the F^- concentration quantified in this study was 0.258–1.0 mg/l. At all sites, the concentration were allowable according to the WHO desirable limit and the water could be utilized for drinking after a disinfection step.

Based on the outcomes of every physicochemical parameter, the hydrochemistry of the analysed samples show that the order of the main cations is presented as $Fe^{2+} > B^+$ of mean abundance, while anions are found in the sequence of mean abundance of $Cl^- > SO_4{}^{2-} > NO_3{}^- > F^-$. This demonstrated how ion exchange, mineral dissolution, and rock weathering control the chemical composition of surface water (Brhane et al. 2024). Therefore, it was possible to clearly view the geospatial maps of the various water quality parameters along the river stretch from the Fig. 19.5(a–t). The aforementioned research reveals that TC and TKN have higher values than recommended by WHO guidelines. This is likely due to the differing sources of geogenic and anthropogenic causes. For example, it is not possible to determine with precision how much the river water quality deviates from the drinking standard only on the values of these metrics. As a result, this study was carried out to evaluate the river water quality using entropy and GIS. The pollution sources were located using F-AHP, and the results suggested the necessary treatment steps needed to make the water fit for human consumption.

19.6 Pollution Source Identification Using Entropy Model

When it comes to classic WQI approaches, they lead us to believe that a surface water body's problematic situation cannot be indicated by its analysis, monitoring, or index development alone. As a result, the entropy approach can be used to easily identify, in further analysis, the relative pollution levels of the sampling locations in relation to the drinking water quality standard. Consequently, another mechanism of decision making namely, F-AHP is suggested to use the following equations, which are covered in the methodology section, to compare their rating processes and

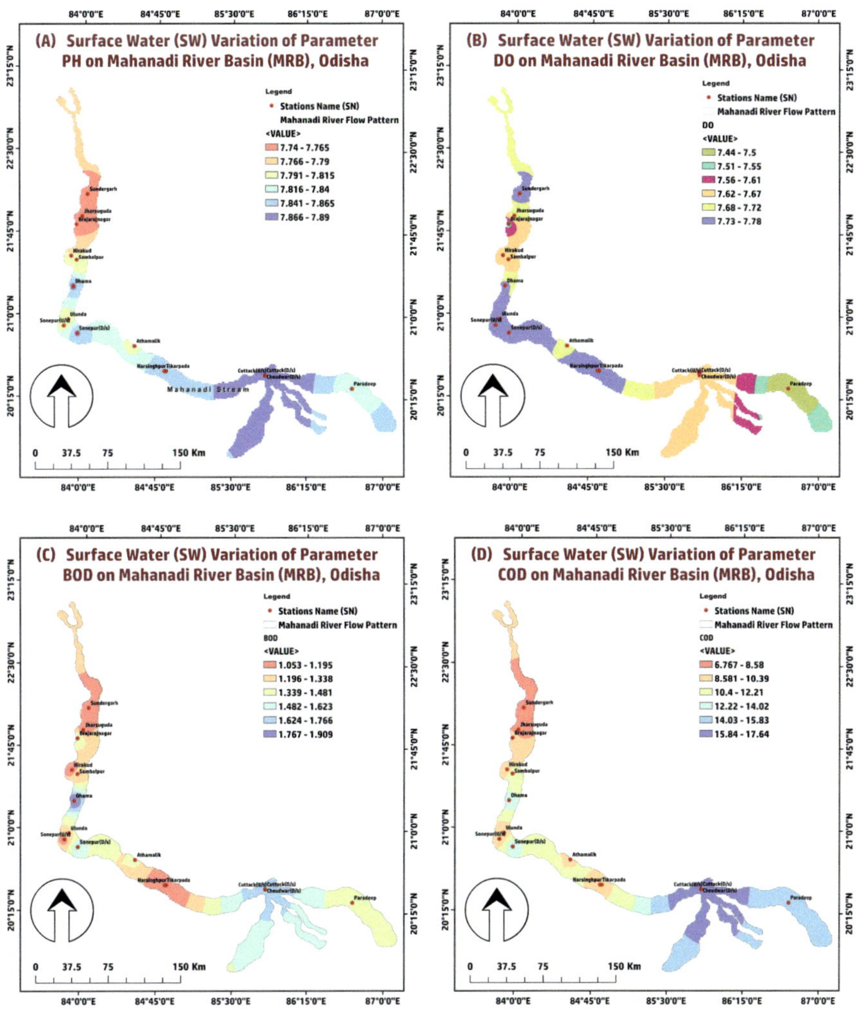

Fig. 19.5 Parameters showing spatial distribution of **a** pH, **b** DO, **c** BOD, **d** COD, **e** Alkalinity, **f** EC, **g** SAR, **h** TDS, **i** TSS, **j** TH, **k** TKN, **l** NH_3-N, **m** Free NH_3, **n** TC, **o** B^+, **p** Fe^{2+}, **q** Cl^-, **r** SO_4^{2-}, **s** NO_3^- and **t** F

averaged rating. During the sampling period (Table 19.1), E-WQI values for each sample (Fig. 6a), ranged from 14.97 to 1065.20, that depicts excellent to extremely poor categories. It is exhibited that river sediment contamination zones can be categorized using the geospatial approach, allowing for improved river and catchment management. The survey data was used to run these interpolations, which were categorized into less affected and higher affected areas. Following the interpolation of the river's WQI readings in Fig. 6b, this stretch was separated into five groups. It is clarified that about 81.25% of the tested locations, that displayed excellent conditions,

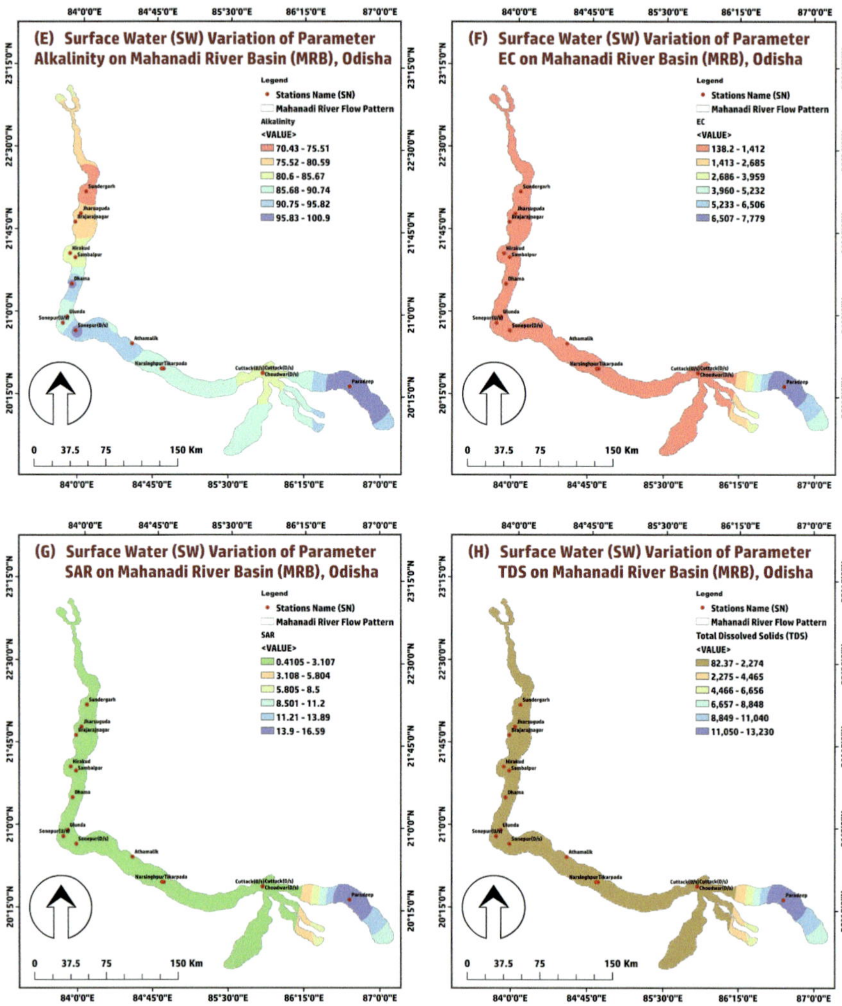

Fig. 19.5 (continued)

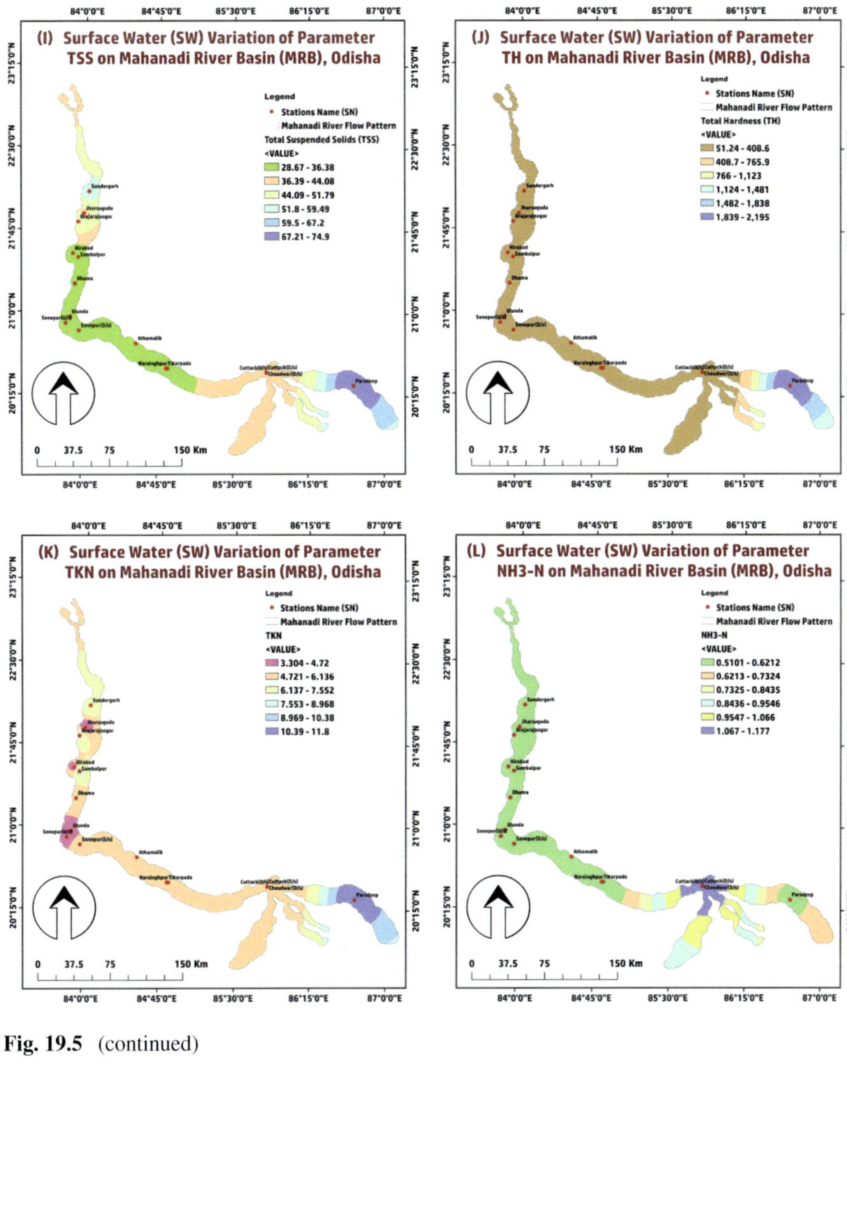

Fig. 19.5 (continued)

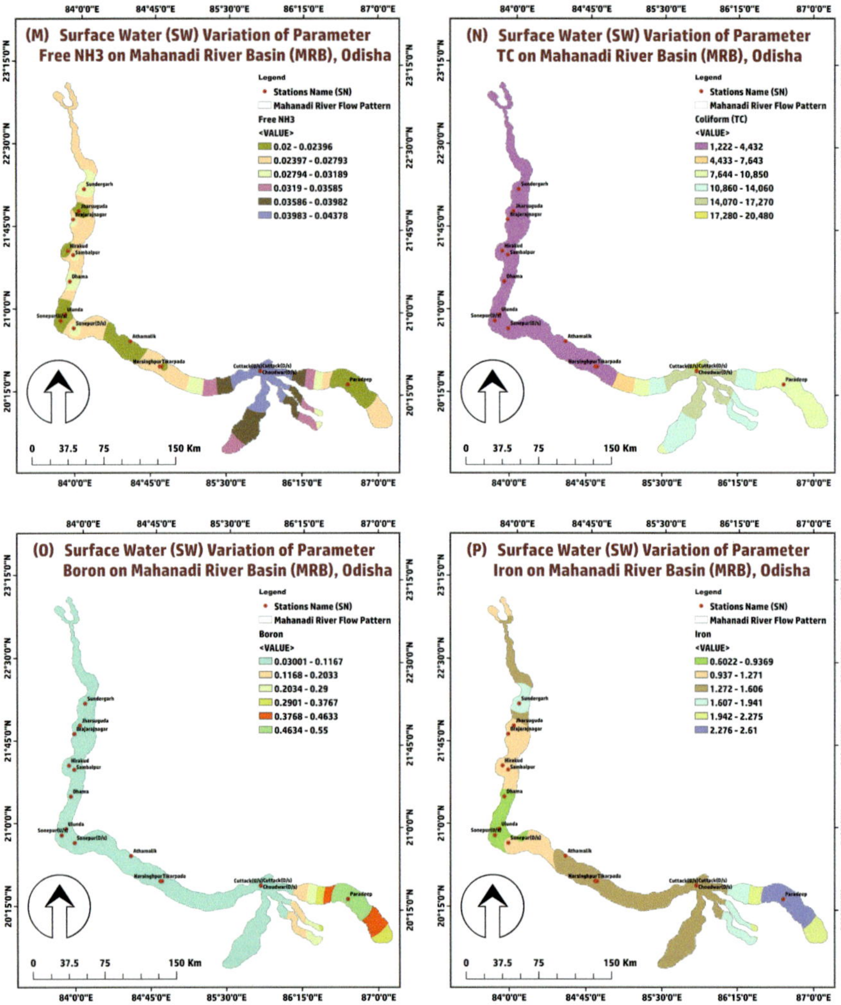

Fig. 19.5 (continued)

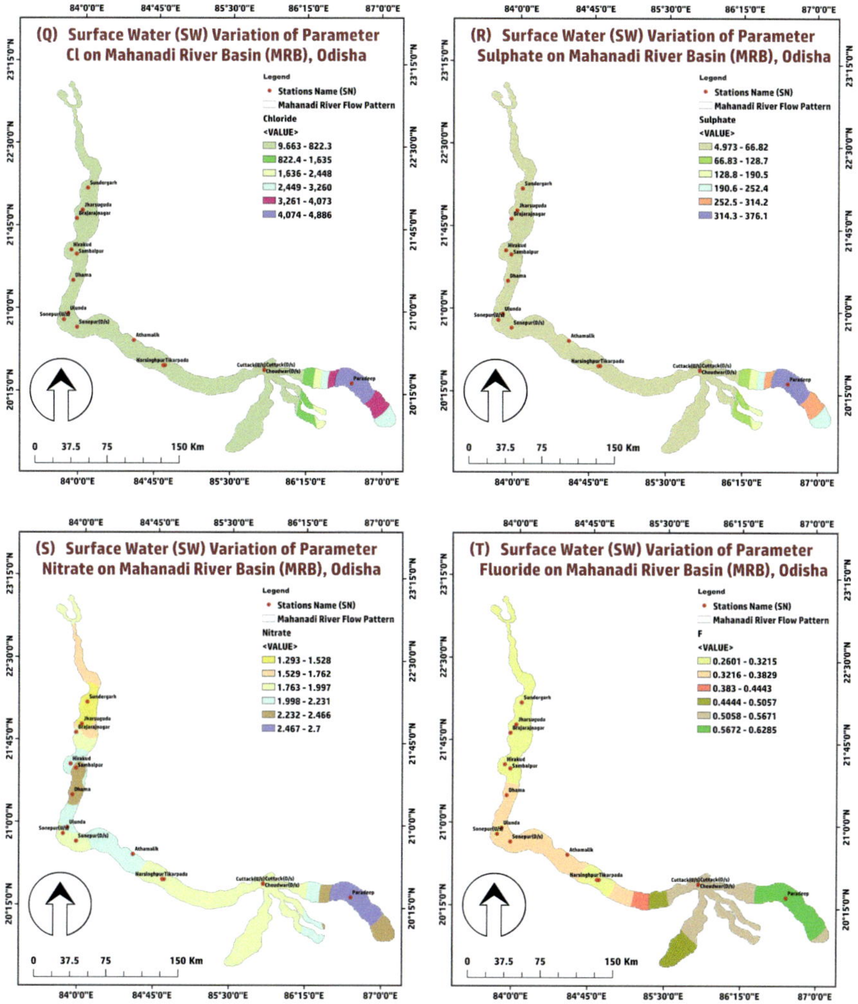

Fig. 19.5 (continued)

12.50% of sites resembles poor conditions and subsequently, 6.25% of principal sites (1 location), expressed under extremely poor water category. The variation in (Fig. 6c) displays the evaluation of the sites' conditions across the entire study region. Thus, it implies that the upstream areas around the sample sites saw an improvement in the quality of the water. Further, stations such as X-(9), is described as an extremely contaminated polluted zone, while X-(8) and (16) were categorized as being in a poor zone since the majority of the region is impacted by agricultural or industrial practices, which are then followed by municipal waste. The spatial extent of the impacted area, which was more exposed to contaminants, was shown by these demarcations in the GIS map. During the study period, it was observed that areas near built-up and industrial areas had the highest concentration of nutrients and were also highly contaminated. Also, elevated values depicted at X-9 site, is because of 'EC, SAR, TDS, TH, Cl^-, SO_4^{2-}, TKN and TC'. This is an example due to analyses and surveys of land use and cover, which indicate that runoff from agricultural areas has had a significant impact, followed by wastewater from homes and businesses. It's thought that agriculture also had an impact on the highly polluted area. The residual stations were classified as being in a moderate-to-low pollution zone because to the influence of agricultural and household runoff. Therefore, for the most part of the research region, the surface water quality falls into the excellent and good categories, making it suitable for both domestic and drinking uses, with the exception of three stations.

Table 19.1 Classification of water type based on E-WQI values and its overall values and its ranking for each site has been computed by F-AHP approach

Sample No (X)	E-WQI Score	Rank (Entropy)	Water type	Fuzzy-AHP WQI score	Rank (FAHP)	Grade of Water (FAHP)
1	15.69	14	Excellent	92.282	13	Good
2	18.4	9	Excellent	196.89	4	Poor
3	16.21	13	Excellent	94.577	12	Good
4	19.88	5	Excellent	95.74	10	Good
5	18.58	8	Excellent	95.63	11	Good
6	19.47	6	Excellent	99.86	9	Good
7	17.61	10	Excellent	194.37	5	Medium
8	196	2	Poor	233.62	2	Very poor
9	1065.2	1	Extremely poor	296.65	1	Unsuitable
10	14.97	16	Excellent	122.83	7	Medium
11	15.08	15	Excellent	176.63	6	Medium
12	16.93	11	Excellent	122.53	8	Medium
13	19.98	4	Excellent	79.443	14	Good
14	16.32	12	Excellent	77.4	15	Good
15	19.45	7	Excellent	53.7	16	Good
16	152	3	Poor	212.71	3	Very Poor

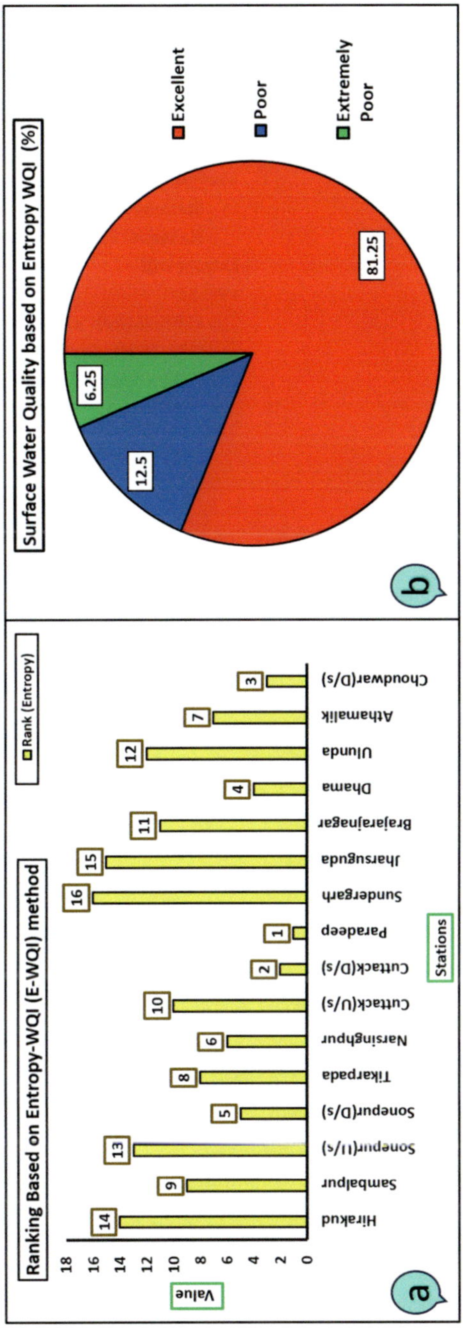

Fig. 19.6 (**a–c**) Entropy water quality Index (E WQI) map for drinking purposes in the area

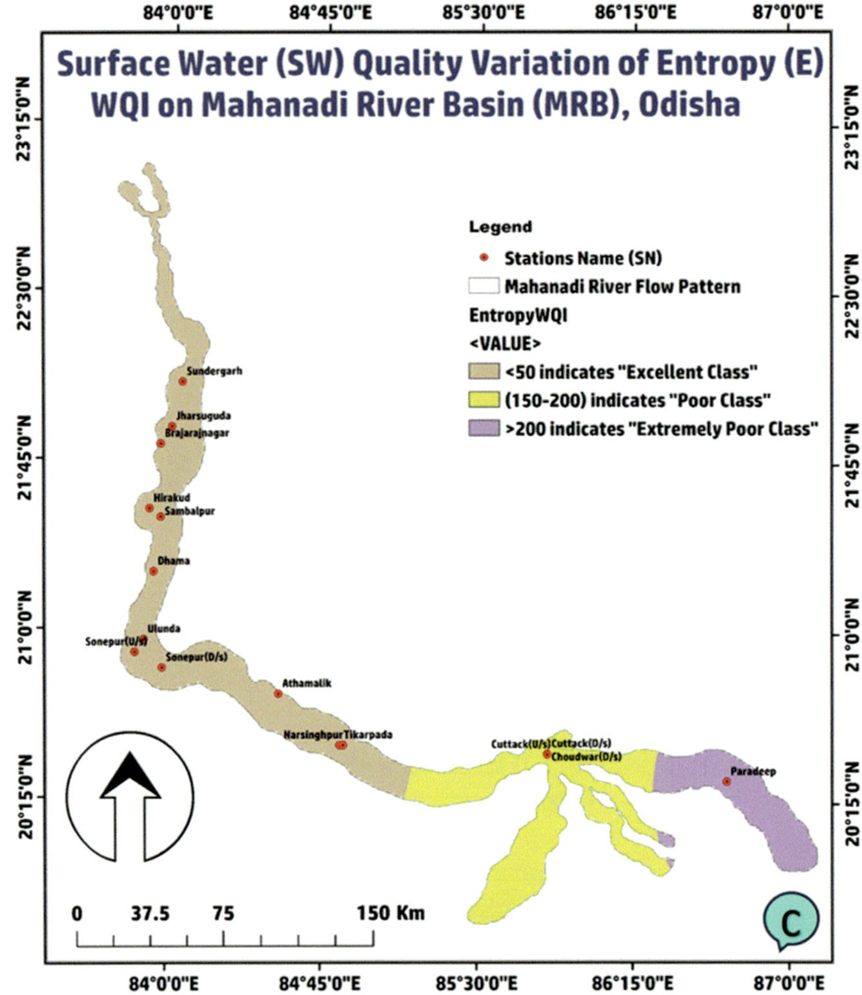

Fig. 19.6 (continued)

19.7 Source Apportionment Using the Fuzzy-Analytic Hierarchy Model

Regarding E-WQI calculations, it should be highlighted that, despite the use of an index, the relative weight and lower quality of each parameter in the WQI grading system are incomprehensible. However, applying normalized weight to every parameter in the F-WQI computations may enhance their impact on WQI grading. It is evident that while the ranking of criteria in the Fuzzy-AHP approach resembles that of AHP, the prioritization of sub-criteria differs significantly. This method is crucial for gathering vital information, providing in-depth knowledge of the topic, and improving comprehension of the current study issue. As such, it is an effective and adaptable method for both qualitative and quantitative research on multi-criteria problems. In a similar vein, the weights are determined in a fuzzy setting, and the ranking criteria and sub-criteria are subsequently applied in accordance with the connections that regulate the Fuzzy-AHP approach, as stated in the methodology section. In addition, by assigning them an overall rating, it also determined the relative pollution level and classified these places as the most polluted. Therefore, the sub-criteria of potential harm to people from the environment, the ability to identify hazardous materials, the potential for taking risks, and the project's implementation costs are given the highest priority; however, the sub-criteria of project management, exposure to natural disasters, and ease of implementation are given lower priority. The findings for the weight values of each qualitative parameter and the well quality grading using F-AHP techniques are shown in Table 19.2.

For instance, the first five fuzzy weights of the parameters were found to be in order: TC > TKN > EC > TDS > Cl^-. It is evident that the parameters with maximum fuzzy weights had the biggest influence on the rating of water quality. Therefore, based on F-AHP grading, it appears that the TC and TKN parameters, with weights of 0.178 and 0.169, respectively, had the biggest impact, whereas the DO parameter, with a weight of 0.002, had the lowest impact. Because of this, it

Table 19.2 Estimation of final weight factor, based on entropy and fuzzy methods

Parameters	Entropy weights	F-AHP weights
pH	0.004	0.052
DO	0.004	0.002
BOD	0.006	0.008
TC	0.057	0.178
TSS	0.007	0.017
Alkalinity	0.004	0.007
COD	0.008	0.035
NH_3-N	0.010	0.021
Free NH_3	0.007	0.025
TKN	0.008	0.169
EC	0.117	0.156
SAR	0.102	0.067
B^+	0.045	0.003
TDS	0.166	0.136
TH	0.094	0.021
Cl^-	0.207	0.117
SO_4^{2-}	0.131	0.097
F^-	0.010	0.008
NO_3^-	0.005	0.112
Fe^{2+}	0.008	0.102

is assumed that the relative weights assigned to each of the qualitative parameters will result in parameters that are higher than the standard mean. In that scenario, the drinking water quality will exhibit the original characteristics of deteriorating water quality, which will have an impact on the location's quality grade. Its reported F-AHP WQI value in Table 19.1 and Fig. 7a, varied in a range of 53.7–296.65, signifying good to unsuitable water category & thus, its standard deviation further, indicates that there is more variance in the dataset. As was previously said, there were some inconsistencies when using E-WQI; however, when comparing the results with the prior method, it was found that, when taking F-AHP WQI into account, over 74% of the research area was graded as good to medium. In order to determine the water quality of the research points, IDW zoning in the ARC GIS environment was used. As explained in the classification as shown by Fig. 7b, 50% of tested locations comes in excellent ($n = 8$) and medium (25%, $n = 4$) classes, whilst 6.25% ($n = 1$) of recorded sites is grouped as poor, which are conferred to unsuitable for drinking uses. Further, 12.5% ($n = 2$) of samples belongs to very poor and finally, unsuitable category holds about 1 station, i.e., 6.25%. Overall, 25% ($n = 4$) of samples are not appropriate for drinking uses and required proper treatment before serving to the local communities. These results could differ greatly, according to the outcomes of grading based on E-WQI after using F-AHP WQI. These alterations are typically visible in sampling points and can be categorized using excellent quality prioritizing. For example, site No. (2) 's E-WQI results place it in the excellent class of water quality, but the F-AHP WQI results place the identical area in the poor category. Once more, site No. (7) has an exceptional grade with an E-WQI value of 17.61, however it ranks fifth according to the F-AHP grading system. Based on the two approaches employed in this investigation, Tables 19.1 and 19.2 present the variations in the qualitative grading of the wells under investigation. Therefore, the quality grading of sample points with WQI was calculated and compared in the current study using the entropy model in conjunction with F-AHP. Additionally, weights of each qualitative parameter in the entropy WQI are obtained using the Shannon approach; on the other hand, weights in the F-AHP are the results of data standardization. Furthermore, it was evident from the F-AHP WQI results that they had superior convergence to the E-WQI approach. The majority of those stations provided the best illustration of the disparity in the qualitative grading of the wells using the methodologies under investigation, as shown in Table 19.1. Considering this, the outcomes in the studied region, observed that the location X-8 (233.62), 19 (212.71), and 2 (196.89) had the second, third and fourth highest F-AHP WQI indicating it as poor/very poor/unsuitable water class, thereby a higher value in three places is noted that follows the trend of higher SO_4^{2-}, TDS, Cl^-, TKN and Conductivity. In addition, place like X-9 (296.65) is located in an area that is very polluted., with the overall rank of 1, due to increased concentrations of the parameters namely, EC, SAR, TDS, TH, Cl^-, SO_4^{2-}, TKN and TC, which were likewise the highest of all the areas and greater than their desired concentration. Because treatment plants cannot handle the volume of sewage generated in cities, residential wastewater from these locations is dumped into the river. According to information from the geospatial map in Fig. 7c, untreated sewage wastes that are dumped into the river downstream may be the main source of pollution in the area around the river. Since river water is the main source of water for homes, farms, and industries, it can deteriorate and have negative effects on crops, aquatic habitat, human health, and even microbial diversity, which can cause ecological imbalances. This makes river water the ultimate source of pollution in both cases. Additionally, a high concentration of both organic and inorganic chemicals can be found in textile effluents. Regarding the in-depth explanation above, the body of research shows that these substances, when combined with the freshwater system, will change the surface water's physical and chemical composition, which could ultimately have an impact on the surface water's quality. Furthermore, because of the excessive extraction of surface water brought on by rapid population increase, commercial activity, and other factors, salt water incursion is a regular occurrence in coastal aquifers. Finally, locations with excellent and poor water quality were identified by contouring the retrieved F-AHP score findings using geostatistical approaches. Combining the two models is therefore helpful in locating possible sources of pollution and in periodically checking the quality of the surface water in the river basin; as a result, it is an essential duty in water resource conservation and monitoring.

19.8 Source Apportionment Using the Geostatistical Model

Here, in later case of discussion, nineteen sampling points were employed in this study to assess the water quality and calculate semi-variogram models for each of the parameters. Geospatial techniques like Kriging and IDW predict for an unobserved site by utilizing neighbouring known qualities and interpolating at the spatial variability of point attributes. Various models, including spherical, circular, exponential, and gaussian models, were employed and determined to be the most suitable for analysing spatial variability with respect to entropy and fuzzy (F)-WQI values. These two models exhibit the characteristics of probabilistic models with minimal error, whereby the linear interpolation approach predicts the value of the unobserved position. Using geospatial statistical approaches, kriging was used in this work to assess the dependability of various surface water quality metrics. The nugget/sill ratio was then used to establish the geographical dependence for water parameters. Regarding conventional categorization, a recorded value of nugget/sill ratio i.e., $>$ 0.75, is regarded as high and shows a significant spatial dependence; a index value ranged as 0.25–0.75, displays moderate correlation and ultimately, a lower value of $<$ 0.25, indicated a poor/weak spatial correlation. An attempt is made to evaluate the prediction performance of these models in order to study the kriging models and four standard statistical indices such as mean standardized error (MSE), root mean square error (RSME), average standard error (ASE), and root mean square standardized error (RMSSE), were determined. Based on outcomes, as stated in Fig. 19.8(a, b), the Gaussian model was believed to be the best match for semi variogram model, for both E-WQI and F-AHP WQI dataset. In Table 19.3, the recorded readings of RMSE and ASE is estimated to be 10.27 and 6.45 for E-WQI and 181.15 and 72.31

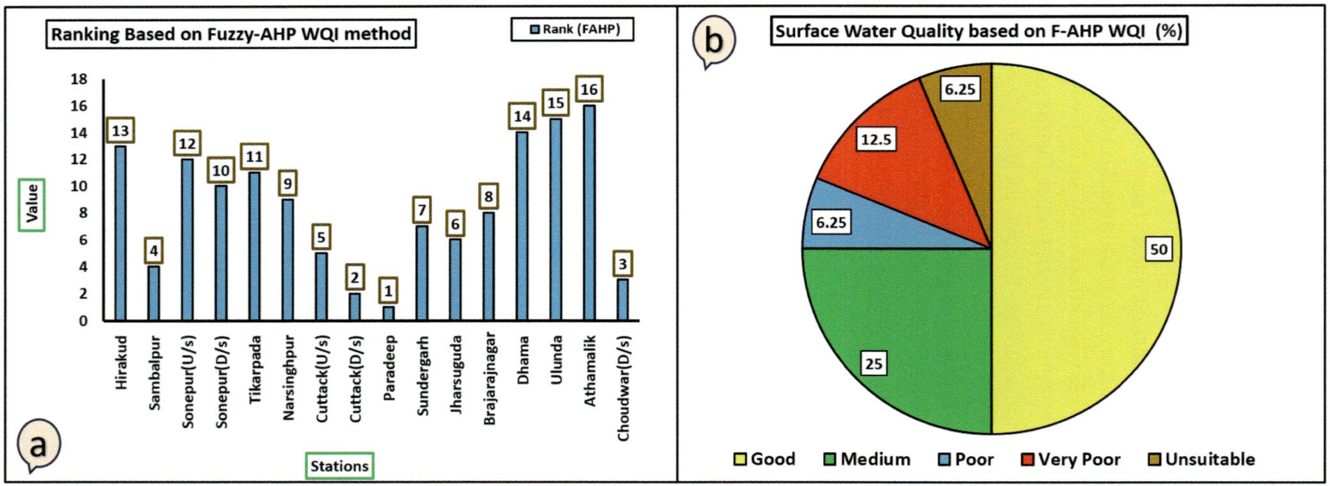

Fig. 19.7 (**a–c**) Fuzzy-AHP water quality Index (WQI) map for drinking purposes in the area

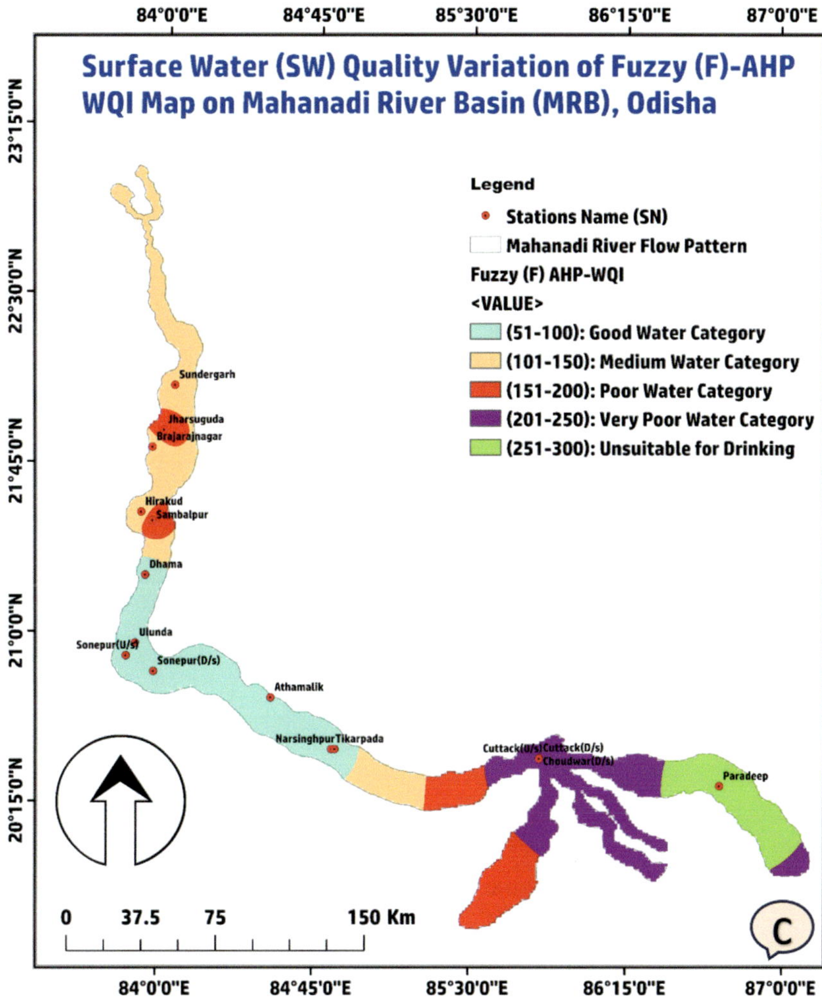

Fig. 19.7 (continued)

for F AHP-WQI respectively. With this, the data elaborated as per Gaussian model, were found as the least, among all other technical models. Also, the tested value of nugget/sill ratio is expressed as 0.11 and 0.05 in the current work, that has been clearly explained for both the indices. The acquired maps make it evident that the majority of the measurements in the X-(9) location had bigger values, which are indicative of more pollution. The factory operations next to the river and the urban waste water discharged into the river could be the cause. Additionally, both indices show a weak spatial correlation that suggests the release of chemical fertilizers and industrial waste water, both of which are probably lowering the quality of the water. Therefore, similar semi-variogram model capabilities have also been documented in

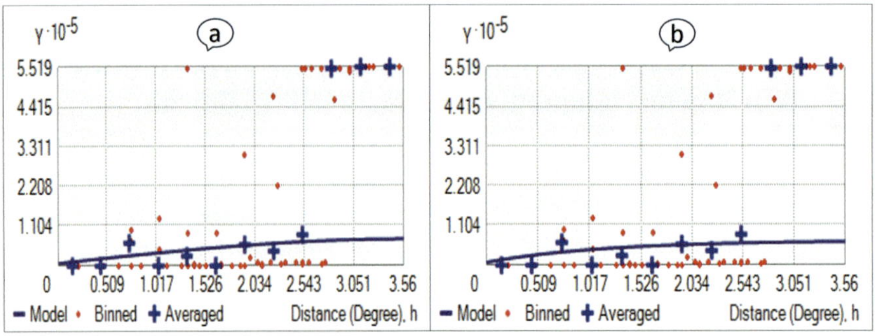

Fig. 19.8 The best-fit gaussian models namely, **a** E-WQI and **b** F-AHP WQI

Table 19.3 Description of the best-fitted variogram model developed for the surface water quality parameters

	Best-fit model	Nugget	Sill	Nugget/sill	ASE	RMSE	MSE	RMSSE
E-WQI	Circular	95.67	91.54	1.04	11.077	13.634	– 0.032	1.076
	Spherical	99.85	76.51	1.33	11.277	13.661	– 0.029	1.066
	Exponential	115.21	83.39	1.37	12.438	13.864	– 0.027	1.142
	Gaussian	43.57	361.45	0.11	6.450	10.27	– 0.108	0.977
F – AHP WQI	Circular	4564	73,663	0.08	124.060	224.413	– 0.121	1.075
	Spherical	6342.2	66,090	0.09	127.451	224.866	– 0.121	1.052
	Exponential	7283.0	59,900	0.12	146.863	229.989	– 0.108	1.411
	Gaussian	2147.2	124,816	0.05	72.31	181.15	– 0.197	0.981

a study based on the evaluation of surface water quality in case of Sylhet district, Bangladesh (Islam et al. 2017).

19.9 Conclusions

Multi-Criteria Decision-Making (MCDM) models are gaining prominence as a viable method for surface water management. Its ability to recognize complex relationships across large datasets and make reliable predictions opens novel opportunities for investigating surface water potential zones and offers useful information for monitoring water quality. In this chapter, an attempt has been made to furnish drinking water in Mahanadi River Basin (MRB), Odisha, India, to determine whether it is appropriate for human consumption, considering 20 water quality (WQ) parameters collected monthly from sixteen water sources. Time frame taken into consideration is 3 years (2020–2023). Entropy-based water quality index (E-WQI) and MCDMs like Fuzzy-AHP method was applied to compute the results. The spatial map was created in ArcGIS 10.5 using the interpolation method (IDW). When the results were compared to the ranges or values provided by standard organizations (WHO) for evaluating WQ, it became clear that the DO was in good condition and the river water was slightly alkaline. With the exception of TC and TKN, the majority of the parameters assessed were found to be within the WHO guidelines' permissible range. High TC and TKN is likely because of peculiar geological features, which render this surface water not suitable for human consumption. But according to E-WQI data, the water quality is contaminated at X-(8) (196), 9 (1065.20) and 16 (152) and it is not fit for drinking purpose, even though the local people are drinking. It is exhibited that at high E-WQI readings at St. (9) were due to the high value of TC, TKN, EC, TDS, Cl^-, SO_4^{2-}, NO_3^- and Fe^{2+}. However, the land use and cover of the region have a significant impact on surface and subsurface conditions including soil erosion, soil moisture, soil fertility, surface runoff, infiltration, interception, evapotranspiration, and other entities. Therefore, seasonal fluctuations, agricultural practices, and other human-induced activities could be blamed for this. The semi-variogram modelling result demonstrates that the Gaussian model fits the E-WQI and FAHP-WQI datasets the best. Furthermore, MCDMs conducted the geographical evaluation of the relative water quality in relation to physicochemical characteristics. These methods are employed to assess the water's purity and suitability for human consumption. According to F-AHP framework, 50% of water samples represent "good water", 25% indicate "medium water", and 6.25% and 12.5% represents "poor and very poor water" and finally, 6.25% indicates "unsuitable water". The situation was thought to be due to overexploitation, leaching of ions, direct discharge of effluents and agricultural impact. Also, the developed fuzzy model was applied on the dataset and ranked the site i.e., X-9 ($P^* = 296.65$) as the most contaminated sampling point on the degree of performance score, followed by 2nd i.e., (X-8, $P^* = 233.62$), 3rd i.e., (X-16, $P^* = 212.71$) and 4th (X-2, $P^* = 196.89$). The main contributors are industrial discharges, human waste, and fertilizers used in agriculture. Furthermore, it was stated that because these locations are on the outskirts of an industrial area, they are primarily impacted by excessive concentrations of various parameters as well as by the effects of strong human activity and industry. Considering the severe health risks, treatment of the surface water has been essentially required to be made safe for drinking purposes. It is also advisable to include alternative sources of potable water as well, so that the health and safety of the community may be safeguarded. Ultimately, the research has validated the viability and dependability of entropy, fuzzy, and GIS techniques for data processing and interpretation related to surface water quality assessment. Since this is the first study in the river basin to show results that show significant discrepancies between the indices at different sampling locations, it was clear from the water quality data set that this study improved our understanding of how to classify the water quality in terms of physicochemical parameters. The outcome of this study sends a red alert for action to prevent further contamination and adverse health effects from long-term exposure to nitrogen parameter in the surface water and to protect the community. Ultimately, in light of these studies, the local government has been advised to use surface water and ground water jointly while closely monitoring and controlling areas with poor surface water quality to assure the resource's sustainable usage. This study will be beneficial for policy-makers and research groups in the quick evaluation of surface water quality. For its scope and recommendations, this work could help future researchers in human health risk assessment and also, recommends that proper waste disposal methods should be followed, sewage lines should be periodically inspected to avoid leakages, and awareness programs for farmers should be implemented highlighting the use of organic fertilizers instead of synthetic fertilizers.

Acknowledgements The author is thankful to State Pollution Control Board (SPCB), Odisha, for his help in the analysis of water samples. The author is also thankful to the honourable reviewer for his scientific comments and suggestions for the improvement of the scientific content of the manuscript. The author is also grateful to the Editor-in-Chief and handling editor for their encouragement and support. In addition, the work space and instrumentation for the current manuscript was supported by Rani-Prativa Das, Research Group, Ruchika Square, Bhubaneswar, Odisha (India).

Author Contributions All authors contributed to the present study through data collection, material preparation, analysis, core findings. Particularly, Abhijeet Das prepared methodology, Data collection and analysis, core findings and conclusions, writing review-editing and supervision.

References

Abdo HG, Almohamad H, Al Dughairi AA, Al-Mutiry M (2022) GIS-based frequency ratio and analytic hierarchy process for forest fire susceptibility mapping in the western region of Syria. Sustaintability 14(8):4668

Abdulkerim E, Fufa F, Takala W (2022) Identification of groundwater recharge site using geographical information system and remote sensing: case study of Sude district, Oromia, Ethiopia. Environ Earth Sci 81(2):56. https://doi.org/10.1007/s12665-022-10170-w

Agarwal B, Pokhriyal S, Vats S, Sharma V, Rawat P, Bajaj M (2023) Crop prediction using ensemble learning. In: 2023 5th International conference on inventive research in computing applications (ICIRCA). IEEE, pp 90–95

Ahmadi H, Kaya OA, Babadagi E, Savas T, Pekkan E (2020) GIS-based groundwater potentiality mapping using AHP and FR models in Central Antalya. Turkey Environ Sci Proc 5(1):11. https://doi.org/10.3390/iecg2020-08741

Anteneh ZL, Alemu MM, Bawoke GT, Kehali AT, Fenta MC, Desta MT (2022) Appraising groundwater potential zones using geospatial and multi-criteria decision analysis (MCDA) techniques in Andasa-Tul watershed, upper Blue Nile basin, Ethiopia. Environ Earth Sci 81:1–20

Anusha BN, Babu KR, Kumar BP, Kumar PR, Rajasekhar M (2022) Geospatial approaches for monitoring and mapping of water resources in semi-arid regions of Southern India. Environ Chall 8(100569):100569. https://doi.org/10.1016/j.envc

APHA (American Public Health Association) (2017) Standard methods for the examination of water and wastewater, 23rd edn. American Public Health Association, Washington, DC

Asmare D, Tesfa C, Zewdie MM (2023) A GIS-based landslide susceptibility assessment and mapping around the Aba Libanos area, northwestern Ethiopia. Appl Geomatics. https://doi.org/10.1007/s12518-023-00499-7

Barroso GR, Pinto CC, Gomes LNL, Oliveira SC (2024) Assessment of water quality based on statistical analysis of physical-chemical, biomonitoring and land use data: Manso river supply reservoir. Sci Total Environ 912:169554

Basharat Ul M, Khan JA, Khalil U et al (2022) Ensuring earthquake proof development in a swiftly developing region through neural network modeling of earthquakes using nonlinear spatial variables. Buildings. https://doi.org/10.3390/buildings12101713

Bassi N (2023) Use of water accounts based indicators for water management in India. River 2(3):307–325

Brhane GK, Mekonen HS (2024) Identifying controlling factors for the evolution of shallow groundwater chemistry of Ellala catchment, northern Ethiopia. Sustain Water Resour Manag 10(2):1–20

Callow JN, Hipsey MR, Vogwill RIJ (2020) Surface water as a cause of land degradation from dryland salinity. Hydrol Earth Syst Sci 24:717–734. https://doi.org/10.5194/hess-24-717-2020

Chauhan SS, Trivedi MK (2023) Assessment of surface water quality index of gwalior-chambal region using fuzzy-based approach. J Inst Eng India Series A 104(1):1–17

Choi JR, Sung JH, Kang DH, Choo KS, Kim BS (2023) A study on the priority selection method for underground dam installation considering humanities and social factors using fuzzy analytic hierarchy process in Korea. Water 15(18):3296

da Silva Monteiro L, de Oliveira-Júnior JF, Ghaffar B et al (2022) Rainfall in the urban area and its impact on climatology and population growth. Atmos Basel. https://doi.org/10.3390/atmos13101610

Daba AW, Qureshi AS (2021) Review of soil salinity and sodicity challenges to crop production in the lowland irrigated areas of Ethiopia and its management strategies. Land 10(12):1377. https://doi.org/10.3390/land10121377

Das A (2022a) Multivariate statistical approach for the assessment of water quality of Mahanadi basin, Odisha. Mater Today Proceed 65:A1–A11

Das A (2022b, August) Water criteria evaluation for drinking purposes in Mahanadi river basin, Odisha. In: International conference on trends and recent advances in civil engineering. Singapore: Springer Nature Singapore. pp 237–260

Das A (2022c, December). Surface water quality modelling using water quality index (WQI) and geographic information system (GIS) on the Mahanadi Basin, Odisha. In: International conference on water technologies. Singapore: Springer Nature Singapore. pp 21–46

Das A (2023a) Anthropogenic effects on surface water quality assessment in Baitarani River Basin, Odisha using GIS and MCDM techniques. Eng Res Trans 5:37–64

Das A (2023b) Identification of surface water contamination zones and its sources on Mahanadi River, Odisha using entropy-based WQI and MCDM Techniques. Eng Res Trans 4:67–92

Das A (2023c) Assessment of potability of surface water and its health implication in Mahanadi Basin, Odisha. Materials today: proceedings

Das A (2023d) Modelling of surface water quality and spatial mapping in Mahanadi Basin, Odisha. J Mines Metals Fuels, 71(9)

Das A (2023e) Characterization of surface water quality using water evaluation indices, EDAS and geo-statistics in Brahmani River Basin (BRB), Odisha, India. In: Hydraulic and civil engineering technology VIII. IOS Press. pp 826–834

Das A (2023f) Abundance, characterization, understanding water quality degradation and strategies for pollutant reduction in Mahanadi River Basin, Odisha. https://doi.org/10.3233/ATDE230857

Das A (2023g, December) A comparative assessment and decision-making approach of water quality evaluation of Mahanadi Basin, Odisha. In: Materials science forum. Trans Tech Publications Ltd. Vol 1111, pp 159–176

Das A (2023h) Assessing surface water quality for drinking water supply using hybrid GIS-based water quality index (WQI) in Mahanadi River Basin (MRB), Odisha, India. AIJR Proceedings, pp 1–21

Das A (2024a) Assessment of surface water quality using entropy-WQI, fuzzy-topsis analysis, irrigation indices and spatial interpolation approaches in Mahanadi River Basin, Odisha India. J Environ Nanotechnol 13(1):182–212

Das A (2024b) Water quality assessment using water quality index (WQI) under GIS framework in Brahmani Basin, Odisha. In: International conference innovation in smart and sustainable infrastructure. Springer, Singapore. pp 131–150

Devkota KP, Devkota M, Rezaei M, Oosterbaan R (2022) Managing salinity for sustainable agricultural production in salt-affected soils of irrigated dry lands. Agric Syst 198:103390. https://doi.org/10.1016/j.agsy.2022.103390

Feng W, Qian H, Panpan X, Hou K (2020) Hydrochemical characteristic of groundwater and its impact on crop yields in the Baojixia irrigation area. China. Water 12:1443. https://doi.org/10.3390/w12051443

Fentahun A, Mechal A, Karuppannan S (2023) Hydrochemistry and quality appraisal of groundwater in Birr River Catchment, Central Blue Nile River Basin, using multivariate techniques and water quality indices. Environ Monit Assess 195:655. https://doi.org/10.1007/s10661-023-11198-6

Gao M, Chen X, Singh SK, Wei L (2022) An improved method to estimate the rate of change of streamflow recession and basin synthetic recession parameters from hydrographs. J Hydrol 604:127254. https://doi.org/10.1016/j.jhydrol.2021.127254

Gautam U, Tiwari V, Tripathi VK (2022) Evaluation of groundwater quality of Prayagraj city using entropy water quality index (EWQI) and new integrated water quality index (IWQI). Sustain Water Resour Manag 8(2):57

Gupta S (2022) AHP-based multi-criteria decision-making for forest sustainability of lower Himalayan foothills in northern circle, India—a case study. Environ Monit Assess 194(12):1–11. https://doi.org/10.1007/s10661-022-10510-0

Hagos YG, Andualem TG (2021) Geospatial and multicriteria decision approach of groundwater potential zone identification in Cuma sub-basin Southern Ethiopia. Heliyon 7(9):e07963. https://doi.org/10.1016/j.heliyon.2021.e07963

Hassani A, Azapagic A, Shokri N (2021) Global predictions of primary soil salinization under changing climate in the 21st century. Nat Commun 12:6663. https://doi.org/10.1038/s41467-021-26907-3

Ifediegwu SI (2021) Assessment of groundwater potential zones using GIS and AHP techniques: a case study of the Lafia district Nasarawa State Nigeria. Appl Water Sci. https://doi.org/10.1007/s13201-021-01556-5

Islam ARM, Shuanghe S, Bodrud-Doza M, Atiqur MR, Samiran D (2017) Assessment of trace elements of groundwater and their spatial distribution in Rangpur district, Bangladesh. Arab J Geosci 10(4). https://doi.org/10.1007/s12517-017-2886-3

Jin L, Whitehead PG, Bussi G, Hirpa F, Taye MT, Abebe Y, Charles K (2021) Natural and anthropogenic sources of salinity in the Awash river and Lake Beseka (Ethiopia): modelling impacts of climate change and lake-river interactions. J Hydrol Reg Stud 36:100865. https://doi.org/10.1016/j.ejrh.2021.100865

Kaur L, Rishi MS, Singh G, Nath S (2020) Groundwater potential assessment of an alluvial aquifer in Yamuna subbasin (Panipat region) using remote sensing and GIS techniques in conjunction with analytical hierarchy process (AHP) and catastrophe theory. Ecol Indic 110:1–19. https://doi.org/10.1016/j.ecolind.2019.105850

Khan MYA, ElKashouty M, Tian F (2022) Mapping groundwater potential zones using analytical hierarchical process and multicriteria evaluation in the Central Eastern Desert Egypt. Water 14(7):1041. https://doi.org/10.3390/w14071041

Laks I, Walczak Z, Walczak N (2023) Fuzzy analytical hierarchy process methods in changing the damming level of a small hydropower plant: Case study of Rosko SHP in Poland. Water Resour Ind 29:100204

Laltu AD (2023) Drinking suitability and source apportionment of physical parameters in surface water in Mahanadi Basin, Odisha. Odisha April 14

Li H, Wang D, Singh VP, Wang Y, Wu J, Wu J (2021) Developing an entropy and copula-based approach for precipitation monitoring network expansion. J Hydrol 598:126366. https://doi.org/10.1016/J.JHYDROL.2021.126366

Mahammad S, Islam A, Shit PK (2022) Geospatial assessment of groundwater quality using entropy-based irrigation water quality index and heavy metal pollution indices. Environ Sci Pollut Res, 1–24

Manzar MS, Benaafi M, Costache R, Alagha O, Mu'azu ND, Zubair M, Abba SI (2022) New generation neurocomputing learning coupled with a hybrid neuro-fuzzy model for quantifying water quality index variable: a case study from Saudi Arabia. Eco Inform 70:101696

Masoud M, El Osta M, Alqarawy A et al (2022) Evaluation of groundwater quality for agricultural under different conditions using water quality indices, partial least squares regression models, and GIS approaches. Appl Water Sci 12:244. https://doi.org/10.1007/s13201-022-01770-9

Mebrahtu TK, Banning A, Girmay EH et al (2021) The effect of hydrogeological and hydrochemical dynamics on landslide triggering in the central highlands of Ethiopia. Hydrogeol J 29:1239–1260. https://doi.org/10.1007/s10040-020-02288-7

Mohamed NA, Wachemo AC, Karuppannan S, Duraisamy K (2022) Spatio-temporal variation of groundwater hydrochemistry and suitability for drinking and irrigation in Arba Minch Town, Ethiopia: An integrated approach using water quality index, multivariate statistics, and Gis. Urban Clim. 46:101338. https://doi.org/10.1016/j.uclim.2022.101338

Mohan U, Krishnakumar A (2022) Geospatial distribution of groundwater quality using entropy water quality index and statistical assessment: A study from a tropical climate river basin. Environ Qual Manage 32(2):269–285

Mukhopadhyay R, Sarkar B, Jat HS, Sharma PC, Bolan NS (2020) Soil salinity under climate change: Challenges for sustainable agriculture and food security. J Environ Manage 280:111736

Naves A, Samper J, Pisani B, Mon A, Dafonte J, Montenegro L (2021) Hydrogeology and groundwater management in a coastal granitic area with steep slopes in Galicia (Spain). Hydrogeol J 29(8):2655–2669

Pandey HK, Tiwari V, Kumar S, Yadav A, Srivastava SK (2020) Groundwater quality assessment of allahabad smart city using GIS and water quality index. Sustain Water Resour Manag. https://doi.org/10.1007/s40899-020-00375-x

Panneerselvam B, Ravichandran N, Kaliyappan SP, Karuppannan S, Bidorn B (2023) Quality and health risk assessment of groundwater for drinking and irrigation purpose in semi-arid region of India using entropy water quality and statistical techniques. Water 15(3):601

Pati SS, Nayak S, Mishra S, Panda BS, Mahala SS, Mohanty SK, Murugesan K (2023) A comprehensive study of the estuary sea environment in the Bay of Bengal, near the Mahanadi River confluence. Discov Water 3(1):20

Pattanayak AA, Swain S, Behera RR, Sharma SD, Panda CR, Mohanty PK (2024) Variability in water quality of two meso-tidal estuaries of Odisha, East Coast of India. J Mar Syst 241:103919

Rajasekhar M, Sudarsana RG, Sreenivasulu Y, Siddi RR (2019) Delineation of groundwater potential zones in semi-arid region of Jilledubanderu river basin, Anantapur District, Andhra Pradesh, India using fuzzy logic, AHP and integrated fuzzy-AHP approaches. HydroResearch 2:97–108. https://doi.org/10.1016/j.hydres.2019.11.006

Rajput P, Sinha MK, Nigam GK (2022) Strategic evaluation of water quality monitoring network using GIS-AHP model in a large river system. In: Water resources management and sustainability. Singapore: Springer Nature Singapore. pp. 281–303

Ranganathan M, Karuppannan S, Murugasen B, Brhane GK, Panneerselvam B (2022) Assessment of groundwater prospective zone in Adigrat town and its surrounding area using geospatial technology. In Climate change impact on groundwater resources, pp 387–405. https://doi.org/10.1007/978-3-031-04707-7_21

Rashidi F, Sharian S (2022) A comparative analysis of three multi-criteria decision-making methods for land suitability assessment. Environ Monit Assess 194(9):1–15

Rawat KS, Jeyakumar L, Singh SK, Tripathi VK (2019) Appraisal of groundwater with special reference to nitrate using statistical index approach. Groundw Sustain Dev 8:49–58. https://doi.org/10.1016/J.GSD.2018.07.006

Roy S, Bose A, Mandal G (2022) Modeling and mapping geospatial distribution of groundwater potential zones in Darjeeling Himalayan region of India using analytical hierarchy process and GIS technique. Model Earth Syst Environ 8(2):1563–1584. https://doi.org/10.1007/s12518-022-00468-6

Sachdeva S, Kumar B (2020) Comparison of gradient boosted decision trees and random forest for groundwater potential mapping in Dholpur (Rajasthan) India. Stoch Environ Res Risk Assess 35(2):287–306. https://doi.org/10.1007/s00477-020-01891-0

Samal P, Singarasubramanian SR, Manoj MC, Srivastava J, Dsouza N, Balakrishna K, Ali S (2023) Heavy metal contamination assessment and its associated human health risk evaluation in the Mahanadi River sediments, India. Int J Environ Sci Technol 20(10):10673–10694

Sarkar K, Majumder M (2020) Application of AHP-based water quality index for quality monitoring of peri-urban watershed. Environ Dev Sustain. https://doi.org/10.1007/s10668-020-00651-y

Savari M, Yazdanpanah M, Rouzaneh D (2022) Factors affecting the implementation of soil conservation practices among Iranian farmers. Sci Rep 12:8396. https://doi.org/10.1038/s41598-022-12541-6

Semy K, Romeo M (2021) Quality assessment of Tsurang River water affected by coal mining along the Tsurangkong Range, Nagaland India. Appl Water Sci 11(7):1–11. https://doi.org/10.1007/s13201-021-01444-y

Semy K, Singh MR (2021) Quality assessment of Tsurang River water affected by coal mining along the Tsurangkong Range, Nagaland, India. Appl Wat Sci 11(7):115

Şener E, Şener Ş, Davraz A (2018) Groundwater potential mapping by combining fuzzy-analytic hierarchy process and GIS in Beyşehir Lake Basin, Turkey. Arab J Geosci 11:1–21

Shokri-Kuehni SM, Raaijmakers B, Kurz T, Or D, Helmig R, Shokri N (2020) Water table depth and soil salinization: From pore-scale processes to field-scale responses. Water Resour Res 56(2):e2019WR026707. https://doi.org/10.1029/2019WR026707

Shrestha G, Shakya BM (2023) Water in Fi Ltration rate in the Kathmandu valley of Nepal amidst present urbanization and land-use change. H2Open J 6(1):1–14. https://doi.org/10.2166/h2oj.2023.044

Singh VK, Kumar D, Singh SK, Pham QB, Linh NTT, Mohammed S, Anh DT (2021) Development of fuzzy analytic hierarchy process based water quality model of Upper Ganga river basin India. J Environ Manag 284:111985

Singh RB, Patra KC (2024) Integrated PCA–RNN approach for surface water quality assessment in the Mahanadi river system. Int J Environ Sci Technol, pp 1–16

Tadesse T, Sharma PD, Ayele T (2022) Effect of the irrigation interval and nitrogen rate on yield and yield components of onion (Allium cepa L.) at Arba Minch Southern Ethiopia. Adv Agric 2022:4655590. https://doi.org/10.1155/2022/4655590

Tamiru H, Wagari M (2021) Evaluation of data-driven model and GIS technique performance for identification of Groundwater Potential Zones: a case of Fincha Catchment, Abay Basin Ethiopia. J Hydrol Reg Stud 37:100902. https://doi.org/10.1016/j.ejrh.2021.100902

Tarkegn GB, Jury MR (2020) Changes in the seasonality of Ethiopian highlands climate and implications for crop growth. Atmosphere 11:892. https://doi.org/10.3390/atmos11090892

Tolche AD (2021) Groundwater potential mapping using geospatial techniques: a case study of Dhungeta-Ramis sub-basin Ethiopia. Geol Ecol Landsc 5(1):65–80

Trach R, Trach Y, Kiersnowska A, Markiewicz A, Lendo-Siwicka M, Rusakov K (2022) A study of assessment and prediction of water quality index using fuzzy logic and ANN models. Sustainability 14(9):5656

Uddin MG, Nash S, Rahman A, Olbert AI (2023) Performance analysis of the water quality index model for predicting water state using machine learning techniques. Process Saf Environ Prot 169:808–828

Vellaikannu A, Palaniraj U, Karthikeyan S, Senapathi V, Viswanathan PM, Sekar S (2021) Identification of groundwater potential zones using geospatial approach in Sivagangai district, South India Arabian. J Geosci 14(1):8. https://doi.org/10.1007/s12517-020-06316-4
Verma M, Loganathan VA, Bhatt VK (2022) Development of entropy and deviation-based water quality index: case of River Ganga India. Ecol Ind 143:109319
WHO (2017) Guidelines for drinking-water quality. World Health Organization
Wudad A, Biru D, Darcho A, Mohammed E, Bazie F (2021) Delineation of ground water potential zones using GIS and remote sensing techniques in the case of Korahe zone, Somali regional state Ethiopia. Int J Sci Res Publ (IJSRP) 11(8):412–433. https://doi.org/10.29322/ijsrp.11.08.2021.p11653
Xia J, Zeng J (2022) Environmental factors assisted the evaluation of entropy water quality indices with efficient machine learning technique. Water Resour Manage 36(6):2045–2060
Xiao K, Tamborski J, Wang X, Feng X, Wang S, Wang Q, Li H (2022) A coupling methodology of the analytic hierarchy process and entropy weight theory for assessing coastal water quality. Environ Sci Pollut Res, 1–18
Yan J, Chen J, Zhang W (2022) Impact of land use and cover on shallow groundwater quality in Songyuan city, China: a multivariate statistical analysis. Environ Pollut 307(May):119532. https://doi.org/10.1016/j.envpol.2022.119532
Zeng Q, Luo X, Yan F (2022) The pollution scale weighting model in water quality evaluation based on the improved fuzzy variable theory. Ecol Ind 135:108562
Zewdie MM, Yeshanew SM (2023) GIS based MCDM for waste disposal site selection in Dejen town Ethiopia. Environ Sustain Indic 18:100228
Zhang S, Wei J, Li Y, Duan M, Nwankwegu AS, Norgbey E (2021) The influence of seasonal water level fluctuations on the soil nutrients in a typical wetland reserve in Poyang Lake China. Sustainability 13:3846. https://doi.org/10.3390/su13073846
Zhao X, Liu X, Xing Y, Wang L, Wang Y (2022) Evaluation of water quality using a Takagi-Sugeno fuzzy neural network and determination of heavy metal pollution index in a typical site upstream of the yellow river. Environ Res 211:113058

Chapter 20
A Hydrologic Modeling Assessment of Future Water Scarcity in the Baitarani River Basin

Sushree Swagatika Swain, Ashok Mishra, Chandranath Chatterjee, and Morgan C Levy

Abstract Water scarcity, defined as an adverse imbalance between freshwater availability and water demand, is a critical concern in the 21st century, and is exacerbated by climate change and increased uncertainties in hydrological cycles. Rising water demands driven by rapid population growth, urbanization, and industrialization further intensify water scarcity, posing significant threats to sustainable development. Addressing water scarcity is particularly challenging in regions with limited data availability, and data constraints can necessitate the use of a water balance approach to quantify water scarcity. Evaluating the hydrological balance of river basins in historical and projected future time periods requires the use of hydrologic and climatic models, wherein the selection of an appropriate model and modelling approach depends on research objectives, data accessibility, site-specific challenges, costs, and model accuracy. In response to these conditions, this study explores hydrological model-based estimates of water scarcity risk in the absence of sector-specific water demand data. Specifically, this study evaluates streamflow, 'blue' and 'green' water balance components derived from streamflow, and a risk-based measure of water scarcity calculated from those water balance components in the Baitarani river basin (13,000 km^2), a basin influenced by climate and land use change. This analysis relies on historical observed climate and hydrologic data (1974–2018) from the Indian Meteorological Department (IMD) and Central Water Commission (CWC), modelled future climate (2020–2064) from the Coupled Model Intercomparison Project 6 (CMIP6), and the Soil Water Assessment Tool (SWAT) hydrologic model.

S. S. Swain (✉)
Scripps Institution of Oceanography, University of California San Diego, San Diego, CA, USA
e-mail: ssswain@ucsd.edu

A. Mishra · C. Chatterjee
Agricultural and Food Engineering Department, Indian Institute of Technology Kharagpur, Kharagpur, India
e-mail: amishra@agfe.iitkgp.ac.in; cchatiit@gmail.com

M. C. Levy
Scripps Institution of Oceanography and School of Global Policy and Strategy, University of California San Diego, San Diego, CA, USA
e-mail: mclevy@ucsd.edu

A. Nanda et al. (eds.), *Navigating the Nexus*, Water Science and Technology Library 102, https://doi.org/10.1007/978-3-031-76532-2_20

This study finds that Baitarani basin water scarcity has decreased in recent decades (1994–2018) relative to earlier decades (1974–1993), and that water scarcity risk will also decrease in the future (2020–2064). Nevertheless, findings remain uncertain due to limited model structural representation of hydrologic processes that may change alongside land use and climate change drivers. Even so, this study offers valuable insights for parsimoniously evaluating basin-scale water availability and scarcity, aiding in the development of adaptive strategies for sustainable water management in data-scarce river basins.

Keywords Water scarcity · Water availability · Climate change · Blue water flow · Green water flow · Green water storage · CMIP6

20.1 Introduction

The World Economic Forum (WEF) has identified water scarcity as one of the top three global systemic risks of utmost concern, based on a comprehensive global survey encompassing stakeholders from business, academia, civil society, governments, and international organizations (World Economic Forum 2014). Water scarcity, defined as a deficit in the supply of freshwater relative to water demand, is a critical concern in water-stressed regions, as water availability and access have extensive impacts on both human societies and ecosystems (Pedro-Monzonís et al. 2015; Veldkamp et al. 2017; Sun and Zhou 2020). Concerns about water scarcity are amplified in regions with rapid economic development, particularly in emerging economies (Abedzadeh et al. 2020). Increasing water demand due to population growth, industrialization, and agricultural expansion, coupled with climate change and anthropogenic activities, intensify water scarcity challenges (Wada et al. 2011; UN World Water Development Report 2021). Urbanization along with climate change can accelerate water scarcity at local to regional scales (Wu et al. 2023), where water scarcity affects food security, access to safe drinking water, and public health (Taylor 2009). Furthermore, many international and interstate river basins are expected to face water scarcity in coming decades (Beck and Bernauer 2011; Gain and Giupponi 2015), posing additional challenges to transboundary water management (Kryston et al. 2022).

The quantification of "blue" surface—and groundwater and "green" rainwater as sources of available water is important for addressing the impacts of water scarcity (Hoekstra 2016), but water scarcity is determined by both physical and social factors, and their interactions. Thus, understanding and addressing water scarcity also requires assessments that extend beyond simple water supply and demand analyses to include socio-economic factors such as governance systems and policies (Falkenmark et al. 2007). Ideally, water scarcity research considers the complex interactions of climate change, variations in human activities, and their impacts on hydrologic responses. "Socio-hydrologic" research (Sivapalan et al. 2012) that bridges coupled physical and social dynamics that govern water scarcity, however, faces challenges such as data scarcity and anthropogenic feedback (Müller and Levy, 2019). For instance, assessing water scarcity risk under the non-availability of basin-scale sectoral water demands and socio-economic factors limits capacity to execute socio-hydrologic research in data-scarce river basins (Swain et al. 2020). Thus, water scarcity risk research that relies on modelling of physical climate and readily observable proxies for various forms of human activity, such as Land Use/ Land Cover (LULC), are a practical and parsimonious approach to analysing water scarcity given inherent complexity.

Model-based analyses of water scarcity frequently acknowledge the multifaceted influences of climate change and/or LULC dynamics on hydrological systems and provide straightforward management-relevant findings (Ayeni et al. 2015; Chanapathi & Thatikonda, 2020; Dosdogru et al. 2020; Omer et al. 2020; Mechal et al. 2022; Dolgorsuren et al. 2024; Li et al. 2024). Modelling can be used quantify the dependence of hydrological processes on precipitation patterns, temperature regimes, and LULC variations (Chen et al. 2020; Jose et al. 2021; Kayitesi et al. 2022; Gupta et al. 2024), and simulations of change in hydrology enable stakeholders to anticipate shifts in water availability and plan adaptation measures accordingly. A widely used hydrological model for the estimation of hydrological response to climate variability and change, as well as LULC, is the Soil and Water Assessment Tool (SWAT), which is a which is a quasi-distributed, physically-based hydrologic model (Srinivisan et al. 1998; Tan et al. 2020). SWAT model analyses have been used to assess quality and quantity of surface and ground water, the impact of LULC and climate change on hydrologic responses, prevention and control of soil erosion, control of point and non-point source pollution, ecosystem assessment and regional river basin management (Tan et al. 2020; Aloui et al. 2023; Dubey et al. 2023; Uniyal et al. 2023; Zhao et al. 2024). Given that uncertainties in climate projections and LULC scenarios challenge accurate prediction within any hydrologic model (Wagener et al. 2010), including but not limited to SWAT, the integration of risk-based approaches into water scarcity analysis with hydrological model output is necessary (Veldkamp et al. 2017).

While a focus on LULC within hydrologic modelling can simplify representations of human activity, and model input and output uncertainty can be acknowledged to a certain extent using a risk framing, fundamental modelling design challenges remain. Specifically, land use and climate change dynamics are embedded in hydrologic records that represent reference conditions in model experiments. Identifying points of change in hydrologic records over time remains an essential step in many hydrologic modelling efforts, as the selection of "baseline" and "assessment" periods underpin quantifications of change. Thus, use of 'change point', 'break-point', or 'turning point' approaches, which indicate points in time at which time series data split into subsets with distinct statistical properties, can be used to establish comparative time periods (Friedman et al. 2016).

Change points within datasets indicate non-stationarities, a concept widely employed for identifying and evaluating the timing and magnitudes of hydro-climatic change at multiple scales (Xie et al. 2019; Zhou et al. 2019), and can indicate abrupt shifts in variables like precipitation, streamflow, reservoir storage, urbanization or agricultural practices (Ryberg et al. 2020). As a result, change points in long-term hydro-climatic data have been used to evaluate water scarcity across comparison time periods (Rafiei-Sardooi et al. 2022). There are several change point detection methods such as the Bayesian Change-Point (BCP) approach (Wang et al. 2018), parametric change-point analysis (Killick and Eckley, 2014), non-parametric change-point analysis (Haynes et al. 2021), and the trend-based approach (Pettitt test) (Pettitt, 1979). However, selecting suitable change point detection methods and then identifying change points within long-term hydro-climatic data are both processes that are subject to inherent uncertainties and limitations related to data quality and availability (He et al. 2022). At a minimum, approaches that are suitable for hydrologic applications avoid invalid statistical assumptions (i.e., normality and independence) and have the capacity to detect multiple change points (Ryberg et al. 2020). Ideally, approaches can identify sequential trends with varying lengths and slopes (Şen, 2019).

It remains that the accuracy of water scarcity risk estimation and projection at the basin scale alone depends on myriad model structural and forcing data factors: representations of time-varying climate, LULC, hydrologic processes, and human intervention. Nevertheless, applied management-relevant research must proceed with estimation in the absence of one or more of these factors. Thus, the question of what parsimonious hydrologic modelling approaches yield in river basins, despite limitations, is valid insofar as the model results are used to understand the future 'possibility space' (Baldassarre et al. 2019). Thus, based on the above critical appraisal, the present study aims to analyse water availability and water scarcity conditions in long-term historical as well as future scenarios in the Baitarani river basin of Eastern India, a basin that has undergone substantial land use change and faces significant projections of hydroclimatic change (Srivastava and Maity, 2023).

This study does so using a SWAT model simulation approach, and using data inputs that include 45 years of observed historical climate and hydrologic data (1974–2018), historical land cover maps (1985, 2005), and 40 years of projected future climate (2020-2059) from two Shared Socioeconomic Pathway (SSP) scenarios from the Coupled Model Intercomparison Project 6 (CMIP6). The flowchart of the overall methodology is shown in Figure 20.1. This study narrates the development of an applied hydrologic model-driven analysis of climate and land use change in a historical reference period, and climate change (with static land use) in a future period, as well as comparisons across time periods and future scenarios. Results from this study illustrate what is, and is not, typically possible in the case of an applied water scarcity risk analysis in a data-scarce basin. Furthermore, model simulation findings and their interpretation highlight the potentially important but often omitted role of model structural representations within management-focused studies of water resources change.

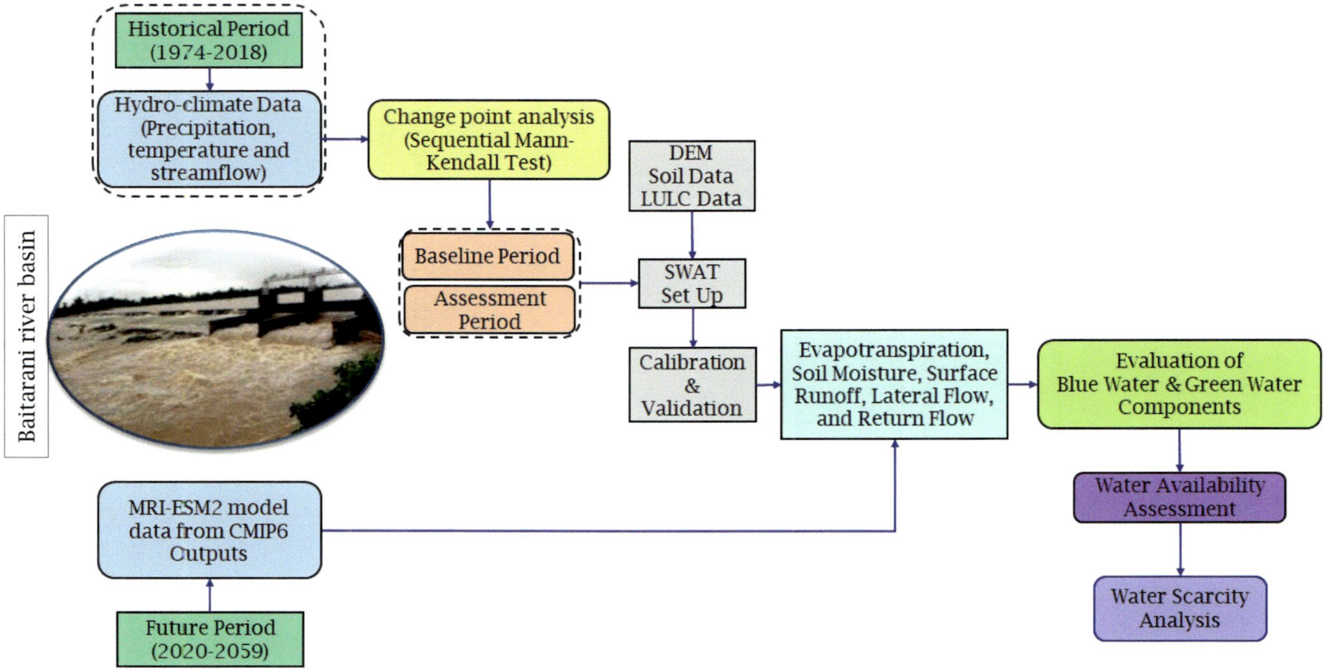

Fig. 20.1 Flowchart for the overall methodology framework of this study.

20.2 Background of the Study Area

The Baitarani river basin is an interstate basin in Eastern India that spans approximately 12,094 km^2, and the river has a travelling length of 360 km before draining into the Bay of Bengal. It extends from 20°30′ N to 22°20′ N in latitude and 85°05′ E to 87°05′ E in longitude (Figure 20.2). Covering primarily the regions of Odisha and a smaller portion of Jharkhand, the Baitarani river basin is a major regional catchment known for its agriculture, fisheries and industrial sectors, and the basin includes a wide variety of land-use types such as agricultural land, water bodies, fallow and wasteland, forested areas and developed/urban land (Visakh et al. 2019). In this basin, the Anandapur streamflow gauging station is located upstream of the Salandi Dam (Figure 20.2), constructed between 1974 and 1983, enabling assessment of variations in unimpaired hydrologic responses with climate change and LULC (Swain et al. 2020). The Salandi Dam is the major water management structure in the Baitarani river basin, storing floodwaters during the monsoon season and supplying water to agriculture and industry during periods of high demand.

The Baitarani river basin falls within a subtropical monsoon climate zone, and the monsoon months (June–September) are the primary source of precipitation, contributing approximately 80% of annual rainfall (Swain et al. 2020). Earlier studies have identified the study area as flood-prone (Rai et al. 2018; Dahm et al. 2019) and simultaneously drought-prone according to projections of increased drought occurrence and magnitudes under future climate change (Ojha et al. 2013; Suman and Maity, 2021). Studying water availability and water scarcity in the context of these divergent and changing extremes poses an illustrative modelling challenge.

20.3 Data Sources

The necessary input data for SWAT consists of the Digital Elevation Model (DEM), LULC maps, soil and meteorological data, and streamflow time series data for model calibration and validation. The details of the data are given as follows:

20.4 Spatial Data Sources

The DEM is sourced from the Shuttle Radar Topography Mission (SRTM), offering a spatial resolution of 30 m × 30 m (Farr et al. 2007; NASA JPL, 2013). Statics derived from soil property maps rely on the Food and Agriculture Organization (FAO) Harmonized World Soil Database (HWSD) v1.2 (Fisher et al. 2008; FAO, 2012), with a spatial scale of 1 km by 1 km. The Decadal LULC Classifications across India dataset provides satellite-derived LULC classifications at a decadal temporal resolution and with a spatial resolution of 100 m by 100 m (Roy et al. 2016). To

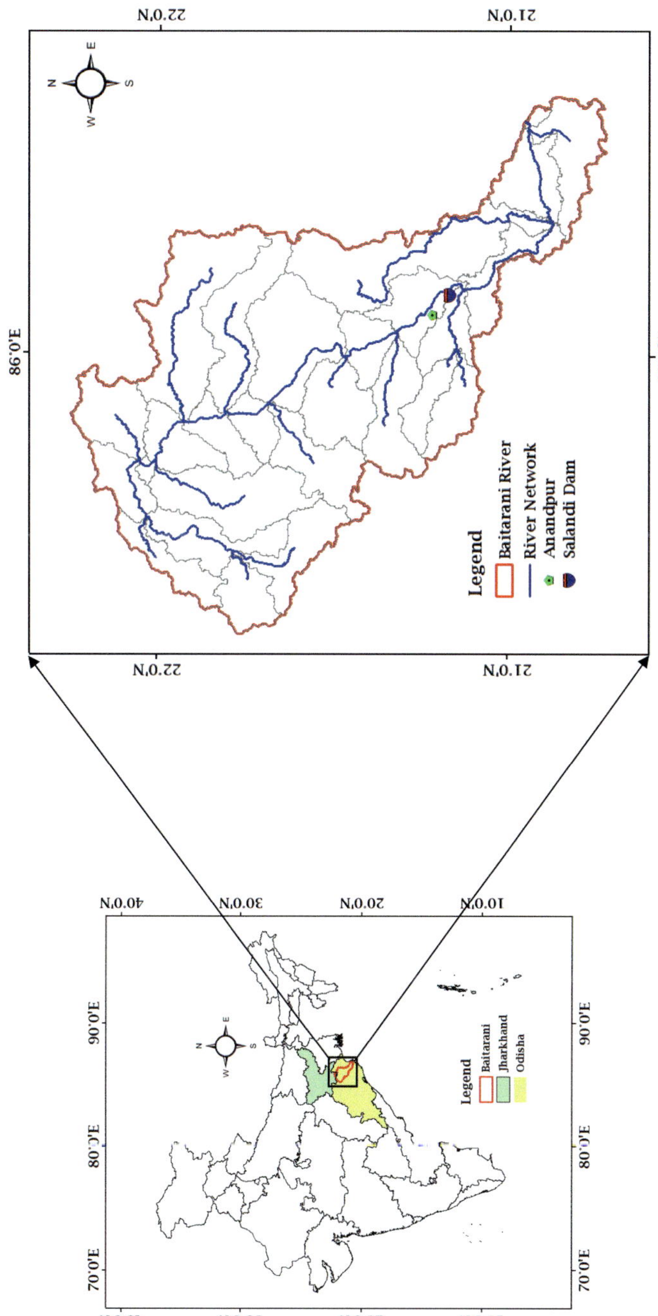

Fig. 20.2 Location and—map of the Baitarani river basin.

analyse the impact of LULC changes during the reference period, representative land cover maps for the years 1985 and 2005 are employed.

20.5 Historical Hydro-Climatic Data Sources

The time period represented by historical data is 1974–2018. The primary observed meteorological data (precipitation, min and max temperature) are obtained on a daily scale from the India Meteorological Department (IMD), Pune (Rajeevan et al. 2006; Pai et al. 2015). The spatial resolution of obtained precipitation data is 0.25°. The obtained 1° gridded temperature data is rescaled to 0.25° resolution through a widely used bilinear interpolation approach (Almazroui et al. 2017, 2020) to maintain spatial consistency with the obtained precipitation data (Swain et al. 2021; Dey et al. 2022). The daily observed streamflow data is collected from the Anandpur (21° 13' 46" N, 86° 7' 30" E) gauging station made available by the Central Water Commission (CWC), Bhubaneswar, and has been extensively used in various applied hydrologic studies (Rai et al. 2018; Visakh et al. 2019; Swain et al. 2023, 2024).

20.6 Future Climatic Data Sources

Global Climate Models (GCMs) are widely employed to simulate and project climate change on both a global and regional scale and are regarded as essential for water resources management research (Fan et al. 2020; Xu et al. 2021). By using standardized simulations within a variety of GCMs, the Coupled Model Intercomparison Project (CMIP) enables quantification of uncertainties in future climate projections across models. GCMs from the sixth phase of CMIP (CMIP6) have delivered substantial enhancements over their previous model generation (CMIP5) in horizontal resolution, physical parameterizations (e.g., cloud representation), the addition of other Earth system processes (like nutrient limitations to the terrestrial carbon cycle) and added components (e.g., ice sheets) (Eyring et al 2016, 2019). A set of socioeconomic and technological development scenarios, collectively known as the Shared Socioeconomic Pathways (SSPs), are developed in CMIP6. SSPs differ from prior CMIP5 Representative Concertation Pathway (RCP) scenarios by describing challenges to adaptation and mitigation, in addition to emissions as presented in the Special Report on Emissions Scenarios (SRES) of the Intergovernmental Panel on Climate Change (IPCC) (O'Neill et al. 2017). The CMIP6 SSP scenarios combine two dimensions, the RCP representation of the severity of climate change (represented by emissions levels), and the SSP narrative characterization of five directions of global societal development (Kriegler et al. 2014). The five SSPs portray alternative pathways of socioeconomic development, such as sustainable development (SSP1), middle-of-the-road development (SSP2), regional rivalry (SSP3), inequality (SSP4) and fossil-fuelled development (SSP5) (Meinshausen et al. 2019).

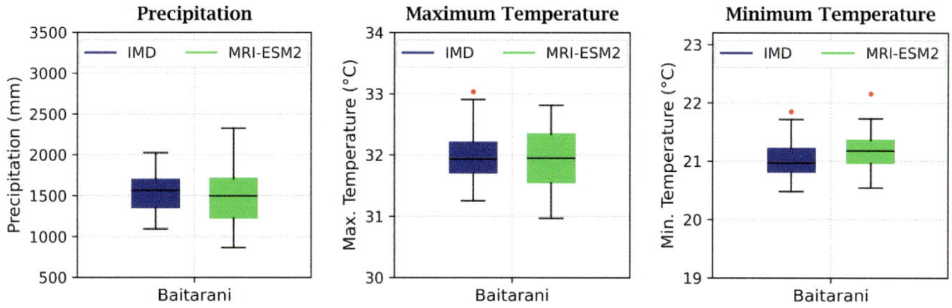

Fig. 20.3 Box and whisker representation of observed IMD and modelled MRI-ESM2 climate variables.

In this study, SSP245 and SSP585 scenarios from CMIP6 are selected to represent future water availability and scarcity under medium and high emission scenarios, respectively. Bias-corrected CMIP6 precipitation, maximum temperature and minimum temperature at a 0.25° × 0.25° spatial resolution and daily temporal resolution come from the Bias Corrected Climate Projections from CMIP6 Models for South Asia dataset (Mishra et al. 2020). These bias-corrected future datasets are widely used for hydro-climatic analysis in the South Asia region (Pandey et al. 2022; Das et al. 2023; Swain et al. 2023). Out of the 13 available GCMs, the MRI-ESM2 model is selected for its superior regional performance, relative to other GCMs, over the modelled historical period (Mahato et al. 2022). Figure 20.3 depicts the annual precipitation and temperature (maximum and minimum) from the observed IMD and modelled MRI-ESM2 spanning the years 1979–2014. There is good agreement between the IMD data and the bias-corrected MRI-ESM2 model data within the study area. Bias-corrected MRI-ESM2 data from the period of 2020–2059 is then used to represent the future period.

20.7 Methodology

20.7.1 Change Point Detection Analysis

This study uses the Sequential Mann–Kendall test for the identification of multiple change points in the historical hydrologic data, and is used to partition the historical period into baseline and assessment periods. The Sequential version of the Mann–Kendall test was initially formulated by Sneyers, 1990, and was subsequently modified and applied in hydrological studies (Douglas et al. 2000; Yang and Tian, 2009; Zhao et al. 2015; Dey and Mishra; 2017; Nourani et al. 2018). Using comparisons between progressive (forward) and retrograde (backward) sequences of ranked time series data, this method detects change points as the onset of trends in those data. The rank of mean annual streamflow magnitudes X_j (where $j = 1, 2, \ldots, n$) are compared to the rank of X_k (where $k = 1, 2, \ldots, j\text{-}1$). The number of events in which the rank of X_k exceeds the rank of X_j is denoted by n_j. The test statistic (T_j) is then calculated based on these counts and given by:

$$T_j = \sum_1^j n_j \ (j = 2, 3, ..., n) \tag{20.1}$$

The test statistics are assumed to be normally distributed. The mean and variance of the test statistics are estimated as:

$$E(T) = \frac{n(n-1)}{4} \tag{20.2}$$

$$\mathrm{Var}(T_j) = \frac{[j(j-1)(2j+5)]}{72} \tag{20.3}$$

The sequential values of the statistic $U(T_j)$ are calculated as:

$$U(T_j) = \frac{T_j - E(T_j)}{\sqrt{\mathrm{Var}(T_j)}} \tag{20.4}$$

which is called the progressive series. Similarly, the retrograde series $U'(T_j)$ is generated in the same way, but with a reversed original time series. The null hypothesis is accepted if $|U(T_j)|$ is less than or equal to $U(T_j)_{1\text{-}\alpha/2}$; where α is the significance level at which the two-sided trend test is performed. The critical value ($U(T_j)_{1\text{-}\alpha/2}$) represents the threshold of the standard normal distribution with a probability greater than $\alpha/2$. A positive $U(T_j)$ indicates a positive trend, while the negative $U(T_j)$ signifies a negative trend, with $U'(T_j)$ reflecting the same pattern as $U(T_j)$. The Sequential Mann–Kendall test in this study is performed at a 5% significance level. The change points in the hydrologic variable (streamflow) are observed at the point where the progressive and retrograde series intersect with each other, and the null hypothesis of there being no trend in the data is rejected for the entire sample if any of the sequential statistics $U(T_j)$ exceed the confidence interval (i.e., +/− 1.96). Here, the intersection point with the greatest sequential statistic value is used to indicate the break point in the long-term streamflow data series. Thus, the study period (i.e. reference period) is partitioned into a baseline period and an assessment period using the year corresponding to the greatest positive value of $U(T_j)$.

20.8 Linear Trend Analysis

The linear trend test is a statistical method frequently used in the hydrology domain to identify and quantify trends in hydro-climatic variables over time, such as precipitation, streamflow, and temperature, providing understanding of long-term changes in water resources and climate patterns (Zhou et al. 2023; Wu et al. 2024). In this study, the linear trend test is performed to evaluate change in annual summaries (e.g., total precipitation, mean streamflow and temperature) of hydro-climatic variables during the reference, baseline and assessment periods, with a significance level set at 5%.

20.9 SWAT Model-Based Scenario Definition

The basin-scale modelling of the study area is done using the SWAT2012 interface (Arnold et al. 2012) with an ArcGIS 10.3-ArcView extension (Wang et al. 2020). SWAT enables representation of various physical hydrological processes, including streamflow, groundwater flow, evapotranspiration, lateral flow and percolation. Streamflow is generated using the SCS curve number method and routed through the Muskingum channel routing method. Evapotranspiration in this study is quantified by the Penman-Monteith method. Subsurface flow components, including groundwater flow, lateral flow and percolation, are assessed through the mass balance of the subsurface system. The conceptualization of hydrological processes in SWAT is presented through the water balance equation, which is formulated as:

$$\mathrm{SW}_t = \mathrm{SW}_o + \sum_{i=1}^{t} (P_i - Q_i - ET_i - G_i - R_i) \quad (20.5)$$

where SW_t is the residual water content in the soil, SW_o is the initial soil water content on ith day, t is time in days, P_i is the precipitation on ith day, Q_i is the surface runoff on ith day, ET_i is the evapotranspiration (ET) on ith day, G_i is the water flowing to the subsurface zone from the soil profile on ith day, and R_i is the return flow on ith day.

SWAT model setups (Table 20.1) are designed around the baseline and assessment time periods that were determined through the change point analysis, and these periods represent alternative climatic and LULC conditions. A warm-up period of 2 years is incorporated for both baseline and assessment periods. The Baitarani river basin is calibrated at the Anandpur gauging station using the SUFI-2 algorithm within the SWAT-CUP interface at a monthly scale (Abbaspour, 2014). Sensitivity analysis of model parameters is conducted using the LH-OAT sampling approach (Van Griensven and Meixner, 2007). The performance of the calibrated and validated models is assessed using three performance indicators: the coefficient of determination (R^2), the Nash-Sutcliffe Efficiency (NSE), and the Percent Bias (PBIAS).

$$R^2 = \left[\sum_{i=1}^{n} (Q_i^o - \overline{Q^o})(Q_i^m - \overline{Q^m})\right]^2 \Bigg/ \sum_{i=1}^{n} (Q_i^o - \overline{Q^o})^2 \sum_{i=1}^{n} (Q_i^m - \overline{Q^m})^2 \quad (20.6)$$

$$\mathrm{NSE} = 1 - \left|\sum_{i=1}^{n} (Q_i^o - Q_i^m)^2 \Bigg/ \sum_{i=1}^{n} (Q_i^o - \overline{Q^o})^2\right| \quad (20.7)$$

Table 20.1 Details of the historical time periods for SWAT modelling.

Time frame	LULC	Climate
Baseline period	1985	1974–1993
Assessment period	2005	1994–2018

$$\text{PBIAS} = \left(\sum_{i=1}^{n} \left(Q_i^m - Q_i^o\right) \times 100\right) \Big/ \sum_{i=1}^{n} Q_i^o \quad (20.8)$$

where $\overline{Q^o}$ is the observed mean streamflow, Q_i^o is the ith observed streamflow, Q_i^m is the ith model-simulated streamflow, $\overline{Q^m}$ is the model-simulated mean streamflow and n is the total number of the data points. The calibrated and validated SWAT model is then used to simulate future hydrology using CMIP6 climate for the period of 2020–2059. Due to the assessed compatibility between historical CMIP6 climate and observed historical climate (Figure 20.2), a SWAT run of historical period CMIP6 climate was deemed unnecessary.

20.10 Blue/ Green Water Availability Analysis

Water availability is assessed using blue and green water components constructed from SWAT outputs. Blue Water Flow (BWF) is calculated by aggregating surface water and groundwater components. Hence, surface runoff/ streamflow, lateral flow and return flow are combined to form the BWF components at a basin scale. The concept of naturally infiltrated water forms the basis of the green water component in a basin, which is estimated using two terms. Firstly, Green Water Storage (GWS) represents the available moisture in the soil. Secondly, the Green Water Flows (GWF) denote the actual evapotranspiration, which is the loss of water from the soil by release to the atmosphere through a combination of transpiration from vegetative and evaporation from water bodies and soil profiles. Therefore, in the present study, BWF and GWS are considered indicators of water availability in that they are used to quantify water scarcity. Further, Water Availability (WA) is defined as the summation of BWF and GWS components (i.e., total water available is surface water and groundwater). It is important to note that the derived estimates for the water balance components are analysed directly from model-simulated outputs due to the lack of available observational data.

20.11 Framework Water Scarcity-Risk Analysis

Water scarcity is usually assessed by comparing sectoral water demands with the available water resources at the basin scale. However, the details of sectoral water demand data are limited in many river basins, including the Baitarani basin. Thus, water scarcity risk assessment is performed in this study based on relative changes in streamflow (ΔQ) and Water Availability (ΔWA) for baseline and assessment periods. It is assumed that there is a state of equilibrium between water demands and water availability for this analysis, which means that the amount of water needed by different sectors (agriculture, industry, households, etc.) is assumed to be balanced by the water available from BWS and GWS. The relative changes in streamflow and water availability are calculated as follows:

$$\Delta Q = \frac{Q_a - Q_b}{Q_b} \tag{20.9}$$

$$\Delta \text{WA} = \frac{\text{WA}_a - \text{WA}_b}{\text{WA}_b} \tag{20.10}$$

The relative change in streamflow (ΔQ) is calculated as the ratio of the difference between the average streamflow for the assessment period (Q_a) and the baseline period (Q_b) to the average streamflow for the baseline period (Q_b). Similarly, the relative change in water availability (ΔWA) is determined as the ratio of the difference between the average water availability for the assessment period (WA_a) and the baseline period (WA_b) to the average water availability for the baseline period (WA_b).

Based on the parameter values outlined above, four distinct risk zones are identified, following the criteria proposed in earlier studies (Garrote et al. 2018; Swain et al. 2020). The criteria for potential water scarcity risk assessment zoning are as follows:

$$\begin{aligned}
\text{Zone1 } (Z_1) &= (\Delta Q > 0,\ \Delta \text{WA} > 0) = \text{Less risk} \\
\text{Zone2 } (Z_2) &= (\Delta Q < 0,\ \Delta \text{WA} > 0) = \text{Moderate risk} \\
\text{Zone3 } (Z_3) &= (\Delta Q > 0,\ \Delta \text{WA} < 0) = \text{High risk} \\
\text{Zone4 } (Z_4) &= (\Delta Q < 0,\ \Delta \text{WA} < 0) = \text{Extreme risk}
\end{aligned}$$

Zone 1 represents low water scarcity risk, wherein water demands exceed water availability, and is represented by positive changes in both streamflow and water availability. Zone 2 signifies a moderate risk of water scarcity due to decreased streamflow (only). Zone 3 indicates a high water scarcity risk due to reduced water availability (only). In zone 4, where both streamflow and water availability decrease, the basin faces an extreme risk of water scarcity and is likely to experience hydrological drought conditions in the region (Goyal et al. 2017). The assessment of water scarcity includes a change point analysis of streamflow, a linear trend analysis of hydro-climate variables, and an evaluation of their impact on water availability.

This impact analysis relies on the calibrated and validated SWAT outputs over the reference (baseline and assessment) period. Water scarcity in the reference period is assessed by comparing water availability and streamflow in the baseline and assessment periods. Then, water scarcity in the future period is assessed by comparing the reference period with SSP245 and SSP 585 hydro-climate variables, respectively.

20.12 Results and Discussions

20.12.1 Change Point Detection Analysis

The Sequential Mann–Kendall test statistic calculated on streamflow data from the Anandapur station in the Baitarani basin identified a total of 11 change points (Table 20.2, Figure 20.4).

Table 20.2 Sequential Mann–Kendal test statistics in the reference period.

Year	Sequential statistic
1978	0.92
1983	2.01
1985	1.63
1989	2.21
1991	1.93
1993	2.32
2002	1.69
2009	1.09
2011	1.11
2012	1.19

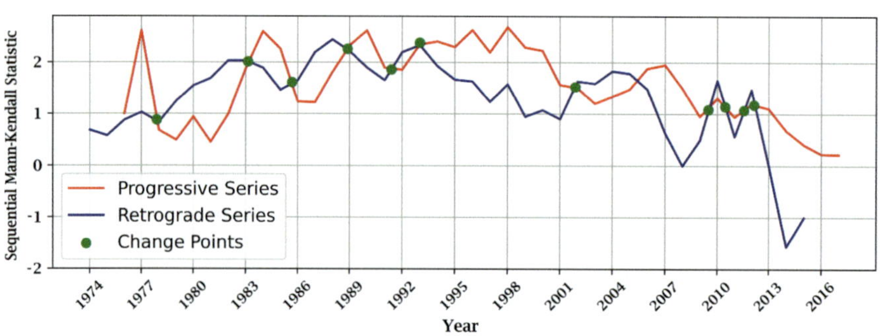

Fig. 20.4 Change points identified by the Sequential Mann–Kendall test of streamflow in the reference period.

Between 2010 and 2013, there are three change points with almost identical sequential statistic values (1.1–1.2). Similarly, in the years 1983 and 1993, sequential statistic values greater than 2 are observed, with the year 1993 exhibiting the greatest value of 2.32. Based on these results, the full reference period of 1974–2018 is divided into two sub-periods: the period before the change point with the greatest corresponding sequential statistic value (1974–1993) is considered to be the baseline period, while the period after (1994–2018) is termed the assessment period.

20.13 Hydro-Climatic and LULC Change Statistics in Reference Period

Comparisons of the trends in the hydro-climatic variables across different periods are illustrated in Figure 20.5. In the reference period from 1974 to 2018, precipitation shows a positive trend ($\Delta = 3.65$ mm/year), though there is no statistical significance (p-value $= 0.2$). Similarly, streamflow has a negligible downward trend ($\Delta = -0.17$ m^3/s/year) during the reference period, which is not statistically significant (p-value $= 0.805$). However, temperature displays a consistent and statistically significant upward trend ($\Delta = 0.01$ °C/year, p-value $= 0.020$) over this period. In the baseline period from 1974 to 1993, precipitation exhibits a significant upward trend ($\Delta = 13.37$ mm/year^2) without any statistical significance (p-value $= 0.102$). Streamflow also demonstrates a significant upward trend ($\Delta = 3.06$ m^3/s/year) during the baseline period, but not significant (p-value $= 0.123$). Similarly, temperature shows no statistically significant trend during this period. Finally, in the assessment period from 1994 to 2018, precipitation displays a slight downward trend ($\Delta = -1.44$ mm/year), though not statistically significant (p-value $= 0.845$). Streamflow, however, reveals a significant downward trend ($\Delta = -3.03$ m^3/s/year), approaching statistical significance (p-value $= 0.088$). Temperature continues its upward trend ($\Delta = 0.02$ °C/year) during the assessment period, and is statistically significant (p-value $= 0.016$). Overall, while precipitation and streamflow exhibit varying trends with statistical insignificance across different periods, temperature consistently displays a significant upward trend, which is especially noticeable in the reference and assessment periods.

The LULC maps of 1985 and 2005 were selected to represent the baseline period and assessment periods, respectively. Summaries of basin LULC classes by time period are shown in Figure 20.6. Between 1985 and 2005, agricultural land increased slightly from 56.22% to 56.89%, indicating continued agricultural expansion. Conversely, deciduous forests experienced a decline from 28.06% to 24.15%, suggesting deforestation or land conversion activities. Fallow land remained relatively stable at 1.35% throughout both periods. Shrubland and mixed forest areas showed modest increases, indicating possible reforestation or natural growth processes. The proportion of water bodies also saw a slight rise from 1.58% to 1.82%, reflecting potential changes in hydrological features. Urban land exhibited a

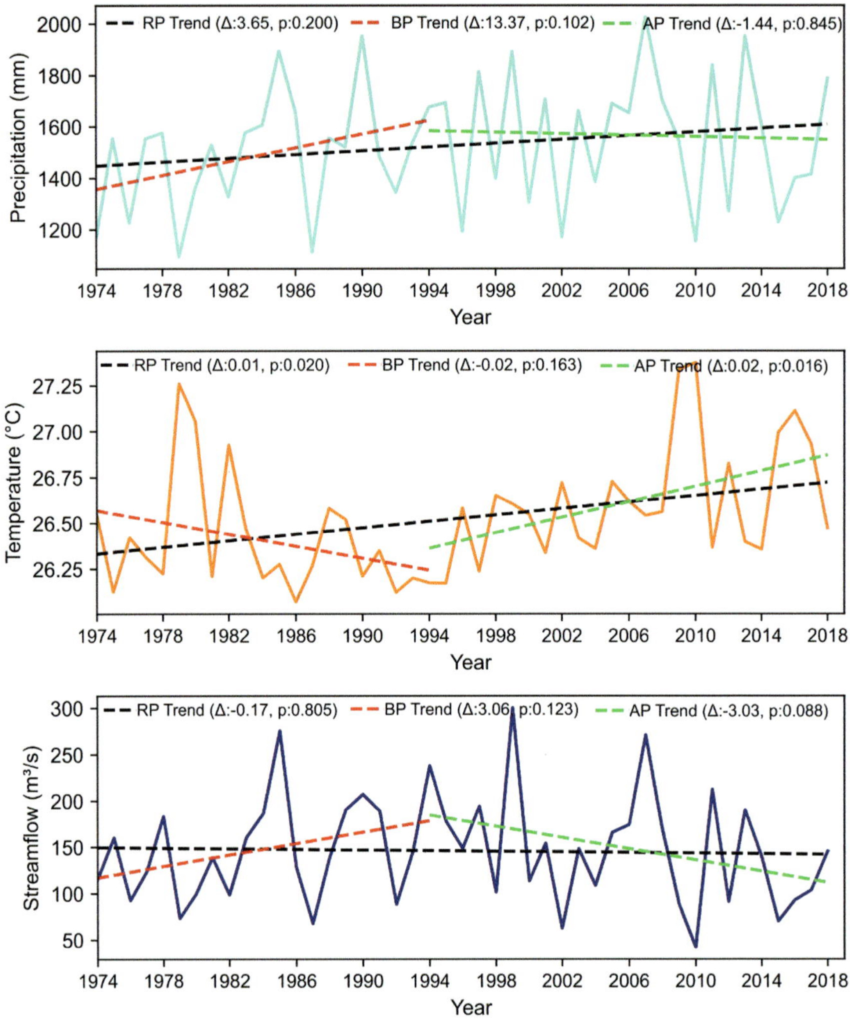

Fig. 20.5 Linear trend variation of annual hydro-climatic variables for different periods.

minor increase from 1.04% to 1.23%, indicating limited urbanization or infrastructure development. These changes highlight shifts in land use patterns over the two decades, with implications for ecosystem health, biodiversity, and water resource management within the study area.

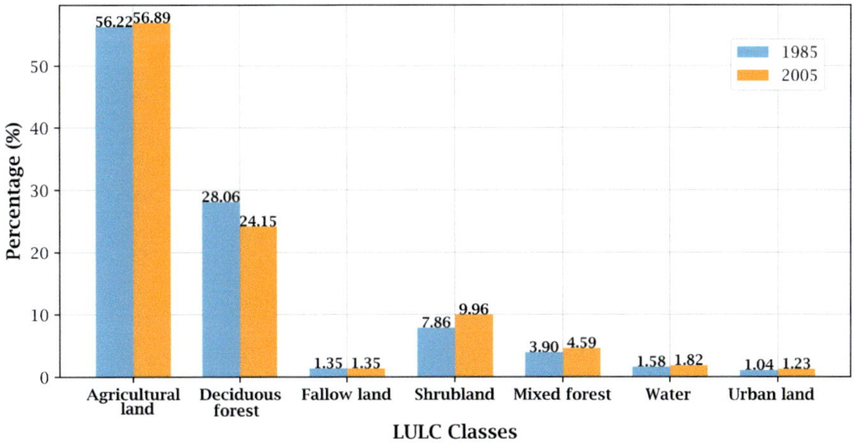

Fig. 20.6 LULC percentages within the Baitarani river basin during baseline (1985) and assessment (2005) years of the reference period.

20.14 SWAT Model Scenario Execution

The two SWAT model setups of baseline and assessment periods, as outlined in Table 20.1, are calibrated and validated using 10 parameters that the model is particularly sensitive to according to earlier literatures (Paul et al. 2019; Visakh et al. 2019; Swain et al. 2020, 2021, 2023). The selected sensitive parameters with suitable ranges and fitted values are detailed in Table 20.3. Sensitive parameters and their ranges are determined through a comprehensive model calibration of the study area. The calibrated peak flow is fine-tuned by adjusting sensitive parameters such as the curve number (CN2.mgt) and available water capacity of the soil layer (SOL_AWC.Sol), drawing from insights gained in earlier studies (Abbaspour et al. 2015). Parameters such as the groundwater delay time (GW_DELAY.gw), threshold depth of water in the shallow aquifer required for return flow to occur (GWQMN.gw), and the groundwater revap coefficient (GW_REVAP.gw) are employed to modify the baseflow of the study area. Changes in LULC are anticipated to influence the variability of sensitive model parameters (Das et al. 2018; Swain et al. 2021), as well as the streamflow response of the basin. The calibration and validation periods are selected based on a ratio of approximately 60:40, respectively. For the baseline period (1974–1993), the model is calibrated from 1974 to 1985 and validated from 1986 to 1993. Similarly, for the assessment period (1994–2018), the model is again calibrated from 1994 to 2008 and validated for the years from 2009 to 2018.

Figure 20.7 illustrates the comparisons of monthly streamflow time series between model simulations and observed streamflow at the Anandapur station. The model results demonstrate an acceptable level of agreement between observed and model-simulated streamflow. The statistical performance indices used to evaluate model performance, including NSE values exceeding 0.5, R2 values above 0.6, and PBIAS

Table 20.3 Best-fit sensitive parameters of model calibration in the study area.

Parameters	Description	Range	Fitted value	
			Baseline period	Assessment period
CN2	Initial SCS runoff number for moisture condition II	(70,90)	80.53	77.30
GW_DELAY	Groundwater delay time (days)	(30,450)	53.94	37.98
ALPHA_BF	Baseflow alpha factor (1/ days)	(0,1)	0.40	0.72
GWQMN	Threshold depth of water in the shallow aquifer required for return flow to occur (mm H_2O)	(0,5000)	2755	2695
CH_K2	Effective hydraulic conductivity in main channel alluvium (mm/h)	(0,200)	11.49	38.70
CH_N2	Manning's n value for the main channel	(0,0.3)	0.07	0.10
ALPHA_BNK	Base flow alpha factor for bank storage (days)	(0,1)	0.47	0.91
SOL_AWC	Available water capacity of the first soil layer (mm H_2O/mm soil)	(0,1)	0.20	0.17
REVAPMN	Threshold of evaporation in the shallow aquifer (mm)	(0,500)	273	19.5
GW_REVAP	Groundwater 'revap' coefficient	(0.02,0.2)	0.11	0.12

values falling within ±25, suggest reliability of the model (Moriasi et al. 2007). The NSE, R^2 and PBIAS values for the calibration period and validation period of the respective model scenarios are outlined in Figure 20.7. During the calibration of the baseline period, the model exhibits an NSE of 0.81 and R^2 of 0.88, indicating a good fit between observed and simulated streamflow values. The PBIAS value of 12.27 suggests a slight overestimation of the observed values. During validation, the model performance improves slightly, with NSE and R^2 values of 0.84 and 0.90, respectively, and a higher PBIAS value of 16.76, indicating a larger overestimation in the validation period. In the calibration period of the assessment period, the model exhibits a slight decrease in performance compared to the baseline period, showing NSE and R^2 values of 0.79 and 0.82, respectively and a negative PBIAS value of − 13.14, indicating an underestimation of observed values. However, in the assessment period validation, the model performance improves, with NSE and R^2 values of 0.83 and 0.88, respectively and a further reduction in PBIAS to − 19.56, indicating a larger underestimation. Overall, the model performance indicators suggest satisfactory performance across different time scales, with some variation in performance metrics between calibration and validation periods and between baseline and assessment periods. Subsequently, the model is simulated for different future SSPs based on the calibrated model parameters of the assessment period.

20.14.1 Assessment of Water Availability

Figure 20.8 presents summaries of estimated blue and green water components (BWF, GWF and GWS), alongside precipitation, across the different time periods and SSPs. In the reference period (baseline and assessment periods combined), the average precipitation is recorded as 247.02 mm, with corresponding values for BWF, GWF, and GWS of 201.055, 21.3 and 24.76 mm, respectively. In the baseline period, there is slightly less precipitation of 245.38 mm, while other components also exhibit minor differences. In the assessment period, precipitation increases very slightly to 248.29 mm, leading to a corresponding slight rise in BWF, GWF and GWS. SSP585 exhibits significantly higher precipitation and BWF, while maintaining the same GWF and only slightly higher GWS compared to the reference period. Both SSP scenarios show similar GWS levels, which are only marginally higher than those in the reference period. This indicates that evaporative and subsurface processes as represented in the SWAT model are unchanged from the reference period, and thus GWS changes only to the extent that increases in BWF propagate through the representation of those processes.

Future SSP simulations only include future climate change in this study, and calibrated SWAT representation of dynamical land surface and subsurface hydrologic processes are unchanged from the reference period. In reality, future LULC dynamics can significantly impact water availability by altering surface runoff, groundwater recharge and evapotranspiration processes, as documented in earlier studies conducted in various regions around the world (Levy et al. 2018; Tankpa et al. 2021; Sinha et al. 2023; Ebodé et al. 2024). For example, the expansion of urban areas or conversion of natural habitats to agricultural land can lead to increased surface runoff and reduced infiltration rates, potentially exacerbating flood risks and reducing groundwater recharge. Conversely, afforestation efforts or restoration of degraded landscapes may enhance water retention and infiltration capacities, mitigating runoff and improving groundwater replenishment (Brown et al. 2005; Zhang et al. 2017). From a modelling perspective, such changes would alter calibration of SWAT model parameters, resulting in change additional to that which is generated by climate alone. It remains that this study's representation of the two baseline and assessment period LULC types, along with their corresponding observed climate, yielded only marginal change in water availability components over the reference period.

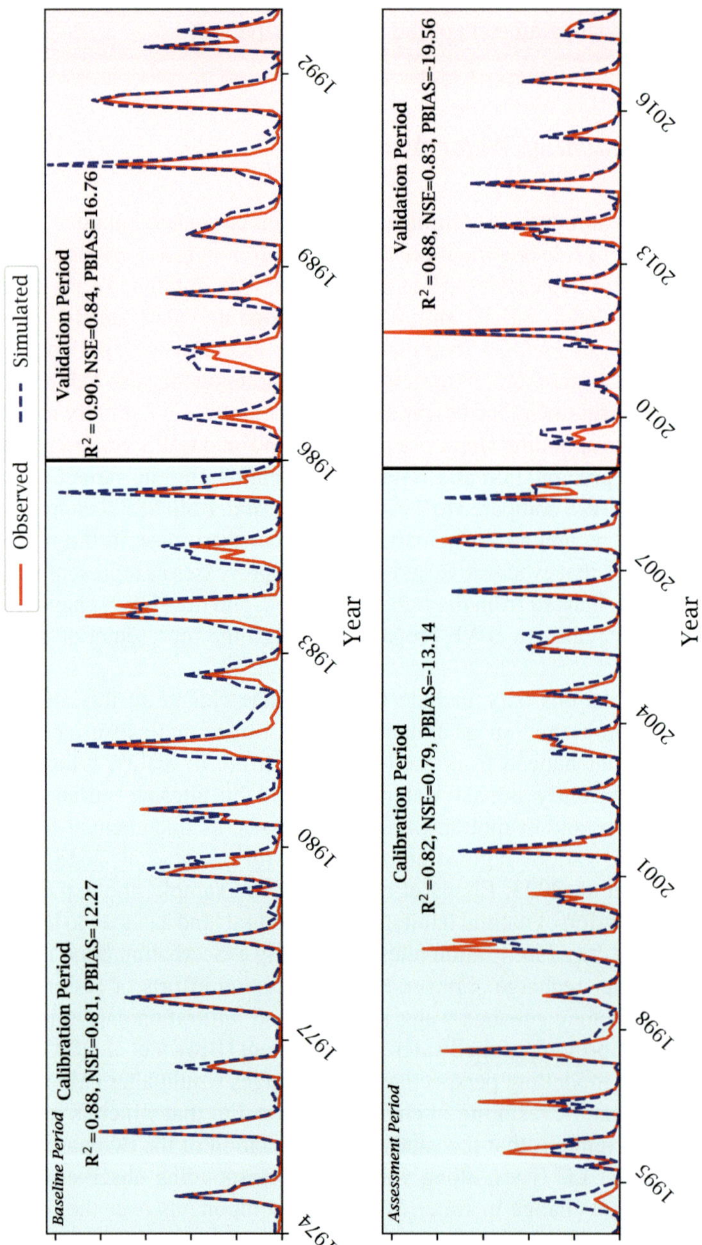

Fig. 20.7 SWAT model performances in the baseline and assessment periods

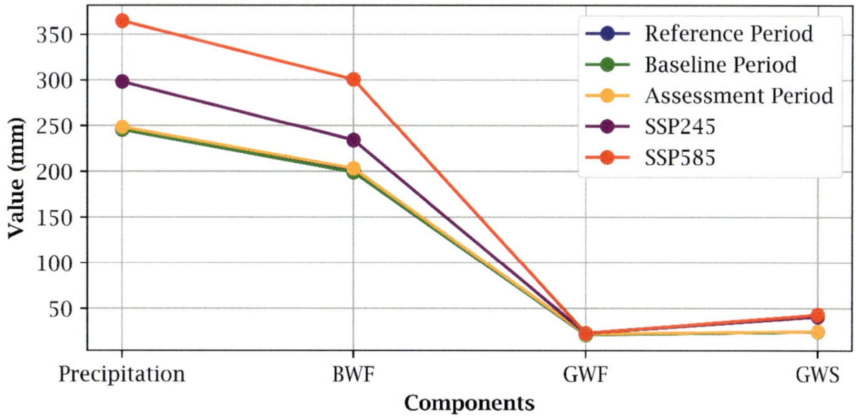

Fig. 20.8 Water availability assessment for different periods.

20.15 Water Scarcity Risk Analysis

The calculation of water scarcity based on streamflow and water availability is assessed at a basin scale and presented in Figure 20.9. The figure represents the changes in water availability (WA) and streamflow (Q) between different periods (reference, baseline, and assessment periods) and scenarios (SSP245 and SSP585), indicating potential shifts in water scarcity risk zoning. In the comparison between the baseline period and the assessment period, there is a marginal increase in water availability (0.02) but a sizable increase in streamflow (0.08). Overall, this generates reduced risk of water scarcity primarily due to enhanced runoff. Comparing the reference period with the SSP245 and SSP585 scenarios (RP-SSP245 and RP-SSP585, respectively), both scenarios show increases in water availability and streamflow. However, the SSP585 scenario exhibits a larger increase in water availability (0.52) compared to SSP245 (0.22); both exhibit relative changes of 0.04 for streamflow.

According to the water scarcity-risk criteria (as defined in subsection 4.4), positive relative changes in both WA and Q, which represents a situation wherein both terms are greater relative to a comparison period, suggests lower water scarcity risk. A positive change in water availability implies an increase in the overall water supply, while a positive change in streamflow indicates higher runoff and potentially improved replenishment of water resources. Thus, in this model simulation experiment, and when manipulating primarily climate variability and change via the use of alternative time periods and CMIP6 SSPs, water scarcity is expected to decrease in the Baitarani river basin in the future.

This conclusion faces several limitations. While the assumption of equilibrium between water demands and water availability simplifies the analysis, the dynamic and variable nature of water supply and demand is influenced by numerous factors such as seasonal variations, population growth, economic development and climate change, which limits understanding of water scarcity risks. This study also uses a

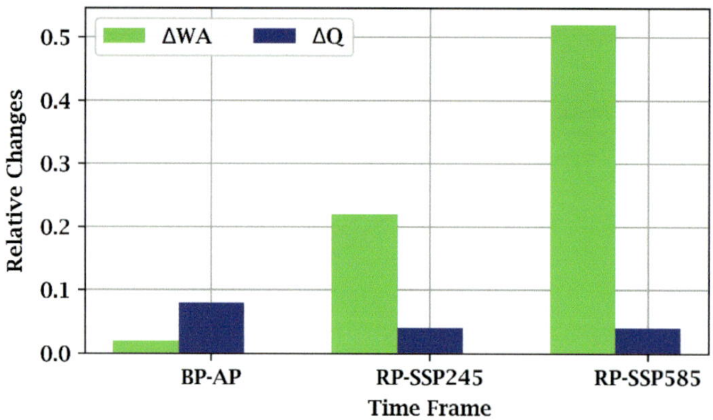

Fig. 20.9 Relative changes in streamflow and water availability for different comparison periods.

hydrologic model whose process representations (e.g., parameterizations) are representative of a static historical period. In contrast to that representation, a study by Garrote et al. (2018) suggested that basin-scale water availability is determined by a combination of streamflow and variable hydrological responses represented by water storage components. Such variable hydrologic responses are not incorporated in this study. One way to do so would be to integrate future LULC scenarios and corresponding parameterizations, along with future climate, into the analysis to assess the combined effects of climate and land use change on hydrologic responses and water scarcity more comprehensively (Leta et al. 2021; Acharya et al. 2023). This would provide valuable insights not only into potential trade-offs and relations between different land management strategies and climate adaptation measures, but also understanding about the extent to which model structural representations are important for understanding future risk of water scarcity.

20.16 Conclusions

This study evaluates modelled water scarcity in an observed historical reference period (1974–2018) and under simulated future climate (2020–2059), alongside representative land use/land cover (LULC) conditions. Using a Sequential Mann–Kendall change point analysis, this study identifies the year (1993) of the most significant change in historical Baitarani river basin streamflow. The subsequent use of this change point analysis to partition the historical reference data into baseline (1974–1993) and assessment periods (1994–2018) represents the use of a data-driven approach to guide otherwise subjective modelling decisions under uncertainty. This study then finds that the SWAT model performs satisfactorily in capturing the dynamic changes in LULC throughout both the baseline and assessment periods. Additionally, modelled blue and green water components yield fluctuations in water availability across the baseline and assessment periods, and in future SSP scenarios. In particular, SSP585 hydrology had notably higher quantities of blue water relative to other time periods, indicating the potentially strong influence of extreme precipitation on water resources in the study region. Building from the analysis of blue and green water components, the analysis of water scarcity indicates that the relative changes in water availability and streamflow are positive for the assessment period, as well as for the SSP245 and SSP585 scenarios, suggesting that the Baitarani basin will face reduced water scarcity risk in the future.

The findings of this study offer valuable insights for making informed decisions and managing water resources effectively in data-limited river basins, especially given the evolving dynamics of climate and changes in land use and land cover. Nevertheless, limitations remain. Here, consideration of future LULC scenarios alongside future climate would enhance the robustness of the model structural representation of hydrologic change, and thereby offer water resource management and planning efforts greater certainty in projected future streamflow and water availability.

Acknowledgement The authors are thankful to the India Meteorological Department (IMD), Pune and the Central Water Commission (CWC), Bhubaneswar for providing the hydro-meteorological datasets required for conducting this study.

References

Abbaspour KC, Rouholahnejad E, Vaghefi S, Srinivasan R, Yang H, Kløve B (2015) A continental-scale hydrology and water quality model for Europe: calibration and uncertainty of a high-resolution large-scale SWAT model. J Hydrol 524:733–752

Abbaspour KC (2014) SWAT-CUP 2012: SWAT calibration and uncertainty programs-a user manual: swiss federal institute of aquatic science and technology

Abedzadeh S, Roozbahani A, Heidari A (2020) Risk assessment of water resources development plans using fuzzy fault tree analysis. Water Resour Manage 34(8):2549–2569

Acharya S, Hori T, Karki S (2023) Assessing the spatio-temporal impact of landuse landcover change on water yield dynamics of rapidly urbanizing Kathmandu valley watershed of Nepal. J Hydrol Reg Stud 50:101562

Almazroui M, Nazrul Islam M, Saeed S, Alkhalaf AK, Dambul R (2017) Assessment of uncertainties in projected temperature and precipitation over the Arabian Peninsula using three categories of Cmip5 multimodel ensembles. Earth Syst Environ 1:1–20

Almazroui M, Saeed S, Saeed F, Islam MN, Ismail M (2020) Projections of precipitation and temperature over the South Asian countries in CMIP6. Earth Syst Environ 4:297–320

Aloui S, Mazzoni A, Elomri A, Aouissi J, Boufekane A, Zghibi A (2023) A review of soil and water assessment tool (SWAT) studies of mediterranean catchments: applications, feasibility, and future directions. J Environ Manage 326:116799

Arnold JG, Moriasi DN, Gassman PW, Abbaspour KC, White MJ, Srinivasan R, Santhi C, Harmel RD, Van Griensven A, VanLiew MW, Kannan N, Jha MK (2012) SWAT: model use, calibration, and validation. Trans ASABE 55(4):1491–1508

Ayeni AO, Kapangaziwiri E, Soneye ASO, Engelbrecht FA (2015) Assessing the impact of global changes on the surface water resources of southwestern Nigeria. Hydrol Sci J 60(11):1956–1971

Beck L, Bernauer T (2011) How will combined changes in water demand and climate affect water availability in the Zambezi river basin? Glob Environ Chang 21(3):1061–1072

Brown AE, Zhang L, McMahon TA, Western AW, Vertessy RA (2005) A review of paired catchment studies for determining changes in water yield resulting from alterations in vegetation. J Hydrol 310(1–4):28–61

Chanapathi T, Thatikonda S (2020) Investigating the impact of climate and land-use land cover changes on hydrological predictions over the Krishna river basin under present and future scenarios. Sci Total Environ 721:137736

Chen Q, Chen H, Zhang J, Hou Y, Shen M, Chen J, Xu C (2020) Impacts of climate change and LULC change on runoff in the Jinsha River Basin. J Geog Sci 30(1):85–102

Dahm RJ, Sperna Weiland FC, Singh UK, Lal M, Marchand M, Singh SK, Singh MP (2019) Assessment of future rainfall for the Brahmani–Baitarani river basin–practical implications of limited data availability. J Water Climate Change 10(4):782–798

Das P, Behera MD, Patidar N, Sahoo B, Tripathi P, Behera PR, Krishnamurthy YVN (2018) Impact of LULC change on the runoff, base flow and evapotranspiration dynamics in eastern Indian river basins during 1985–2005 using variable infiltration capacity approach. J Earth Syst Sci 127(2):1–19

Das J, Das S, Umamahesh NV (2023) Population exposure to drought severities under shared socioeconomic pathways scenarios in India. Sci Total Environ, 161566

Dey P, Mishra A (2017) Separating the impacts of climate change and human activities on streamflow: a review of methodologies and critical assumptions. J Hydrol 548:278–290

Dey A, Remesan R (2022) Multimodel quantification of green and blue water components and its error propagations through parameter transferability approach across input choices. J Hydrol 607:127579

Di Baldassarre G, Sivapalan M, Rusca M, Cudennec C, Garcia M, Kreibich H, Blöschl G (2019) Sociohydrology: scientific challenges in addressing the sustainable development goals. Water Resour Res 55(8):6327–6355

Dolgorsuren SE, Ishgaldan B, Myagmartseren P, Kumar P, Meraj G, Singh SK, Kanga S, Almazroui M (2024) Hydrological responses to climate change and land-use dynamics in central Asia's semi-arid regions: an SWAT model analysis of the Tuul River Basin. Earth Syst Environ

Dosdogru F, Kalin L, Wang R, Yen H (2020) Potential impacts of land use/cover and climate changes on ecologically relevant flows. J Hydrol 584:124654

Douglas EM, Vogel RM, Kroll CN (2000) Trends in floods and low flows in the United States: impact of spatial correlation. J Hydrol 240(1–2):90–105

Dubey SK, Kim J, Her Y, Sharma D, Jeong H (2023) Hydroclimatic impact assessment using the SWAT model in India—State of the art review. Sustainability 15(22):15779

Ebodé VB, Onana JYN, Boyomo TMS (2024) Water resources availability in the Mefou basin, Cameroon: under current and future climate, and land use and land cover. Sustain Water Resour Manag 10(2):78

Eyring V, Bony S, Meehl GA, Senior CA, Stevens B, Stouffer RJ, Taylor KE (2016) Overview of the Coupled Model Intercomparison Project Phase 6 (CMIP6) experimental design and organization. Geosci Model Dev 9(5):1937–1958

Eyring V, Cox PM, Flato GM, Gleckler PJ, Abramowitz G, Caldwell P, Williamson MS (2019) Taking climate model evaluation to the next level. Nat Clim Chang 9(2):102–110

Falkenmark M, Berntell A, Jägerskog A, Lundqvist J, Matz M, Tropp H (2007) On the verge of a new water scarcity: a call for good governance and human ingenuity. SIWI Policy Brief. SIWI

FAO (2012) Harmonized world soil database (version 1.2). Food and Agriculture Organization, Rome, Italy, and IIASA, Laxenburg, Austria

Fan X, Duan Q, Shen C, Wu Y, Xing C (2020) Global surface air temperatures in CMIP6: historical performance and future changes. Environ Res Lett 15(10):104056

Farr TG et al (2007) The shuttle radar topography mission. AGU J

Fisher MJ et al (2008) Harmonized world soil database v1.2

Friedman D, Schechter J, Baker B, Mueller C, Villarini G, White KD (2016) US army corps of engineers nonstationarity detection tool user guide. US Army Corps of Engineers, Washington, DC

Gain AK, Giupponi C (2015) A dynamic assessment of water scarcity risk in the Lower Brahmaputra River Basin: an integrated approach. Ecol Ind 48:120–131
Garrote L, Iglesias A, Granados A (2018) Country-level assessment of future risk of water scarcity in Europe. Proceed Int Assoc Hydrol Sci 379:455–462
Goyal MK, Gupta V, Eslamian S (2017) Hydrological drought: water surface and duration curve indices. In: Handbook of drought and water scarcity. CRC Press. pp 45-71
Gupta V, Syed B, Pathania A, Raaj S, Nanda A, Awasthi S, Shukla DP (2024) Hydrometeorological analysis of July-2023 floods in Himachal Pradesh, India. Natural Hazards, 1-26
Haynes K, Killick R, Fearnhead P, Eckley I (2021) Changepoint. np. Methods for nonparametric changepoint detection. R package version 1.0. 3
He Q, Wang M, Liu K, Li K, Jiang Z (2022) GPRChinaTemp1km: a high-resolution monthly air temperature data set for China (1951–2020) based on machine learning. Earth Syst Sci Data 14(7):3273–3292
Hoekstra AY (2016) A critique on the water-scarcity weighted water footprint in LCA. Ecol Ind 66:564–573
Jose DM, Makhdumi W, Dwarakish GS (2021) Hydrological Modelling to Study the Impacts of Climate and LULC Change at Basin Scale: A Review. In: Jha R, Singh VP, Singh V, Roy LB, Thendiyath R (eds) Water resources management and reservoir operation: hydraulics, water resources and coastal engineering. Springer International Publishing, pp 13–26
Kayitesi NM, Guzha AC, Mariethoz G (2022) Impacts of land use land cover change and climate change on river hydro-morphology—a review of research studies in tropical regions. J Hydrol 615:128702
Killick R, Eckley IA (2014) Changepoint: an R package for changepoint analysis. J Stat Softw 58:1–19
Kriegler E, Edmonds J, Hallegatte S, Ebi KL, Kram T, Riahi K, Van Vuuren DP (2014) A new scenario framework for climate change research: the concept of shared climate policy assumptions. Clim Change 122:401–414
Kryston A, Müller MF, Penny G, Bolster D, Tank JL, Mondal MS (2022) Addressing climate uncertainty and incomplete information in transboundary river treaties: a scenario-neutral dimensionality reduction approach. J Hydrol 612:128004
Leta MK, Demissie TA, Tränckner J (2021) Hydrological responses of watershed to historical and future land use land cover change dynamics of Nashe watershed Ethiopia. Water 13(17):2372
Levy MC, Lopes AV, Cohn A, Larsen LG, Thompson SE (2018) Land use change increases streamflow across the arc of deforestation in Brazil. Geophys Res Lett 45(8):3520–3530
Li Y, Huang Y, Li Y, Zhang H, Fan J, Deng Q, Wang X (2024) Spatiotemporal heterogeneity in meteorological and hydrological drought patterns and propagations influenced by climatic variability, LULC change, and human regulations. Sci Rep 14(1):5965
Mahato PK, Singh D, Bharati B, Gagnon AS, Singh BB, Brema J (2022) Assessing the impacts of human interventions and climate change on fluvial flooding using CMIP6 data and GIS-based hydrologic and hydraulic models. Geocarto Int 37(26):11483–11508
Mechal A, Takele T, Meten M, Deyassa G, Degu Y (2022) A modeling approach for evaluating the impacts of Land Use/Land Cover change for Ziway Lake watershed hydrology in the Ethiopian Rift. Model Earth Syst Environ 8(4):4793–4813
Meinshausen M, Nicholls Z, Lewis J, Gidden MJ, Vogel E, Freund M, Wang HJ (2019) The SSP greenhouse gas concentrations and their extensions to 2500. Geosci Model Dev Discuss 2019:1–77
Mishra V, Bhatia U, Tiwari AD (2020) Bias corrected climate projections from CMIP6 models for South Asia. Zenodo
Moriasi DN, Arnold JG, Van Liew MW, Bingner RL, Harmel RD, Veith TL (2007) Model evaluation guidelines for systematic quantification of accuracy in watershed simulations. Trans ASABE 50(3):885–900
Müller MF, Levy MC (2019) Complementary vantage points: Integrating hydrology and economics for sociohydrologic knowledge generation. Water Resour Res 55(4):2549–2571

NASA JPL (2013) Shuttle radar topography mission (SRTM)
Nourani V, Danandeh Mehr A, Azad N (2018) Trend analysis of hydroclimatological variables in Urmia lake basin using hybrid wavelet Mann–Kendall and Şen tests. Environ Earth Sci 77(5):1–18
O'Neill BC, Kriegler E, Ebi KL, Kemp-Benedict E, Riahi K, Rothman DS, Solecki W (2017) The roads ahead: Narratives for shared socioeconomic pathways describing world futures in the 21st century. Glob Environ Chang 42:169–180
Ojha R, Nagesh Kumar D, Sharma A, Mehrotra R (2013) Assessing severe drought and wet events over India in a future climate using a nested bias-correction approach. J Hydrol Eng 18(7):760–772
Omer A, Elagib NA, Zhuguo M, Saleem F, Mohammed A (2020) Water scarcity in the Yellow River Basin under future climate change and human activities. Sci Total Environ 749:141446. https://doi.org/10.1016/j.scitotenv.2020.141446
Pai DS, Sridhar L, Badwaik MR, Rajeevan M (2015) Analysis of the daily rainfall events over India using a new long period (1901–2010) high resolution (0.25× 0.25) gridded rainfall data set. Climate Dyn 45(3):755–776
Pandey D, Tiwari AD, Mishra V (2022) On the occurrence of the observed worst flood in Mahanadi River basin under the warming climate. Weather Climate Extremes 38:100520
Paul PK, Gaur S, Kumari B, Panigrahy N, Mishra A, Singh R (2019) Diagnosing credibility of a large-scale conceptual hydrological model in simulating streamflow. J Hydrol Eng 24(4):04019004
Pedro-Monzonís M, Solera A, Ferrer J, Estrela T, Paredes-Arquiola J (2015) A review of water scarcity and drought indexes in water resources planning and management. J Hydrol 527:482–493. https://doi.org/10.1016/j.jhydrol.2015.05.003
Pettitt AN (1979) A non-parametric approach to the change-point problem. J Roy Stat Soc: Ser C (Appl Stat) 28(2):126–135
Rafiei-Sardooi E, Azareh A, Joorabian Shooshtari S, Parteli EJR (2022) Long-term assessment of land-use and climate change on water scarcity in an arid basin in Iran. Ecol Model 467:109934. https://doi.org/10.1016/j.ecolmodel.2022.109934
Rai PK, Dhanya CT, Chahar BR (2018) Coupling of 1D models (SWAT and SWMM) with 2D model (iRIC) for mapping inundation in Brahmani and Baitarani river delta. Nat Hazards 92:1821–1840
Rajeevan M, Bhate J, Kale JD, Lal B (2006) High resolution daily gridded rainfall data for the Indian region: analysis of break and active monsoon spells. Curr Sci, 296-306
Roy PS, Meiyappan P, Joshi PK, Kale MP, Srivastav VK, Srivasatava SK, Krishnamurthy YVN (2016) Decadal land use and land cover classifications across India, 1985, 1995, 2005. ORNL DAAC, Oak Ridge, Tennessee, USA
Ryberg KR, Hodgkins GA, Dudley RW (2020) Change points in annual peak streamflows: Method comparisons and historical change points in the United States. J Hydrol 583:124307
Şen Z (2019) Partial trend identification by change-point successive average methodology (SAM). J Hydrol 571:288–299
Shuttle Radar (n.d.) Topography mission 1 arc-second global (digital object identifier (DOI) number: /https://doi.org/10.5066/F7PR7TFT). Accessed: 2020-02-01
Sinha RK, Eldho TI, Subimal G (2023) Assessing the impacts of land use/land cover and climate change on surface runoff of a humid tropical river basin in Western Ghats, India. Int J River Basin Manag 21(2):141–152
Sivapalan M, Savenije HH, Blöschl G (2012) Socio-hydrology: a new science of people and water. Hydrol Process 26(8):1270–1276
Sneyers R (1990) On the statistical analysis of series of observations (No. 551.501. 3 SNE)
Srinivasan R, Arnold JG, Jones CA (1998) Hydrologic modelling of the United States with the soil and water assessment tool. Int J Water Resour Dev 14(3):315–325
Srivastava A, Maity R (2023) Unveiling an Environmental Drought Index and its applicability in the perspective of drought recognition amidst climate change. J Hydrol 627:130462

Suman M, Maity R (2021) Assessment of basin-wise future agricultural drought status across India under changing climate. J Water Climate Change 12(6):2400–2421
Sun C, Zhou X (2020) Characterizing hydrological drought and water scarcity changes in the future: a case study in the Jinghe River Basin of China. Water 12(6):1605
Swain SS, Mishra A, Sahoo B, Chatterjee C (2020) Water scarcity-risk assessment in data-scarce river basins under decadal climate change using a hydrological modelling approach. J Hydrol 590:125260
Swain SS, Mishra A, Chatterjee C, Sahoo B (2021) Climate-changed versus land-use altered streamflow: A relative contribution assessment using three complementary approaches at a decadal time-spell. J Hydrol 596:126064
Swain SS, Kumar SB, Mishra A, Chatterjee C (2023) Sensitive or resilient catchment?: A Budyko-based modeling approach for climate change and anthropogenic stress under historical to CMIP6 future scenarios. J Hydrol 622:129651
Swain SS, Mishra A, Chatterjee C (2024) Time-varying evaluation of compound drought and hot extremes in machine learning-predicted ensemble CMIP5 future climate: a multivariate multi-index approach. J Hydrol Eng 29(2):04024001
Tan ML, Gassman PW, Yang X, Haywood J (2020) A review of SWAT applications, performance and future needs for simulation of hydro-climatic extremes. Adv Water Resour 143:103662
Tankpa V, Wang L, Awotwi A, Singh L, Thapa S, Atanga RA, Guo X (2021) Modeling the effects of historical and future land use/land cover change dynamics on the hydrological response of Ashi watershed, Northeastern China. Environ Dev Sustain 23:7883–7912
Taylor R (2009) Rethinking water scarcity: the role of storage. EOS Trans Am Geophys Union 90(28):237–238
UN World Water Development Report (2021) UN-Water. Retrieved February 5, 2024, from https://www.unwater.org/publications/un-world-water-development-report-2021
Uniyal B, Kosatica E, Koellner T (2023) Spatial and temporal variability of climate change impacts on ecosystem services in small agricultural catchments using the Soil and Water Assessment Tool (SWAT). Sci Total Environ 875:162520
Van Griensven A, Meixner T (2007) A global and efficient multi-objective auto-calibration and uncertainty estimation method for water quality catchment models. J Hydroinf 9(4):277–291
Veldkamp TIE, Wada Y, Aerts JCJH, Döll P, Gosling SN, Liu J, Masaki Y, Oki T, Ostberg S, Pokhrel Y, Satoh Y, Kim H, Ward PJ (2017) Water scarcity hotspots travel downstream due to human interventions in the 20th and 21st century. Nat Commun 8(1):15697
Visakh S, Raju PV, Kulkarni SS, Diwakar PG (2019) Inter-comparison of water balance components of river basins draining into selected delta districts of Eastern India. Sci Total Environ 654:1258–1269
Wada Y, van Beek LPH, Viviroli D, Dürr HH, Weingartner R, Bierkens MFP (2011) Global monthly water stress: 2. Water demand and severity of water stress. Water Resour Res, 47(7)
Wagener T, Sivapalan M, Troch PA, McGlynn BL, Harman CJ, Gupta HV, Kumar P, Rao PSC, Basu NB, Wilson JS (2010) The future of hydrology: an evolving science for a changing world. Water Resour Res, 46(5)
Wang Q, Xu Y, Wang Y, Zhang Y, Xiang J, Xu Y, Wang J (2020) Individual and combined impacts of future land-use and climate conditions on extreme hydrological events in a representative basin of the Yangtze River Delta China. Atmos Res 236:104805
Wang X, Erdman C, Emerson JW (2018) Bayesian analysis of change point problems
World Economic Forum (2014) WEF Global Risks
Wu C, Liu W, Deng H (2023) Urbanization and the emerging water crisis: identifying water scarcity and environmental risk with multiple applications in urban agglomerations in Western China. Sustainability 15(17):17
Wu J, An P, Zhao C, Wei Z, Lan T, Li X, Wang G (2024) Effects of multi-year droughts on the precipitation-runoff relationship: an integrated analysis of meteorological, hydrological, and compound droughts. J Hydrol, 131064

Xie P, Gu H, Sang YF, Wu Z, Singh VP (2019) Comparison of different methods for detecting change points in hydroclimatic time series. J Hydrol 577:123973
Xu Z, Han Y, Tam CY, Yang ZL, Fu C (2021) Bias-corrected CMIP6 global dataset for dynamical downscaling of the historical and future climate (1979–2100). Sci Data 8(1):293
Yang Y, Tian F (2009) Abrupt change of runoff and its major driving factors in Haihe River Catchment China. J Hydrol 374(3–4):373–383
Zhang M, Liu N, Harper R, Li Q, Liu K, Wei X, Liu S (2017) A global review on hydrological responses to forest change across multiple spatial scales: Importance of scale, climate, forest type and hydrological regime. J Hydrol 546:44–59
Zhao J, Huang Q, Chang J, Liu D, Huang S, Shi X (2015) Analysis of temporal and spatial trends of hydro-climatic variables in the Wei River Basin. Environ Res 139:55–64
Zhao J, Zhang N, Liu Z, Zhang Q, Shang C (2024) SWAT model applications: from hydrological processes to ecosystem services. Sci Total Environ, 172605
Zhou C, van Nooijen R, Kolechkina A, Hrachowitz M (2019) Comparative analysis of nonparametric change-point detectors commonly used in hydrology. Hydrol Sci J 64(14):1690–1710
Zhou J, Deitch MJ, Grunwald S, Screaton E (2023) Do the Mann–Kendall test and Theil-Sen slope fail to inform trend significance and magnitude in hydrology? Hydrol Sci J 68(9):1241–1249

Chapter 21
Summary Chapter

Aliva Nanda, Pankaj Kumar Gupta, Vivek Gupta, Prakash Kumar Jha, and Swatantra Kumar Dubey

Abstract This comprehensive book examines the intricate relationships between hydrology, agriculture, pollution, and climate change across multiple regions. The text begins by highlighting the crucial role of headwater catchments in regulating water flow, carbon storage, and ecosystem health. Through various case studies and research findings, it explores three main themes: hydrological changes in catchment areas, agricultural systems and water management, and pollution remediation strategies. The first section analyzes streamflow regime shifts in headwater catchments and presents projections for water availability in regions like the Indus basin, incorporating CMIP6 climate scenarios. The agricultural focus emphasizes innovative approaches such as coconut husk mulch for water conservation and examines climate extremes' impact on crop productivity, particularly in Indian agricultural systems. A significant portion addresses water pollution challenges, particularly in agricultural and industrial contexts. The text details various remediation technologies for contaminated groundwater systems, including both ex-situ and insitu methods, with special attention to simulation-optimization models for aquifer restoration. The book also explores the complex dynamics of fertilizer leaching and heavy metal contamination

A. Nanda (✉) · V. Gupta
School of Civil and Environmental Engineering, Indian Institute of Technology Mandi, Mandi, Himachal Pradesh, India
e-mail: aliva@iitmandi.ac.in

V. Gupta
e-mail: vivekgupta@iitmandi.ac.in

P. K. Gupta
Center for Rural Development and Technology, Indian Institute of Technology Delhi, New Delhi, India
e-mail: pk3gupta@uwaterloo.ca

P. K. Jha
Mississippi State University, Starkville, MS, USA
e-mail: pj442@msstate.edu

S. K. Dubey
Environmental Engineering, Seoul National University of Science and Technology, Seoul, Korea (Republic of)
e-mail: swatantratech1@gmail.com

A. Nanda et al. (eds.), *Navigating the Nexus*, Water Science and Technology Library 102, https://doi.org/10.1007/978-3-031-76532-2_21

in industrial corridors, presenting case studies from regions like Tamil Nadu. The final sections integrate advanced technological solutions, including precision agriculture, ERA5-Land reanalysis data, and Bayesian Model Averaging for streamflow prediction. The book concludes by addressing policy perspectives on transboundary water conflicts and water quality assessment, emphasizing the need for integrated approaches to tackle contemporary environmental challenges.

Keywords Hydrology · Policy · Agriculture · Climate change · Pollution

Headwater catchments play a crucial role in governing the hydrologic cycle of both upstream and downstream regions. Headwater catchments control the timing and volume of water flow downstream. They store and slowly release water, helping to moderate the flow during both wet and dry periods. This buffering effect reduces the risk of downstream flooding during heavy rains and ensures a steady supply of water during droughts. Further, headwater catchments often store significant amounts of carbon, primarily in the form of organic matter in soils and vegetation. This carbon storage can help mitigate climate change by sequestering carbon and preventing its release into the atmosphere. Overall, headwater catchments are integral components of watershed ecosystems. Their functions extend far beyond their small size, influencing water quality, habitat availability, flow regulation, and nutrient cycling downstream. Protecting and conserving headwater catchments is essential for maintaining the health and resilience of entire watersheds. Thus, the first theme of our book focuses on the hydrology of headwater catchments in changing climatic conditions. Mehan et al., discuss the role of headwater catchments in altering streamflow regimes. Chapter 1 explores strategies for adaptation and mitigation, emphasizing the need for interdisciplinary collaboration and effective policy measures. By providing a comprehensive overview of the causes and impacts of streamflow regime shifts in headwater catchments, this chapter is an essential resource for researchers, practitioners, and policymakers seeking to better understand and address this critical issue in a changing environment. Nazeer et al., discuss the future water availability of the Indus basin in changing climatic scenarios. They analyzed the future water availability of the Hunza River Basin (sub-basins of the UIB) under the Coupled Model Intercomparison Project Phase 6 (CMIP6) climatic projections. The 12–32% annual precipitation increase is expected by the end of the twenty-first century, relative to the baseline precipitation in the Hunza River Basin. Shifts in precipitation as snow and increased rainfall are expected. Policymakers and stakeholders could use the results of this study for reservoir development, modifying hydropower and reservoir development policy, sustainable agriculture, and efficient water resources management. Further, Yaka et al., simulate the changes in glacier mass and volume over time, focusing on the initial state of the Chhota Shigri Glacier. Their study examined the sensitivity of the Chhota Shigri Glacier through a series of experiments conducted from 2003 to 2019 using in-situ WGMS and geodetic data. Sharma et al., analyzed the LULC change on runoff using Variable Infiltration Capacity (VIC) model. Their study emphasizes the importance of hydrological modelling in understanding water balance components and simulating hydrological responses under LULC alterations. Also, the adoption of appropriate workflows for assembling multiple datasets and extraction techniques and analysis of the time-series impact of urbanization on the watershed's runoff potential.

Chapter 5 investigates the synergistic effects of using coconut husk mulch (CHM) on the hydrothermal regime, growth characteristics, and water consumption of onion crops in environments with limited water resources. Field trials assessed the effectiveness of CHM in retaining soil moisture and optimizing irrigation schedules for onion cultivation. The study found that CHM significantly reduced maximum soil temperatures, enhanced plant development, and decreased overall irrigation requirements compared to conditions without mulch or with other mulch types. Specifically, at Maximum Allowable Depletion (MAD) levels of 30% to 70%, CHM reduced the need for 1–2 irrigations, demonstrating its potential to conserve water and improve crop productivity under varying moisture conditions. These findings underscore the practical benefits of CHM in sustainable water management and agricultural production, highlighting its role in creating favorable soil conditions and optimizing water use efficiency for onion cultivation in water-limited regions. Chapter 6 explores the role of climate extremes on crop productivity. Analyzing the relationship between climate extremes and crop yields, positive correlations were observed with precipitation indices, which boost plant growth and soil microbial activities, enhancing productivity. The findings of this study offer valuable insights for decision-makers aiming to formulate effective adaptation and mitigation policies to address the impacts of climate change on crop yield in Bihar. More studies are needed to understand how these climate extremes affect plants' phenological stages and yield. Chapter 7 discusses the Pathways to accelerate agricultural production system technologies in India. They integrated appropriate agronomic, chemical, mechanical, and agrostological techniques customized for soil management and the conservation of green and blue water across various basins and agroecological zones. These proposed practices are further expanded toward value chain development, ensuring nutritional and economic benefits. Chapter 8 explains the impact of climate change on crop yield and its consequences for the water-energy-food nexus in an Indian river basin. The interconnection of water, energy, and food (WEF Nexus) underscores the need to explore how alterations in food security, driven by climate change-induced crop yield changes, may reverberate across water and energy security. This study addresses a notable gap by utilizing the CMIP6-GCM to estimate climate change effects on crop yield, explicitly focusing on the Kangsabati River basin.

Agricultural return flow, often referred to as agricultural runoff, is a critical issue in water management and environmental sustainability. It involves the movement of excess water, which could be irrigation water, rainwater, or melted snow, that flows off agricultural fields back into surface or groundwater systems. This flow carries with it a variety of substances, including fertilizers, pesticides, sediments, and other agricultural byproducts, which can have significant impacts on soil and water quality. Paul and Bhatia (Chap. 9) address rice farming pollution, focusing on preventing and remedying water and soil contamination due to agricultural practices. Various engineering, biological, and ecological methods are suggested for treating polluted river water. The global impact of pollution on ecosystems and public health is highlighted, emphasizing the need for international cooperation. Innovations in rice farming, such as high-yielding varieties, mechanization, advanced irrigation practices, biotechnology, integrated pest management, and crop diversification, are discussed for their potential to enhance sustainability and productivity. Additionally, the adverse effects of chemical inputs like pesticides, herbicides, and fertilizers on soil and water and soil erosion due to rice cultivation are explored, advocating for sustainable and eco-friendly farming practices. Furthermore, it discusses the importance of water in crop cultivation. Water is crucial for rice growth, but continuous flooded irrigation leads to significant water loss, nutrient depletion, and groundwater pollution. The limited expansion of rice fields is due to high water demand, labour-intensive farming, and soil degradation, with rice's water productivity being lower than other crops. Waterlogging and salinization further hinder agricultural productivity, requiring sustainable practices. Effective strategies include precision farming, organic farming, integrated pest management, cover cropping, conservation tillage, and efficient irrigation management. These methods aim to optimize resource use, enhance soil health, and promote sustainable rice production amidst growing climate challenges and water scarcity.

Groundwater pollution sources include agricultural waste, fertilizers, and sewage, with pollutants like pesticides and nitrates causing health problems. Nigam et al. (Chap. 10) reviewed the technologies for remediation of contaminated groundwater systems including ex-situ methods like chemical precipitation and in-situ methods like bioremediation and phytoremediation, which uses plants to absorb and degrade contaminants. Effective groundwater management and ongoing research are essential for maintaining water quality and availability. Improving in-situ treatments can enhance pollutant degradation rates, making the process more efficient and cost-effective. Effective management and prevention of groundwater degradation are crucial due to the complexity and expense of treatment. Developing advanced groundwater purification technologies is essential for sustainable water quality and economic stability.

Restoring contaminated groundwater systems is an expensive and long-term process. Remediation techniques depend on the hydrogeological properties of the aquifer, the type and quantity of the contaminant, and the potential for future contamination. Predicting the future and fate of the contaminant, considering factors influencing the remediation process, and minimizing remediation costs for a desired duration are crucial for designing effective management and remediation plans. Das and Eldho (Chap. 11) used Simulation–Optimization (SO) for aquifer restoration and management. An SO model typically combines a numerical simulation tool for solving the groundwater contaminant transport equation and an optimization algorithm for selecting the best values of decision variables to minimize remediation costs or duration. The simulation model is called during every fitness evaluation of the optimization model. A groundwater contaminant transport simulation model solves the governing transport equation to predict concentration distribution over time. Analytical models provide exact solutions with simplifying assumptions but are impractical for real cases. They discussed that numerical models like Finite Difference Method (FDM), Finite Element Method (FEM), and Finite Volume Method (FVM) are successful but rely on a background mesh, complicating the automation and refinement process for accuracy. Meshless methods, which limit the use of a mesh, significantly improve the discretization and adaptive analysis of the aquifer. Meshless models have been used for contaminant transport studies, including the weak-form Element Free Galerkin Method (EFGM). These models have been developed for various applications, such as Monte Carlo Smoothed Particle Hydrodynamics, Point Collocation Method, Radial Basis Function Collocation Method, and Local Radial Point Interpolation Method. The Meshless Local Petrov Galerkin (MLPG) model, first used for groundwater problems, has been successfully applied in pump and treat (PAT) remediation. This study presents an SO model for PAT remediation design, based on MLPG simulation and DE optimization, aiming to minimize remediation costs and improve financial resource usage. The model's accuracy and efficiency make it an efficient alternative to mesh-based methods. The MLPG-DE-PAT model effectively reduces TDS concentration in aquifers to acceptable levels within specified durations. They have highlighted that It offers better accuracy, efficient remediation, and simpler applications compared to mesh-based methods. They found that after 3 years, the concentration distribution shows significant improvement, ensuring potable groundwater. This model is a valuable tool for designing optimal PAT remediation plans, minimizing costs, and enhancing groundwater restoration efforts.

Although groundwater remediation is achievable, the vadose zone contamination posses a complex problem. Kumar et al. (Chap. 12) evaluate the effect of rainfall dynamics on nitrogen, phosphorus, and potassium (NPK) concentration at various soil depths in cotton-growing regions of South India. Cotton is India's most valuable fibre and cash crop, playing a significant role in the country's manufacturing and agricultural economies. It supplies the cotton textile industry with the basic raw material (cotton fibre) and is a direct source of income for 6 million farmers. The cotton trade and manufacturing employ 40–50 million people. India has ten major cotton-growing states, divided into three zones: the north zone (Punjab, Haryana, and Rajasthan), the central zone (Madhya Pradesh, Maharashtra, and Gujarat), and the south zone (Andhra Pradesh, Telangana, Karnataka, and Tamil Nadu). The average cotton yield in these regions is 1.8 quintal per hectare, close to the national average. Crop yields vary from year to year due to changes in rainfall and nutrient supply in the soil caused by climate change. The Intergovernmental Panel on Climate Change (IPCC) defines climate change as a long-term change in climatic conditions identified by changes in the mean or variability of its properties. The United Nations Framework Convention on Climate Change (UNFCCC) attributes climate change to human activity that alters the global atmosphere, beyond natural climate variability. Climate change significantly affects rainfall patterns, which in turn impacts groundwater recharge and nutrient transport, crucial for plant development. India is the world's second-largest fertilizer producer after China and ranks first in fertilizer imports.

Fertilizer use in India increased from 78 thousand tons in 1965–66 to 26.5 million tonnes in 2009–2010. Reducing fertilizer use is not viable due to the priority of optimizing food output for food and nutritional security. Improving fertilizer usage through balanced nutrient distribution, good planning, and water conservation can reduce fertilizer leaching beyond the root zone of crops. Various models, including HYDRUS-1 and HYDRUS-2, are used to study nitrogen, phosphorus, and potassium leaching. Kumari et al. (2019) investigated nitrate–N load and groundwater vulnerability in the Madhubani area of Bihar, showing higher vulnerability in the eastern part due to alluvial sandy lithology and a shallow groundwater table. Gupta et al. (2020) evaluated nitrate vulnerability in groundwater in three districts of Bihar using the HYDRUS 1D model, indicating high nitrate-leaching vulnerability in the southern part due to Gangetic kankar in the subsurface. A computer simulation model, Hydrus2D/3D, was used to simulate NPK distribution at various soil depths, considering rainfall dynamics. The Hydrus-2D model, which solves Richards' equation for water flow and the convection–dispersion equation for solute transport, is critical for predicting environmental impacts associated with human activities such as the overuse of fertilizer and irrigation. The study analyzes soil data for a tropical crop, cotton, and its impact on nitrogen and potassium (NPK) transport in the soil column. It finds that rainfall dynamics significantly influence solute transport and distribution, emphasizing the need for considering rainfall variability in fertilizer application and crop nutrient availability. Further research should explore long-term effects and environmental factors.

Likewise, Loganathan (Chap. 13) investigates groundwater contamination by heavy metals in Tamil Nadu's northern industrial corridors, specifically focusing on Ambur, Ranipet, and Vaniyambadi. The findings reveal significant contamination by Cadmium (Cd) and Chromium (Cr), with levels exceeding WHO limits, posing substantial health risks to local populations. The contamination levels were classified using the Contamination Factor (CF) method, indicating considerable to high contamination, particularly in the Ranipet and Ambur regions. In Ambur, a strong positive correlation was found among Cr, Cd, and Co, suggesting these metals likely originate from similar sources such as tannery wastewater. In Ranipet, a strong correlation among all five heavy metals indicates significant anthropogenic contributions from improper waste discharge. Health risk assessments for Cr and Cd were conducted, focusing on ingestion pathways for infants, children, and adults. Carcinogenic risk assessments (CDI and ILCR) indicate that all three industrial corridors pose carcinogenic risks, with children and infants in Ranipet being the most affected. Hazard Quotient (HQ) results suggest that Cd has a greater impact on children and infants, while Cr significantly affects all age groups in Ambur. The study underscores the urgent need for effective waste management and industrial effluent treatment to mitigate groundwater pollution and protect public health. The research provides critical baseline data for policymakers and stakeholders to implement corrective measures and improve groundwater quality management in these industrial areas. The findings highlight severe contamination of groundwater with toxic heavy metals, emphasizing the necessity for immediate action to prevent further health risks and environmental damage.

The complex climatic problems required advanced tools to tackle the ongoing climatic issues. Chapter 14 uses the VIC model to assess the impact of climate change in the Beas River basin. This research introduces a novel methodological framework for hydrological modelling, ensuring that regional climate change assessments are both precise and actionable for policymakers. Their findings highlight stark differences in hydrological impacts under various Shared Socioeconomic Pathways, emphasizing the need for adaptive strategies in water resource management to accommodate anticipated climatic shifts. Chapter 15 discusses the precision agriculture technology for climate resiliency. This chapter explores how precision farming techniques and technologies are used to manage water resources and tackle the challenges brought by climate change in agriculture. Precision farming relies on automation, sensors, drones, and/or telemetry systems to monitor soil conditions, water availability, and weather patterns. Further, Chap. 16 uses ERA5-Land reanalysis data for analyzing the water budgeting and the impact of climatic extremes on it. Chapter 17 analyzed streamflow prediction using Bayesian Model Averaging (BMA) to simulate monthly discharge in the Godavari basin, using projections from five Global Climate Models (GCMs) and five Global Hydrological Models (GHMs). The study found BMA outperformed traditional averaging methods in accuracy and reliability. BMA effectively captured seasonal flow fluctuations and provided a robust framework for uncertainty analysis. The findings highlight BMA's potential in enhancing streamflow projections in water resource management.

The soil–water, agriculture, and climate change are inextricably linked. These systems are interdependent, and decisions made in one sector can significantly impact the others. Policymakers are increasingly recognizing the importance of this nexus in addressing a range of challenges, including climate change, food security, and sustainable development. Policymakers can identify and implement policies that improve resource use efficiency and reduce negative trade-offs by considering the interactions between different sectors. To increase resiliency and promote equity under climate change, research, and development-oriented policies should go together. This theme discusses policy perspectives on integrated landscape management, payment for ecosystem services, climate-smart agriculture, and market-based instruments to encourage sustainability. Chapter 18 discusses transboundary water conflicts for effective management of water resources for regional stability and international cooperation. AI and ML technologies enhance water resource management by enabling predictive modeling, optimizing resource allocation, and fostering transparency and collaboration among nations. This chapter explores AI and ML's role in resolving water conflicts, using case studies to demonstrate their potential in predicting water scarcity and promoting cooperation. Chapter 19 assesses the surface water quality of the Mahanadi River Basin in Odisha for drinking purposes using entropy-based WQI and reliability-based MCDM models, specifically Fuzzy-AHP. Analysis of samples from 16 locations over 2020–2023 revealed varying water quality, with some sites showing poor or very poor conditions due to high levels of TKN, coliform, and other contaminants. The integrated approach aids in identifying areas needing restoration and optimizing water quality monitoring programs. Chapter 20 discusses issues related to water scarcity, which is driven by climate change, population growth, and urbanization. This study uses hydrological models to estimate water scarcity risk in the Baitarani river basin, relying on historical and projected climate data. Findings indicate a decrease in water scarcity from 1994–2018 and predict further decline until 2064, though uncertainties remain due to model limitations. The research provides insights for sustainable water management in data-scarce regions.

We believe the content of this book will give a broader perspective to all our readers on Hydrology, Agriculture, Pollution, and Climate Change.

References

Gupta PK, Kumari B, Gupta SK, Kumar D (2020) Nitrate-leaching and groundwater vulnerability mapping in North Bihar, India. Sustain Water Resour Manage 6:1–12

Kumari B, Gupta PK, Kumar D (2019) In-situ observation and nitrate-N load assessment in Madhubani District, Bihar, India. J Geol Soc India 93:113–118

GPSR Compliance

The European Union's (EU) General Product Safety Regulation (GPSR) is a set of rules that requires consumer products to be safe and our obligations to ensure this.

If you have any concerns about our products, you can contact us on ProductSafety@springernature.com

In case Publisher is established outside the EU, the EU authorized representative is:

Springer Nature Customer Service Center GmbH
Europaplatz 3
69115 Heidelberg, Germany

Batch number: 08057028

Printed by Printforce, the Netherlands